INTRODUCTORY FLUID MECHANICS
for
Physicists and Mathematicians

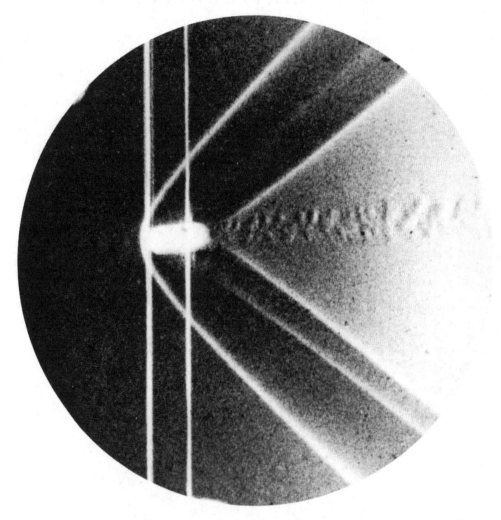

Frontispiece The first observation of shock waves generated by the passage of a bullet through air by Mach and Salcher(1887). Behind the primary shock can be seen the following rarefaction. Note also the turbulet wake leaving the rear of the bullet. Since the bullet was blunt the shock is slightly detached, although this cannot be clearly seen. The two vertical lines are for timing. The photograph was taken using a shadowgraph technique.

INTRODUCTORY FLUID MECHANICS
for
Physicists and Mathematicians

Geoffrey J. Pert

Department of Physics, University of York, UK

A John Wiley & Sons, Ltd., Publication

This edition first published 2013
© 2013 John Wiley & Sons Ltd.

Registered office
John Wiley & Sons Ltd, The Atrium, Southern Gate, Chichester, West Sussex, PO19 8SQ, United Kingdom

For details of our global editorial offices, for customer services and for information about how to apply for permission to reuse the copyright material in this book please see our website at www.wiley.com.

Library of Congress Cataloging-in-Publication Data applied for.

Hardback ISBN: 9781119944843

A catalogue record for this book is available from the British Library.

Typeset in 11/13.5pt Computer Modern by Laserwords Private Limited, Chennai, India
Printed and bound in Malaysia by Vivar Printing Sdn Bhd

This book is dedicated
with grateful thanks and appreciation to
my doctoral supervisor
John Pain
who introduced me to the science of
shock waves and fluid dynamics
and who provided help and support
in times of stress
(even when we attacked the *Tirpitz*)

Contents

Preface   xvii

1 Introduction   1
   1.1   Fluids as a State of Matter   1
   1.2   The Fundamental Equations for Flow of a Dissipationless Fluid   3
   1.3   Lagrangian Frame   4
         1.3.1   Conservation of Mass   6
         1.3.2   Conservation of Momentum–Euler's Equation   7
         1.3.3   Conservation of Angular Momentum   8
         1.3.4   Conservation of Energy   8
         1.3.5   Conservation of Entropy   8
   1.4   Eulerian Frame   8
         1.4.1   Conservation of Mass–Equation of Continuity   8
         1.4.2   Conservation of Momentum   9
         1.4.3   Conservation of Angular Momentum   10
         1.4.4   Conservation of Energy   11
         1.4.5   Conservation of Entropy   11
   1.5   Hydrostatics   12
         1.5.1   Isothermal Fluid–Thermal and Mechanical Equilibrium 12
         1.5.2   Adiabatic Fluid–Lapse Rate   12
         1.5.3   Stability of an Equilibrium Configuration   15
   1.6   Streamlines   16
   1.7   Bernoulli's Equation: Weak Form   16
   1.8   Polytropic Gases   17
         1.8.1   Applications of Bernoulli's Theorem   19
                 1.8.1.1   Vena Contracta   19
                 1.8.1.2   Flow of gas along a pipe of varying cross-section 20
         Case study 1.I   Munroe Effect–Shaped Charge Explosive . .   22

2 Flow of Ideal Fluids   25
   2.1   Introduction   25
   2.2   Kelvin's Theorem   26

2.2.1   Vorticity and Helmholtz's Theorems   27
        2.2.1.1   Simple or rectilinear vortex   29
        2.2.1.2   Vortex sheet   29
2.3   Irrotational Flow   31
      2.3.1   Crocco's Equation   31
2.4   Irrotational Flow–Velocity Potential and the Strong Form of
      Bernoulli's Equation   32
2.5   Incompressible Flow–Streamfunction   33
      2.5.1   Planar Systems   33
      2.5.2   Axisymmetric Flow–Stokes Streamfunction   34
2.6   Irrotational Incompressible Flow   35
      2.6.1   Simply and Multiply Connected Spaces   37
2.7   Induced Velocity   38
      2.7.1   Streamlined Flow around a Body Treated
               as a Vortex Sheet   41
2.8   Sources and Sinks   42
      2.8.1   Doublet Sources   43
               2.8.1.1   Doublet sheets   45
      2.8.2   Flow Around a Body Treated as a Source Sheet   45
      2.8.3   Irrotational Incompressible Flow Around a Sphere . . .   48
      Case study 2.I   Rankine Ovals   49
2.9   Two-Dimensional Flow   51
      2.9.1   Irrotational Incompressible Flow   51
2.10 Applications of Analytic Functions in Fluid Mechanics   52
      2.10.1 Flow from a Simple Source and a Simple Vortex   52
               2.10.1.1 Free vortex   53
               2.10.1.2 Two-dimensional doublets and vortex loops .   55
      2.10.2 Flow Around a Body Treated as a Sheet of Complex
               Sources and Doublets   55
      Case study 2.II   Application of Complex Function Analysis to
               the Flow around a Thin Wing   59
      2.10.3 Flow Around a Cylinder with Zero Circulation   62
      2.10.4 Flow Around a Cylinder with Circulation   64
      2.10.5 The Flow Around a Corner   66
2.11 Force on a Body in Steady Two-Dimensional Incompressible
      Ideal Flow   66
2.12 Conformal Transforms   69
Appendix 2.A   Drag in Ideal Flow   70
      2.A.1   Helmholtz's Flow and Separation   71
      2.A.2   Lines of Vortices   72
               2.A.2.1   Single infinite row of vortices   72

2.A.2.2   Two parallel symmetric rows of vortices . . .   73
2.A.2.3   Two parallel alternating rows of vortices . . .   73

3   Viscous Fluids   **75**
3.1   Basic Concept of Viscosity   75
3.2   Differential Motion of a Fluid Element   76
3.3   Strain Rate   76
3.4   Stress   77
3.5   Viscous Stress   78
    3.5.1   Momentum Equation   79
    3.5.2   Energy Equation   79
    3.5.3   Entropy Creation Rate   80
3.6   Incompressible Flow–Navier–Stokes Equation   80
    3.6.1   Vorticity Diffusion   81
    3.6.2   Couette or Plane Poiseuille Flow   81
3.7   Stokes' or Creeping Flow   82
    3.7.1   Stokes' Flow around a Sphere   82
        3.7.1.1   Oseen's correction   85
        3.7.1.2   Proudman and Pearson's solution   85
        3.7.1.3   Lamb's solution for a cylinder   86
3.8   Dimensionless Analysis and Similarity   86
    3.8.1   Similarity and Modelling   88
    3.8.2   Self-similarity   89
Appendix 3.A   Buckingham's Π Theorem and the Complete Set of
Dimensionless Products   90

4   Waves and Instabilities in Fluids   **93**
4.1   Introduction   93
4.2   Small-Amplitude Surface Waves   94
    4.2.1   Surface Waves at a Free Boundary of a Finite Medium   96
        4.2.1.1   Capillary waves   96
        4.2.1.2   Gravity waves   97
        4.2.1.3   Transmission of energy   98
    Case study 4.I   The Wake of a Ship–Wave Drag   99
        4.I.i   Two-dimensional wake, Kelvin wedge   100
4.3   Surface Waves in Infinite fluids   102
    4.3.1   Surface Wave at a Contact Discontinuity   102
    4.3.2   Rayleigh–Taylor Instability   103
4.4   Surface Waves with Velocity Shear Across a
Contact Discontinuity   104
4.5   Shallow Water Waves   106

4.6   Waves in a Stratified Fluid 108
4.7   Stability of Laminar Shear Flow 112
4.8   Nonlinear Instability 115

5 **Turbulent Flow**   **117**
5.1   Introduction 117
        5.1.1   The Generation of Turbulence 119
5.2   Fully Developed Turbulence 121
5.3   Turbulent Stress–Reynolds Stresses 126
5.4   Similarity Model of Shear in a Turbulent Flow–von Karman's
        Hypothesis 127
5.5   Velocity Profile near a Wall in Fully Developed
        Turbulence–Law of the Wall 127
5.6   Turbulent Flow Through a Duct 129
        5.6.1   Prandtl's Distribution Law 130
        5.6.2   Von Karman's Distribution Law 130
        Case study 5.I   Turbulent Flow Through
                       a Horizontal Uniform Pipe 132
                5.I.i   Blasius wall stress correlation 135
        Appendix 5.A   Prandtl's Mixing Length Model 136

6 **Boundary Layer Flow**   **139**
6.1   Introduction 139
6.2   The Laminar Boundary Layer in Steady Incompressible
        Two-Dimensional Flow–Prandtl's Approximation 141
6.3   Laminar Boundary Layer over an Infinite Flat Plate–Blasius's
        Solution 144
6.4   Laminar Boundary Layer–von Karman's Momentum Integral
        Method 146
        6.4.1   Application to Boundary Layers with an Applied
                   Pressure Gradient 149
6.5   Boundary Layer Instability and the Onset
        of Turbulence–Tollmein–Schlichting Instability 151
6.6   Turbulent Boundary Layer on a Flat Smooth Plate 152
        6.6.1   Turbulent Boundary Layer–Power Law Distribution . 154
6.7   Boundary Layer Separation 156
        6.7.1   Viscous Flow Over a Cylinder 159
6.8   Drag 161
        Case study 6.I   Control of Separation in Aerodynamic
                       Structures 163
6.9   Laminar Wake 163

6.10   Separation in the Turbulent Boundary Layer 166
        6.10.1 Turbulent Wake 168
Appendix 6.A   Singular Perturbation Problems and the Method of
        Matched Asymptotic Expansion 169

7 Convective Heat Transfer   **175**
        7.1   Introduction 175
        7.2   Forced Convection 176
               7.2.1   Empirical Heat Transfer Rates from a Flowing Fluid . 178
                        7.2.1.1   Heat transfer from a fluid flowing along a pipe 178
                        7.2.1.2   Heat transfer from a fluid flowing across a pipe 179
                        7.2.1.3   Heat exchanger design 180
                        7.2.1.4   Logarithmic mean temperature 181
               7.2.2   Friction and Heat Transfer Analogies in Turbulent Flow 182
                        7.2.2.1   Reynolds analogy 182
                        7.2.2.2   Prandtl–Taylor correction 183
                        7.2.2.3   Von Karman's correction 184
                        7.2.2.4   Martinelli's correction 186
                        7.2.2.5   Colburn's modification 188
        7.3   Heat Transfer in a Laminar Boundary Layer 189
               7.3.1   Boundary Integral Method 190
        7.4   Heat Transfer in a Turbulent Boundary Layer on a Smooth
               Flat Plate 193
        7.5   Free or Natural Convection 194
               7.5.1   Boussinesq Approximation 196
               7.5.2   Free Convection from a Vertical Plate 198
                        7.5.2.1   Similarity analysis 198
                        7.5.2.2   Boundary layer integral approximation 199
               7.5.3   Free Convection from a Heated Horizontal Plate 201
               7.5.4   Free Convection between Parallel Horizontal Plates . . 202
                        7.5.4.1   Rayleigh–Bénard instability 204
               7.5.5   Free Convection around a Heated Horizontal Cylinder 206
               Case study 7.I   Positive Column of an Arc 206

8 Compressible Flow and Sound Waves   **209**
        8.1   Introduction 209
        8.2   Propagation of Small Disturbances 211
               8.2.1   Plane Waves 212
               8.2.2   Energy of Sound Waves 213
        8.3   Reflection and Transmission of a Sound Wave at an Interface . 214
        8.4   Spherical Sound Waves 215

8.5   Cylindrical Sound Waves 217

9 Characteristics and Rarefactions   219
9.1   Mach Lines and Characteristics 219
9.2   Characteristics 221
      9.2.1   Uniqueness Theorem 222
      9.2.2   Weak Discontinuities 223
      9.2.3   The Hodograph Plane 223
      9.2.4   Simple Waves 223
9.3   One-Dimensional Time-Dependent Expansion 224
      9.3.1   The Centred Rarefaction 226
      9.3.2   Reflected Rarefaction 228
      9.3.3   Isothermal Rarefaction 230
9.4   Steady Two-Dimensional Irrotational Expansion 231
      9.4.1   Characteristic Invariants 232
      9.4.2   Expanding Supersonic Flow around a Corner 235
      9.4.3   Flow around a Sharp Corner–Centred Rarefaction . . 235
              9.4.3.1   The complete Prandtl–Meyer flow 238
              9.4.3.2   Weak rarefaction 239

10 Shock Waves   241
10.1   Introduction 241
10.2   The Shock Transition and the Rankine–Hugoniot Equations . 242
       10.2.1 Rankine–Hugoniot Equations for a Polytropic Gas . . 243
              10.2.1.1 Strong shocks 244
10.3   The Shock Adiabat 245
       10.3.1 Weak Shocks and the Entropy Jump 248
10.4   Shocks in Real Gases 250
10.5   The Hydrodynamic Structure of the Shock Front 254
       10.5.1 Polytropic Gas Shocks 256
              10.5.1.1 Shocks supported by heat transfer 260
       10.5.2 Weak Shocks 261
10.6   The Shock Front in Real Gases 264
10.7   Shock Tubes 267
       10.7.1 Shock Tube Theory 269
10.8   Shock Interaction 271
       10.8.1 Planar Shock Reflection at a Rigid Wall 271
              10.8.1.1 Collision between two planar shocks 274
       10.8.2 Overtaking Interactions 275
              10.8.2.1 Shock overtaking a shock 276
              10.8.2.2 Shock–rarefaction overtaking 276

10.8.2.3 Shock interaction with a contact surface . . . 276
10.9 Oblique Shocks 277
10.9.1 Large Mach Number 281
10.9.2 The Shock Polar 282
10.9.3 Supersonic Flow Incident on a Body 285
10.10 Adiabatic Compression 287
Appendix 10.A An Alternative Approach to the General Conservation Law Form of the Fluid Equations 290
10.A.1 Hyperbolic Equations 290
10.A.2 Formal Solution 291
10.A.3 Discontinuities 292
10.A.4 Weak Solutions 293

11 Aerofoils in Low-Speed Incompressible Flow   295
11.1 Introduction 295
11.1.1 Aerofoils 296
11.2 Two-Dimensional Aerofoils 298
11.2.1 Kutta Condition 299
11.3 Generation of Lift on an Aerofoil 301
11.4 Pitching Moment about the Wing 302
11.5 Lift from a Thin Wing 304
11.6 Application of Conformal Transforms to the Properties
of Aerofoils 309
11.6.1 Blasius's Equation 309
11.6.2 Conformal Mapping of a Circular Cylinder 310
11.6.3 The Lift and Pitching Moment of Aerofoils Generated
by Transformations of a Circle 312
11.7 The Two-Dimensional Panel Method 314
11.8 Three-Dimensional Wings 315
11.8.1 Velocity at the Wing Surface 318
11.8.2 The Force on the Wing 319
11.8.3 Prandtl's Lifting Line Model–Downwash Velocity . . . 320
11.8.4 Lift and Drag as Properties of the Wake 323
Case study 11.I Calculation of Lift and Induced Drag for
Three-Dimensional Wings 327
11.I.i   Wing loading 327
11.I.ii   Elliptic loading 329
11.9 Three-Dimensional Panel Method 330
Appendix 11.A Evaluation of the Principal
Value Integrals 331
Appendix 11.B The Zhukovskii Family of Transformations 332

11.B.1 Zhukovskii Transformation 333
  11.B.1.1 Transformation of a circle to a streamlined
           symmetric body 333
  11.B.1.2 Transformation of a circle to a streamlined
           asymmetric body 333
11.B.2 Karman–Treffetz Transformation 334
11.B.3 Von Mises Transformation 336
11.B.4 Theodorsen's Solution for an Arbitrary Profile 338

12 Aerofoils in High-Speed Compressible Fluid Flow   341
12.1 Introduction 341
12.2 Linearised Theory for Two-Dimensional Flows:
      Subsonic Compressible Flow around a Long Thin
      Aerofoil – Prandtl–Glauert Correction 344
      12.2.1 Improved Compressibility Corrections 347
12.3 Linearised Theory for Two-Dimensional Flows:
      Supersonic Flow about an Aerofoil – Ackeret's Formula 347
12.4 Drag in High-Speed Compressible Flow 350
      12.4.1 Swept Wings 350
      12.4.2 Drag in Supersonic Flow 351
      12.4.3 Transonic Flow 351
12.5 Linearised Theory of Three-Dimensional Supersonic Flow –
      von Karman Ogives and Sears–Haack Bodies 354
      12.5.1 Whitcomb Area Rule 358
      Case study 12.I Hypersonic Wing 359

13 Deflagrations and Detonations   363
13.1 Introduction 363
      13.1.1 Deflagrations 363
             13.1.1.1 Propagating burn 364
             13.1.1.2 Deflagration propagating in a closed tube . . 367
      13.1.2 Detonations 367
13.2 Detonations, Deflagrations and the Hugoniot Plot 368
      13.2.1 The Structure of a Deflagration 373
             13.2.1.1 The Shvab–Zel'dovich model of a deflagration 374
             13.2.1.2 Detonations as deflagrations initiated
                      by a shock 375
      13.2.2 Chapman–Jouget Hypothesis 376
      Case study 13.I   Deflagrations and Detonations in
             Laser–Matter Breakdown 377
             13.I.i   Solid targets 378

13.I.i.a   High-intensity irradiation –
deflagration model 379

13.I.i.b   Low-intensity irradiation –
self-regulating model 380

13.I.ii   Gaseous targets 381

**14 Self-similar Methods in Compressible Gas Flow
and Intermediate Asymptotics**   **383**

14.1   Introduction 383

14.2   Homogeneous Self-similar Flow of a Compressible Fluid 386

14.2.1   General Homogeneous Expansion of a Compressible
Gas 386

14.2.1.1   Adiabatic flow 389

14.2.1.2   Isothermal flow 389

14.2.2   Homogeneous Adiabatic Compression 390

14.2.2.1   Homogeneous collapse of spheres 390

14.2.2.2   Homogeneous collapse of shells 393

14.3   Centred Self-similar Flows 395

14.4   Flow Resulting from a Point Explosion in Gas – Blast Waves . 397

14.5   Adiabatic Collapse of a Sphere 402

14.6   Convergent Shock Waves – Guderley's Solution 407

14.6.1   Compression of a Shell and Collapse of Fluid
into a Void 412

Case study 14.I   The Fluid Dynamics of Inertial Confinement
Fusion 414

14.I.i   Basic principles 414

14.I.i.a   Hydrodynamic compression 415

Problems   **417**

Solutions   **427**

Bibliography   **455**

Index   **463**

Preface

Every physicist is familiar with the two revolutions in thought which took place at the turn of the nineteenth and twentieth centuries. These revolutions in modern physics in the guise of relativity and quantum physics form essential components of today's undergraduate physics courses. Despite the fact that it has played a critical role in the development of the modern world, underlying the design of aircraft and industrial plants, the third contemporary revolution in fluid dynamics is less familiar, yet it was also necessary to eliminate paradoxes inherent in nineteenth-century physics. Fluid mechanics has become a neglected subject in the modern physics curriculum. The reason for this is twofold. Firstly, the subject is an area of 'classical physics', now generally regarded by students as 'old fashioned', yet one area, namely turbulence, remains one of the most intractable problems in physics at the forefront of complexity. Secondly, much of the field was developed by engineers and applied mathematicians. Until recently physicists have been less involved. In recent years this has changed as environmental physics, and plasma physics, have increasingly required a working knowledge of fluid dynamics. More generally, when physicists are expected to act as 'troubleshooters' in many fields of applied physics, a basic knowledge of fluid mechanics is needed.

The dominant problem in nineteenth-century fluid mechanics was drag (the resistance of the moving flow on a stationary body) in the flow of fluids with low viscosity. Stokes (1851) had solved the problem of the drag force experienced by a sphere in a flow dominated by viscosity. However, it was well known that fluids of low viscosity also generated drag, which had been measured and analysed notably by Froude in 1871, who measured the drag due to a ship's motion. The revolution in thought referred to above followed two seminal papers. The first by Reynolds (1883) introduced turbulence, namely that fluid motion could be disorganised and chaotic. The second was by Prandtl (1904), in which he introduced not only the concept of the boundary layer, but also flow separation (in only 10 pages!). These two papers opened the way to the resolution of this central problem of nineteenth-century fluid mechanics: that is, the drag force exerted by a moving fluid on a body immersed in it.

The development of the theory of compressible flow also dates from the late nineteenth-century, most importantly with the introduction of characteristics by Riemann (1860). Waves of discontinuity (shock waves) were propounded by Rankine (1870) and Hugoniot (1887) between 1870 and 1890, but were generally discounted on theoretical grounds despite observations, e.g. by Mach and Salcher (1887) and Boys (1893), in back-lit photographs of the motion of supersonic projectiles. It was not until the work of Rayleigh (1910), Taylor (1910), and more fully Becker (1922), demonstrated a narrow 'shock layer' supported by dissipation in a thin zone between two regions of dissipationless flow, that it became accepted that shock waves in compression existed, stabilised by thermal conduction and viscosity.

Both the problems associated with the boundary layer and shock wave stemmed from a lack of understanding of the role of viscosity in nearly inviscid fluids. It was assumed that because the viscosity of most fluids was known to be small, the flow could be accurately described by the 'ideal fluid' equations in which the viscosity is equal to zero. In fact, even though the viscosity is small, viscous forces are strong in regions of large velocity gradient, respectively adjacent to the surface of a solid body and in the shock 'discontinuity'. The equations for an ideal fluid are therefore an 'asymptote' of those for a real fluid in the limit when the viscosity tends to zero. The singularities in the inviscid fluid equations, which are associated with the difference between zero viscosity and infinitesimally small viscosity, consequently disappear in a full treatment. More recently, new mathematical methods, based on the generalisation of the theory of functions through distributions, have allowed the singularities in ideal flow to be formally treated directly (Lax, 1954).

One further development in physics at the turn of the nineteenth to the twentieth century has played a major role in fluid mechanics. This was the evolution of dimensional analysis by Rayleigh (1899) and later Buckingham (1914). This led to the development of the methods of similarity and modelling, which are widely used to analyse experimental measurements.

This book is intended to give a pedagogical summary of the physics of fluid flow. Thus it builds on several classic texts, which cover specific aspects of the field in more detail than is possible here. In appropriate places the reader is referred to these for further study. The book includes both the applied mathematical development, which underlies much of the subject, and results from the more empirical engineering approach. Although lacking the rigour of the former, the latter are of equal importance to the working physicist. Unfortunately limitations of space have also prevented the discussion of two topics, which should have been included, were the book to be inclusive. These are, firstly, flows in a rotating environment, important in geophysics and meteorology, and, secondly, computational fluid dynamics. A full discussion of modern developments in turbulence was also not deemed appropriate, and a more engineering type of approach has been adopted.

Experimental fluid mechanics is a very visual subject. Many beautiful illustrations of effects described in this book have been published. Rather than extensive reproductions of these photographs, the reader is referred to relevant pictures in van Dyke (1982) or the classic photographs of Prandtl and Tietjens (1957). The 'internet' also provides an easily accessible source of much illustrative material.

Ideal inviscid fluid flow provides the basis for solving many problems of practical importance involving waves and instabilities, and underlying boundary layer theory. Its mathematical tractability allows it to be used for many of the methods of computational fluid dynamics. It is therefore important to give some background to the analytic treatment of ideal fluid mechanics, which occupies Chapters 1 and 2. Viscosity is introduced in Chapter 3, together with a brief account of dimensional analysis. Waves and instabilities, which underlie many aspects of fluid dynamics, are discussed in Chapter 4. Turbulence is introduced in Chapter 5 and is treated predominately phenomenologically, with only a brief nod to modern analysis. Boundary layers in Chapter 6 cover the basic Prandtl approach, treating only simple problems, and development into separation and drag. Chapter 7 gives a brief account of the engineering approach to heat transfer.

Compressible fluids and the characteristic problems associated with them fill the remainder of the book. Chapter 8 gives a brief introduction to sound waves. Rarefaction flow, treated by the method of characteristics, occupies Chapter 9. Compression and shock waves are introduced in Chapter 10 studying both the foundations of the theory of shock waves and their application. The behaviour of fluid flow around aerofoils and wings occupies Chapters 11 and 12, the first for subsonic and the second for transonic and supersonic flight. Detonation and deflagration associated with flames and explosives are treated in Chapter 13. The book is concluded with Chapter 14 describing the application of self-similar methods applied to example problems in compressible flow.

It is expected that the book will be used by final year undergraduates and postgraduates in physics and applied mathematics. A good working knowledge of vector calculus and the functions of a complex variable is therefore assumed. Dimensional analysis is an essential tool of the fluid dynamicist, and some knowledge is expected. A brief introduction to Buckingham's Π theorem is given for those who have not previously met this approach.

The basic development of the subject forms the main text. Sections in small type are intended for reference rather than parts of the main development of the text. However, examples of applications are included either as case studies or as problems, whose solutions may be found at the end of book. Some specific points of mathematical development or of historical interest appear as appendices.

Chapter 1

Introduction

1.1 Fluids as a State of Matter

A standard dictionary definition of a fluid is

> *a substance whose particles can move about*
> *with freedom – a liquid or gas.*

Whilst this formulation encapsulates our general concept of a fluid, it is not entirely satisfactory as a scientific basis for the understanding of such materials. More formally within the context of *fluid mechanics* the fluid is seen as an isotropic, locally homogeneous, macroscopic material whose particles are free to move within the constraints established by the dynamical laws of continuum physics. The requirement that the fluid be a continuum implies that if a volume of fluid is successively subdivided into smaller elements, each element will remain structurally similar to its parent, and that this process of subdivision can be carried out down to infinitesimal volumes. Under these conditions several useful macroscopic concepts may be defined:

Fluid particle   a fictitious particle fixed within the fluid continuum and moving with the velocity of the flow, and representing an average over a large number of microscopic particles.

Fluid point   fixed in the fluid moving with the flow velocity. A fluid particle is always situated at the same fluid point.

Infinitesimal volume   within the continuum of the fluid, and large compared with microscopic scales, but small compared with macroscopic ones.

Introductory Fluid Mechanics for Physicists and Mathematicians, First Edition. Geoffrey J. Pert.
© 2013 John Wiley & Sons, Ltd. Published 2013 by John Wiley & Sons, Ltd.

In fact of course the fluid is not a continuum in the strict mathematical sense used above. The fluid is made up of discrete microscopic particles, namely molecules, which are distributed randomly with a distribution of velocities characteristic of the fluid in thermal equilibrium, typically given by the Maxwell–Boltzmann distribution in a gas. Fortunately, at the densities at which most experiments are conducted, the intermolecule separation is extremely small and very much less than the laboratory scale. It is therefore possible to average over small volumes which contain a very large number of particles, yet are very small on the laboratory scale, and allow us to recover the continuum approximation. In this manner we obtain terms which characterise the fluid as a bulk material. Typical of these average quantities are:

Density   number or mass of particles per unit volume.

Temperature   average energy of the random motion per particle in thermal equilibrium.

Pressure   average momentum flow associated with the random motion per unit area.

Flow velocity   mean velocity of the molecules averaging out the random motion.

The role of collisions amongst the particles plays an important role in defining irreversibility through the loss of correlation between the particles. Particles collide on average after a distance equal to the mean free path, and time after the collision interval. Since fluid mechanics assumes the fluid particles are in thermal equilibrium and randomly distributed, this condition requires that spatial and temporal averages be taken to include a large number of collisions, i.e. the laboratory-scale length is large compared with the mean free path and time to the collision interval. In practice this is not normally a restrictive condition. The effects of the collisions on fluid transport (momentum and energy) are thereby averaged over the thermal distribution to yield bulk properties of the material, namely viscosity and thermal conduction respectively. Consequently (ideal) fluid motion without viscosity or thermal conduction is dissipationless, entropy generation being due to viscosity and thermal conduction.

Within the continuum theory it is implicitly assumed that locally the fluid is in thermal equilibrium, although the temperature may vary globally through the flow. As a result the thermodynamics of bulk materials may be applied locally in the flow to calculate the pressure from the density and temperature (say) using the equation of state of the fluid. More generally the quantities and relations of equilibrium thermodynamics may be applied in the flow.

The flow of a basic fluid may be calculated using Newtonian mechanics, classical thermodynamics and the values of viscosity and thermal conductivity. From the above discussion, the conditions under which this theory may be applied are:

> *Laboratory-scale lengths must be large compared with the intermolecule separation and mean free path.*
> *Characteristic laboratory times must be large compared with the collision interval.*
> *The fluid must locally be in thermal equilibrium.*

The theory may be readily extended to relativistic mechanics and also to include additional dissipative terms, e.g. due to radiation. However, under normal laboratory conditions these are not required. Astrophysical systems provide examples of flows where more general approaches may be required.

Provided the above conditions are met, it is relatively straightforward to show that the fluid dynamical equations (to be obtained later) may be directly derived from the governing kinetic theory of the molecules.

1.2   The Fundamental Equations for Flow of a Dissipationless Fluid

The basic equations of fluid mechanics stem from simple concepts of conservation applied to mass, momentum and energy. These are completed by the thermodynamic equation of state of the material, in which the flow is to be calculated. The equations are, of course, complemented by the boundary conditions in an appropriate form, depending on the nature of the problem. In any problem, we seek to find five variables: three velocity components ($\mathbf{v}$) and two thermodynamic state variables, e.g. density ρ and pressure p, as functions of space $\mathbf{r}$ and time t. In many problems the actual number of variables required is reduced, either by symmetry to a restricted number of spatial dimensions or by a specified thermodynamic state, e.g. constant entropy or constant temperature. The problem is often further simplified by the restriction to *steady flow*, when there is no time variation.

Initially we will consider only dissipationless or *ideal* flow where the entropy of a fluid particle remains constant, i.e. viscosity and thermal conduction are neglected, deferring the treatment of flows in which viscosity plays a role until later; many important systems are treatable within the inviscid limit. We may quite generally identify two different conditions of flow involving the entropy of the fluid: *adiabatic flow* when the specific entropy of a fluid particle is constant in time; and *isentropic flow* where the specific entropy of each fluid particle has the same initial value. Many flows are both isentropic and adiabatic, e.g. the ideal steady flow of a fluid, whose specific entropy on entry is everywhere constant.

The basic equations may be formulated in two complementary ways:

- In the frame of the laboratory–the Eulerian frame. In this frame the co-ordinates are fixed in space and time. The derivatives used are the usual partial derivatives

$$\frac{\partial}{\partial t}\bigg|_{\mathbf{r}} \quad \text{and} \quad \nabla|_{t}$$

- In the frame of the moving particle–the Lagrangian frame. In this system the spatial variation seen by the particle due to its motion is absorbed into the time derivative

$$\frac{\mathrm{d}}{\mathrm{d}t} = \frac{\partial}{\partial t} + \mathbf{v} \cdot \nabla \tag{1.1}$$

This system is often easier to set up, but becomes more complicated when the dissipative terms, viscosity and thermal conduction, are important.

However, the two systems are entirely equivalent and each may be easily derived from the other. They may also be used mixed if required. The actual choice of which to use will depend on the nature of the problem.

1.3 Lagrangian Frame

The Lagrangian frame of reference considers the fluid from the point of view of an observer on a fluid particle. Since many methods of calculation in computational fluid mechanics use the Lagrangian approach, we give a brief formal introduction to these methods. A fluid particle may be conveniently identified by a co-ordinate set, which is fixed on the particle, namely $\mathbf{\Lambda} = (\lambda, \mu, \nu)$, i.e. a triad of numbers. For example, these may be the initial position of the particle $\mathbf{r}_0 = (x_0, y_0, z_0)$. The position, velocity and thermodynamic state of the particle are therefore functions of time alone. Conceptually this leads to a simple set of kinematic and dynamic relations governing the motion of the particle, namely

$$\frac{\mathrm{d}\mathbf{r}}{\mathrm{d}t} = \mathbf{v} \qquad \text{and} \qquad \frac{\mathrm{d}\mathbf{v}}{\mathrm{d}t} = \frac{\mathbf{F}}{m} \tag{1.2}$$

where $\mathbf{F}$ is the force acting on and m the mass of the fluid particle. The particle has a finite size expressed by the increments in the Lagrangian co-ordinates $\delta\Lambda_i = (\delta\lambda, \delta\mu, \delta\nu)$, and whose volume is given by the Jacobian

$$\delta V = \frac{\partial(x,y,z)}{\partial(\lambda,\mu,\nu)} \, \delta\lambda \, \delta\mu \, \delta\nu \tag{1.3}$$

which can be expressed as[1]

$$J = \frac{\partial(x,y,z)}{\partial(\lambda,\mu,\nu)} = \frac{1}{N!} \left\{ \varepsilon_{ijk}\, \varepsilon_{lmm} \frac{\partial x_i}{\partial\Lambda_l} \frac{\partial x_j}{\partial\Lambda_m} \frac{\partial x_k}{\partial\Lambda_n} \right\}$$
$$= \left[\frac{\partial\mathbf{r}}{\partial\Lambda_1} \wedge \frac{\partial\mathbf{r}}{\partial\Lambda_2} \right] \cdot \frac{\partial\mathbf{r}}{\partial\Lambda_3} \tag{1.4}$$

where ε_{ijk} is the perturbation symbol[2]

$$\varepsilon_{ijk} = \begin{cases} 1 & \text{if } (i \neq j \neq k) \text{ are in the sequence (1,2,3)} \\ -1 & \text{if } (i \neq j \neq k) \text{ are in the sequence (1,3,2)} \\ 0 & \text{otherwise} \end{cases} \tag{1.5}$$

Spatial derivatives of quantities associated with the fluid particles, e.g. thermodynamic variables, are directly calculated in a Lagrangian framework. The gradient of a scalar quantity $f(\lambda,\mu,\nu)$, which is defined on the fluid particle, is often required in the inertial Eulerian frame. Such terms are obtained by the use of the total differential for the variable f and Cramer's rule to solve

---

[1]We make extensive use of the index notation for vectors, where the vector is represented by a general component in a Cartesian co-ordinate system

$$\mathbf{A} \equiv (A_x, A_y, A_z) \equiv A_i \qquad \text{where} \quad i = 1, 2, 3$$

Sums over the indices are represented by Einstein's repeated index summation notation. Thus for example a scalar product is

$$\mathbf{A} \cdot \mathbf{B} = A_i B_i \equiv \sum_{i=1}^{N} A_i B_i$$

A repeated index indicates summation of that index over the full range of values N. The summation rule also applies to the elements of matrices.

[2]The expansion of the determinant of an $N \times N$ matrix $A = a_{ij}$ may be written as

$$\det A = \varepsilon_{ijk} a_{i1}\, a_{j2}\, a_{k3} = \varepsilon_{lmn} a_{1l} a_{2m} a_{3n} = \frac{1}{N!} \varepsilon_{ijk}\, \varepsilon_{lmn} a_{il}\, a_{jm}\, a_{kn}$$

the divisor $N!$ appearing because the first index l may be chosen in N different ways, the second m in $N-1$ ways, etc.

the resulting set of simultaneous equations.[3] Using the subscript notation and Einstein's repeated index summation rule gives

$$\nabla f|_i \equiv \frac{\partial f}{\partial x_i} = \frac{1}{N!\,J}\left\{\varepsilon_{ijk}\,\varepsilon_{lmn}\,\frac{\partial f}{\partial \Lambda_l}\,\frac{\partial x_j}{\partial \Lambda_m}\,\frac{\partial x_k}{\partial \Lambda_n}\right\}$$ (1.7)

where $N = 3$ is the dimensionality. The calculation of vector operators in Eulerian space, grad, div and curl, follows directly.

Taking the time derivative of the Jacobian, remembering that $\mathbf{v}_i = \mathrm{d}r_i/\mathrm{d}t$ and using equation (1.7) we obtain

$$\frac{\mathrm{d}J}{\mathrm{d}t} = J\,\nabla \cdot \mathbf{v}$$ (1.8)

Since the mass δm of the particle is constant,

$$\rho\,J = \delta m/\delta\lambda\,\delta\mu\,\delta\nu = \rho_0\,J_0$$ (1.9)

where the initial density is ρ_0 and the Jacobian J_0. This is the Lagrangian form for the conservation of mass. The specific volume of the particle is clearly related to the Jacobian through

$$V = \frac{1}{\rho} = \frac{J}{J_0}V_0$$

and making use of equation (1.8) we obtain the more familiar form of the Lagrangian mass conservation equation

$$\frac{\mathrm{d}\rho}{\mathrm{d}t} + \rho\nabla \cdot \mathbf{v} = 0$$

1.3.1 Conservation of Mass

This equation may be derived in a more direct manner by considering the change in volume of a fluid particle ΔV with constant mass $\Delta m = \rho\Delta V$ and

---

[3]Cramer's rule solves the non-singular set of simultaneous equations

$$a_{ij}\,x_i = b_j$$

by forming a set of determinants for the matrix elements $D = \det a_{ij}$ and those formed by progressively replacing the ith column by the column 'vector' b_j, namely

$$D_i = \begin{vmatrix} a_{i'j} & \text{when} & i' \neq i \\ b_j & & i' = i \end{vmatrix}$$

The solution is then

$$x_i = D_i/D$$ (1.6)

surface ΔS as it moves through the fluid. In a time δt the volume increases by

$$
\begin{aligned}
\delta(\Delta V) &= \int_{\Delta S} \mathbf{v} \cdot d\mathbf{S}\, \delta t \\
&= \int_{\Delta V} \nabla \cdot \mathbf{v}\, dV\, \delta t
\end{aligned}
$$

by Gauss's theorem. As ΔV is small, the rate of dilation is

$$
\dot{\Theta} = \frac{1}{\Delta V} \lim_{\delta t \to 0} \left\{ \frac{\delta(\Delta V)}{\delta t} \right\} = \nabla \cdot \mathbf{v} = -\frac{1}{\rho}\frac{d\rho}{dt}
$$

since the mass of the particle is constant. Hence we obtain

$$
\frac{d\rho}{dt} + \rho \nabla \cdot \mathbf{v} = 0 \tag{1.10}
$$

If the density of the fluid particle remains constant, i.e. the fluid is incompressible,

$$
\nabla \cdot \mathbf{v} = 0 \tag{1.11}
$$

and the rate of dilation is zero, or alternatively the volume of a fluid element is constant.

1.3.2 Conservation of Momentum–Euler's Equation

We consider the change of momentum of the fluid particle as a result of the forces applied to it. The total force is due to the pressure exerted inwards over the surface of the particle

$$
-\int_{\Delta S} p\, d\mathbf{S} = -\int_{\Delta V} \nabla p\, dV \approx -\nabla p\, \Delta V
$$

and gravity $\Delta m\, \mathbf{g}$ where $\mathbf{g}$ is the acceleration due to gravity. Hence using Newton's second law of motion we obtain Euler's equation:

$$
\frac{d\mathbf{v}}{dt} = -\frac{1}{\rho} \nabla p + \mathbf{g} \tag{1.12}
$$

The preceding equation for flow in an inertial frame must be modified to include the Coriolis and centrifugal forces in a rotating frame

$$
\frac{d\mathbf{v}}{dt} + 2\,\mathbf{\Omega} \wedge \mathbf{v} + \mathbf{\Omega} \wedge (\mathbf{\Omega} \wedge \mathbf{r}) = -\frac{1}{\rho} \nabla p + \mathbf{g} \tag{1.13}
$$

where $\mathbf{\Omega}$ is the angular velocity of rotation.

1.3.3   Conservation of Angular Momentum

Angular momentum is not frequently used in fluid mechanics. However, as in mechanics, it must be conserved in the absence of external torques. We may obtain the relationship governing its variation directly by taking the vector product of the radius vector with Euler's equation (1.12). Thus the angular momentum per unit mass of a fluid particle is

$$\frac{d(\mathbf{r} \wedge \mathbf{v})}{dt} = \mathbf{r} \wedge \frac{d\mathbf{v}}{dt} = -\frac{1}{\rho}\mathbf{r} \wedge \nabla p \, \mathbf{r} \wedge \mathbf{g} \qquad (1.14)$$

1.3.4   Conservation of Energy

Since the fluid is dissipationless, the energy equation takes the particularly simple form of the first law of thermodynamics for an adiabatic change:

$$\frac{d\epsilon}{dt} = \frac{p}{\rho^2}\frac{d\rho}{dt} = -\frac{p}{\rho}\nabla \cdot \mathbf{v} \qquad (1.15)$$

where ϵ is the specific internal energy (per unit mass).

1.3.5   Conservation of Entropy

If dissipation in the fluid is negligible, i.e. the flow of an *ideal fluid*, the entropy of a fluid element is constant. Therefore

$$\frac{ds}{dt} = 0 \qquad (1.16)$$

where s is the specific entropy (per unit mass).

1.4   Eulerian Frame

We turn now to the equations in Eulerian form.

1.4.1   Conservation of Mass–Equation of Continuity

Consider a fluid of density ρ moving with velocity $\mathbf{v}$ in a closed volume V stationary in the Eulerian frame with bounding surface S. The mass of fluid enclosed in V is $\int_V \rho \, dV$. Thus the rate of increase of mass in V is

$$\int_V \frac{\partial \rho}{\partial t} \, dV$$

This mass gain must be balanced by a mass flow rate into V through S. Since the mass flow rate out through an element $\mathrm{d}\mathbf{S}$ is $\rho\mathbf{v}\cdot\mathrm{d}\mathbf{S}$, we obtain

$$\int_V \frac{\partial\rho}{\partial t}\,\mathrm{d}V = -\int_S \rho\,\mathbf{v}\cdot\mathrm{d}\mathbf{S}$$

$$= -\int_V \nabla\cdot(\rho\,\mathbf{v})\,\mathrm{d}V \qquad (1.17)$$

Hence, since V is arbitrary, the integrands must be equal, i.e.

$$\frac{\partial\rho}{\partial t} + \nabla\cdot(\rho\,\mathbf{v}) = 0 \qquad (1.18)$$

namely, the equation of continuity.

This equation has the characteristic form of a conservation equation, i.e.

$$\frac{\partial}{\partial t}[\text{Quantity per unit volume}] \quad + \quad \mathrm{div}[\text{Flux of quantity}]$$

$$= \quad [\text{Input per unit volume per unit time}]$$

$$\{ \quad = \quad 0 \text{ in this case}\}$$

where the flux is the quantity flowing per unit time through unit area normal to the flow (see Appendix 10.A). In this case the flux $\mathbf{j} = \rho\,\mathbf{v}$.

1.4.2   Conservation of Momentum

Momentum introduces a complication in that momentum itself is a vector. The momentum flux is a tensor. We therefore work in Cartesian components using the general notation, i.e. u_i for the ith component ($i = 1, 2, 3\,[x, y, z]$). We also use the Einstein summation convention for a repeated index, namely $a_i\,b_i = \sum_{i=1}^{3} a_i b_i = \mathbf{a}\cdot\mathbf{b}$. The total momentum in V is thus $\int_V \rho\,v_i\,\mathrm{d}V$, and the flow of momentum leaving through $\mathrm{d}\mathbf{S}$ is $\rho\,v_i\,v_j\,\mathrm{d}S_j$. The sources of momentum in V are the forces: the internal force due to the hydrostatic pressure $-p\,\mathrm{d}S_i$ (minus sign since pressure acts inwards) and the external force due to gravity ρg_i per unit volume.

The momentum balance equation for V is thus

$$\int_V \frac{\partial(\rho\,v_i)}{\partial t}\,\mathrm{d}V = -\int_S \rho\,v_i\,v_j\,\mathrm{d}S_j - \int_S p\,\mathrm{d}S_i + \int_V \rho\,g_i\,\mathrm{d}V$$

$$= \int_V \rho\,g_i - \frac{\partial}{\partial x_j}[\rho\,v_i\,v_j + p\,\delta_{ij}]\,\mathrm{d}V \qquad (1.19)$$

where δ_{ij} is the Kronecker delta.

Hence, as before, since V is arbitrary

$$\frac{\partial(\rho\,v_i)}{\partial t} + \frac{\partial(\rho\,v_i\,v_j + p\,\delta_{ij})}{\partial x_j} = \rho\,g_i \qquad (1.20)$$

This has the general form noted above since $\partial a_j/\partial x_j = \nabla \cdot \mathbf{a}$. The momentum flux $\Gamma_{ij} = \rho\, v_i\, v_j + p\,\delta_{ij}$ includes the internal force, in this case pressure alone. By Newton's second law of motion, this force corresponds to an impulse which transfers momentum within the fluid body, but is conserved overall, as one part of the fluid exerts an equal and opposite force (and therefore momentum transfer) on another. The external force, gravity, corresponds to a source term, which is not conserved.

1.4.3 Conservation of Angular Momentum

Angular momentum obeys a conservation law in the absence of external torques. However, the form is a little more difficult to establish than for linear momentum. As with linear momentum we expect that the angular momentum flux will be a second-order tensor. We may obtain the relations directly by considering the conservation of angular momentum in an arbitrary volume V with surface S. The rate of change of the total angular momentum in V must be balanced by the flow of angular momentum through the surface S due to transport and to the torques exerted on the fluid at S by the pressure and internally by any volume force. Thus

$$\frac{\partial}{\partial t}\int_V \rho\,(\mathbf{r}\wedge\mathbf{v})\,dV = -\int_S [\rho\,\mathbf{r}\wedge\mathbf{v}]\,\mathbf{v}\cdot d\mathbf{S} - \int_S p\,\mathbf{r}\wedge\,d\mathbf{S} + \int_V \rho\,\mathbf{r}\wedge\mathbf{g}\,dV$$

As before, noting that V is arbitrary and using Gauss's theorem,[4] we obtain

$$\frac{\partial}{\partial t}[\rho\,\mathbf{r}\wedge\mathbf{v}] + \frac{\partial}{\partial x_j}[\rho\,(\mathbf{r}\wedge\mathbf{v})_i\,v_j] + \mathbf{r}\wedge\nabla p = \mathbf{r}\wedge(\rho\,\mathbf{g}) \qquad (1.21)$$

which may also be obtained by taking the vector product of $\mathbf{r}$ with equation (1.20). This is clearly not in conservation law form due to the pressure term. This can, however, be written as the divergence of a second-order tensor as follows:

$$\mathbf{r}\wedge\nabla p = \varepsilon_{ijk}\,x_j\,\frac{\partial p}{\partial x_k} = -\varepsilon_{ijk}\,\frac{\partial}{\partial x_j}(p\,x_k)$$

where ε_{ijk} is the perturbation symbol defined in equation (1.5).

---

[4]This result is obtained from the vector identities

$$\nabla\cdot(\mathbf{A}\wedge\mathbf{B}) = \mathbf{B}\cdot\nabla\wedge\mathbf{A} = \mathbf{A}\cdot\nabla\wedge\mathbf{B} \qquad \text{and} \qquad \nabla\wedge(\phi\,\mathbf{A}) = \phi\,\nabla\wedge\mathbf{A} - \mathbf{A}\wedge\nabla\phi$$

Hence using Gauss's theorem when $\mathbf{a}$ is an arbitrary constant vector

$$\mathbf{a}\cdot\int p\,\mathbf{r}\wedge d\mathbf{S} = \int \mathbf{a}\wedge(p\mathbf{r})\cdot d\mathbf{S} = \int \nabla\cdot(\mathbf{a}\wedge p\mathbf{r})\,dV = \mathbf{a}\cdot\int \mathbf{r}\wedge\nabla p\,dV$$

from which, since $\mathbf{a}$ is arbitrary, equation (1.21) follows.

Hence the total angular momentum flux is

$$\Xi_{ij} = \rho v_i \left(\varepsilon_{jk\ell}\, x_k\, v_\ell \right) - \left(\varepsilon_{ijk}\, p\, x_k \right)$$

1.4.4   Conservation of Energy

The total energy per unit mass of the fluid includes both internal energy ϵ and kinetic energy $\frac{1}{2}v^2$. The work done on the fluid is due to the pressure force on the surface and to gravity. Thus

$$\int_V \frac{\partial}{\partial t} \left[\rho \left(\epsilon + \frac{1}{2}v^2 \right) \right] dV = - \int_S \left[\rho \left(\epsilon + \frac{1}{2}v^2 \right) \right] \mathbf{v} \cdot d\mathbf{S}$$

$$- \int_S p\,\mathbf{v} \cdot d\mathbf{S} + \int_V \rho\,\mathbf{g} \cdot \mathbf{v}\, dV + \int_V W\, dV$$

where W is the energy deposited by external sources per unit volume per unit time. Hence we obtain by the use of Gauss's theorem and the arbitrary nature of the volume V, as before,

$$\frac{\partial}{\partial t} \left[\rho \left(\epsilon + \frac{1}{2}v^2 \right) \right] + \nabla \cdot \left[\rho \left(h + \frac{1}{2}v^2 \right) \mathbf{v} \right] = W + \rho\,\mathbf{g} \cdot \mathbf{v} \qquad (1.22)$$

where $h = \epsilon + p/\rho$ is the specific enthalpy (per unit mass). The energy flux is thus $(h + \frac{1}{2}v^2)\,\mathbf{v}$ and includes in the enthalpy, h, a term for the work done by one section of the fluid on another corresponding to the internal forces. The work done by the external force is not conserved.

Since gravity is a conservative force, which is constant in time, we may include the gravitational field in overall fluid energy $(\epsilon + \frac{1}{2}v^2 + U)$ per unit mass where $\mathbf{g} = -\nabla U$ defines the gravitational potential. Using the equation of continuity (1.18) and the time invariance of the gravitational field, U, we obtain

$$\frac{\partial}{\partial t} \left[\rho \left(\epsilon + \frac{1}{2}v^2 + U \right) \right] + \nabla \cdot \left[\rho \left(h + \frac{1}{2}v^2 + U \right) \mathbf{v} \right] = W \qquad (1.23)$$

1.4.5   Conservation of Entropy

In an ideal fluid, where entropy is conserved, entropy may also be written as a conservation law. From equations (1.16) and (1.18) we obtain

$$\frac{\partial}{\partial t}\,(\rho\, s) + \nabla \cdot (\rho\, s\, \mathbf{v}) = 0 \qquad (1.24)$$

1.5 Hydrostatics

Consider the situation where the fluid is at rest. It follows from Euler's equation (1.12) (or directly) that the pressure force and gravity must balance, i.e.

$$\nabla p = \rho \mathbf{g} \qquad (1.25)$$

Since gravity only acts in the vertical direction (z, measured upwards) this equation takes the simple form when the density is constant

$$p = -\rho g z + \text{const}$$

$\rho g z$ is known as the hydrostatic head.

If the thermodynamic condition of the fluid is predetermined, the pressure and density are related by the appropriate equation of state and are both functions of the vertical height only. The fluid is therefore stratified. We may identify two important cases.

1.5.1 Isothermal Fluid–Thermal and Mechanical Equilibrium

If the fluid is isothermal, i.e. the temperature T is constant everywhere, the system is in thermal equilibrium. The equilibrium condition is written in terms of the thermodynamic potential per unit mass Φ given by

$$d\Phi = -s\,dT + V\,dp$$

where s is the specific entropy (per unit mass) and $V = 1/\rho$ the specific volume. Hence equation (1.25) integrates to

$$\Phi + gz = \text{const} \qquad (1.26)$$

throughout the fluid.

This result is recognised as the standard result from thermodynamics for a system in thermal equilibrium in an external field.

1.5.2 Adiabatic Fluid–Lapse Rate

If the fluid is isentropic, the specific entropy of the fluid is everywhere constant. Since the enthalpy is

$$dh = T\,ds + V\,dp$$

equation (1.25) takes the form

$$h + gz = \text{const} \qquad (1.27)$$

Alternatively, making use of the thermodynamic relations

$$\left.\frac{\partial T}{\partial p}\right|_s = \left.\frac{\partial V}{\partial s}\right|_p = \left.\frac{\partial T}{\partial s}\right|_p \left.\frac{\partial V}{\partial T}\right|_p = \frac{\alpha V T}{c_p} \tag{1.28}$$

where α the coefficient of volume expansion and c_p the heat content at constant pressure (per unit mass),

$$\alpha = \left.\frac{1}{V}\frac{\partial V}{\partial T}\right|_p \quad \text{and} \quad c_p = \left.\frac{\partial h}{\partial T}\right|_p = \left.T\frac{\partial s}{\partial T}\right|_p \tag{1.29}$$

The temperature varies as a function of the height alone

$$\frac{\mathrm{d}T}{\mathrm{d}z} = \left.\frac{\partial T}{\partial p}\right|_s \frac{\mathrm{d}p}{\mathrm{d}z} = \frac{\alpha V T}{C_p}\frac{\mathrm{d}p}{\mathrm{d}z} = -\frac{\alpha T}{c_p}g \tag{1.30}$$

The rate at which the temperature decreases with height in the atmosphere, namely $\Gamma = -\mathrm{d}T/\mathrm{d}z$, and the corresponding density and pressure changes, are known as the adiabatic lapse rate. A very simple alternative derivation of this result is useful. Starting from the second $T\mathrm{d}S$ equation of thermodynamics we may write directly that

$$T\mathrm{d}s = c_p\mathrm{d}T - \left.T\frac{\partial V}{\partial T}\right|_p \mathrm{d}p = 0 \tag{1.31}$$

since $\mathrm{d}s = 0$ for an adiabatic change. Hence using equation (1.25) we obtain equation (1.30).

For a perfect gas obeying the ideal gas laws, $\alpha = 1/T$ and the lapse rate takes the particularly simple form $\Gamma = g/c_p$.

In atmospheric physics several distinct lapse rates are identified:

1. **Dry adiabatic lapse rate** is the rate of decrease of temperature with height of dry (unsaturated) air under adiabatic conditions. Since air is approximately a perfect gas the lapse rate is given by $g/c_p \approx 9.8\,\mathrm{K/km}$, where in dry air $c_p = 1004$ J/kg/K and $g = 9.81\,\mathrm{m/s}^2$. The dry adiabatic lapse rate is independent of the height.

2. **The moist saturated adiabatic lapse rate** is the rate at which the air temperature decreases as it rises when maintained at its dew point (i.e. saturated with water vapour). It is significantly smaller than the dry lapse rate due to the latent heat released by the water vapour as it condenses, forming a liquid cloud, and thereby raising the temperature. Unlike the dry lapse rate the saturated value varies with height, typically having a value of about $5\,\mathrm{K/km}$.

An approximate expression for the moist adiabatic lapse rate may be derived by including the latent heat released as the air rises by modifying equation (1.31) to include the condensation

$$T \, ds = c_p \, dT - T \frac{\partial V}{\partial T}\bigg|_p \, dp + L \, dr = 0 \qquad (1.32)$$

where r is the *specific humidity* or *mixing ratio* (mass ratio of water vapour to dry air) and $L = 2.453 \times 10^6$ J/kg the latent heat of vaporisation.

The variation in the mixing ratio as the temperature and pressure change may be approximately calculated from the application of Dalton's law of partial pressures

$$r = \epsilon \frac{p_v}{(p - p_v)} \approx \epsilon \frac{p_v}{p} \qquad (1.33)$$

where p_v is the water vapour partial pressure and $\epsilon = R_a/R_v = 287/462$ is the ratio of the gas constants for air and water vapour respectively, which can be expressed in terms of the ratio of molecular masses, $\epsilon = M_v/M_a = 18.015/28.964 = 0.622$. The saturated water vapour pressure p_v is given by the Clausius–Clapeyron equation

$$\frac{dp_v}{dT} = \frac{L}{(V_v - V_{lT})} \approx \frac{L p_v}{R_v T^2} \qquad (1.34)$$

where the specific volume of the vapour V_v is much larger than that of the liquid V_l, and the vapour behaves as an ideal gas. Hence we obtain

$$\begin{aligned}
\frac{1}{r}\frac{dr}{dz} &\approx \frac{1}{p_v}\frac{dp_v}{dz} - \frac{1}{p}\frac{dp}{dz} \\
&\approx \frac{L}{R_v T^2}\frac{dT}{dz} + \frac{1}{p}\rho g \\
&= -\frac{L}{R_v T^2}\Gamma + \frac{1}{R_a T}g \qquad (1.35)
\end{aligned}$$

Substituting for dp/dz and dr/dz we obtain an approximate expression for the lapse rate of air saturated with water vapour

$$\Gamma = \frac{g}{c_p}\frac{\left[1 + \dfrac{L\,r}{R_a\,T}\right]}{\left[1 + \dfrac{L^2\,r\,\epsilon}{c_p\,R_a\,T^2}\right]} \qquad (1.36)$$

The mixing ratio r varies with temperature as the saturated vapour pressure given by the integral of equation (1.34). Since the latent heat is almost constant

$$\ln\left(\frac{p_v}{p_0}\right) = \frac{L}{R_v}\left(\frac{1}{T_0} - \frac{1}{T}\right)$$

where (p_0, T_0) is a suitably chosen initial (saturated) condition, e.g. at the triple point of water where $p_0 = 611.73$ Pa, $T_0 = 273.16$ K. This equation taken with equation (1.33) may be used to calculate the *dew point*: that is, the temperature to which a given parcel of air (with known specific humidity and pressure) must be cooled before it starts to condense. However, tabular and graphical representations are used to make this a much easier task in practice.

3. **Environmental lapse rate** is the measured rate of decrease of temperature with height in the atmosphere.

1.5.3   Stability of an Equilibrium Configuration

Although the system may be in mechanical equilibrium, it may not be stable. If the configuration is unstable, *convection* currents are set up within the fluid, which tend to mix the fluid to establish a uniform temperature.

We may derive the condition for the equilibrium to be stable by considering the effect of a small displacement of a fluid element ξ in the direction z. The system is stable if its response is to restore the perturbation to its original position. Thus let us suppose that the fluid element has a specific volume $V(p, s)$ at its equilibrium position z and let us suppose that the specific volume changes adiabatically in response to a pressure change to p' at $(z + \xi)$ to $V(p', s)$.[5] The fluid will displace an equal volume of fluid whose pressure p' and entropy s' correspond to the equilibrium values at $(z + \xi)$. If the displaced fluid element is heavier than the one it displaces, it will tend to sink and the equilibrium will be restored. Thus a necessary condition for stability is

$$V(p', s') - V(p', s) \approx \left.\frac{\partial V}{\partial s}\right|_p \frac{ds}{dz}\xi = \frac{\alpha T}{\rho c_p}\frac{ds}{dz}\xi > 0 \qquad (1.37)$$

where we have made use of the thermodynamic relations as before, equation (1.28).

Since the constants in the above inequality are all positive, the condition for stability reduces to

$$\frac{ds}{dz} > 0 \qquad (1.38)$$

that is, the entropy increases with height.

The limiting case of stability, that the entropy is constant with height, corresponds to the adiabatic lapse rate, equation (1.30). Applying these results to the atmosphere, we can see that it is unstable if the temperature falls less

---

[5]We may imagine that the change takes place sufficiently slowly to be reversible with viscosity and thermal conduction negligible.

rapidly than the adiabatic lapse rate going to higher altitudes, i.e. if the environmental lapse rate is greater than the adiabatic one

$$\frac{\mathrm{d}T}{\mathrm{d}z} > \frac{\mathrm{d}T}{\mathrm{d}z}\Big|_{\text{lapse}} = -\frac{\alpha T}{c_p} g \qquad (1.39)$$

In practice convection tends to reduce temperature gradients to the adiabatic lapse rate where the atmosphere is marginally stable. The adiabatic lapse rate may therefore be used to give an approximation to the variation of temperature and pressure with altitude.

The relationship of the adiabatic and the environmental lapse rates plays an important role in determining the generation of upward thermals. If a parcel of air is unsaturated and rises, being unstable it ascends, cooling at the dry lapse rate, until the dew point is reached and water vapour starts to condense. This is approximately the level of the cloud base. At this point the lapse rate is decreased to the moist saturated vapour rate, causing the parcel to ascend more rapidly and leading to the formation of rain. In an extreme case the rapidly rising stream of air leads to the formation of a characteristic thunder cloud.

1.6   Streamlines

A streamline is a line whose tangent is everywhere parallel to the flow. In Cartesian co-ordinates its equation is

$$\frac{\mathrm{d}x}{v_x} = \frac{\mathrm{d}y}{v_y} = \frac{\mathrm{d}z}{v_z} \qquad (1.40)$$

In steady flow, but *not* in non-steady flow, the streamlines are the particle paths (streaklines). The surface of a body immersed in a flow must be a streamline, since there is no flow through it. In streamlined flow around a body, the neighbouring streamlines closely parallel the surface from entry to exit. This is in contrast to the flow around a non-streamlined body, where the streamline touching the surface may separate and leave the neighbourhood of the body.

A closely related concept is the tube of flow, namely a region of flow whose walls are streamlines, and thus parallel to the flow. Consequently no flow can take place through the wall of the tube. The total flux of any quantity through any cross-section of a tube of flow is therefore constant.

1.7   Bernoulli's Equation: Weak Form

Bernoulli's equation is a direct consequence of the equations for a dissipationless fluid. It has a simple form, which makes it suitable for many applications, particularly for order of magnitude estimates.

In steady, dissipationless flow, the flow at any point on a streamline satisfies the following simple relation:

$$h + \frac{1}{2}v^2 + U = \text{const} \tag{1.41}$$

If the fluid is incompressible, $\rho = \text{const}$, then the enthalpy h is replaced by p/ρ. Under the above conditions this equation is known as the *weak form of Bernoulli's equation*. Other forms will be seen to occur under different conditions.

The proof follows from Euler's equation (1.12) as follows. In steady flow $\partial/\partial t \equiv 0$, and thus

$$(\mathbf{v} \cdot \nabla)\mathbf{v} = -\frac{1}{\rho}\nabla p - \nabla U$$
$$\nabla(1/2v^2) - \mathbf{v} \wedge (\nabla \wedge \mathbf{v}) = -\nabla(h + U) \tag{1.42}$$

since the fluid is dissipationless, the entropy change $\mathrm{d}s = 0$ and $\mathrm{d}h = (1/\rho)\,\mathrm{d}p$. Hence integrating along the streamline and noting that $\mathbf{v} \wedge (\nabla \wedge \mathbf{v})$ is perpendicular to $\mathbf{v}$, we obtain equation (1.41).

Alternatively, if we consider a narrow tube of flow surrounding a streamline of cross-section δS, normal to the flow, then in steady flow the total mass flow $\rho v \, \delta S = \text{const}$, the total energy flow

$$\rho v \left(h + 1/2v^2 + U\right) \delta S = \text{const}$$

and we again obtain Bernoulli's equation (1.41).

The two proofs are based on momentum and energy flow respectively–a consequence of the fact that in ideal flow the equation of state has only one independent thermodynamic variable, because, due to the dissipationless nature of the flow, the entropy of a fluid particles is constant.

Bernoulli's equation enables the definition of a useful quantity expressing the total energy available in the flow, which is the value of the constant along the streamline in equation (1.41):

$$\frac{1}{2}v_{\text{max}}^2 = h + \frac{1}{2}v^2 + U \tag{1.43}$$

v_{max} being the limit speed, i.e. the velocity that the fluid would acquire if both the pressure and gravitational potential were zero.

1.8   Polytropic Gases

A useful representation of many real gases is the polytropic gas, whose equation of state is the ideal gas law, and whose adiabatic equation of state is the familiar expression

$$\frac{p}{\rho^{\gamma}} = \text{const} \tag{1.44}$$

where γ is the ratio of specific heats, or adiabatic index. A large number of gases behave as polytropic gases with an appropriate value of the adiabatic index γ. For these materials the specific internal energy and specific enthalpy are given by

$$\epsilon = \frac{1}{(\gamma - 1)}\frac{p}{\rho} \qquad \text{and} \qquad h = \frac{\gamma}{(\gamma - 1)}\frac{p}{\rho} \tag{1.45}$$

For future reference the specific entropy (per unit mass) is given by

$$s = c_V \ln\left(\frac{p}{\rho^{\gamma}}\right) + s_0 \tag{1.46}$$

where c_V is the specific heat per unit mass.

The limit speed in polytropic gases can be expressed in terms of the critical velocity, when the flow velocity equals the local sound speed $c = \sqrt{\partial p/\partial \rho|_s} = \sqrt{\gamma p/\rho}$. From Bernoulli's equation (1.41) we obtain

$$v_* = c_* = \sqrt{\frac{(\gamma - 1)}{(\gamma + 1)}}\, v_{\max} \tag{1.47}$$

Another set of quantities, which are often useful in calculations with polytropic gases, is the stagnation sound speed, pressure and density. These are specified by the condition that the flow velocity is zero. They are generally defined by the stagnation sound speed

$$c_0 = \sqrt{\frac{(\gamma - 1)}{2}}\, v_{\max} \tag{1.48}$$

with p_0 and ρ_0 obtained through the equation of state in the form

$$\frac{c}{c_0} = \left(\frac{p}{p_0}\right)^{(\gamma-1)/2\gamma} = \left(\frac{\rho}{\rho_0}\right)^{(\gamma-1)/2} = \left\{\frac{1 + (\gamma - 1)M_0^2/2}{1 + (\gamma - 1)M^2/2}\right\}^{1/2} \tag{1.49}$$

where $M = v/c$ is the Mach number.

The other class of material, which is important in dissipationless flows, is incompressible, where the density of a fluid particle is constant, $\rho = \text{const}$. Liquids are the obvious examples of this condition. However, as we shall see, gases also behave in this way when their flow speed is much less than the sound speed (subsonic flows). For incompressible flow the integral

$$\frac{1}{\rho}\int dp = \int \frac{dp}{\rho} = \frac{p}{\rho}$$

is used in applications such as Bernoulli's equation. We can therefore replace $h \Rightarrow p/\rho$ in these cases.

1.8.1   Applications of Bernoulli's Theorem

1.8.1.1   Vena Contracta

We consider the flow from a reservoir containing an incompressible fluid through a small hole in one of the walls of area S_2. The pressure in the fluid in the reservoir far from the hole is approximately constant p_1 and the flow velocity $v_1 \approx 0$. Bernoulli's equation for the flow speed through the hole v_2 where the pressure is atmospheric, p_2, is

$$\frac{1}{2}\cancelto{0}{v_1^2} + \frac{p_1}{\rho} = \frac{1}{2}v_2^2 + \frac{p_2}{\rho} \tag{1.50}$$

However, in steady flow, the momentum flux through the hole must balance that in the reservoir. Thus the momentum balance in direction i is given by

$$\int_S (\rho\, v_i\, v_j + p\, \delta_{ij})\, \mathrm{d}S_j = \int_{(S-S_2)} (\rho\, v_i\, v_j + p)\, \mathrm{d}S_j + \int_{S_2} (\rho\, v_i\, v_{2x} + p_2)\, \mathrm{d}S_x = 0$$

where S is the surface area of the reservoir including the hole and x is the direction of the normal to the area of the hole. We assume that the pressure over the wall (excluding the hole) is approximately constant and equal to p_1. By symmetry we may assume that the components in the y and z directions cancel. Similarly the integral over the inner surface contains area elements which cancel except over the projection of the hole on to the internal surface. Therefore

$$\int_{S_2} (\rho\, \cancelto{0}{v_1^2} + p)\, \mathrm{d}S = \left(\rho\overline{v_{2x}^2} + p_2 \right) S_2 \tag{1.51}$$

where $\overline{v_{2_x}^2}$ is the mean square velocity at the hole in the direction normal to its area x. Comparing equation (1.50) with equation (1.51) we see that $\overline{v_{2_x}^2} \neq v_2^2$. This is due to flow near the wall, where the velocity is non-parallel through the hole. As a result, after leaving the hole the flow continues to converge, reaching a minimum cross-section when all the streamlines are approximately parallel, and the velocity is approximately $v_2^2 \approx v_{2_x}^2$. The minimum area–vena contracta–is thus approximately $S_{\min} \approx \frac{1}{2}S_2$. In fact experiment gives a value of $S_{\min} \approx 0.624\, S_2$. The difference is accounted for mainly by pressure variations in the fluid near the hole.

If a tube is inserted into the fluid–Borda's mouthpiece–so that the fluid enters the exiting flow well away from the wall, the pressure and velocity in the neighbourhood of the entry correspond to the uniform symmetric value, as assumed above. As a result the area reduction is found to be very nearly the predicted value $\frac{1}{2}$.

1.8.1.2   Flow of gas along a pipe of varying cross-section

Gas obeying the polytropic equation of state moves steadily along a pipe of decreasing cross-section from a reservoir at pressure p_0 and density ρ_0 to an exit at pressure p_1. We assume that the speed of flow across a cross–section is constant–the hydraulic approximation. Since the flow is adiabatic we may apply Bernoulli's equation to the flow

$$h + \frac{1}{2} v^2 = \frac{\gamma}{(\gamma - 1)} \frac{p}{\rho} + \frac{1}{2} v^2 = \frac{\gamma}{(\gamma - 1)} \frac{p_0}{\rho_0}$$

where the initial speed v_0 is assumed to be very small, i.e. stagnation.

Introducing the sound speed and noting that the adiabatic equation of state (1.44) is appropriate,

$$c = \sqrt{\left(\frac{\gamma p}{\rho} \right)} = c_0 \left(\frac{\rho}{\rho_0} \right)^{(\gamma-1)/2}$$

Bernoulli's equation can be rewritten as

$$\frac{1}{2} v^2 = \frac{1}{(\gamma - 1)} (c_0^2 - c^2)$$

The mass flux is therefore

$$j = \rho v = \sqrt{\left[\frac{2}{(\gamma - 1)} \right]} \rho_0 \left(\frac{c^2}{c_0^2} \right)^{1/(\gamma-1)} \sqrt{(c_0^2 - c^2)}$$

Differentiating j with respect to c^2 we find a turning point at $c_0^2 - c^2 = 1/2 (\gamma - 1) c^2$. Since $j \to 0$ as $v \to 0$ and $c \to c_0$, this turning point must be a maximum. At the maximum, the flow speed equals the sound speed (sonic flow):

$$v = v_* = c = c_* = \sqrt{\frac{2}{\gamma + 1}} c_0 \qquad (1.52)$$

the critical speed.

Since the initial velocity $v_0 \approx 0$ the critical pressure and density at the sonic point are

$$p_* = \left[\frac{2}{(\gamma + 1)} \right]^{\gamma/(\gamma-1)} p_0 \qquad \text{and} \qquad \rho_* = \left[\frac{2}{(\gamma + 1)} \right]^{1/(\gamma+1)} \rho_0 \qquad (1.53)$$

Figure 1.1 shows the characteristic parameters of the flow plotted as fractions of those at the critical point, where the flow velocity equals the sound speed, together with the local Mach number $M = v/c$.

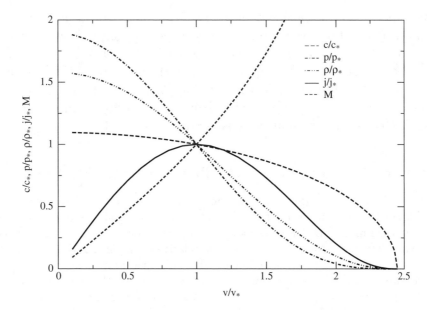

Figure 1.1: Characteristic parameters of the flow through a convergent/divergent nozzle.

Provided the external pressure p_{ext} exceeds the critical pressure p_*, the flow in the pipe is determined by the conditions at the exit, $p_1 = p_{ext}$.

The discharge (total mass flow) jS must remain constant, whilst the flux j cannot increase beyond the critical value, even if the cross-section of the pipe continues to decrease. Therefore the flux must adjust itself to be a maximum at the minimum cross-section, i.e. at the exit $p_1 = p_*$ when the external pressure p_{ext} is less than p_*. The discharge is then

$$\rho_* v_* S_{min}$$

If the external pressure p_{ext} is less than the critical p_*, the flow is said to be *choked*, and there must be an additional expansion external to the pipe. In a uniformly converging pipe, the flow cannot become supersonic even if the external pressure is very low. To achieve a supersonic flow it is necessary to allow the flow to expand after the sonic point so that the increasing speed can be accommodated by a decreasing flux. This is accomplished by terminating the converging section by a throat of minimum cross-section followed by a diverging section. Such a pipe is a *de Laval nozzle*, and may be used to produce a supersonic jet of gas. The discharge in such a nozzle is determined by the critical flux at the throat.

The nozzle with fixed walls is an example of a tube of flow, since in each case the boundary condition of no flow through the wall must be upheld. The flow parameters along such a tube must therefore be identical to those derived above. In particular, in the neighbourhood of the sonic point, where $M = 1$, the flux is nearly constant even though the flow speed changes. The cross-section of the tube of flow is therefore nearly constant in this locality. Hence, within the ideal (dissipationless) flow approximation, the flow can neither expand nor contract transversely in the transonic region where $M \approx 1$ –behaviour which Busemann called a *streampipe*. This result turns out to be important in the design of aircraft at near sonic speeds (see Section 12.5.1).

Case study 1.I Munroe Effect–Shaped Charge Explosive

The collapse of lined cavities has long been known to produce a high-velocity jet and a slower moving slug from the liner material. The jet can penetrate steel plate. Originally used in mining it was developed as an anti-tank weapon during the Second World War. The basic theory of the effect is relatively simple (Birkhoff *et al.*, 1948). A wedge-shaped block of explosive, lined with a thin metal layer (liner), is detonated from the apex end. The detonation wave moving through the explosive causes an inward implosion. The metal layer is fluidised by the intense pressure generated by the detonation, and driven towards the axis with velocity v_0. As a result the apex moves along the axis as the detonation proceeds through the block, Figure 1.2(a). To an observer moving with the apex it appears that material flows steadily down the arms and leaves as a jet (forward) and slug (backward) along the axis, Figure 1.2(b). Provided the detonation moves with constant velocity through the block u_0, the flow in this frame of reference is steady. Therefore we may apply Bernoulli's theorem to the collision of the two streams from each arm, and thus to the incoming and outgoing flows. Following the impulsive pressure pulse immediately after the detonation, the pressure falls rapidly and the fluid moves freely along the arms, so that the pressure on both the fluid in the arms and that moving along the axis is approximately constant on impact at the apex. Hence it follows from Bernoulli's theorem that the flow speeds in both axial flows, backwards (slug) and forward (jet), are equal to that of the incoming flows v' in the apex frame. The velocities of the slug and jet in the laboratory frame are then obtained by transforming back from the apex frame.

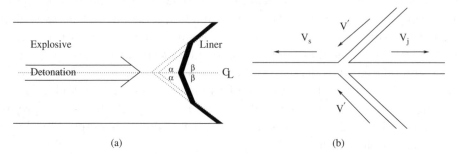

(a)   (b)

Figure 1.2: The geometrical arrangement of the flow system for the shaped charge detonation with a wedge: (a) shows the arrangement in the laboratory frame, and (b) that seen by an observer moving with the apex.

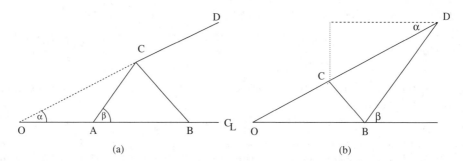

Figure 1.3: The geometrical arrangement of the flow system for the shaped charge detonation with a wedge in the laboratory frame. OCD is the original position of the liner. OAB is the axis, i.e. the path of the apex. $AC = v'(t' - t)$ is the position of the liner at time t, apex position A. $BC = v_0(t' - t)$ is the path of the liner particles from t to t', when they reach the apex. Thus the apex moves from A to B in time t to t' and $AB = u(t' - t)$. In time t to t' the detonation has moved from C to D and $CD = u_0(t' - t)/\cos(\alpha)$, and the liner lies along BD. The relevant angles are $\angle OCA = \angle BDC = (\beta - \alpha)$, $\angle ACB = \angle BCD = \angle CBD = [\pi - (\beta - \alpha)]/2$ and $\angle ABC = [\pi - (\beta + \alpha)]/2$, and the dynamics described by $\triangle ABC$ and $\triangle ABD$. (a) Detonation at C at time t. (b) Detonation at D at time t'.

The analysis of the problem in the laboratory frame involves the geometry of the flow. Let α be the half angle of the wedge, and β the angle of the flow the fluidised liner makes with the centre line, Figure 1.3. We assume that in the laboratory frame, the fluidised liner velocity bisects the initial and accelerated planes of the layer. In Figure 1.3(a) we see the situation where the detonation has reached the point C at time t. The line AC represents the line of the fluidised layer at this time, A being at the apex at time t. AC also represents the line of the flow in the apex frame. The fluid itself moves along the line BC bisecting the angle $\angle ACD$ in the laboratory frame, reaching the axis at time t'. Thus AB is the movement of the apex in time $t' - t$, namely $u(t' - t)$, and BC the flow of the fluid over the same time $v_0(t' - t)$. Since $\angle ACB = [\pi - (\beta - \alpha)]/2$ and $\angle BAC = \beta$, the velocity of the apex

$$u = v_0 \frac{\cos[\frac{1}{2}(\beta - \alpha)]}{\sin \beta}$$

Furthermore, since the velocity of the fluid along the arms in the apex frame $\mathbf{v'} = \mathbf{v_0} - \mathbf{u}$, we see that it is given by the third side of the $\triangle ABC$, namely

$$v' = v_0 \frac{\cos[\frac{1}{2}(\beta + \alpha)]}{\sin \beta}$$

since $\angle ABC = [\pi - (\beta + \alpha)]/2$. Alternatively AC is the line of flow in the apex frame for fluid starting at t and arriving at the apex at t', thus $AC = v'(t' - t)$ and we obtain the same result for v'.

It now remains to calculate the value of β from the speed of the detonation through the block, u_0. Referring to Figure 1.3(b), the detonation reaches point D at time t'.

$\triangle BCD$ is isosceles since $\angle BCD = \angle CBD = [\pi - (\beta - \alpha)]/2$, so that

$$\sin\left[\frac{1}{2}(\beta - \alpha)\right] = \frac{v_0 \cos \alpha}{2u_0} \qquad \text{and} \qquad \beta = \alpha + 2\arcsin\left(\frac{v_0 \cos \alpha}{2u_0}\right)$$

In the apex frame, the component of momentum along the axis after collision must balance that before

$$m_s v_s - m_j v_j = m_0\, v'\cos \beta \tag{1.54}$$

where m_s, m_j and $m_0 = m_s + m_j$ are the masses of the slug, jet and liner respectively, and the angle of the incoming flow to the axis is β. The laboratory velocities of the slug and the jet respectively are given by the transformation of the velocities $\mp v'$ from the apex frame back into the laboratory frame:

$$v_s = u - v' \qquad \text{and} \qquad v_j = u + v' \tag{1.55}$$

Thus we obtain the velocity of the slug and the jet in the laboratory frame in terms of the experimental parameters $(u_0,\, v_0,\, \alpha)$

$$
\begin{aligned}
v_s &= v_0 \left\{ \frac{\cos[\frac{1}{2}(\beta - \alpha)]}{\sin \beta} - \frac{\cos[\frac{1}{2}(\beta + \alpha)]}{\sin \beta} \right\} = 2\,\frac{\sin(\beta/2)\,\sin(\alpha/2)}{\sin \beta}\, v_0 \\[2mm]
v_j &= v_0 \left\{ \frac{\cos[\frac{1}{2}(\beta - \alpha)]}{\sin \beta} + \frac{\cos[\frac{1}{2}(\beta + \alpha)]}{\sin \beta} \right\} = 2\,\frac{\cos(\beta/2)\,\cos(\alpha/2)}{\sin \beta}\, v_0
\end{aligned}
\tag{1.56}
$$

respectively, and the masses are

$$m_s = m_0\,\frac{(1 + \cos \beta)}{2} \qquad \text{and} \qquad m_j = m_0\,\frac{(1 - \cos \beta)}{2} \tag{1.57}$$

Chapter 2

Flow of Ideal Fluids

2.1 Introduction

In this chapter we examine the classical theory of inviscid (dissipationless) fluid flow, most commonly for incompressible fluids, often called *ideal flow*. This was developed in the second half of the nineteenth century, largely following potential theory, by workers such as Helmholtz, Kirchoff, Kelvin and Rayleigh. The application of harmonic functions, solutions of Laplace's equation, derived in the development of electromagnetic theory led to mathematical results of great elegance, but sadly often of only limited applicability, for treating a variety of flows, particularly in two dimensions.

As we shall see in Chapter 6 ideal (inviscid) flow is unusual in that it is essentially an asymptotic solution of the fluid dynamical equations. This arises since inviscid flow is normally *not* the limit solution of the full viscous equations subject to the appropriate boundary conditions, as the viscosity tends to zero. It can be clearly seen in the role of the boundary layer in Chapter 6, where even the presence of an infinitesimal degree of viscosity is sufficient to generate a very thin boundary layer and hence drag. This may give rise to separation, but always enables the resolution of the essential non-uniqueness of inviscid flow, which arises in the treatment of the flow around the surface of a solid body as in Section 2.3.

Nonetheless inviscid solutions have many important practical applications:

1. Flow around streamlined bodies, e.g. aerofoils. In this case the problem of non-uniqueness is resolved by an additional empirical condition, which frequently allows accurate representations of the flow found in practice.

Introductory Fluid Mechanics for Physicists and Mathematicians, First Edition. Geoffrey J. Pert.
© 2013 John Wiley & Sons, Ltd. Published 2013 by John Wiley & Sons, Ltd.

Much of the theory of aerodynamics in consequence has been constructed around inviscid flow models.

2. Waves and instabilities. Both surface and internal waves in liquids are well described by the inviscid equations as viscosity plays little role in their behaviour, and there is no flow around an external body.

3. Geophysical and meteorological flows are often well described by the inviscid equations, when the problem does not involve the flow at the surface of a body. We shall not discuss these rather specialised flows.

4. Most compressible supersonic flows are effectively dissipation-less (Chapter 8). Only shock waves involve significant dissipation (Chapter 10).

In this chapter we identify the underlying concepts of ideal flow, and refer to Lamb (1932) and Milne-Thomson (1968), where the flow of inviscid, ideal fluids is treated in detail and extensive analytic solutions are given.

2.2   Kelvin's Theorem

We define the circulation around a closed loop L as

$$\Gamma = \oint_L \mathbf{v} \cdot d\mathbf{l} \tag{2.1}$$

Kelvin's theorem states that in the flow of a dissipationless fluid, the circulation around a closed loop fixed in the fluid, i.e. always comprising the same fluid particles, is constant:

$$\frac{d\Gamma}{dt} = 0 \tag{2.2}$$

We define a parameter on the loop fixed in the fluid λ, e.g. the distance around the loop to a particular fluid particle at time $t = 0$. Then using Euler's equation (1.12) we may write

$$
\begin{aligned}
\frac{d}{dt}\oint_L \mathbf{v} \cdot d\mathbf{l} &= \oint_L \frac{d\mathbf{v}}{dt} \cdot d\mathbf{l} + \oint_L \mathbf{v} \cdot \frac{d}{d\lambda}\left(\frac{d\mathbf{l}}{dt}\right) d\lambda \\
&= \oint_L -\left(\frac{1}{\rho}\nabla p + \nabla U\right) \cdot d\mathbf{l} + \oint_L \mathbf{v} \cdot \frac{d\mathbf{v}}{d\lambda} d\lambda
\end{aligned}
$$

since the loop is made up of fluid particles $\mathbf{v} = d\mathbf{l}/dt$. Furthermore, since entropy is constant, or the flow incompressible, the pressure is a function of the density alone and we may define some function $H(\rho)$ such that $\nabla p/\rho = \nabla H$:

$$\frac{d\Gamma}{dt} = \oint_L -(\nabla H + \nabla U) \cdot d\mathbf{l} + \oint_L \frac{1}{2}\frac{d}{d\lambda}(v^2)\, d\lambda = 0 \tag{2.3}$$

thereby establishing Kelvin's theorem. The theorem is valid for incompressible fluids provided the density is constant for all particles. If there is a discontinuity, e.g. at a boundary between two dissimilar fluids, the theorem fails for loops taken across the boundary, since for these loops $\int \mathrm{d}p/\rho \neq p/\rho$ everywhere.

The loop of fluid can be defined quite generally within the fluid. For example, the loop may be taken around a solid body without invalidating the theorem, provided the flow on the loop is everywhere inviscid, i.e. it does not touch the surface. As the flow proceeds, the fluid loop must continue to enclose the body, so that Kelvin's theorem remains valid around the loop.

2.2.1   Vorticity and Helmholtz's Theorems

We define the vorticity as

$$\boldsymbol{\zeta} = \nabla \wedge \mathbf{v} \tag{2.4}$$

Vorticity has several useful properties which were first enunciated by Helmholtz in 1858. The vorticity is clearly related to the presence of rotational elements in the flow. Indeed we shall see later that vorticity is related to the solid body rotation of small fluid elements, the angular velocity being given as $\boldsymbol{\omega} = \frac{1}{2}\boldsymbol{\zeta}$ (Section 3.2). It has the important property, which follows from Kelvin's and Stokes' theorems, that

$$\int_S \boldsymbol{\zeta} \cdot \mathrm{d}\mathbf{S} = \oint_L \mathbf{v} \cdot \mathrm{d}\mathbf{l} = \Gamma \tag{2.5}$$

where S is an area enclosed by the loop L. Clearly S is to some extent arbitrary, being only bounded by L.

The generation of vorticity is easily calculated by taking the curl of Euler's equation (1.12)

$$\frac{\partial \boldsymbol{\zeta}}{\partial t} - \nabla \wedge (\mathbf{v} \wedge \boldsymbol{\zeta}) = 0 \tag{2.6}$$

Since

$$\nabla \wedge (\mathbf{v} \wedge \boldsymbol{\zeta}) = (\boldsymbol{\zeta} \cdot \nabla)\mathbf{v} - (\mathbf{v} \cdot \nabla)\boldsymbol{\zeta} - \boldsymbol{\zeta}(\nabla \cdot \mathbf{v}) + \mathbf{v}(\nabla \cdot \boldsymbol{\zeta})$$

then, making use of the equation of continuity,

$$\left(\frac{\partial}{\partial t} + \mathbf{v} \cdot \nabla\right)\frac{\boldsymbol{\zeta}}{\rho} = \left(\frac{\boldsymbol{\zeta}}{\rho} \cdot \nabla\right)\mathbf{v} \tag{2.7}$$

This equation represents the convection of vorticity. It implies that lines of vorticity[1] and particle paths are tied together.

---

[1]Lines and tubes of vorticity are defined in a similar manner to streamlines and tubes of flow. Thus a line of vorticity is a line tangent to the vorticity. A tube of vorticity is a tube whose walls are everywhere lines of vorticity.

Consider a tube of vorticity. We define the strength of the tube as the circulation around the tube $\Gamma = \int \zeta \cdot d\mathbf{S}$ integrated across a cross-section. Since $\nabla \cdot \zeta = 0$, Gauss's theorem shows that the strength of a tube of vorticity is everywhere constant. Thus considering an infinitely narrow tube of vorticity, it follows that lines of vorticity must be either continuous or terminate at the boundaries of the flow.

The circulation around any loop initially on the wall of a vortex tube is zero, since no vorticity passes through it. As the fluid moves, the circulation around the loop fixed in the fluid remains zero by equation (2.5). Since this is true of every such loop, each of which always comprises the same fluid particles, the totality of such loops must continue to form the wall of the same vortex tube. Therefore the tube always comprises the same fluid particles. In particular if we consider an infinitesimal tube of vorticity enclosing a vortex line, it is clear that the fluid particles in their motion are tied to a specific line of vorticity.

These results are encapsulated in Helmholtz's theorems of vortex motion:[2]

    i. *The vorticity represents the solid body rotation of infinitely small fluid elements.*

    ii. *The intensity of a vortex tube is constant everywhere along the tube. Consequently vortex lines are continuous, either closed or terminating only at infinity.*

    iii. *The vortex lines are tied to the fluid motion, i.e. a vortex line always comprises the same fluid particles.*

    iv. *The intensity of a vortex tube remains constant throughout the motion.*

The first and second theorems are statements of the identity div curl $= 0$, and the definition of vorticity. They are therefore always valid even if viscosity is present. Hence we obtain the important result that lines of vorticity either form closed loops, or terminate at infinity or on the surface of solid bodies. Consequently *vorticity is generated in the flow at the surface of solid bodies.* In viscous flow, we shall find that lines of vorticity are diffused but not destroyed, Section 3.6.1.

In dissipationless flow, lines of vorticity have the remarkable properties that the vortex tube is continuous, that the circulation around it is constant in space and time, and that it cannot be destroyed. Vorticity is persistent in

---

[2]**Historical note** Many authors only quote two or three of the theorems. In fact Helmholtz (1858) gives all four. Helmholtz's proof of the third theorem is based on equation (2.7), but is not rigorous (see Lamb, 1932). Kelvin in his paper (Thomson, 1869) gives alternative derivations as above, based on his theorem of circulation. Kelvin's proof of the third theorem is more satisfactory.

dissipationless fluids, and can only be diffused by viscosity. This remarkable property, exemplified by the persistence of smoke rings, led to speculation in the late nineteenth century that atoms were perhaps in some way composed from vortex rings.

From Helmholtz's third theorem a vortex tube must coincide with a stream tube. A short length $\mathbf{\Delta\ell}$ of a narrow stream tube of width ΔS, which moves with the fluid, must satisfy the kinematic relation

$$\frac{\mathrm{d}\,(\mathbf{\Delta\ell})}{\mathrm{d}\,t} = (\mathbf{\Delta\ell} \cdot \nabla)\,\mathbf{v}$$

The similarity of this equation with equation (2.7) suggests the relation $\zeta \sim \rho\,\Delta\ell$. Therefore if the length of the tube increases, its vorticity must increase. Since the circulation is constant, the cross-sectional area ΔS must decrease. The mass contained in the volume $\rho\,\Delta S\,\Delta\ell$ of the tube therefore remains constant, consistent with the equation of continuity.

2.2.1.1   Simple or rectilinear vortex

Consider a narrow tube of flow, within which there is non-zero vorticity, of radius ϵ, directed along the line. The circulation around a line is clearly $\Gamma = \pi\epsilon^2\zeta$. As the radius of the filament is reduced to zero, the circulation remains constant, and the filament forms the core of a simple vortex. The axis of the vortex may be defined as normal to the plane of the vortex in the direction prescribed by the sense of the circulation and the 'corkscrew rule', corresponding to the direction of the vorticity ζ through the element.

2.2.1.2   Vortex sheet

Consider an infinitesimal layer of flow across which the tangential component in the velocity changes discontinuously from $\mathbf{v}_- = \mathbf{v} - \frac{1}{2}\,\delta\mathbf{v}$ on one side to $\mathbf{v}_+ = \mathbf{v} + \frac{1}{2}\,\delta\mathbf{v}$ on the other, i.e. there is a discontinuity in the velocity tangential to the surface of $\delta\mathbf{v}$. The normal component of velocity is continuous across the surface. The velocity of flow 'in the sheet' is clearly $\mathbf{v} = \frac{1}{2}\,(\mathbf{v}_- + \mathbf{v}_+)$. The sheet may be envisaged as an assembly of infinitesimally narrow simple vortices arranged within the surface. Such a tangential discontinuity is known as a *vortex sheet* and plays an important role in the ideal flow theory of aerofoil sections.

The *surface vorticity* (or strength) of the sheet per unit surface area is defined in terms of the velocity difference across it to be

$$\boldsymbol{\gamma} = \hat{\mathbf{n}} \wedge \delta\mathbf{v} = \hat{\mathbf{n}} \wedge (\mathbf{v}_+ - \mathbf{v}_-) \tag{2.8}$$

where $\hat{\mathbf{n}}$ is the unit vector normal to the sheet in the direction from the lower surface $\ominus$ to the upper one $\oplus$. The surface vorticity vector is clearly tangential to the sheet and normal to the velocity difference. $\gamma\,\delta w$ represents the circulation around a loop of width δw embedded in the sheet. Thus as noted earlier the sheet may also be visualised as a series of infinitely thin contiguous vortices lying in the sheet with axes in the direction of the surface vorticity vector.

The relationship of the strength of the vortex sheet to vorticity may be seen by imagining the sheet to have a finite width ϵ. The total vorticity contained in an element of surface δS of the sheet is therefore

$$\int_{\delta V} \zeta\,\mathrm{d}V = \int_{\delta S} \hat{\mathbf{n}} \wedge (\mathbf{v}_+ - \mathbf{v}_-)\,\mathrm{d}S$$
$$= \gamma\,\delta S$$

where $\delta V = \epsilon\,\delta S$ is the element of volume, and we have made use of Gauss's theorem in the form[3]

$$\int_V (\nabla \wedge \mathbf{A})\,\mathrm{d}V = - \int_S \mathbf{A} \wedge \mathrm{d}\mathbf{S}$$

so that $\gamma = \epsilon\,\zeta$ as $\epsilon \to 0$ and $\zeta \to \infty$ whilst γ remains constant.

Since

$$v_2{}^2 - v_1{}^2 = (\mathbf{v}_2 - \mathbf{v}_1) \cdot (\mathbf{v}_2 + \mathbf{v}_1)$$

we have three alternatives if the speed is the same on both sides of the vortex sheet:

1. $(\mathbf{v}_+ - \mathbf{v}_-) = 0$, the trivial solution with the velocities equal on both sides of the sheet.
2. $(\mathbf{v}_+ + \mathbf{v}_-) = 0$, the velocity in the sheet is zero; the sheet is at rest.
3. $(\mathbf{v}_+ - \mathbf{v}_-) \perp (\mathbf{v}_+ + \mathbf{v}_-)$, the surface vorticity is perpendicular to the flow velocity in the sheet. This will be found to be the case associated with the wake behind an aerofoil.

---

[3]This equation is obtained by considering the scalar product with an arbitrary constant vector $\mathbf{B}$ and application of the identity

$$\nabla.(\mathbf{A} \wedge \mathbf{B}) = \mathbf{B} \cdot \nabla \wedge \mathbf{A} - \mathbf{A} \cdot \nabla \wedge \mathbf{B}$$

Therefore

$$\mathbf{B} \cdot \int_S \mathbf{A} \wedge \mathrm{d}\mathbf{S} = - \int_S \mathbf{A} \wedge \mathbf{B} \cdot \mathrm{d}\mathbf{S} - \int_V \nabla \cdot (\mathbf{A} \wedge \mathbf{B})\,\mathrm{d}V = -\mathbf{B} \cdot \int_V (\nabla \wedge \mathbf{A})\,\mathrm{d}V$$

Since $\mathbf{B}$ is arbitrary the integral follows.

Vortex sheets are always unstable (see Section 4.4). An infinite vortex sheet will develop by growing sinusoidal perturbations, limited only by a nonlinear phase in which a series of isolated vortices form. A finite sheet tends to roll up towards its ends to form a pair of separated rectilinear vortices.

2.3   Irrotational Flow

It follows from equation (2.5) that if the circulation on a streamline is initially zero, it will remain so for all time, provided the streamline does not touch the surface of a body–a condition known as *irrotational flow*. If this requirement is violated the loop is broken as one part goes along one side of the surface and the other part along the other. Kelvin's theorem therefore no longer holds. This streamline thus separates the region of zero vorticity from one where the vorticity may be finite. As a consequence a *tangential discontinuity* may form along the streamline, across which the tangential component of velocity changes discontinuously, but the pressure and normal velocity component are continuous. The discontinuity itself forms a 'vortex sheet' due to the velocity difference, which is equivalent to an infinitely thin layer of vortices of appropriate strength per unit area. In practice, viscosity will eliminate the discontinuity, introducing vorticity and possibly turbulence into the neighbouring flow across the boundary.

The spaces enclosed by the different segments of these streamlines may contain vorticity. A consequence is that flow around a body may not be mathematically unique in a dissipationless (ideal) fluid, as the flow may separate along the surface and the body streamline may leave the surface at an arbitrary point. As we shall see in dissipational fluids this non-uniqueness is removed by the inclusion of viscosity (see Chapter 6). In practice in calculations with ideal flow, this problem is often resolved by additional knowledge of the expected flow from experiment.

In general, due to viscosity, no flow which includes a body surface can be irrotational everywhere. Nonetheless it does form a good approximation to the flow around streamlined bodies, provided some additional experimental guidance is available to ensure uniqueness. These flows form an important class of flows, particularly in aerodynamics.

2.3.1   Crocco's Equation

In the above discussion it has been implicitly assumed that the fluid is everywhere homogeneous. In practice this may not be correct throughout the flow field. We therefore examine under what conditions the condition of irrotationality may be maintained.

In steady flow along a streamline Bernoulli's equation (1.41) may be written as

$$H = h + \frac{1}{2}v^2 = \frac{1}{2}v_{\text{max}}^2 = \text{const} \qquad (2.9)$$

From Euler's equation (1.12) we have

$$
\begin{aligned}
\mathbf{v} \wedge \boldsymbol{\zeta} &= \nabla\left(\frac{1}{2}v^2\right) + \frac{1}{\rho}\nabla p \\
&= \nabla H - T\nabla s \qquad (2.10)
\end{aligned}
$$

where s is the entropy per unit mass and T the temperature, and from the first law of thermodynamics

$$\mathrm{d}h = T\mathrm{d}s + \frac{1}{\rho}\mathrm{d}p$$

Hence we conclude that irrotational flow implies and requires that the fluid be both *homo-energetic* (H constant everywhere) and *homo-entropic* (S constant everywhere).

2.4   Irrotational Flow–Velocity Potential and the Strong Form of Bernoulli's Equation

If the flow is irrotational, such that $\nabla \wedge \mathbf{v} = 0$, then the velocity may be expressed in terms of a potential $\mathbf{v} = \nabla\phi$, the *velocity potential*. Inserting this result into Euler's equation (1.12) we obtain

$$\frac{\partial}{\partial t}\nabla\phi + \nabla\left(\frac{1}{2}v^2\right) = -\nabla(h + U)$$

and integrating over space

$$\frac{\partial\phi}{\partial t} + h + \frac{1}{2}v^2 + U = f(t)$$

where $f(t)$ is an arbitrary function of time. Since the potential ϕ is only defined in space, we include the function $f(t)$ in ϕ to give the *strong form of Bernoulli's equation*:

$$\frac{\partial\phi}{\partial t} + h + \frac{1}{2}v^2 + U = 0 \qquad (2.11)$$

Comparison with the earlier form of Bernoulli's equation (1.41) shows that the inclusion of the additional term $\partial\phi/\partial t$ yields a form which is time dependent and valid throughout the flow, not just along a streamline, but is only applicable in irrotational flow.

Since $\int \mathbf{v} \cdot d\boldsymbol{\ell}$ increases along a streamline, it follows from Stokes' theorem that, if the potential is single valued (*acyclic flow*),[4] the streamlines must start and finish on the surface of a body or at infinity. The velocity potential increases continuously along the streamline; for if $\delta\boldsymbol{\ell}$ is an interval along the streamline

$$\delta\phi = \nabla\phi \cdot \delta\boldsymbol{\ell} > 0 \qquad (2.12)$$

since $\nabla\phi = \mathbf{v}$ is parallel to the streamline $\delta\boldsymbol{\ell}$, unless the flow velocity is zero and the potential constant.

2.5   Incompressible Flow–Streamfunction

If the flow is incompressible, $\nabla \cdot \mathbf{v} = 0$, the velocity may be expressed as a vector potential $\mathbf{v} = \nabla \wedge \boldsymbol{\psi}$. In a general three-dimensional system, this quantity may find little application as, being a vector, it does not reduce the dimensionality of the problem. However, if the system has a symmetry so that it may be reduced to two dimensions in either Cartesian planar symmetry or cylindrical axisymmetry, the vector has only a single term in the direction perpendicular to the plane of the co-ordinates: in the z direction in a planar (x, y) system; and in the azimuthal (ϕ) direction in the axisymmetric (z, ϱ) system.[5]

2.5.1   Planar Systems

In this case we define the streamfunction ψ in Cartesian co-ordinates (x, y)

$$v_x = \frac{\partial\psi}{\partial y} \qquad \text{and} \qquad v_y = -\frac{\partial\psi}{\partial x} \qquad (2.13)$$

The streamfunction has two important properties:

1. Along a streamline $dx/v_x = dy/v_y$

$$d\psi = \frac{\partial\psi}{\partial x}\,dx + \frac{\partial\psi}{\partial y}\,dy = -v_y\,dx + v_x\,dy = 0 \qquad (2.14)$$

and the streamfunction is constant. This has the important consequence that the surface of a body immersed in the flow must be a streamline.

---

[4]Since the circulation is

$$\Gamma = \oint_C \nabla\phi \cdot d\boldsymbol{\ell} = [\phi]_C$$

multiple values of the potential will occur if the circulation in any closed circuit along the path taken C is non-zero. This may occur if the space is not simply connected, i.e. one in which all circuits can be reduced to one another without crossing a boundary. We return to this point later.

[5]Note the difference between the symbol ϱ representing the distance from the axis and ρ the density.

2. The total mass flow per unit width through a line segment between the points ① and ② is given by the difference in the values of the streamfunction at these points ψ_1 and ψ_2 respectively. The mass flow per unit width is

$$J = \int_1^2 \rho \mathbf{v} \cdot d\mathbf{s} = \rho \int_1^2 \left\{ \frac{\partial \psi}{\partial x} \, dx + \frac{\partial \psi}{\partial y} \, dy \right\} = \rho \, (\psi_2 - \psi_1) \qquad (2.15)$$

since the area element $d\mathbf{s} = \hat{\mathbf{i}} \, dy - \hat{\mathbf{j}} \, dx$ is normal to the line element $d\mathbf{l} = \hat{\mathbf{i}} \, dx + \hat{\mathbf{j}} \, dy$, $\hat{\mathbf{i}}$ and $\hat{\mathbf{j}}$ being unit vectors in the directions x and y respectively.

2.5.2 Axisymmetric Flow–Stokes Streamfunction

From the definition of velocity in terms of a vector potential in cylindrical co-ordinates (z, ϱ) we have

$$v_z = \frac{1}{\varrho} \frac{\partial \psi}{\partial \varrho} \qquad \text{and} \qquad v_\varrho = -\frac{1}{\varrho} \frac{\partial \psi}{\partial z} \qquad (2.16)$$

The streamfunction is conventionally set to zero on the axis.

As in the planar case, the stream function has two important properties:

1. Along a streamline

$$\frac{1}{\varrho} d\psi = \frac{1}{\varrho} \left\{ \frac{d\psi}{dz} dz + \frac{d\psi}{d\varrho} d\varrho \right\} = -v_\varrho \, dz + v_z \, d\varrho = 0 \qquad (2.17)$$

which is the equation of a streamline. Therefore the streamfunction is constant on a streamline. The surface of a body immersed in the flow must be a streamline.

2. Consider a surface bounded by the axis and the circle ①; then the total mass flux passing through the surface is

$$J = \int_0^1 \rho \mathbf{v} \cdot d\mathbf{s} = 2 \pi \rho \int_0^1 \varrho \left\{ (v_z \, d\varrho - v_\varrho \, dz) \right\} = 2 \pi \, (\psi_1 - \psi_0) \qquad (2.18)$$

since the area element is normal to the line element and

$$ds_z = -2 \pi \varrho \, d\varrho \qquad \text{and} \qquad ds_\varrho = 2 \pi \varrho \, dz$$

It follows from the conservation of mass that the total flux is independent of the surface chosen along a streamline and, furthermore, that if the flow is continued along the streamline the total flux is constant through successive rings.

A uniform flow with velocity U parallel to the axis has streamfunction

$$\psi = \frac{1}{2}U\varrho^2 \tag{2.19}$$

2.6   Irrotational Incompressible Flow

If the flow is both irrotational and incompressible

$$\nabla \cdot \mathbf{v} = \nabla^2 \phi = 0 \tag{2.20}$$

so that the velocity potential satisfies Laplace's equation, independent of time. The methods of potential theory, developed in the latter half of the nineteenth century to solve problems in electromagnetism, thus become applicable to calculating irrotational incompressible flows. These applications are particularly important in view of the wide range of solutions to Laplace's equation available from classical potential theory. The equation also has some important properties such as uniqueness, which make it especially attractive for solving fluid dynamical problems. As a result it has formed the basis for the classical theory of aerodynamics, and still continues to fill an important role in the design of aircraft (Chapter 11). When viscosity is non-zero, but weak, the irrotational solution is often the asymptotic form of the solution far from the body, which is matched to the boundary layer solution near the surface (Chapter 6) using the method of matching asymptotics.

One of the most important properties of the solutions, known as *harmonic functions*, is due to the linearity of the equation, namely the principle of superposition. Thus if ϕ_1 and ϕ_2 are two different solutions of the equation, then $\phi = \phi_1 + \phi_2$ is also, but satisfying different boundary conditions to each of its parents. Fluid mechanics makes extensive use of this principle.

In an acyclic region where the velocity potential is single valued and satisfies Laplace's equation, there can be neither a maximum nor a minimum value. For if this were the case we should be able to construct a small spherical surface S to surround the extremum over which the normal velocity v_r would be approximately constant and non-zero. This is forbidden by Gauss's theorem since

$$\int_V \nabla^2 \phi \, \mathrm{d}V = \int_S \mathbf{v} \cdot \mathrm{d}\mathbf{s} = \int r^2 \, v_r \, \mathrm{d}\Omega = 0$$

where the surface S has radius r.

In a similar manner it follows that $v^2 = (\nabla\phi)^2$ cannot have a maximum, for each component of velocity v_i must individually satisfy Laplace's equation, and cannot therefore have either a maximum or minimum. Hence considering the squared velocity $v^2 = v_x{}^2 + v_y{}^2 + v_z{}^2$ we see that the flow speed v cannot have a maximum. A minimum, namely a stagnation point, may, of course, occur.

From Green's theorem and Laplace's equation, it follows quite generally that

$$\int_V (\nabla \phi)^2 \, \mathrm{d}V = - \int_S \phi \nabla \phi \cdot \mathrm{d}\mathbf{S} \tag{2.21}$$

From these preceding results we may draw some important conclusions:

1. If the potential is constant on all boundary surfaces, the potential is constant throughout the space, and the velocity is zero. This is established in a number of ways since the potential must increase along all streamlines (2.12); or alternatively no maxima/minima of potential can occur; or directly from equation (2.21).

2. If the normal gradient of the potential at every surface is zero, the potential is everywhere constant and no flow occurs. For, from (2.12), it follows that no streamline can start or leave the boundary as required since the flow is acyclic; or directly from (2.21) since the integrand on the left side is always positive.

3. If the boundary consists partly of surfaces over which either a constant potential (S) or zero normal velocity (Σ) is specified, the potential is constant and no flow occurs. For no streamline can leave or enter Σ and none can pass from one point to another on S.

Consider the solution ϕ, which is itself the difference between two solutions ϕ_1 and ϕ_2 each of which satisfies the identical boundary conditions of the problem, specified in one of the following forms:

1. Potential given at every point on the boundary–*Dirichlet boundary conditions*.

2. Normal gradient of potential given on the boundary–*Neumann boundary conditions*.

3. Potential specified on one part of the boundary and the normal component of velocity on the other–*Mixed boundary conditions*.

So if either ϕ or $\mathrm{d}\phi/\mathrm{d}n$ is zero on the boundary, we establish that $\phi = 0$ and hence that $\phi_1 = \phi_2$. Thus we obtain the well-known uniqueness theorem:

> *In an acyclic irrotational incompressible flow, the potential is uniquely defined by Dirichlet, Neumann or mixed boundary values applied to all the bounding surfaces.*

2.6.1   Simply and Multiply Connected Spaces

We define a *reducible* curve in space as a closed loop which may be continuously reduced to a point without crossing any internal boundary.

In a *simply connected space* all closed loops are reducible. Three-dimensional space containing only closed surfaces (i.e. bodies) is simply connected.

A *multiply connected space* contains closed loops, which are not reducible. Examples of multiply connected spaces are toroids, e.g. anchor rings, and infinitely long cylinders. Thus a two-dimensional space, which is essentially that of infinitely long cylinders normal to the two-dimensional plane, is multiply connected.

The order of multiplicity n identifies the number of independent irreducible curves. Thus a two-dimensional space of n cylindrical bodies is an n-fold connected space. We may reduce the connectivity by 1 from n to $n - 1$ by inserting a barrier joining two unconnected bodies.

In a simply connected region the potential is single valued as any curve is reducible and may therefore be bounded by a surface which lies outside any boundary. Then from Stokes' theorem

$$\Gamma = \oint \mathbf{v} \cdot d\boldsymbol{\ell} = \int \boldsymbol{\zeta} \cdot d\mathbf{S} = 0$$

In a multiply connected space, the circulation taken around a body cannot be a priori guaranteed to be single valued as Stokes' theorem cannot be applied. Hence the potential may no longer be single valued, due to non-zero circulation, Γ. Since

$$\oint_C \nabla \phi \cdot d\boldsymbol{\ell} = [\phi]_C = \Gamma \tag{2.22}$$

it follows that ϕ is undetermined to an additive term $\kappa \Gamma$ where κ is an integer representing the number of rotations about the centre of circulation. Γ must be set as an additional boundary condition. If the circulation is non-zero the flow is *cyclic*.

We can easily show that cyclic flows are unique provided additional boundary conditions, namely the values of the circulation around the n centres of circulation, are set. Consider two solutions satisfying the boundary conditions above and the equality of all circulations, then the flow $\phi_1 - \phi_2$ is acyclic, and the preceding theorems above apply. Therefore $\phi_1 - \phi_2 = \text{const}$ as before, and the flows are identical.

We therefore generalise the uniqueness theorem to include multiply connected spaces:

> In a cyclic irrotational incompressible flow, the potential is uniquely defined by Dirichlet, Neumann or mixed boundary values applied to all the bounding surfaces, together with the values of the circulation around all centres of circulation in space.

2.7   Induced Velocity

The condition for incompressible flow is that $\nabla \cdot \mathbf{v} = 0$. This allows the velocity to be expressed in terms of a vector potential $\mathbf{v} = \nabla \psi$.

If the flow is also irrotational, $\nabla \wedge \mathbf{v} = 0$ and therefore

$$\nabla \wedge (\nabla \wedge \psi) = \nabla(\nabla \cdot \psi) - \nabla^2 \psi = 0 \tag{2.23}$$

However, $\nabla \cdot \psi$ is undefined and may without loss of generality be set to zero, to yield Laplace's equation for ψ.

Similarly if the velocity is expressed in terms of the velocity potential $v = \nabla \phi$, we have

$$\nabla^2 \phi = \nabla^2 \psi = 0 \tag{2.24}$$

In general the velocity field is specified by the two quantities, the scalar and vector potentials, the latter being associated with vorticity. Indeed there is a direct relationship between the vector potential and the vorticity. We assume for the present that the flow is at rest at infinity, and that the space is simply connected. Consider the integral

$$\psi = \frac{1}{4\pi} \int \zeta(\mathbf{r}') \frac{1}{|\mathbf{r} - \mathbf{r}'|} dV' \tag{2.25}$$

We take the Laplacian derivative of this term

$$\nabla^2 \psi = \frac{1}{4\pi} \int \zeta(\mathbf{r}') \nabla^2 \left(\frac{1}{|\mathbf{r} - \mathbf{r}'|} \right) dV' \tag{2.26}$$

noting that

$$\nabla^2 \left(\frac{1}{|\mathbf{r} - \mathbf{r}'|} \right) = 0 \qquad \mathbf{r} \neq \mathbf{r}'$$

and that there is a singularity at $\mathbf{r} = \mathbf{r}'$. We therefore exclude a small region in this neighbourhood by an infinitesimal sphere of radius ϵ centred on $\mathbf{r}'$. Using Gauss's theorem to transform the volume integral to one over the bounding surface at ∞, whose contribution is zero, and the sphere of radius ϵ we obtain

$$\nabla^2 \psi = -\zeta(\mathbf{r}) \tag{2.27}$$

Since

$$\zeta = \nabla \wedge \mathbf{v} = \nabla \wedge (\nabla \wedge \psi) = \nabla(\nabla \cdot \psi) - \nabla^2 \psi$$

and since $\nabla \cdot \psi$ may without error be set to zero, we see that equation (2.25) yields the relation between the vector potential and the vorticity.

Furthermore, since the velocity is the curl of the vector potential we have

$$
\begin{aligned}
\mathbf{v}(\mathbf{r}) &= -\frac{1}{4\pi} \int \boldsymbol{\zeta}(\mathbf{r}') \wedge \nabla \left(\frac{1}{|\mathbf{r} - \mathbf{r}'|} \right) dV' \\
&= \frac{1}{4\pi} \int \boldsymbol{\zeta}(\mathbf{r}') \wedge \nabla' \left(\frac{1}{|\mathbf{r} - \mathbf{r}'|} \right) dV' \\
&= \frac{1}{4\pi} \int \boldsymbol{\zeta}(\mathbf{r}') \wedge \left(\frac{(\mathbf{r} - \mathbf{r}')}{|\mathbf{r} - \mathbf{r}'|^3} \right) dV'
\end{aligned}
\tag{2.28}
$$

This velocity component is known as the *induced velocity*. It is not, however, a velocity induced by the vorticity but rather the velocity field which is consistent with the prescribed vorticity field. The two fields are mutually consistent, neither one nor the other being the origin or consequence of the other.

Biot–Savart law–the rectilinear vortex

Consider the induced velocity at $\mathbf{r}$ resulting from an element $\delta\boldsymbol{\ell}'$ along the axis of a vortex filament of strength Γ specified by its circulation at $\mathbf{r}'$:

$$
\mathbf{v}(\mathbf{r}) = \frac{1}{4\pi} \Gamma \, \delta\boldsymbol{\ell}' \wedge \left(\frac{(\mathbf{r} - \mathbf{r}')}{|\mathbf{r} - \mathbf{r}'|^3} \right)
\tag{2.29}
$$

The direction of the velocity is normal to both $\delta\boldsymbol{\ell}'$ and $(\mathbf{r} - \mathbf{r}')$. Thus if we have a uniform straight element which subtends angles θ_1 and θ_2 to the perpendicular from $\mathbf{r}$ to the filament at the ends, the induced velocity is

$$
\mathbf{v} = \frac{1}{4\pi r} \Gamma \left(\cos \theta_1 + \cos \theta_2 \right) \hat{\boldsymbol{\ell}} \wedge \hat{\mathbf{r}}
\tag{2.30}
$$

where $\mathbf{r}$ is the perpendicular distance vector from the point of measurement *to* the filament, and $\hat{\boldsymbol{\ell}}$ and $\hat{\mathbf{r}}$ are the unit vectors in the direction of the vortex axis and radius respectively. The velocity vector is normal to the plane containing the filament and the point of measurement, or alternatively the flow rotates in circles about the filament, clearly in the same sense as the circulation, i.e. in the sense defined by the 'corkscrew rule' and the axis.

The potential is multi-valued due to the circulation. Thus taking the zero of the azimuthal angle from some appropriate plane and in the same sense of rotation as the circulation, the potential measured from the zero of the azimuthal angle φ is

$$
\phi(\varphi) - \phi(0) = \frac{1}{4\pi} \Gamma \left(\cos \theta_1 + \cos \theta_2 \right) \varphi
\tag{2.31}
$$

In particular, if the filament is infinite $v = \Gamma/2\pi r$, a result which can easily be deduced from Stokes' theorem, and the potential jump for a single rotation $\phi_+ - \phi_- = \Gamma$.

Induced velocity from a vortex sheet

The induced velocity at a point $\mathbf{r}$ due to an element $\delta S'$ at $\mathbf{r}'$ of a vortex sheet is easily shown to be

$$\mathbf{v}(\mathbf{r}) = \frac{1}{4\pi}\,\gamma(\mathbf{r}') \wedge \left(\frac{(\mathbf{r}-\mathbf{r}')}{|\mathbf{r}-\mathbf{r}'|^3}\right)\,\delta S' \tag{2.32}$$

where γ is the surface vorticity of the sheet, i.e. the circulation associated with unit surface area of the sheet.

Relationship with magnetostatics

In incompressible flow the two vector equations

$$\nabla \wedge \mathbf{v} = \boldsymbol{\zeta} \qquad \text{and} \qquad \nabla \cdot \mathbf{v} = 0 \tag{2.33}$$

may be recognised as the fundamental equations of magnetostatics, with the velocity taking the place of the induction ($\mathbf{v} \equiv \mathbf{B}$) and the vorticity that of the current density ($\boldsymbol{\zeta} \equiv \mu_0\mathbf{j}$). Thus we expect that there will be a direct relationship between the velocity and vorticity equivalent to the Biot–Savart law in magnetostatics, as in equation (2.28) above.

There is, however, a clear distinction with magnetostatics in that, as we noted earlier, the induced velocity is not generated by the vorticity, in contrast to magnetostatics where the current is directly the source of the magnetic field.

Induced velocity due to a vortex loop

Consider a loop S with constant circulation Γ centred at the origin $r = 0$. The induced velocity at the point $\mathbf{r}$ is directly obtained from the Biot–Savart law (2.29) by integration around the loop ℓ bounding S

$$\mathbf{v}(\mathbf{r}) = \oint \frac{1}{4\pi|\mathbf{r}-\mathbf{r}'|^3}\Gamma\,\mathrm{d}\boldsymbol{\ell}' \wedge (\mathbf{r}-\mathbf{r}') \tag{2.34}$$

where $\mathbf{r}'$ are points on the loop.

It follows from Stokes' theorem that

$$\oint_\ell \mathbf{B} \wedge \mathrm{d}\boldsymbol{\ell}\Big|_i = \int_S \frac{\partial B_j}{\partial x_j}\,\mathrm{d}S_i - \int_S \frac{\partial B_i}{\partial x_j}\,\mathrm{d}S_j$$

where S is the surface bounded by the loop ℓ.

Making use of the results that $\nabla(1/|\mathbf{r}-\mathbf{r}'|) = -\nabla'(1/|\mathbf{r}-\mathbf{r}'|)$ and $\nabla^2(1/r') = 0$ except at $r = 0$, where ∇ and ∇' are the gradient operators with respect to $\mathbf{r}$ and $\mathbf{r}'$ respectively, we obtain

$$\mathbf{v}(\mathbf{r}) = -\frac{\Gamma}{4\pi}\nabla\int\frac{(\mathbf{r}-\mathbf{r}')}{|\mathbf{r}-\mathbf{r}'|^3}\cdot\delta\mathbf{S}' = -\frac{\Gamma}{4\pi}\nabla\Omega \tag{2.35}$$

where Ω is the solid angle subtended by the loop ℓ at the point $\mathbf{r}$. The potential is therefore

$$\phi = -\frac{\Gamma}{4\pi}\Omega \tag{2.36}$$

and the velocity components are readily evaluated.

Applying this result to a small loop δS we obtain

$$v_r = \frac{\Gamma\,\delta S}{2\pi r^3}\cos\theta \qquad\qquad v_\theta = \frac{\Gamma\,\delta S}{4\pi r^3}\sin\theta \qquad\qquad v_\phi = 0 \tag{2.37}$$

where θ is the angle between the radius vector $\mathbf{r}$ and the normal to the loop. The product $\Gamma\,\delta S$ defines the strength of an infinitesimal loop.

The potential resulting from the velocity field is easily seen to have a discontinuity in the plane of the loop. Thus taking a circuit from below the plane of the loop to above, around and not enclosing the loop

$$\phi_+ - \phi_- = -\frac{\Gamma}{4\pi}(\Omega_+ - \Omega_-) = \Gamma \tag{2.38}$$

since $\Omega_\pm = \mp 2\pi$ for points lying just above and just below the plane of the loop respectively. This result is clearly in conformity with the fact that the loop is a multiply connected space.

A finite loop L of circulation Γ enclosing a surface S may be imagined to be tiled with a set of adjacent touching infinitesimal loops, such that every loop is everywhere in contact with either a neighbour or the bounding loop, each loop having circulation Γ. The contribution to the circulation at the line of contact of two infinitesimal loops is zero as the two are equal in magnitude but with reversed direction and therefore cancel out. Around the boundary each small loop makes a contribution to the total circulation, which is therefore Γ. The resulting field may then be obtained directly by integration over the individual loops.

2.7.1 Streamlined Flow around a Body Treated as a Vortex Sheet

Irrotational incompressible flow requires the solution of Laplace's equation subject to the boundary conditions on the surface of the body and the incoming flow at infinity. The solution of this equation is uniquely determined by the boundary conditions alone. Thus although the boundary conditions may be applied in different ways, the solution remains the same. It may usefully be applied in the case of the flow over the surfaces of a streamlined body by replacing the surface by a vortex sheet of varying strength. The necessary boundary condition to be satisfied is that the flow is everywhere tangential to the surface. The velocity difference across the sheet parallel to the surface defines the surface vorticity

$$\gamma = \mathbf{v}_{2\|} - \mathbf{v}_{1\|} \quad \text{and} \quad \mathbf{v}_{2\perp} - \mathbf{v}_{1\perp} = 0 \tag{2.39}$$

where $\mathbf{v}_{2\|}$ and $\mathbf{v}_{1\|}$ are the velocity components parallel to the sheet, and $\mathbf{v}_{2\perp}$ and $\mathbf{v}_{1\perp}$ the components along the normal to the sheet, respectively.

If the interior of the body is considered to be replaced by the same material as that external to it, and if the incoming flow is at rest at infinity, we may apply the equation for the induced velocity to the flow to account for the perturbation introduced into the flow by the body. To satisfy this condition we may consider the body to be moving into stationary fluid with speed $\mathbf{U}$. The induced velocity expression will now hold. The interior of the body is therefore fluid moving with the same velocity $\mathbf{U}$, but enclosed by a vortex sheet whose strength matches the above boundary condition.[6] Since there are no sources of mass or vorticity inside the body sheet, it follows from the uniqueness theorem that the flow velocity inside the body must be constant and the internal surface an equipotential.

Transforming back to the rest frame of the body, we may add an additional velocity $-\mathbf{U}$ to those calculated from the induced velocity equation to obtain the resultant $\mathbf{v}' - \mathbf{U}$, where $\mathbf{v}'$ is the induced velocity. Thus we see that the induced velocity may be treated simply as the incremental velocity obtained by replacing the surface of the body by a vortex sheet of appropriate strength. The fluid inside the surface is at rest, as follows from the appropriate boundary condition, since the normal velocity at the surface is everywhere zero.

It is clear from the uniqueness theorem given earlier that if the normal gradient (velocity) condition on the boundary and the velocity at infinity are specified then the complete velocity field is uniquely determined. This result is independent of the way in which those boundary conditions are established. Therefore the flow is the same for both body and vortex sheet. Since the surface strength of the sheet is given by the tangential velocity at the surface of the body, it is therefore uniquely determined, when the sheet coincides with the surface of the body.

The determination of the surface vorticity in two-dimensional flow is straightforward (page 58). In three dimensions the satisfaction of a Helmholtz condition, namely that vortex lines are continuous, makes the use of vortex sheets complex. Similar alternative approaches, but using source and doublet sheets (Section 2.8.2), are more straightforward and preferred.

2.8   Sources and Sinks

Consider the flow with a singularity at the origin, where mass flows outwards at a rate ρm per unit time, m the volume flow rate being known as the strength of the source. Since the flow must be spherically symmetric it is obvious that

---

[6]An appropriate pressure distribution is applied over the surface to maintain this configuration: namely, that associated with the aerodynamic forces.

the total volume flux through a sphere of radius r must be m so that the radial velocity is

$$v_r = \frac{m}{4\pi r^2} \tag{2.40}$$

Such a flow is known as a simple source. If m is negative, fluid is extracted from the flow at a rate $-m$ per unit time and we have a sink.

The potential of this flow is clearly

$$\phi = -\frac{m}{4\pi r} \tag{2.41}$$

By considering the z component of the velocity $v_z = v_r \cos\theta$ it is easily shown that the Stokes streamfunction is

$$\psi = -\frac{m}{4\pi} \cos\theta \tag{2.42}$$

where $\theta = \arctan(\varrho/z)$ is the angle that the radius vector makes with the axis.

Source sheets

As with vorticity it is often convenient to identify a continuous distribution of sources along the surface of an infinitesimally thin sheet. Thus we suppose that a source of strength $\sigma\,\delta S$ is embedded in the infinitesimal element of area δS. The velocity tangential to the sheet is clearly zero, but the normal component $v_n = \frac{1}{2}\sigma$, as is easily seen by calculating the volume flow through a cylinder with faces δS above and below the surface.

As with the vortex sheet we may represent the surface of a solid body by a source sheet. The boundary condition to be applied is that the total velocity normal to the sheet is zero. It is made up of the contribution from all elements of the sheet given by the expression above, namely

$$v_n = \iint \sigma(\mathbf{r}') \frac{(\mathbf{r} - \mathbf{r}')}{|\mathbf{r} - \mathbf{r}'|} \cdot \hat{\mathbf{n}}\, dS' = 0 \tag{2.43}$$

where $\hat{\mathbf{n}}$ is the unit normal at $\mathbf{r}$. This condition determines the function $\sigma(\mathbf{r}')$ on the surface in a similar way to strength of the vortex sheet.

2.8.1 Doublet Sources

A important extension of these flows is provided by a combination of a source and a sink of equal strength, so that the outflow from one is extracted by the other. Thus letting the source/sink combination be separated by $\mathbf{d}$, the potential is

$$\phi = -\frac{m}{4\pi \left|\mathbf{r} - \frac{1}{2}\mathbf{d}\right|} + \frac{m}{4\pi \left|\mathbf{r} + \frac{1}{2}\mathbf{d}\right|} \tag{2.44}$$

At large distances, relative to the separation $r \gg d$, the potential has the form

$$\phi \approx -\frac{m\,d}{4\pi r^2}\cos\theta$$

where θ is the angle between $\mathbf{r}$ and $\mathbf{d}$.

Taking the limit $d \to 0$ in such a way that $m\,d$ remains finite, we obtain the potential for a doublet located at the origin

$$\phi = -\frac{\mathcal{M}}{4\pi r^2}\cos\theta \tag{2.45}$$

where $\mathcal{M} = m\,\mathbf{d}$ is the doublet strength, and the axis is taken along the line of the doublet.

The velocity components are easily found:

$$
\begin{aligned}
v_r &= \frac{\mathcal{M}}{2\pi r^3}\cos\theta \\
v_\theta &= \frac{\mathcal{M}}{4\pi r^3}\sin\theta \\
v_\phi &= 0
\end{aligned}
\tag{2.46}
$$

We note that on the axis $\theta = 0$ the flow is directed along that axis, and on the normal to the axis $\theta = \pi/2$ the flow is again parallel to the axis. The streamlines thus form a series of loops in the azimuthal plane. Comparing the velocity profile above with that for an infinitesimal vortex loop (2.37), we see that the two are identical provided the strength of the doublet $\mathcal{M}$ is equal to the strength of the vortex loop $\Gamma\,\delta S$. However, there is an important difference between doublets and vortex loops. In the case of the doublet, the potential passing from the negative to the positive source is equal to the negative of that one passing around the dipole. The total potential difference for a closed path passing through the origin is therefore zero, in contrast to that for a vortex ring where it is equal to the circulation.

The Stokes streamfunction for a doublet aligned along the axis is easily found from that of a pair of sources a very small distance d apart to be

$$\psi = \frac{\mathcal{M}}{4\,\pi r}\sin^2\theta \tag{2.47}$$

since the angle of separation between the two radius vectors is

$$\theta_+ - \theta_- = \delta\theta = d\,\sin\theta/r$$

2.8.1.1   Doublet sheets

As with sources we may have doublet sheets, with doublet density (strength per unit area) μ. The doublet strength of the element of the sheet δS is therefore $M = \mu \, \delta S$. The potential of a doublet sheet S, bounded by the line L, of uniform strength μ at a point P where the sheet subtends a solid angle Ω can be derived from equation (2.45)

$$\phi = -\frac{1}{4\pi} \int \mu \, \frac{1}{r^2} \cos\theta \, \mathrm{d}S = -\frac{1}{4\pi} \int \mu \, \mathrm{d}\Omega = -\mu \, \Omega \qquad (2.48)$$

where r is the distance from the element $\mathrm{d}S$ to the point of measurement, θ the angle between the normal to the surface and the line r, and $\mathrm{d}\Omega$ the element of solid angle subtended at P by $\mathrm{d}S$. Clearly this result is independent of the surface provided it is bounded by L and subtends the same solid angle at P.

A doublet sheet is equivalent to a vortex sheet where the sheet is made up of infinitesimal vortex loops of circulation $\Gamma = \mu$.

2.8.2   Flow Around a Body Treated as a Source Sheet

Consider an acyclic flow with zero velocity at infinity. We show that the flow around a finite body can be considered as a set of sources and doublets distributed over the internal boundaries.

If ϕ and ψ are two harmonic functions (solutions of Laplace's equation), then it follows from Green's theorem[7] that.

$$\oiint_S \{\phi \, \nabla\psi - \psi \, \nabla\phi\} \cdot \mathrm{d}\mathbf{S} = 0 \qquad (2.49)$$

where S is the boundary of the flow.

Defining r to be the distance from the point of measurement P, which is internal to the surface S, $\psi = r^{-1}$ is a well-behaved harmonic function except at $r = 0$. We may consider the solution excluding a small region surrounding P. Integrating over the surface of this region S',

$$\oiint_{S'} \phi \nabla \left(\frac{1}{r}\right) \mathrm{d}S = -\phi_P \int \frac{1}{r^2} r^2 \mathrm{d}\Omega = -4\pi\phi_P$$

---

[7]Green's theorem is a direct consequence of Gauss's theorem applied to products of the two scalar functions ϕ and ψ

$$\iiint_V \{\phi \, \nabla^2\psi - \psi \, \nabla^2\phi\}\mathrm{d}V = \iint_S \{\phi \, \nabla\psi - \psi \, \nabla\phi\} \cdot \mathrm{d}\mathbf{S}$$

where S is the bounding surface of the volume V.

where ϕ_P is the value of ϕ at P. Integrating over the surfaces S and S' when P lies in the interior of the flow,

$$\phi_P = - \underbrace{\iint_S \frac{1}{r} \frac{\partial \phi}{\partial n}\, \mathrm{d}S}_{\text{Source term}} + \underbrace{\iint_S \phi \frac{\partial}{\partial n}\left(\frac{1}{r}\right)\, \mathrm{d}S}_{\text{Doublet term}} \qquad (2.50)$$

The contribution from the element $\mathrm{d}S$ through the first term scales as r^{-1}, i.e. as a source at the element, whose strength is proportional to its area (equation 2.41). Similarly the second term has the form of the potential from a doublet element directed normal to the element $\mathrm{d}S$ (equation 2.45). Hence the flow may be considered to be the result of a set of sources of strength $\partial \phi / \partial n$ per unit area[8] and doublets of strength ϕ per unit area, where $\mathbf{n}$ is the surface normal distributed over the boundary surface. However, the distributions of sources and doublets are not unique, being just one of an infinite set, as will be shown by the following equation (2.52). The distributions of sources and doublets are arranged to satisfy the boundary conditions on the flow on the surface and at infinity. The flow ϕ', external to S, is arbitrary and usually unphysical, often taking place in the fluid which is used to replace the interior of the solid body.

If the point P is external to the surface S, i.e. does not lie in the volume of the flow, it follows that the origin $r = 0$ is no longer a singularity for the integrand and therefore that the integral over the infinitesimal surface surrounding P is no longer required. The volume integration implied in Green's theorem therefore yields a zero value

$$0 = - \iint_S \frac{1}{r} \frac{\partial \phi}{\partial n}\, \mathrm{d}S + \iint_S \phi \frac{\partial}{\partial n}\left(\frac{1}{r}\right)\, \mathrm{d}S \qquad (2.51)$$

Suppose the flow is divided by the surface S into two acyclic regions and let ϕ be the potential in the region which contains P, and ϕ' the potential in the region not containing P; then ϕ satisfies equation (2.50) and ϕ' equation (2.51).[9] Hence, adding we obtain

$$\phi_P = - \iint_S \frac{1}{r}\left(\frac{\partial \phi}{\partial n} + \frac{\partial \phi'}{\partial n'}\right)\, \mathrm{d}S + \iint_S (\phi - \phi') \frac{\partial}{\partial n}\left(\frac{1}{r}\right)\, \mathrm{d}S \qquad (2.52)$$

where n' is the boundary normal in the region external to P. The function ϕ' is determined by the unknown values of ϕ' and $\partial \phi' / \partial n'$ on the boundary.

The principal application of these results is to the calculation of the flow around a solid surface, where we imagine the flow perturbation generated by the body to be the result of a series of sources and dipoles. Their strengths are determined by the requirement to satisfy the boundary conditions at the surface and at infinity. To match this condition with the above results we imagine that the surface of the body is replaced by a discontinuity with the fluid internal and external to the surface having

---

[8]Compare this relationship with electrostatics where the surface charge density is proportional to the normal electric field intensity.

[9]Note that S may be a surface of discontinuity so that ϕ and ϕ' are independent.

independent properties. The potentials ϕ and ϕ' determine the source and doublet strengths, and match the boundary conditions.

At present the flow external to P is undetermined and the problem is consequently not unique. However, some specific conditions are used to limit the nature of the sources and dipoles, and in doing so make the problem unique:

1. Let $\phi' = \phi$ on the boundary. The tangential component of the velocity is continuous across the boundary, but the normal component discontinuous

$$\phi_P = - \iint\limits_S \frac{1}{r} \left(\frac{\partial \phi}{\partial n} + \frac{\partial \phi'}{\partial n'} \right) dS \qquad (2.53)$$

   and we have a distribution of sources alone with density $(\partial\phi/\partial n + \partial\phi'/\partial n')$.

2. Let $\partial\phi/\partial n = -\partial\phi'/\partial n'$ on the boundary. The normal velocity is continuous across the boundary, but the tangential component discontinuous

$$\phi_P = \iint\limits_S (\phi - \phi') \frac{\partial}{\partial n} \left(\frac{1}{r} \right) dS \qquad (2.54)$$

   and we have a distribution of doublets of density $(\phi - \phi')$.

It can be shown that the representations in terms of sources alone or doublets alone are each unique, whereas the representation in terms of both sources and doublets together is indeterminate.

The relation of the above discussion to the preceding one concerning the representation of the body surface by a vortex sheet (Section 2.7.1) is fairly obvious. The surface which separates one region of flow from another may be considered to be the surface of the body, and also to include the wake. The unprimed system is therefore the external flow around the body, and the primed one the internal flow imagined when the body is replaced by fluid. The wake, being infinitely thin, must consist of doublets alone. The set of sources considered in the earlier section, being vortex rings, is essentially equivalent to doublets.

Thus far this calculation considers the body moving into stationary fluid with a speed $-\mathbf{U}$, as in Section 2.7.1. We may therefore add a potential ϕ_∞ associated with the change of frame to one where the fluid is moving and the body stationary, with velocity $\mathbf{U}$. The potential, calculated in the preceding equations, represents the perturbation, which is introduced by the body, to the incoming flow, ϕ_∞. The position variable $\mathbf{r}$ may be expressed in a more convenient form by introducing the position of the point P, namely $\mathbf{r}$, and the position of the surface element dS', namely $\mathbf{r}'$, separately. Therefore we change $\mathbf{r} \to \mathbf{r}' - \mathbf{r}$. The velocity potential thus becomes

$$\phi(\mathbf{r}) = \phi_\infty(\mathbf{r}) - \iint\limits_{S'} \sigma(\mathbf{r}') \frac{1}{|\mathbf{r}' - \mathbf{r}|} dS' + \iint\limits_{S'} \mu(\mathbf{r}') \frac{\partial}{\partial n'} \left(\frac{1}{|\mathbf{r}' - \mathbf{r}|} \right) dS' \qquad (2.55)$$

where $\sigma(\mathbf{r}')$ and $\mu(\mathbf{r}')$ are the source and doublet densities at the point $\mathbf{r}'$ on the surface respectively, and $\phi_\infty(\mathbf{r})$ the potential at P due to the flow at infinity. As noted earlier the distribution of sources and doublets together with the external applied flow must be such that all the boundary conditions are satisfied. If either sources or doublets only are considered the distribution is unique, but if both are included there is an infinite set of possible solutions. The satisfaction of the boundary conditions is achieved in the manner outlined in Section 2.7.1 and forms the basis of the calculation technique known as the panel method, described later in section 11.9.

2.8.3   Irrotational Incompressible Flow Around a Sphere

The solution for the flow around a sphere may be obtained very simply by considering the Stokes streamfunction resulting from an incoming flow of velocity U and a doublet of strength $\mathcal{M}$

$$\psi = \frac{1}{2}U\,r^2\sin^2\theta + \frac{\mathcal{M}}{4\,\pi\,r}\sin^2\theta \tag{2.56}$$

The streamline $\psi = 0$ is therefore found on the spherical surface $r = R$ if $\mathcal{M} = -2\,\pi\,U\,R^3$. The solution represents an isolated internal flow inside the sphere with fluid flowing from one side of the doublet to the other. Outside the surface, the flow is entirely composed of the incoming fluid. The two are separated by the surface of the sphere.

The velocity of the flow is

$$v_r = U\cos\theta + \frac{\mathcal{M}}{2\pi r^3}\cos\theta$$

$$v_\theta = -U\sin\theta + \frac{\mathcal{M}}{4\pi r^3}\sin\theta \tag{2.57}$$

$$v_\phi = 0$$

As expected, the normal velocity is zero on the surface $r = R$, which is therefore made up entirely of streamlines running along the lines of constant azimuthal angle. The flow around the surface of a sphere is characterised by the normal velocity being everywhere zero, so that no fluid flows into the interior. The points at which the incoming and outgoing streamlines meet the surface are *stagnation points* where the total velocity is zero. It is a general result that when the flow is perpendicular to the surface there can be no normal component of the velocity, and by symmetry the tangential component must be zero also.

It follows from the uniqueness theorem that the flow is determined by the boundary conditions, not by the way in which they are satisfied. Thus the above flow is the solution for the irrotational incompressible flow around a sphere due to an incoming flow of velocity U.

The tangential, and therefore total, velocity on the surface of the sphere is therefore

$$v_\theta = -\frac{3}{2}U\sin\theta \tag{2.58}$$

The sign is negative since θ is measured from the exit point on the surface. Hence from Bernoulli's theorem, the pressure on the sphere is

$$p = p_0 + \frac{1}{2}\rho U^2 \left(1 - \frac{9}{4}\sin^2\theta \right) \tag{2.59}$$

where p_0 is the pressure in the incoming flow at infinity.

The pressure is a maximum on the axis where the flow joins ($\theta = \pi$) and leaves ($\theta = 0$) the sphere and the flow velocity is zero, and has a minimum at the top ($\theta = \pi/2$) and bottom ($\theta = 3\pi/2$). The excess pressure maximum is simply the stagnation pressure $\frac{1}{2}\rho U^2$ and the minimum $-\frac{5}{8}\rho U^2$.

If $p_0 < \frac{5}{8}\rho U^2$ the minimum pressure is calculated to be negative, which is clearly unphysical. This results in *cavitation* where the flow breaks away from the surface. We will defer further discussion of this effect until later in connection with the flow around a cylinder.

Case study 2.I Rankine Ovals

When the separation of the sources d is finite (Figure 2.1), the streamfunction becomes

$$\psi = \frac{1}{2}U R^2 \sin^2\theta + \frac{m}{4\pi}(\cos\theta_+ - \cos\theta_-) \tag{2.60}$$

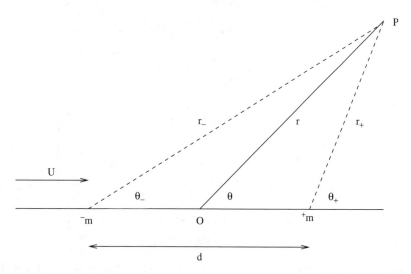

Figure 2.1: Combination of a sink and a source of equal strength m on the axis forming a doublet with separation d. An incoming flow of uniform velocity has been added.

As for the sphere, the streamline $\psi = 0$ forms a closed surface, but now of an oval form provided $md < 0$. By reversing the signs of the incoming velocity U and source strengths m it is easily seen that the oval is symmetric about the centre line $\theta = \pi/2$, and, by reversing the signs of the angles θ, about the axis $\theta = 0$. The points on the surface where $\theta = 0$ and $\theta = \pi$ are therefore stagnation points. It is simple to see from dimensional analysis that these surfaces possess geometrical similarity, the shape depending on the value of the dimensionless parameter $U\,d^2/|m|$. The surfaces are known as *Rankine ovals*. If the shape parameter is small, $U\,d^2/|m| \ll 1$, the ovals collapse to a sphere.

The distance along the axis z_0 of the stagnation points is easily found by requiring the total velocity to be zero at z_0. Calculating the velocity on axis from the values for two sources at distances $z_0 - d$ and $z_0 + d$ and the incoming flow U, we find z_0 as the solution of the quartic equation

$$\left[z_0{}^2 - \left(\frac{1}{2}d\right)^2 \right]^2 + \frac{m\,d}{2\pi U}\, z_0 = 0 \tag{2.61}$$

and the width of the body ϱ_0 is given by the condition $\psi = 0$ at $\theta = \pi/2$.

$$\varrho_0{}^2 \sqrt{(d/2)^2 + \varrho_0{}^2} + \frac{md}{2\pi U} = 0 \tag{2.62}$$

Figure 2.2 shows the variation of the length z_0 and width ϱ_0 plotted as ratios of the source/sink separation d with the scaling parameter $Ud^2/|m|$. The fractional dependence of z_0/d and ϱ_0/d expresses the geometrically similar nature of the flows, which depend solely on the parameter $Ud^2/|m|$ as predicted. It can be seen that for small values of the scaling parameter the oval reduces to a sphere. For large values the length is equal to the separation d and the width decreases as $(\pi Ud^2/|m|)^{-1/2}\,d$.

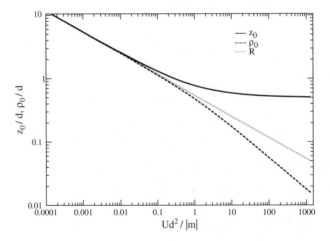

Figure 2.2: Plots of the length and width of a Rankine oval expressed as ratios of the source/sink separation d as functions of the dimensionless parameter $Ud^2/|m|$. The radius of the equivalent sphere R is also shown.

This approach using a uniform flow from infinity and pairs of balanced source/sink combinations may be used to generate the flow around a range of closed bodies, a method known as *distributed sources*. In particular, if the flow consists of a uniform incoming flow with a point source and a line sink with total mass flow, of zero, the body has a blunt front tapering towards the rear typical of the form of an airship.

If the source and sink are not balanced so that there is a net inflow (or outflow) in the system the 'body' streamline extends to infinity generating an open system, but with the fluid from the external flow separated from that of the source/sink combination.

2.9 Two-Dimensional Flow

Two-dimensional flows play an important role in many physical situations. They correspond to the flow around bodies of large aspect ratio, where the length normal to the dominant plane of the flow is very large, and the velocity in this normal direction is approximately zero. As we have already seen in connection with Laplace's equation, the system differs from the three-dimensional case by being necessarily multiply connected.

2.9.1 Irrotational Incompressible Flow

The case of irrotational flow where $\nabla \wedge \mathbf{v} = 0$ introduces a particularly important class of incompressible flows in two dimensions where

$$\nabla \wedge (\nabla \wedge \psi) \rightarrow \nabla^2 \psi = 0 \tag{2.63}$$

and both the velocity potential and the streamfunction satisfy Laplace's equation. In fact the two are closely related, for

$$
\begin{aligned}
v_x &= \frac{\partial \phi}{\partial x} = \frac{\partial \psi}{\partial y} \\
v_y &= \frac{\partial \phi}{\partial y} = -\frac{\partial \psi}{\partial x}
\end{aligned}
\tag{2.64}
$$

These equations will be recognised as the Cauchy–Riemann relations for the real and imaginary parts of the complex function $w = \phi + \imath \psi$ of the complex number $z = x + \imath y$. Thus we conclude that the solution of irrotational incompressible flow in two dimensions is given by the real or imaginary parts of an analytic function, *the complex potential*, which is subject to the appropriate boundary conditions. The complex velocity is given by

$$v = v_x - \imath v_y = \frac{dw}{dz} \tag{2.65}$$

The integral around a contour C is

$$\oint_C \frac{dw}{dz} dz = \oint_C (v_x \, dx + v_y \, dy) + \imath (v_x \, dy - v_y \, dx) = \Gamma + \imath Q \tag{2.66}$$

where Γ is the circulation around C and ρQ the mass outflow through C. The integral may be readily performed using Cauchy's theorem

$$\Gamma + \imath Q = 2\pi\imath \sum_{\text{poles}} \text{residues of } \frac{dw}{dz} \qquad (2.67)$$

where the sum is taken of the residues over all the poles inside C, which are therefore the source and vortex singularities.

The lines of constant potential (equipotentials) and streamfunction (streamlines) are clearly orthogonal. Furthermore, if w is replaced by $\imath w$, the potential and the streamfunctions are interchanged since $\phi + \imath \psi \to \imath \phi - \psi$.

2.10   Applications of Analytic Functions in Fluid Mechanics

There are a number of important applications of the methods of complex functions to problems in two-dimensional fluid mechanics. We examine a few of those of particular relevance.

2.10.1   Flow from a Simple Source and a Simple Vortex

Consider the complex potential expressed in polar form

$$w = a \, \log(z) \qquad \text{or} \qquad z = \exp\left(\frac{w}{a}\right) \qquad (2.68)$$

If a is real, the potential and the streamfunction are given by

$$\phi = a \ln(r)$$
$$\psi = a\theta$$

and the equipotentials (lines of constant ϕ) are circles of radius $\exp(\phi/a)$ and the streamlines radii at angles ψ/a. The flow is therefore that from a source at the origin $r = 0$ along a line normal to the two-dimensional plane. The mass output is easily calculated to be $2\pi\rho a$ per unit length. This flow therefore represents a simple source at the origin of strength $2\pi a$ per unit length, and is the two-dimensional form of that considered earlier (Section 2.8).

If a is imaginary $(-\imath b)$, the equipotentials and streamlines are interchanged, $\phi = b\theta$ and $\psi = -b \ln r$. Thus the streamlines are circles, the flow rotating anti-clockwise around the origin with speed $v_\theta = b/r$. The anti-clockwise circulation is therefore

$$\Gamma = \oint \mathbf{v} \cdot d\mathbf{l} = \int_0^{2\pi} b \, d\theta = 2\pi b \qquad (2.69)$$

The flow $w = -\imath\,(\Gamma\,/\,2\,\pi)\,\log z$ is therefore a simple or rectilinear vortex of circulation Γ centered at the origin. The induced velocity is $\Gamma/2\pi r$, in agreement with our earlier derivation for the rectilinear vortex (Section 2.8.1).

More generally if a is complex

$$\frac{\mathrm{d}w}{\mathrm{d}z} = \frac{a}{z} \tag{2.70}$$

has a pole of residue a at the origin. Therefore there is a source of strength $Q = 2\pi\,\Re(a)$ and a vortex of circulation $\Gamma = -2\pi\,\Im(a)$. In this general case the flow spirals outwards, the flow being most easily obtained as a summation of the above two solutions:

$$\phi = \frac{Q}{2\pi}\ln r - \frac{\Gamma}{2\pi}\theta \quad\text{and}\quad \psi = \frac{Q}{2\pi}\theta - \frac{\Gamma}{2\pi}\ln r \tag{2.71}$$

where ρQ is the mass outflow per unit length from the source. It is readily seen that the streamlines obey the equation

$$\ln r = \frac{\Gamma}{Q}\theta - \psi \tag{2.72}$$

which is the equation of a family of equiangular spirals with angle $\arctan(\Gamma/Q)$ between the tangent and the radius vector.

In two-dimensional flow, the vortex sheet becomes an assembly of infinitely small rectilinear vortices distributed continuously with a total circulation $\gamma(\ell)\,\delta\ell$ in a length $\delta\ell$ along the sheet. The axis of each vortex is perpendicular to the plane of the flow. The total circulation around the sheet is therefore

$$\Gamma = \oint \gamma\,\mathrm{d}\ell \tag{2.73}$$

and the induced velocity at a point $\mathbf{r}$ is

$$v(\mathbf{r}) = \oint \frac{\gamma(\mathbf{r}')}{2\pi|\mathbf{r} - \mathbf{r}'|}\,\mathrm{d}\ell' \tag{2.74}$$

perpendicular to $(\mathbf{r} - \mathbf{r}')$, where $\mathbf{r}'$ is the position vector of the element $\mathrm{d}\ell'$, the sense being anti-clockwise if $\gamma > 0$.

2.10.1.1   Free vortex

The simple vortex with complex potential given by $w = -\imath\,(\Gamma\,/\,2\,\pi)\,\log z$ cannot exist in a practical situation due to the singularity at the origin. In a real environment vortices avoid this unphysical behaviour in one of two ways:

Tied vortex   The flow rotates about a solid body. This occurs in many important situations of which probably the most important is the flow around aerofoils, where the circulation leads to the generation of lift (Chapter 11). The centre of the vortex lies within the solid and the vortex is not able to move freely.

Free vortex   The flow rotates about a central core of rotational flow. The vortex may move freely under the influence of flow generated externally. There are many examples such as tropical storms (hurricanes, typhoons, cyclones), tornadoes, smoke rings, fluid draining through a hole, etc.

The simplest model of free vortex flow is the Rankine vortex, which gives a good representation of many flows found in nature. There is a central core of radius R with constant vorticity ζ surrounded by an irrotational flow outside. Clearly the system is axisymmetric. The rotational core therefore has a velocity profile given by Stokes' theorem $2\pi r v_\theta = \pi r^2 \zeta$. The velocity at the core/surround interface must be continuous and therefore the circulation of the surrounding fluid is $\Gamma = \pi \zeta R^2$. The velocity profile of the vortex is therefore

$$v_\theta = \begin{cases} \dfrac{1}{2} \zeta\, r & \text{if } r \leq R \\[2mm] \dfrac{1}{2} \zeta\, \dfrac{R^2}{r} & \text{otherwise} \end{cases} \tag{2.75}$$

Within the core where $v_\theta = \omega r$, there is rotation with a constant angular velocity $\omega = \frac{1}{2}\zeta$. The pressure distribution is determined by a balance between the pressure gradient and the centrifugal acceleration. Outside the core the flow is irrotational and the strong form of Bernoulli's equation hold:

$$p = \begin{cases} p_0 + \dfrac{1}{8}\rho\, \zeta^2\, r^2 & \text{if } r \leq R \\[2mm] p_\infty - \dfrac{1}{8}\rho\, \zeta^2\, \dfrac{R^4}{r^2} & \text{otherwise} \end{cases} \tag{2.76}$$

where p_0 is the pressure on axis and p_∞ the pressure at infinity. At the boundary $r = R$ the pressures must be equal and therefore

$$p_0 = p_\infty - \frac{1}{4}\rho\, \zeta^2 R^2 \tag{2.77}$$

A free vortex in a moving fluid moves with the fluid, the rotation being superimposed on the normal flow velocity. A tornado consists of a vortex tube of hot moist air in contact with and rising up from the earth to contact the cloud base, where it forms a 'funnel cloud'. The upward-rising air acquires circulation and the lines of vorticity become extended, increasing the vorticity. The uprising column becomes the core of a vortex, which moves differentially between its base where the wind velocity is small and up to its top where the wind velocity is high, as shown by 'tipping' of the funnel cloud. The core vortex tube comprising the tornado becomes extended following Helmholtz's third law, as the lines of vorticity are tied to the fluid motion. Since the vortex tube must coincide with a stream tube, the area of the tube decreases as its length increases in accordance with the requirement of mass conservation. As the area decreases, the vorticity and therefore the velocity increase in accordance with Helmholtz's fourth theorem. The funnel cloud is formed within the vortex core by condensation in the lowered pressure.

2.10.1.2   Two-dimensional doublets and vortex loops

As in three dimensions two equal and opposite sources (source and sink) form a doublet, in this case a line doublet. The complex potential is easily shown to be given by

$$w = -\frac{\mathcal{M}}{2\pi z}$$

$$\phi = -\frac{\mathcal{M}}{2\pi}\frac{\cos\theta}{r} \qquad\qquad \psi = \frac{\mathcal{M}}{2\pi}\frac{\sin\theta}{r} \qquad\qquad (2.78)$$

where $\mathcal{M}$ is the doublet strength per unit length. As earlier (Section 2.8.1), the doublet strength is a vector directed along the line from the negative sink to the positive source. The angle θ is that between the radius vector and the doublet strength. The velocity components are

$$v_r = \frac{\mathcal{M}}{2\pi}\frac{\cos\theta}{r^2} \qquad\qquad v_\theta = \frac{\mathcal{M}}{2\pi}\frac{\sin\theta}{r^2} \qquad\qquad (2.79)$$

In this case also we may consider the case of a pair of vortices of opposing circulations $\pm\Gamma = \pm 2\pi b$ separated by a small distance d, the pair forming the two-dimensional arms of an infinitesimal vortex loop closed at infinity. The vortex strength is therefore Γd.

If d is imaginary, so that the displacement is in the y direction, the loop strength is $(-\imath b/2\pi)(\imath d)$ equivalent to a doublet of strength $\mathcal{M} = \Gamma d$ with axis in the x direction. More generally it is clear that if d is considered to be complex the doublet and vortex pair are identical with the axis of one normal to that of the other. In general both flows may be treated simply as components of one with the axis in an appropriate direction. This result is in accordance with our earlier analysis for three-dimensional singularities.

2.10.2   Flow Around a Body Treated as a Sheet of Complex Sources and Doublets

In section Section 2.8.2 we considered the representation of the acyclic flow external to a three-dimensional body in terms of a distribution of sources and doublets, or vortex rings. In two dimensions the problem is complicated by the cyclic nature of the flow, and the consequent need for the circulation to be specified. Although the problem may be attacked by an extension of the methodology used earlier, it is more convenient to make use of the methods of complex variables and in particular the Cauchy integral theorem.

It is well known that the Cauchy integral theorem applies only to functions that are analytic within the contour over which the integral is performed. This requires that the function be single valued everywhere, in addition to the conditions of continuity and the existence of the complex derivative. Since the flow is cyclic, the complex potential is not single valued, increasing by Γ for a rotation around the body, and we require a calculation of the flow variables, e.g. potential, external to the body. The contour of integration must therefore be chosen to contain the external point P within it, but not involve a complete rotation around the body. Such a contour is shown in Figure 2.3 where the integral is taken nearly round the body C_1 lying on its surface, before being taken along a path to infinity, C_2, and completed by a loop at infinity, C_3, before returning by a path, C_4, parallel to, but separated by an infinitesimal distance from, C_2. Thus the complete contour $C = C_1 + C_2 + C_3 + C_4$ and contains the point P.

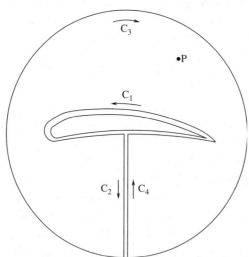

Figure 2.3: Contour for integrating the flow at an external point P around a body generating cyclic flow made up of a set of elements C_1 to C_4.

As in the three-dimensional case, we consider initially the case where the flow at infinity is at rest. At this stage the body is considered to be moving through the fluid with speed $-\mathbf{U}$. We may then bring the body to rest by adding a further external flow with potential ϕ_∞ representing an incoming flow with speed $\mathbf{U}$.

Consider the integral containing the integrand formed by the term $z^{-1}\,dw/dz$ containing the complex velocity over C, where z is taken with origin at the point P, and which therefore is a pole with residue $dw/dz|_P$. The component integrals

$$\oint_{C_2} \frac{dw}{dz}z^{-1}dz = -\oint_{C_4} \frac{dw}{dz}z^{-1}dz \qquad \text{and} \qquad \oint_{C_3} \frac{dw}{dz}z^{-1}dz = 0$$

since $(v_x, v_y)/r \to 0$ faster than $1/r$ at infinity. The integral around C is therefore determined by the residue at P which is described in a clockwise sense. Therefore it follows from the Cauchy integral theorem that

$$\oint_{C_1} \frac{dw}{dz}z^{-1}dz = \oint_C \frac{dw}{dz}z^{-1}dz = -\imath\,2\pi \left.\frac{dw}{dz}\right|_P$$

This equation may be cast into a more familiar form by moving the origin to a convenient location, so that if z is the position of the point P and z' that of an element of the surface of the body,

$$\left.\frac{dw}{dz}\right|_z = \imath\,\frac{1}{2\pi} \oint_{C_1} \left.\frac{dw}{dz}\right|_{z'} \frac{1}{(z - z')}\,dz' \tag{2.80}$$

and, by integration, the potential

$$w(z) = \imath\,\frac{1}{2\pi} \oint_{C_1} \left.\frac{dw}{dz}\right|_{z'} \ln(z - z')\,dz' \tag{2.81}$$

These results generate the analytic continuation of the complex velocity from the surface into the external flow. The equations represent the velocity and potential distributions due to a set of sources and vortices distributed around the body, with complex strength $a = \imath/(2\pi)\,dw/dz$ per unit length. Defining the velocity components $v_{\parallel}$ along the surface in an anti-clockwise sense and $v_{\perp}$ along the outward normal to the surface, it is easily shown that

$$\frac{dw}{dz}\,dz = (v_{\parallel} - \imath\,v_{\perp})\,ds$$

where ds is an element of the path around the surface. *The strength of the source per unit length is $v_{\perp}$ and the vortex strength per unit length $v_{\parallel}$.* If the body surface is impermeable, the normal component of velocity is zero, $v_{\perp} = 0$, and the surface is simply a vortex sheet. The circulation around the body is easily seen to be the total contribution of the vortices, i.e. the imaginary part of the source.

A similar analysis can be performed for the function w/z to give an equivalent result for the analytic continuation of the complex potential

$$w(z) = i \frac{1}{2\pi} \oint_{C_1} w(z') \Big|_{z'} \frac{1}{(z - z')} \, dz' \qquad (2.82)$$

which represents the flow in terms of a distribution of complex doublets and vortex pairs.

In general we may form the solution for the flow by a sum of these two solutions, i.e. in terms of an arbitrary sum of complex sources and doublets. The equivalence of these results to those obtained earlier in three dimensions in Section 2.8.2 is immediately obvious. In the above analysis we followed a derivation based on the properties of complex functions. However, an alternative, but entirely equivalent, formulation using Green's theorem and the potential of a simple line source is also possible following the methods of Sections 2.8.2.

As noted above, the flow at this stage $w(z)$ is calculated in a frame in which the fluid is at rest at infinity. We may transform to the frame in which the body is at rest by adding the additional potential w_∞ due to an incoming flow of velocity $\mathbf{U}$ to obtain

$$w(z) = w_\infty(z) + \frac{1}{2\pi} \oint_{C_1} (\sigma(z') - i\gamma(z')) \ln(z - z') dz' \qquad (2.83)$$

where $\sigma(z)$ is the source density and $\gamma(z)$ the surface vorticity. As noted earlier for a solid surface, $\sigma(z)$ is zero, although it will seen that for some purposes a finite value may be appropriate. The flow described by equations (2.80), (2.81) and (2.82) represents the perturbation induced in the flow by the body.

It is clear from the above result that we may consider the surface of the body as a vortex sheet, as was done earlier. The distribution of the surface vorticity may then be directly calculated as follows. The sheet is represented by a finite distribution of N rectilinear vortices. The induced velocity due to each vortex may be calculated at each vortex point in turn, to which must be added that due to the external incoming flow ϕ_∞. Since the normal component of the velocity on the surface must be zero ($v_\perp = 0$), this condition gives rise to a set of N simultaneous equations, balancing the induced velocity from the vortices against that due to the incoming flow. However, the equations are singular, and they are therefore not independent. This is a consequence of the fact that the total mass flow through the surface is zero, and results from the sum of the products of the normal component of the flow velocity with the length of the element. One equation is therefore redundant and must be replaced appropriately, in this case by the sum of the tangential velocities representing the prescribed circulation. The singularity of the resulting algebraic equations is a consequence of the multi-connectivity in two-dimensional flow, which can only be resolved by specifying the circulation around the body.

The direct application of complex function analysis to the properties of a thin wing is given in the following case study.

Case study 2.II Application of Complex Function Analysis to the Flow around a Thin Wing

In contrast to three dimensions, we cannot normally have a 'free' vortex sheet, i.e. not bound to the surface of a body, in two dimensions. This is due to the fact that the necessary tangential velocity discontinuity is not consistent with the requirements of continuity of the normal velocity component and pressure across the sheet required by Bernoulli's equation. This condition will be relaxed if there is a contact discontinuity of composition or entropy.

We consider the case of a wing inclined at an angle of attack α to an incoming flow of velocity U along the x axis. The wing is thin if the thickness of the wing profile t is much smaller than the chord c, the distance between the leading and trailing edges, and both the angular deflection along the wing θ and the angle of attack α are small. The wing then makes only a small perturbation to the incoming flow. The wing profile may be considered to be made up of two components, one symmetric representing the thickness of the wing along the x direction $y_\pm = \pm g(x)$, where $+$ refers to the upper surface and $-$ to the lower, and the other the shape of the mean line (camber) and the angle of attack $y = h(x) - \alpha x$. The full profile is therefore

$$y_\pm = f_\pm(x) = \pm g(x) + h(x) - \alpha x$$

Since the equations of fluid mechanics are additive for irrotational incompressible flow, we may write the perturbation potential in terms of a symmetric component ϕ_s associated with the thickness and a second anti-symmetric component ϕ_a with the camber and the angle of attack. As the wing is thin, the normal component of the perturbation velocity is approximately given by the y derivative of the relevant potential, and the angle the surface makes with the incoming flow is approximately given by dy/dx. Consequently the condition for zero net flux through the wing may be written as

$$\lim_{y \to 0\pm} \frac{\partial \phi_s}{\partial y} = u_y^s(x, 0\pm) = \pm U \frac{dg}{dx}$$

$$\lim_{y \to 0\pm} \frac{\partial \phi_a}{\partial y} = u_y^a(x, 0\pm) = U \left[\frac{dh}{dx} - \alpha \right] \qquad (2.84)$$

so that ϕ_s is symmetric and ϕ_a anti-symmetric with respect to the x axis, α being the angle of attack. The perturbation velocities exhibit corresponding symmetries

$$
\begin{aligned}
u_x^s(x, -y) &= u_x^s(x, y) \quad : \quad u_y^s(x, -y) = -u_y^s(x, y) \\
u_x^a(x, -y) &= -u_x^a(x, y) \quad : \quad u_y^a(x, -y) = u_y^a(x, y)
\end{aligned}
\qquad (2.85)
$$

Because ϕ_s and ϕ_a are both solutions of Laplace's equations satisfying their respective boundary conditions (2.84), we obtain the source and vorticity distributions from the thickness and camber functions. Since the wing is thin we may treat it as a combination of source and vorticity sheets, the strength of the sheets being given by the change in the appropriate velocity discontinuities across the wing:

$$
\begin{aligned}
\sigma_s &= \lim_{y \to 0+} \frac{\partial \phi_s}{\partial y} - \lim_{y \to 0-} \frac{\partial \phi_s}{\partial y} = 2u_{y0} \\
\gamma_s &= \lim_{y \to 0+} \frac{\partial \phi_s}{\partial x} - \lim_{y \to 0-} \frac{\partial \phi_s}{\partial x} = 0 \\
\sigma_a &= \lim_{y \to 0+} \frac{\partial \phi_a}{\partial y} - \lim_{y \to 0-} \frac{\partial \phi_a}{\partial y} = 0 \\
\gamma_a &= \lim_{y \to 0+} \frac{\partial \phi_s}{\partial x} - \lim_{y \to 0-} \frac{\partial \phi_a}{\partial x} = 2u_{x0}
\end{aligned}
\tag{2.86}
$$

where $u_{x0} = U(\mathrm{d}h/\mathrm{d}x - \alpha)$ and $u_{y0} = U\,\mathrm{d}g/\mathrm{d}x$ are the perturbation velocities at the wing surface. Clearly the sources are associated with the symmetric perturbation due to the thickness of the wing, and the vorticity with the anti-symmetric one due to the camber and inclination of the mean line. It is the latter which gives rise to lift.

The velocity determined by the symmetric potential function ϕ_s is calculated by the direct application of equation (2.80) taking the contour C_1 to be forward along the bottom surface and backward along the top of the wing

$$
\left.\frac{\mathrm{d}w}{\mathrm{d}z}\right|_s = \frac{1}{2\pi} \int_0^c \frac{\left(u_y^s(x',0+) - u_y^s(x',0-)\right)}{(z - x')}\,\mathrm{d}x' = \frac{U}{\pi} \int_0^c \frac{1}{(z - x')} \frac{\mathrm{d}g}{\mathrm{d}x}\,\mathrm{d}x'
\tag{2.87}
$$

making use of the surface boundary condition (2.84). Hence we see that the symmetric flow is determined independently from the anti-symmetric one and depends solely on the thickness profile of the wing section.

The velocity may be also written in terms of the source distribution

$$
\left.\frac{\mathrm{d}w}{\mathrm{d}z}\right|_s = \frac{1}{2\pi} \int_0^c \frac{\sigma(x')}{(z - x')}\,\mathrm{d}x' = \frac{1}{\pi} \int_0^c \frac{u_{y0}(x')}{(z - x')}\,\mathrm{d}x'
\tag{2.88}
$$

Separating the real and imaginary parts,

$$
\begin{aligned}
u_x^s(x, y) &= \frac{1}{\pi} \int_0^c \frac{(x - x')\,u_{y0}(x')}{(x - x')^2 + y^2}\,\mathrm{d}x' \\
u_y^s(x, y) &= \frac{1}{\pi} \int_0^c \frac{y\,u_{y0}(x')}{(x - x')^2 + y^2}\,\mathrm{d}x'
\end{aligned}
$$

In the limit as the point of measurement tends to the surface, $y \to 0+$, the second integral receives contributions only from x' near x, and the integral tends to the expected result $\lim(y \to 0+)u_y^s(x, y) \to u_{y0}$. However, the integral for u_x^s at the aerofoil is less simple. Divide the integral into three parts

$$u_x^s(x, 0\pm) = \frac{1}{\pi} \int\limits_{0}^{(x'-\delta)} \frac{(x-x')\, u_{y0}(x')}{(x-x')^2 + y^2} \mathrm{d}x' + \frac{1}{\pi} \int\limits_{(x'-\delta)}^{(x'+\delta)} \frac{(x-x')\, u_{y0}(x')}{(x-x')^2 + y^2} \mathrm{d}x'$$
$$+ \frac{1}{\pi} \int\limits_{(x'+\delta)}^{c} \frac{(x-x')\, u_{y0}(x')}{(x-x')^2 + y^2} \mathrm{d}x'.$$

Consider the second integral: as $\delta \to 0$, $u_{y0}(x') \approx u_{y0}(x)$ and may be treated as constant in the integral. The remaining integrand is odd in $(x - x')$ and the integral therefore zero. The velocity $u_y^s(x, 0\pm)$ is therefore given by the Cauchy principal value[10]

$$u_x^s(x, 0\pm) = u_{x0}(x) = \frac{1}{\pi} \fint_0^c \frac{u_{y0}(x')}{(x-x')} \mathrm{d}x' = \frac{1}{\pi} \fint_0^c \frac{\sigma_s(x')}{(x-x')} \mathrm{d}x' = U\frac{\mathrm{d}g}{\mathrm{d}x} \tag{2.89}$$

The solution of this integral equation allows the source distribution $\sigma(x)$ to be evaluated knowing the symmetric component of the wing section profile $g(x)$.

Turning now to the anti-symmetric components, which are determined by the vorticity function,

$$\left.\frac{\mathrm{d}w}{\mathrm{d}z}\right|_a = -\frac{1}{2\pi \imath} \int_0^c \frac{\gamma_a(x')}{(z-x')} \mathrm{d}x \tag{2.90}$$

whose imaginary part gives the normal velocity on axis. As above, the integral on axis is singular and in a similar manner can be shown to be equal to the Cauchy principal value:

$$u_y^s(x, 0\pm) = u_{y0}(x) = -\frac{1}{2\pi} \fint_0^c \frac{\gamma_a(x')}{(x-x')} \mathrm{d}x' = U\left[\frac{\mathrm{d}h}{\mathrm{d}x} - \alpha\right] \tag{2.91}$$

This integral equation, known as Glauert's equation, forms the basis of the theory of thin wings. It allows the vorticity distribution $\gamma(x)$ to be obtained knowing the functional form of the asymmetric component of the profile $h(x)$. The equation may be integrated directly, but the more usual approach, through a Fourier series approximation for the vorticity, will be outlined in Section 11.5.

We cannot approach the calculation of the velocity of the anti-symmetric components using the Cauchy integral theorem as in equation (2.87) as simply as for the symmetric ones since v_y^a is symmetric across the wing. To do so we need to find a function with which we can multiply the normal component of the velocity in order to generate an anti-symmetric function in y which may then be treated as u_y^s. Such a term is $\zeta(z) = \sqrt{z/(z-c)}$. We take the positive root on the x axis for $x > c$, $\zeta = \sqrt{x/(x-c)}$. For $0 < x < c$, $\zeta(x, 0\pm) = \pm\imath\sqrt{x/(x-c)}$. For $x < 0$ on the x axis,

---

[10]The Cauchy principal value of an integral, whose integral contains a singularity at $x' = x$, is defined by

$$\fint_0^c \sim \mathrm{d}x' \equiv \lim_{\delta \to 0} \left[\int_0^{(x-\delta)} + \int_{(x+\delta)}^c\right] \sim \mathrm{d}x'$$

$\zeta(z)$ is real and negative, $\zeta = -\sqrt{x/(x-c)}$. Thus ζ has a branch cut from $x = 0$ to $x = c$. Substituting the function $\zeta(z)\,u_y^a$ for u_y^s in equation (2.87) we obtain

$$
\begin{aligned}
\left.\frac{dw}{dz}\right|_a &= \frac{1}{2\pi\imath}\,\zeta^{-1}(z)\int_0^c \frac{(\zeta(x',0+) - \zeta(x',0-))}{(z-x')}\,u_y^a(x,0)\,dx' \\[2mm]
&= \frac{U}{\pi\imath}\sqrt{\frac{(z-c)}{z}}\int_0^c \frac{1}{(z-x')}\left(\frac{dh}{dx} - \alpha\right)\sqrt{\frac{x'}{c-x'}}\,dx' \qquad (2.92)
\end{aligned}
$$

We may calculate the circulation about the wing by integrating the velocity around a contour at large distances from the wing, $|z| \gg c$. The circulation is obtained from the residue at the origin $z = 0$, since there is no source of vorticity in the flow external to the wing:

$$
\frac{dw}{dz} \approx -\imath\frac{U}{\pi z}\int_0^c \left(\frac{dh}{dx} - \alpha\right)\sqrt{\frac{x'}{c-x'}}\,dx'
$$

$$
\Gamma = 2U\int_0^c \left(\frac{dh}{dx} - \alpha\right)\sqrt{\frac{x'}{c-x'}}\,dx' \qquad (2.93)
$$

The circulation is therefore determined solely by the camber (mean) line of the wing and the angle of attack. For a flat plate, $h(x) \approx 0$ and the circulation has the value $\Gamma = -\pi\,c\,\alpha\,U$. Our choice of the function $\zeta(z)$ ensures that the velocity at the trailing edge $x = c$, equation (2.92), is finite. This condition (the Kutta condition), necessary for a well behaved flow around the wing section, is discussed in Chapter 11. The velocity at the leading edge $(x = 0)$, however, tends to infinity. This is due to the sharp edge at the leading edge, which is avoided in real wing sections.

2.10.3   Flow Around a Cylinder with Zero Circulation

Consider the complex potential established by the superposition of a uniform flow with a doublet at the origin, as suggested by the solution for the flow around a sphere:

$$
w(z) = U\left(z + \frac{R^2}{z}\right) \qquad (2.94)
$$

The real and imaginary parts are easily identified

$$
\phi + \imath\psi = \left\{Ux\left[1 + \frac{R^2}{(x^2 + y^2)}\right] + \imath Uy\left[1 - \frac{R^2}{(x^2 + y^2)}\right]\right\}
$$

Clearly $\psi = 0$ on the surface $(x^2 + y^2) = R^2$, i.e. on the surface of a cylinder, and on the x axis, $y = 0$, which is therefore a streamline. As $x \to \pm\infty$, $\phi \to Ux$ and $v_x = \partial\phi/\partial x = U$ and $v_y = \partial\phi/\partial y = 0$. Thus there is an incoming and outgoing flow of velocity U parallel to the x axis. The solution represents the flow around a cylinder of radius R with incoming velocity U.

The velocity at any point on the surface identified by its polar co-ordinates r, θ is easily calculated

$$
\begin{aligned}
v_r &= \frac{\partial \phi}{\partial r} = U \cos\theta \left(1 - \frac{R^2}{r^2}\right) \\
v_\theta &= \frac{1}{r}\frac{\partial \phi}{\partial \theta} = -U \sin\theta \left(1 + \frac{R^2}{r^2}\right)
\end{aligned}
\tag{2.95}
$$

and $v_r = 0$ if $r = R$, i.e. on the surface. The pressure on the surface may be calculated from Bernoulli's equation:

$$
\begin{aligned}
p &= p_0 + \frac{1}{2}\rho\left(U^2 - v^2\right) \\
&= p_0 + \frac{1}{2}\rho U^2 \left[1 - 4\sin^2\theta\right]
\end{aligned}
\tag{2.96}
$$

The pressure maxima at $\theta = 0$ and π result from the flow being brought to rest. At these points the streamline from the free stream joins the surface. By symmetry the velocity at this point must be zero, a *stagnation point* where the velocity is zero, as can be checked from equation (2.95). The pressure $p = p_0 + \frac{1}{2}\rho U^2$ is the result of the ambient pressure plus the kinetic pressure resulting from the flow being brought to rest. The pressure minima at $\theta = \pi/2$ and $3\pi/2$ result from the acceleration of the fluid as the streamlines are displaced by the surface, in accordance with the conservation of mass. If $p_0 < \frac{3}{2}\rho U^2$, equation (2.96) shows that the pressure would be negative, which is clearly non-physical. In fact a cavity will form before this would occur, around which the flow will contour, a phenomenon known as *cavitation*. The cavity will in fact be filled with vapour.

If we compare the values of the differential pressure minima on a sphere, $-\frac{5}{8}\rho U^2$, with those on a cylinder, $-\frac{3}{2}\rho U^2$, we notice that the latter are much larger in magnitude. This is due to the *relieving effect* associated with three-dimensional space. It is a consequence of the fact that fluid has a much greater space to move into in three dimensions than in two, with the result that the maximum speed in the latter case is increased from $\frac{3}{2}U$ to $2U$. This effect is symptomatic of the difference between three- and two-dimensional flow.

Clearly $p(\theta) = p(\pi - \theta) = p(\pi + \theta) = p(-\theta)$, and pressure is symmetrically distributed over the surface. Since the only force on the cylinder is due to pressure, the net force is therefore zero, i.e. contrary to experience there is no drag force exerted in the direction of flow. This is a consequence of the 'ideal fluid' approximation, which does not allow dissipation, which would be generated by drag. More generally it is an example of *d'Alembert's paradox* which states that there is no drag in steady flow of an ideal fluid when the body is immersed in the fluid.

2.10.4 Flow Around a Cylinder with Circulation

As we have seen, the flow in two dimensions is only fully specified if we include the circulation about the cylinder. Therefore we add a vortex flow to that already considered

$$w(z) = U\left(z + \frac{R^2}{z}\right) - i\frac{\Gamma}{2\pi}\ln\left(\frac{z}{R}\right) \tag{2.97}$$

The potential and the streamfunction have the values

$$\phi + i\psi = \left\{Ux\left[1 + \frac{R^2}{(x^2 + y^2)}\right] - \frac{\Gamma}{2\pi}\theta\right\} + i\left\{Uy\left[1 - \frac{R^2}{(x^2 + y^2)}\right] - \frac{\Gamma}{2\pi}\ln\left(\frac{r}{R}\right)\right\}$$

The velocities are easily calculated

$$v_r = \frac{\partial\phi}{\partial r} = U\cos\theta\left(1 - \frac{R^2}{r^2}\right) \quad \text{and} \quad v_\theta = \frac{1}{r}\frac{\partial\phi}{\partial\theta} = -U\sin\theta\left(1 + \frac{R^2}{r^2}\right) + \frac{\Gamma}{2\pi r} \tag{2.98}$$

The stagnation points occur on the surface of the cylinder $r = R$ when the azimuthal velocity is zero, i.e.

$$\sin\theta = \frac{\Gamma}{4\pi UR}$$

which has solutions when $\Gamma \leq 4\pi R$, and the stagnation points occur on the top half of the cylinder for positive circulation Γ. In this condition the incoming flow is sufficiently strong to overcome the effect of the circulation on the surface.

For the case $\Gamma > 4\pi UR$, it is clear from symmetry arguments that if a stagnation point occurs, it must lie on the line $\theta = \pi/2$ away from the surface where

$$v = v_\theta = -\left(1 + \frac{R^2}{r^2}\right)U - \frac{\Gamma}{2\pi r}$$

Hence there are stagnation points when

$$r = \frac{\Gamma}{4\pi U} \pm \sqrt{\left(\frac{\Gamma}{4\pi U}\right)^2 - R^2}$$

One stagnation point lies outside the cylinder and is clearly physical. The second occurs inside the cylinder and is clearly non-physical. It is associated with the parent flows, from which the potential was constructed, namely a doublet and a uniform flow, together with the circulation.

The total velocity on the surface is just the tangential component v_θ so that the pressure is obtained from Bernoulli's equation

$$p = p_0 + \frac{1}{2}\rho U^2 \left[1 - \left(2\sin\theta + \frac{\Gamma}{2\pi UR} \right)^2 \right] \qquad (2.99)$$

The condition for the development of cavitation is weakened on the vacuum side of the cylinder at $\theta = \pi/2$ by the presence of circulation to $p_0 < \frac{1}{2}\rho U^2 \left\{ [(\Gamma/2\pi UR) + 2]^2 - 1 \right\}$.

The solution is quite arbitrary in that the circulation is arbitrary and any value will give a realistic flow around the cylinder. This is in conformity with our conclusions regarding uniqueness in two-dimensional flow.

The forces on the cylinder are easily calculated as they are due solely to the pressure on the surface

$$
\begin{aligned}
\mathbf{F} &= -\oint p \, \mathrm{d}\mathbf{S} \\
&= -\oint \left(\hat{\mathbf{i}} \cos\theta + \hat{\mathbf{j}} \sin\theta \right) p R \, \mathrm{d}\theta \\
&= -\int_0^{2\pi} \left[p_0 + \frac{1}{2}\rho U^2 \left(1 - \frac{\Gamma^2}{4\pi^2 U^2 R^2} \right) \right] - \frac{1}{2}\rho U^2 \left[4\sin^2\theta - 2\frac{\Gamma}{\pi UR}\sin\theta \right] \\
&\quad \times \left(\hat{\mathbf{i}} \cos\theta + \hat{\mathbf{j}} \sin\theta \right) R \, \mathrm{d}\theta \\
&= 0\,\hat{\mathbf{i}} - \rho U \Gamma \,\hat{\mathbf{j}} \qquad (2.100)
\end{aligned}
$$

The drag force, in the direction of the incoming flow, is therefore zero, as predicted by d'Alembert's paradox. There is also a lift force in the $-y$ direction. This is due to the pressure difference established between the upper surface $(+y)$ and the lower one $(-y)$ due to the slowing and speeding up of the flow on these surfaces respectively, which results from the circulation; the pressure change is then accounted for by Bernoulli's equation. The form of the lift

$$L = -\rho U \Gamma \qquad (2.101)$$

is an example of a more general result, to be derived later in Section 2.11, namely the Kutta–Zhukovskii lift formula.

The lift from a rotating body is familiar behaviour due to this effect. This behaviour, known as the *Magnus effect*, was first observed in the motion of rotating cannon balls in the eighteenth century. Today we see it in the swerve imparted to spinning tennis or golf balls. It is also considered of some importance in the flight of spinning missiles.

2.10.5   The Flow Around a Corner

Consider the complex potential

$$w = Az^n \qquad (2.102)$$

Using polar co-ordinates (r, θ)

$$\phi + \imath \psi = Ar^n \left[\cos(n\theta) + \imath \sin(n\theta) \right]$$

Clearly $\psi = 0$ if $\theta = k\pi/n$, where k is integral. Thus if $n > 1$ the potential represents the flow inside a corner of angle π/n, and if $n < 1$ then around the outside. The velocities are

$$
\begin{aligned}
v_r &= \frac{\partial \phi}{\partial r} &&= nAr^{n-1}\cos(n\theta) \\
v_\theta &= \frac{1}{r}\frac{\partial \phi}{\partial \theta} &&= -nAr^{n-1}\sin(n\theta)
\end{aligned}
\qquad (2.103)
$$

If $n > 1$ the flow lies inside the corner, v_r, $v_\theta \to 0$, as $r \to 0$, and the flow is well behaved. However, if $n < 1$ the flow lies around the outside of the corner v_r, $v_\theta \to \infty$, as $r \to 0$. This is clearly non-physical, and indeed requires the pressure $p \to -\infty$. The flow will, as a result, separate at the corner with the flow below the line of separation becoming undefined. The introduction of a weak viscosity leads to the formation of a boundary layer (Chapter 6) which separates due to the large adverse velocity gradient at the corner (Section 6.7). The separated flow forms a vortex which curls around the corner. In practice the resultant flow is likely to be confused and turbulent[11]

2.11   Force on a Body in Steady Two-Dimensional Incompressible Ideal Flow

As we have seen in Section 2.10.4, the flow around an asymmetric streamlined body generates a lift force normal (y direction) to the incoming direction of flow. This is a result of the pressure differential developed on the body by virtue of Bernoulli's equation as the flow over the (upper) more extended side of the body is more accelerated than that over the (lower) less extended side. The mass flux in the stream tubes being is increased on the upper side to accommodate the reduced cross-section. Consequently the pressure on the upper side is reduced more than that on the lower one, and a net force is generated. The body is immersed in the fluid, no separation of the flow over the surface must

---

[11]In compressible flow this condition is relaxed and flow around the outside of a corner can occur without separation provided the angle is not too large (Section 9.4.2).

occur, i.e. the body must be streamlined, or the flow is no longer well behaved and the lift may be destroyed. Separation only occurs along the same line on both the upper and lower surfaces (*trailing edge*).

We initially consider only the two-dimensional flow about a section, i.e. a transverse element dz of a body of large span and uniform section. The flow is continuous with separation only at the trailing edge. The forces on the body may be easily calculated by considering the flow of momentum onto the surface. The force is then the rate at which momentum is given to the surface, which in turn is equal to minus the rate at which fluid gains momentum from the surface. This is determined by the momentum flux at the surface:

$$F_i = -\int_S (p\,\delta_{ij} + \rho\, v_i\, v_j)\, dS_j$$

where $\mathbf{v}$ is the flow velocity and S is the surface of the body.[12]

In steady flow the divergence of the momentum flux tensor $\operatorname{div}\Gamma = \partial\Gamma_{i,j}/\partial x_j = 0$ and it follows from Gauss's theorem that the surface S may be replaced by any enclosing the section (and excluding any additional singularities). We therefore extend the surface of integration to one at very large distances from the body. As the distance tends to infinity the perturbation introduced by it tends to zero, and the flow reduces to that of the incoming flow. Thus we write $\mathbf{v} = \mathbf{U} + \mathbf{u}$, where $\mathbf{U}$ is the incoming flow velocity and $\mathbf{u}$ the (small) perturbation velocity. Defining co-ordinate directions x in the direction of the incoming flow, and y perpendicular to it defined by an anti-clockwise rotation, and considering only two-dimensional bodies, the area element $d\mathbf{S}$ can be written in terms of the line element $d\mathbf{l}$ per unit width. Hence taking $d\mathbf{S}$ outwards and $d\mathbf{l}$ anti-clockwise[13] (Figure 2.4), $dS_x = dl_y\, dz = dy\, dz$ and $dS_y = -dl_x\, dz = -dx\, dz$.

---

[12]Since there is no mass flux through any element of the surface $\mathbf{v} \cdot d\mathbf{S} = v_j\, dS_j = 0$, the lift force may be simply expressed as an integral of the pressure over the surface $F_i = -\int_S p\, dS_i$.

[13]The sense of rotation is taken in accordance with the normal mathematical form, as used for example in complex forms, through the small angle $x \to y$. Thus on a normal plot positive rotation is anti-clockwise. With the axes as defined, i.e. the x direction going from the leading to the trailing edge, this does not conform with conventional aeronautical practice, where the sense is clockwise. As a result the signs of terms associated with rotation, namely circulation or vorticity (lift and moment), are opposite to those in many standard aeronautical engineering texts. However, this convention allows the direct use of the Cauchy integral theorem for terms such as circulation in terms of residues. To conform to aeronautical practice, we introduce a sign change in the definition of coefficients of the lift and the pitching moment when necessary. There should be no confusion provided the reader is aware which convention is being used.

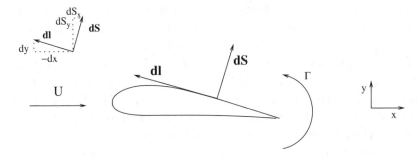

Figure 2.4: Sketch of the geometry of the flow around the body surface.

Since the surface of integration may be taken far from the wing, we may neglect terms of second order in $\mathbf{u}$. Bernoulli's equation becomes

$$p \;+\; \tfrac{1}{2}\rho\,(U_i + u_i)^2 \;=\; p_0 + \tfrac{1}{2}\rho U^2$$
$$\therefore p \;\approx\; p_0 - \rho\,U_i\,u_i \;=\; p_0 - \rho U\,u_x$$

Since the surface is closed, $\oint \rho\,\mathbf{u}\cdot d\mathbf{S} = 0$ and $\oint p_0\,d\mathbf{S} = 0$. Evaluating the forces to first order in $\mathbf{u}$ we obtain the following:

1. **Drag force** parallel to the incoming flow

$$F_x = -\oint (p + \rho U\,u_x)\,dS_x + U \oint \rho\,(u_x\,dS_x + u_y\,dS_y) \approx -\oint p_0\,dS_x = 0$$
$$(2.104)$$

This result validates d'Alembert's paradox for ideal flow, but is only valid if the flow is inviscid and steady state. The proof is easily extended to three dimensions provided no discontinuity occurs.

2. **Lift force** normal to incoming flow

$$F_y = -\oint (p\,dS_y + \rho U\,u_y\,dS_x) \approx -\oint p_0\,dS_y + \rho U \oint \left(u_x\,dS_y - \oint u_y\,dS_x \right)$$
$$= -\rho U \oint (u_x\,dl_x + u_y\,dl_y)\,dz = -\rho U \oint \mathbf{u}\cdot d\mathbf{l}\,dz = -\rho U\,\Gamma\,dz$$
$$(2.105)$$

where

$$\Gamma = \oint \mathbf{u}\cdot d\mathbf{l}$$
$$(2.106)$$

is the circulation around the body, taken anti-clockwise. Since the flow is assumed to be ideal, the circulation is independent of the contour around the body which is used to evaluate it.

This analysis is restricted to the case where the flow at large distances from the body can be considered as a perturbation, i.e. the flow around the body must be continuous. This excludes cases where a region of independent flow, known as the wake, is separated by vortex sheets from the incoming flow. In flow around a streamlined body, where the streamlines run smoothly around the surface, such a discontinuity is forbidden in two dimensions by Bernoulli's theorem. This analysis is easily extended in three dimensions. However, in three dimensions a vortex sheet with discontinuous tangential velocity is allowed even for streamlined bodies. As we shall see (Section 11.8.4) this leads to a drag force as the perturbation within the localised extended vortex sheet is not necessarily small.

2.12   Conformal Transforms

Conformal transformations relate a one-to-one mapping from one two-dimensional space to another. Thus to each point $P' \equiv (\xi, \eta)$ in one space ζ there is a corresponding point $P \equiv (x, y)$ in the second z. The transformation is therefore represented by a pair of functions $x(\xi, \eta)$ and $y(\xi, \eta)$. In terms of the complex numbers $\zeta = \xi + i\eta$ and $z = x + iy$, this transformation is represented by a complex function $z = f(\zeta)$. It is assumed for the purposes of the mapping that within the space to be mapped $z = f(\zeta)$ is single valued, contains no singularities and its inverse $\zeta = f^{-1}(z)$ is similarly well behaved.

If the function $f(\zeta)$ is analytic, so that the derivative $f'(\zeta) = dz/d\zeta$ is well defined, the transformation is conformal. Consider two small elements $d\zeta_1$ and $d\zeta_2$ in the space of ζ, which become dz_1 and dz_2 in that of z. Then since the derivative is independent of direction,

$$dz_1 = \frac{dz}{d\zeta} d\zeta_1 \qquad \text{and} \qquad dz_2 = \frac{dz}{d\zeta} d\zeta_2 \qquad (2.107)$$

The angles between the elements dz_1 and dz_2, and the elements $d\zeta_1$ and $d\zeta_2$, are given by the difference of their arguments. Therefore since

$$\arg(dz_2) - \arg(dz_1) = \arg\left(\frac{dz_2}{dz_1}\right) = \arg\left(\frac{d\zeta_2}{d\zeta_1}\right) = \arg(d\zeta_2) - \arg(d\zeta_1)$$

the angles between elements are unchanged by the transformation. Furthermore, if we consider the magnitude of the elements during the transformation

$$\frac{|dz_1|}{|d\zeta_1|} = \left|\frac{dz}{d\zeta}\right| = \frac{|dz_2|}{|d\zeta_2|}$$

we notice that small lengths are transformed by a constant scale factor. Hence we see the conformal nature of the mapping, in that infinitesimal areas are transformed so as to maintain geometrical similarity, the scale factor being equal to the derivative of the transformation $f'(\zeta)$. Small rectangular elements are therefore transformed into small rectangular elements but of different size. However, straight lines of finite length are not necessarily transformed into straight lines. Thus the similarity of finite elements is not maintained.

Conformal transformations allow a great extension of known solutions of flow, such as those identified earlier, by mapping from a simple space into one more closely matching the flow to be studied, e.g. by transforming the flow around a cylinder to that around an aerofoil section. Clearly if the complex potential in z space $w(z)$ is analytic, then so is $w(f(\zeta))$ in ζ space and therefore it represents a solution to the corresponding fluid problem. The velocity potential and the streamfunction also transform in the same way. Thus if the streamfunction is constant on a surface S in z, it will also be constant on the transformed surface S' in ζ. Hence if a circle S in z can be transformed into an aerofoil section S' in ζ, we may obtain the flow around the wing by a direct transformation from the solution in the space of the cylinder into the space of the aerofoil.

The complex velocity is transformed by

$$v = v_x - \imath v_y = \frac{dw}{dz} = \frac{dw}{d\zeta}\frac{d\zeta}{dz} = \left[f'(\zeta)\right]^{-1}(v_\xi - \imath v_\eta) \qquad (2.108)$$

Conformal transforms have many applications in solving two-dimensional problems. For example, consider the transform $z = \zeta^n$. Then the wedge space contained by the lines $\theta = 0$ and $\theta = \pi/n$ in z space is transformed to the upper half plane in ζ. Therefore consider a flow parallel to the real axis in ζ, namely $w = U\zeta = Uz^{1/n}$, and we obtain the flow around a corner discussed in Section 2.10.5. Several different flows may be constructed in this manner, e.g. the Helmholtz flows described in the following appendix. Further examples of the approach are to be found in the flow around an ellipse, problems #8 and #9. Further applications of the conformal transform approach will be found in Appendix 11.B.

Appendix 2.A Drag in Ideal Flow

As we have seen, d'Alembert's paradox ensures that there is no drag in ideal flow provided the flow is both:

a. **Continuous** No cavitation or separation has taken place around the surface of the body.

b. **Steady** The flow around the body introduces no oscillations into the flow.

If either of these conditions is not upheld, a net drag force may be expected from a purely irrotational flow. In real flows viscosity plays a subtle role in both the configurations discussed in this appendix and indirectly leads to the drag.

2.A.1 Helmholtz's Flow and Separation

Separation of the flow away from the surface of a body is characteristic of the flow around a surface containing a discontinuity of gradient at a salient edge (the case $n < 1$ in Section 2.10.5). Figure 2.A.1 shows a sketch of the streamlines of the flow incident normally on a flat plate. The flow breaks away at the edge, in principle forming a cavity, but in reality a region of dead, often turbulent, slow-moving flow. Before the advent of boundary layer theory, which resolved d'Alembert's paradox of zero drag in nearly inviscid flow, Kirchoff and Rayleigh (1876) independently calculated the drag in ideal flow behind a two-dimensional flat plate subject to certain restrictive conditions. It was assumed that the pressure at the tangential discontinuity between the incoming and separated fluid is constant, and is determined by the conditions far downstream. The pressure and thus the velocity are therefore equal to the incoming pressure, which is lower than that on the front surface. This pressure differential leads to a drag force in the direction of the incoming flow $D \sim \rho S U^2$ on a plate of cross-section S. Flows of this type with free boundary streamlines over which the pressure is constant are known as *Helmholtz flows* and may be analytically calculated in two dimensions by a conformal transformation method. Kirchoff found that a two-dimensional plate of length ℓ has a drag coefficient

$$C_D = \frac{D}{\frac{1}{2}\rho \ell U^2} = \frac{2\pi}{\pi + 4} \approx 0.88 \qquad (2.A.1)$$

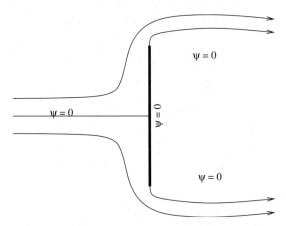

Figure 2.A.1: Sketch of the flow around a plate mounted normal to the incoming flow. The plate is shown by the thick vertical line. The streamline $\psi = 0$ is the free streamline downstream of the plate separating the irrotational incoming flow from the rotational in the cavity.

In fact experiment measures a drag coefficient of approximately 2. Kirchoff's model is unsatisfactory for a number of reasons. It is found experimentally that behind the plate is a region of suction rather than pressure as assumed in the model, hence the increased drag coefficient. In Kirchoff's calculation it is found that the cavity expands parabolically far downstream, whereas in practice vortices form and cavities close after a finite distance. The discontinuity surface is unstable and leads to non-steady mixing and turbulence of the fluid behind the plate. This region of generally turbulent flow is known as the *wake*. The type of drag which is associated with this general structure of flow, leading to a large wake, is known as *form drag*. The general scaling of the drag factor when separation occurs is discussed later in Section 6.8. A full account of Helmholtz flows may be found in Lamb (1932) and Milne-Thomson (1968). Batchelor (1967) gives a thorough discussion of the limitations of Kirchoff's and Rayleigh's theory.

2.A.2 Lines of Vortices

2.A.2.1 Single infinite row of vortices

Consider an infinite row of identical vortices each of circulation Γ separated by a distance a along the x axis. The vortices are therefore at the points $(0,0)$, $(\pm a, 0)$, $(\pm 2a, 0)$, ..., $(\pm na, 0)$, ... The complex potential is the sum of those due to the individual vortices, namely for the $(2n + 1)$ nearest the origin:

$$
\begin{aligned}
w &= -\imath \Gamma \sum_{m=-n}^{n} \log(z \pm ma) = -\imath \Gamma \log \left\{ \prod_{m=-n}^{n} (z \pm ma) \right\} \\
&= -\imath \Gamma \log \left\{ \frac{\pi z}{a} \prod_{m=1}^{n} \left(1 - \frac{z^2}{m^2 \pi^2} \right) \right\} - \imath \Gamma \log \left\{ \frac{a}{\pi} \prod_{m=1}^{n} m^2 a^2 \right\} \quad (2.A.2)
\end{aligned}
$$

Since w is a potential, we may neglect the constant term, and make use of the infinite product expansion of $\sin x$ (Hobson, 1911, §282), namely

$$
\sin x = x \prod_{m=1}^{\infty} \left(1 - \frac{x^2}{m^2 \pi^2} \right) \quad (2.A.3)
$$

to obtain

$$
w = -\imath \Gamma \log \sin \frac{\pi z}{a} \quad (2.A.4)
$$

The complex velocity is

$$
\frac{dw}{dz} = v_x - \imath v_y = -\imath \frac{\Gamma}{2a} \cot \frac{\pi z}{a} \quad (2.A.5)
$$

and hence

$$
\begin{aligned}
v_x &= -\frac{\Gamma}{2a} \left\{ \frac{\sinh(2\pi y/a)}{\cosh(2\pi y/a) - \cos(2\pi x/a)} \right\} \\
v_y &= \frac{\Gamma}{2a} \left\{ \frac{\sin(2\pi x/a)}{\cosh(2\pi y/a) - \cos(2\pi x/a)} \right\}
\end{aligned} \quad (2.A.6)
$$

The individual vortices at $(x = \pm na, y = 0)$ have zero velocity, and the pattern as a unit remains stationary. It is, however, unstable with respect to small departures from exact periodicity (Lamb, 1932, §156).

2.A.2.2 Two parallel symmetric rows of vortices

If a second row of vortices is introduced with the vortices aligned at a separation b, the velocity potential is easily calculated by adding a second term to that obtained above, equation (2.A.3). From equation (2.A.6) velocity of one row of vortices along its line is therefore

$$V = \frac{\Gamma}{2\,a}\,\coth\frac{\pi\,b}{a}\qquad\qquad(2.A.7)$$

The mean velocity in the medial plane is Γ/a. This system is also unstable to small perturbations in the position of the vortices.

2.A.2.3 Two parallel alternating rows of vortices

If a second row of vortices is introduced with the vortices aligned at the mid-points of the other row and separation b (Figure 2.A.2), the velocity potential is easily calculated by adding a second term to that obtained above, equation (2.A.3). From equation (2.A.6) the velocity of one row of vortices along its line therefore

$$V = \frac{\Gamma}{2\,a}\,\tanh\frac{\pi\,b}{a}\qquad\qquad(2.A.8)$$

The mean velocity in the medial plane is also Γ/a.

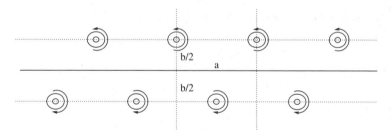

Figure 2.A.2: Sketch of arrangement of the vortices in the Karman vortex street.

This system is unstable except for the particular condition, due to von Karman (Lamb, 1932, §156; Milne-Thomson, 1968, §13.72), that

$$\cosh^2 k\,a = 2 \qquad k\,a = 0.8814 \qquad b/a = k = 0.281$$

It is found experimentally that for a range of Reynolds numbers (to be defined later) vortices leave opposite edges of a cylindrical body placed in a flow. The vortices leave regularly from alternate edges. At a distance behind the body a well-defined vortex stream of definite period, known as the *Karman vortex street*, is formed. At large distances from the body the vortices are damped by viscosity or turbulence. The energy carried away from the body by these vortices appears as a drag force on the body. If the body moves through stationary fluid with velocity U and the vortex street has velocity V, the drag force per unit width is

$$D = \rho\,\Gamma\,k\,(U - 2V) + \rho\,\Gamma^2\,/\,2\,\pi\,a$$

Inserting the values appropriate to the stable Karman vortex street, the corresponding drag coefficient is (Prandtl and Tietjens, 1957, §83)

$$C_D = \left[1.587 \left(\frac{V}{U}\right) - 0.628 \left(\frac{V}{U}\right)^2\right] \frac{a}{d} \qquad (2.A.9)$$

where d is a linear dimension characteristic of the body. At one time Karman vortex streets were thought to be major contributors to the drag from aircraft wings, but are now known to play a negligible role.

Karman vortex streets are, however, an important feature in the design of tall buildings. Vortex shedding as wind blows across large engineering structures can lead to structural instability, in severe cases collapse, as for example in the well-known example of three cooling towers at the Ferrybridge power station in 1965. The phenomenon is now well understood and designs use a number of different measures to avoid the effect. Vortex shedding gives rise to *aeolian tones*, the low-frequency sounds that are heard when wind blows across a string. The vortex shedding frequency f from cylinders is governed by the dimensionless *Strouhal number* in terms of the cylinder diameter d and the flow velocity U, which has the approximate empirical value

$$S = \frac{f\,d}{U} \approx 0.2121 \left(1 - \frac{2.7}{\mathcal{R}}\right) \qquad (2.A.10)$$

in the range of Reynolds number $400 \lesssim \mathcal{R} \lesssim 10^4$.

Chapter 3

Viscous Fluids

3.1 Basic Concept of Viscosity

Viscosity has long been understood as a friction force tending to destroy a velocity gradient. The simple description of this effect, known as Newtonian viscosity, considers the force between neighbouring laminar planes of fluid in differential motion. The flow is assumed to be unidirectional and the velocity gradient normal to the planes containing the flow. The force is proportional to the area of the planes, and to the velocity gradient normal to the planes. Thus the motion is a shear and the force a shearing stress:

$$\frac{F_x}{A} = -\mu \frac{\mathrm{d}v_x}{\mathrm{d}y} \tag{3.1}$$

where x is the direction of flow and y the direction of the gradient. The force F_x is opposed to the direction of flow ($-$ sign) and exerted over an area A. The coefficient μ is known as the first or Newtonian coefficient of viscosity.

This simple result is adequate for treating many simple problems, such as Poiseuille flow through a pipe, but is clearly insufficient to treat complex problems involving more than unidirectional flow. The generally form for viscosity was first given by Navier in 1827 based on a model of dubious validity. The general accepted picture is due to Saint Venant in 1843 and Stokes (1845) following the lines given in this chapter.

Introductory Fluid Mechanics for Physicists and Mathematicians, First Edition. Geoffrey J. Pert.
© 2013 John Wiley & Sons, Ltd. Published 2013 by John Wiley & Sons, Ltd.

3.2   Differential Motion of a Fluid Element

Consider a fluid particle at position $\mathbf{r}$ with velocity $\mathbf{v(r)}$ at time t. A neighbouring particle at $(\mathbf{r} + \delta \mathbf{r})$ has velocity $\mathbf{v}(\mathbf{r} + (\delta \mathbf{r} \cdot \nabla)\mathbf{v})$. The velocity difference is

$$(\delta \mathbf{r} \cdot \nabla)\mathbf{v} = \left[\delta x_j \frac{\partial}{\partial x_j}\right] v_i$$

$$= \frac{1}{2}\delta x_j \left\{\frac{\partial v_i}{\partial x_j} + \frac{\partial v_j}{\partial x_i}\right\} + \frac{1}{2}\delta x_j \left\{\frac{\partial v_i}{\partial x_j} - \frac{\partial v_j}{\partial x_i}\right\}$$

The term

$$\delta x_j \left\{\frac{\partial v_i}{\partial x_j} - \frac{\partial v_j}{\partial x_i}\right\} = \delta \mathbf{r} \wedge (\nabla \wedge \mathbf{v})$$

is a solid body rotation of a fluid element with angular velocity given by the vorticity $\mathbf{z} = \tilde{N}\hat{U}\mathbf{v}$. Further, we define the strain rate tensor

$$\dot{e}_{ij} = \frac{1}{2}\left\{\frac{\partial v_i}{\partial x_j} + \frac{\partial v_j}{\partial x_i}\right\}$$

Thus we obtain as the velocity of the particle at $(\mathbf{r} + \delta \mathbf{r})$

$$\mathbf{v}(\mathbf{r} + \delta \mathbf{r}) = \mathbf{v(r)} + [\delta x_j \, \dot{e}_{ij}] + \frac{1}{2}\delta \mathbf{r} \wedge (\nabla \wedge \mathbf{v}) \qquad (3.2)$$

These terms represent the motion of elements of the fluid as:

1. Uniform translation: namely, $\mathbf{v(r)}$.

2. Solid body rotation with angular velocity given by the vorticity $\frac{1}{2}\zeta = \frac{1}{2}\nabla \wedge \mathbf{v}$: namely, $\frac{1}{2}\delta \mathbf{r} \wedge (\nabla \wedge \mathbf{v})$.

3. Local strain at a rate determined by the strain rate tensor $\dot{e}_{ij}$: namely, $\delta x_j \, \dot{e}_{ij}$.

3.3   Strain Rate

We may identify two different components of the strain rate:

1. The longitudinal or diagonal terms such as $\dot{e}_{xx}$, which represent an extension in the same direction as the length, e.g. measured in direction x. The sum of the longitudinal strain rates in the three directions, i.e. the trace of the corresponding matrix, is

$$\dot{\Theta} = \dot{e}_{ii} = \frac{\partial v_i}{\partial x_i} = \nabla \cdot \mathbf{v} \qquad (3.3)$$

This will be recognised as the rate of dilation $\dot{\Theta}$, which is the fractional rate of change of volume in the fluid.

2. The shear or off-diagonal terms such as $\dot{e}_{xy}$, which represent the differential movement of a length in, for example, the x direction, with respect to variations in the perpendicular direction, e.g. y; that is the angle of shear of the face x over y. Clearly $\dot{e}_{ij} = \dot{e}_{ji}$ so that the strain rate tensor is symmetric.

The change of shape of a fluid element is determined by removing the volume change from the strain to obtain the distortion tensor

$$\dot{D}_{ij} = \dot{e}_{ij} - \frac{1}{3}\dot{\Theta}\,\delta_{ij} \tag{3.4}$$

so that $\dot{D}_{ii} = 0$.

Clearly both the strain rate and the distortion tensors are symmetric.[1]

3.4 Stress

Stress is defined as the force per unit area exerted on an area. Thus two vector directions are associated with the stress, one the direction of the force and the other that of the area. It can be shown that stress is therefore a tensor of rank 2, namely τ_{ij}, where i is associated with the force and j with the area. Thus the force exerted on area $\mathrm{d}S_j$ in direction i is $\tau_{ij}\,\mathrm{d}S_j$.

Viscosity is an internal force due to the interaction between the particles in the fluid. By Newton's third law, the force on a fluid element is balanced by an equal and opposite force on its neighbour. If the stress is constant, the force on the particle is zero, as

$$\int_S \tau_{ij}\,\mathrm{d}S_j = \int_V \frac{\partial \tau_{ij}}{\partial x_j}\,\mathrm{d}V = 0$$

where S is the surface area of the particle and V its volume and using Gauss's theorem. Similarly the net torque should be zero, so that no spontaneous rotation is possible

$$\begin{aligned}(\text{Torque})_i = \int_S \varepsilon_{ijk}\,x_j\,\tau_{k\ell}\,\mathrm{d}S_\ell &= \int_V \varepsilon_{ijk}\,\delta_{j\ell}\,\tau_{k\ell}\,\mathrm{d}V \\ &= -\int_V (\tau_{jk} - \tau_{kj})\,\mathrm{d}V\end{aligned}$$

where ε is the permutation symbol (equation 1.5). Thus stress is also a symmetric tensor $\tau_{ij} = \tau_{ji}$.

---

[1]It is well known from elasticity theory that in an isotropic medium, the strain may be quite generally expressed in terms of two independent quantities: for example, the shear and the dilation. As a result the elastic constants of the medium may be expressed in terms of the shear and bulk moduli.

As with the strain rate we identify two different stresses, the longitudinal (diagonal) stresses, where the force is parallel to the area vector, and the shear (off-diagonal) stresses where the force is tangential to the area.

3.5   Viscous Stress

As we have seen above, viscosity is a force depending linearly on the velocity gradient, or more succinctly the strain rate. Fluids are isotropic, i.e. there is no preferred direction in the fluid, in contrast to plasma where the magnetic field may introduce anisotropy. The most general form of a linear relationship between stress and strain rate involves two scalar constants, the first (or Newtonian) coefficient μ and the second ζ:[2]

$$\sigma_{ij} = 2\mu \left(\dot{e}_{ij} - \frac{1}{3}\dot{\Theta}\delta_{ij} \right) + \zeta\,\dot{\Theta}\,\delta_{ij} \tag{3.5}$$

The term associated with the Newtonian viscosity μ is dependent on the distortion rate rather than the strain rate, and that with second coefficient ζ depends on the dilation only. The factor 2 ensures agreement with the elementary theory of viscosity discussed earlier.

The second coefficient of viscosity is not often significant. It plays an important role in the absorption of sound waves, and in the classical theory of shock wave structure, but is otherwise negligible. Consequently the assumption is often made to set $\zeta = 0$, known as Stokes' hypothesis. This is clearly valid for incompressible flow, but also for monatomic gases where ζ is explicitly zero, being due to rotational relaxation of the molecules.

The total stress exerted on the fluid includes the pressure. Thus

$$\tau_{ij} = \sigma_{ij} - p\,\delta_{ij} \tag{3.6}$$

the minus sign appearing as pressure acts inwards.

The force on an element can be obtained immediately

$$\text{Force} = \int_S \tau_{ij}\,\mathrm{d}S_{ij} = \int_V \frac{\partial \tau_{ij}}{\partial x_j}\,\mathrm{d}V \tag{3.7}$$

i.e. a force $\partial\tau_{ij}/\partial x_j$ per unit volume.

---

[2]There is an exact analogy with the stress/strain relations in the linear theory of elasticity, where the first coefficient corresponds to the shear modulus and the second to the bulk. As with the viscosity only two moduli are independent.

3.5.1   Momentum Equation

Using this force we may easily revise the analysis of Sections 1.3.2 and 1.4.2. Thus following through the previous analysis in the Lagrangian description, Euler's equation (1.12) becomes

$$\frac{dv_i}{dt} = \frac{1}{\rho}\frac{\partial \tau_{ij}}{\partial x_j} + g_i \tag{3.8}$$

and the momentum conservation equation in Eulerian form (1.20) becomes

$$\frac{\partial (\rho v_i)}{\partial t} + \frac{\partial}{\partial x_j}(\rho v_i v_j - \tau_{i,j}) = g_i \tag{3.9}$$

The momentum flux thus includes an additional term $-\sigma_{ij}$, which is due to viscous momentum transfer as viscosity is an internal force. The total momentum flux is therefore $\rho v_i v_j - \tau_{ij}$.

3.5.2   Energy Equation

Similarly we may modify the energy equations to take into account the work done by the viscous force. The total work done at the surface of the fluid element by the stress in the fluid is

$$\text{Work done/unit time} = \int_S v_i \tau_{ij}\, dS_j = \int_V \frac{\partial (v_i \tau_{ij})}{\partial x_j}\, dV$$

In the Lagrangian description, we do not include the kinetic energy since we treat the internal energy separately. Thus we only need the work done by the stress increasing the volume of the element

$$\begin{aligned}
\text{Work done/unit time} &= \frac{1}{\rho}\frac{\partial (\tau_{ij} v_i)}{\partial x_j} - v_i\frac{dv_i}{dt} \\
&= \frac{1}{\rho}\tau_{ij}\frac{\partial v_i}{\partial x_j}
\end{aligned}$$

per unit mass.

There is a further dissipational mechanism of heat transfer which we have not yet included, namely thermal conduction, which has flux $-\kappa\,\nabla T$ where T is the temperature. By including this term in the Lagrangian energy conservation equation (1.15) the final form of the energy equation in Lagrangian form is

$$\frac{d\epsilon}{dt} = \frac{1}{\rho}\tau_{ij}\,\dot{e}_{ij} + \frac{1}{\rho}\nabla\cdot(\kappa\,\nabla T) + g_i v_i \tag{3.10}$$

However, in the Eulerian description we must include the total work done per unit volume in the energy conservation equation in the Eulerian framework, equation (1.23). Hence the equation of energy conservation including thermal conduction takes the form

$$\frac{\partial}{\partial t}\left[\rho\left(\epsilon + \frac{1}{2}v^2 + U\right)\right] + \frac{\partial}{\partial x_j}\left[\rho v_j\left(h + \frac{1}{2}v^2 + U\right) - v_i\,\sigma_{ij} - \kappa\frac{\partial T}{\partial x_j}\right] = 0$$

(3.11)

We note that the viscous term appears as an energy flux $v_i\,\sigma_{ij}$ as is expected for an internal force.

3.5.3   Entropy Creation Rate

The first law of thermodynamics can be written as

$$T\,\mathrm{d}s = \mathrm{d}\epsilon + p\,\mathrm{d}V = \mathrm{d}\epsilon - \frac{p}{\rho^2}\,\mathrm{d}\rho$$

where s is the entropy. Hence we obtain the rate of increase of entropy per unit mass from equation (3.10)

$$\rho\frac{\mathrm{d}s}{\mathrm{d}t} = \frac{1}{T}\left[\sigma_{ij}\,\dot{e}_{ij} + \nabla\cdot(\kappa\,\nabla T)\right]$$

(3.12)

since gravity does not contribute to the entropy generation.

3.6   Incompressible Flow–Navier–Stokes Equation

If the fluid is incompressible $\rho = \mathrm{const}$ and the viscosity is also constant, $\mu = \mathrm{const}$, the terms in the viscous stress depending on the dilation rate $\dot{\Theta} = \nabla\cdot\mathbf{v} = 0$. The viscous stress term simplifies to

$$\frac{\partial\sigma_{ij}}{\partial x_j} = \mu\left(\frac{\partial}{\partial x_j}\frac{\partial v_j}{\partial x_i}^{\,0} + \frac{\partial}{\partial x_j}\frac{\partial v_i}{\partial x_j}\right)$$

$$= \mu\nabla^2\mathbf{v}$$

Hence we obtain from equation (3.8) the Navier–Stokes equation (Stokes, 1845)

$$\frac{\mathrm{d}\mathbf{v}}{\mathrm{d}t} = -\frac{1}{\rho}\nabla p + \frac{\mu}{\rho}\nabla^2\mathbf{v} + \mathbf{g}$$

(3.13)

The term $\nu = \mu/\rho$ is often used and is known as the kinematic viscosity (μ being the dynamic viscosity). Due to the mathematical complexity of this equation, very few analytic solutions exist. Fortunately it is possible to introduce two different limiting cases, which contain much of the essential physics associated with viscosity. In a flow with characteristic velocity U and length ℓ, the quantity $\mathcal{R} = U\ell/\nu$, known as the *Reynolds number*, expresses the relative strength of the viscosity and convective terms in the Navier–Stokes equation. In flows with large Reynolds number, $\mathcal{R} \gg 1$, viscosity is relatively weak, and the flow is nearly ideal, except in a narrow region adjacent to the surface of a body. This condition is known as *boundary layer flow* and is studied in Chapter 6. On the other hand if the Reynolds number is small, $\mathcal{R} \ll 1$, viscosity dominates, a condition known as *Stokes' flow* or 'creeping' flow, described in Section 3.7.

3.6.1   Vorticity Diffusion

Taking the curl of equation (3.13) we obtain

$$\frac{d\zeta}{dt} = \nu \nabla^2 \zeta \tag{3.14}$$

which represents the dissipation of vorticity through an incompressible fluid. Viscosity plays a remarkable twofold role in relation to vorticity, both creating it and subsequently destroying it. Considering an arbitrary Cartesian component ζ_i, the equation reduces to the standard diffusion equation. The vorticity component per unit volume is therefore a conserved quantity with flux $-\nu \nabla \zeta_i$. Vorticity is therefore not annihilated but distributed through the fluid by viscosity, and in so doing gives rise to heating and entropy generation. Thus viscosity causes 'spreading' of vorticity from its source through the fluid. The vortex sheet associated with the wake behind a streamlined body thus broadens at a rate dependent on the viscosity. Consequently the wake behind an aerofoil, where viscosity is weak, remains narrow for a considerable distance downstream.

3.6.2   Couette or Plane Poiseuille Flow

The steady slow flow of an incompressible fluid between two parallel plates with no slip at the walls is easily calculated from the Navier–Stokes equation in two dimensions

$$
\begin{aligned}
v_x \frac{\partial v_x}{\partial x} + v_y \frac{\partial v_x}{\partial y} &= -\frac{1}{\rho}\frac{\partial p}{\partial x} + \frac{\mu}{\rho}\left[\frac{\partial^2 v_x}{\partial x^2} + \frac{\partial^2 v_x}{\partial y^2}\right] \\
v_x \frac{\partial v_y}{\partial x} + v_y \frac{\partial v_y}{\partial y} &= -\frac{1}{\rho}\frac{\partial p}{\partial y} + \frac{\mu}{\rho}\left[\frac{\partial^2 v_y}{\partial x^2} + \frac{\partial^2 v_y}{\partial y^2}\right] \\
\frac{\partial v_x}{\partial x} + \frac{\partial v_y}{\partial y} &= 0
\end{aligned}
\tag{3.15}
$$

Since the total mass flowing along the duct, $\int_{-h}^{h} \rho\, v_x\, dy$, must be constant, we infer that the flow is independent of x. The velocity component v_y is zero at the walls $-h < y < h$. The solution in a steady flow is therefore

$$v_x = -\frac{1}{2\mu}\frac{dp}{dx}\left(h^2 - y^2\right) \qquad v_y = 0 \qquad p = p(x) \qquad \frac{dp}{dx} = \text{const}$$

$$(3.16)$$

where the width of the duct is $2h$ and $-h < y < h$ is measured from the centre line.

3.7 Stokes' or Creeping Flow

Consider the situation when the Reynolds number is small. In this case the viscous terms dominate the convective ones in the Navier–Stokes equation (3.13). Neglecting gravity, the steady state flow of incompressible fluid is therefore given by

$$\nabla p = \nu\, \nabla^2 \mathbf{v} \qquad \text{and} \qquad \nabla \cdot \mathbf{v} = 0 \qquad (3.17)$$

The problem is fully specified by including the boundary conditions at ∞ and on the surface of the body.

These flows, which are laminar, move slowly around the body as the viscosity tries to make the flow become stationary. They are therefore known as creeping flows. They are also known as Stokes' flows after Stokes, who found the first solution to a problem of this type in 1851.

3.7.1 Stokes' Flow around a Sphere

Consider a sphere of radius R with an incoming flow along the x direction with speed U and pressure P. Assuming no slip at the surface of the sphere, the boundary conditions are therefore

$$v_x = v_y = v_z = 0 \quad \text{at } r = R \qquad \text{and} \qquad \left\{ \begin{array}{c} v_x \rightarrow U \\ p \rightarrow P \end{array} \right\} \quad \text{as } x \rightarrow -\infty$$

We try as a solution[3]

$$\mathbf{v} = \nabla\phi + \mathbf{v}' \qquad (3.18)$$

Hence

$$\nabla p = \mu\, \nabla^2 (\nabla\phi) + \mu\, \nabla^2 \mathbf{v}'$$

---

[3]This solution is based on the fact that $\nabla^2 \mathbf{a} = \nabla(\nabla \cdot \mathbf{a})$ if and only if $\mathbf{a} = \nabla\phi$, a result which follows directly from $\nabla \wedge (\nabla \wedge \mathbf{a}) = \nabla(\nabla \cdot \mathbf{a}) - \nabla^2 \mathbf{a}$.

At this stage, our choice of $\mathbf{v}'$ is undetermined; let $\nabla^2 \mathbf{v}' = 0$. Then by integration

$$p = P + \mu \nabla^2 \phi \qquad (3.19)$$

and from the equation of continuity

$$\nabla^2 \phi + \nabla \cdot \mathbf{v}' = 0 \qquad (3.20)$$

Now further assume that only one component of $\mathbf{v}'$ is non-zero, namely v'_x satisfying $\nabla^2 v'_x = 0$ as above. Then the simplest solution is

$$v'_x = \frac{a}{r} \qquad (3.21)$$

and hence

$$\nabla^2 \phi + a\, \frac{\partial(1/r)}{\partial x} = 0$$

which has the solution

$$\phi = -\frac{1}{2}\, a\, \frac{\partial r}{\partial x} \qquad (3.22)$$

where a is an undetermined constant.

This solution for ϕ is a particular integral as we may add any solution of $\nabla^2 \phi$ as a complementary function to obtain the general solution, the appropriate one being determined by the boundary conditions. In particular we require a function such that $\mathbf{v} \to \mathbf{U}$ as $r \to \infty$. Such a function is clearly Ux. Another is $\{b\, \partial(1/r)/\partial x\}$, where b is also an undetermined constant. We introduce both in order to satisfy the boundary conditions on the sphere. Our solution thus far satisfies the governing differential equations, but must also satisfy the boundary conditions. The complete solution is

$$\mathbf{v} = \nabla \left[Ux + b\, \frac{\partial(1/r)}{\partial x} - \frac{a}{2}\, \frac{\partial r}{\partial x} \right] + \frac{a}{r}\, \hat{\mathbf{i}} \qquad (3.23)$$

At the surface of the sphere $r = R$ we require $\mathbf{v} = 0$, i.e. $v_x = v_y = v_z = 0$. But $v_y = v_z = 0$ if $b = -a\, R^3/6$, and $v_x = 0$ if $b = U\, R^3/4$, hence $a = -\frac{3}{2}\, U R$. Thus we have found the solution which obeys the equation of continuity and the Navier–Stokes equation, and which satisfies all the boundary conditions. In component form:

$$v_x = U \left(1 - \frac{3}{4}\frac{R}{r} - \frac{1}{4}\frac{R^3}{r^3} \right) - \frac{3}{4}\frac{UR}{r^3}\left(1 - \frac{R^2}{r^2} \right) x^2$$

$$v_y = -\frac{3}{4}\frac{UR}{r^3}\left(1 - \frac{R^2}{r^2} \right) x\, y \qquad (3.24)$$

$$v_z = -\frac{3}{4}\frac{UR}{r^3}\left(1 - \frac{R^2}{r^2} \right) x\, z$$

and the pressure

$$p = P + \mu \nabla^2 \phi = P - \frac{3}{2} \frac{\mu U R}{r^3} x \qquad (3.25)$$

To calculate the force on the sphere, we neglect the velocity term for the momentum flux, being small, and note that by symmetry only the force in the direction of the incoming flow, x, will be non-zero. The stress over the surface of the sphere is

$$F_i = -\oint p \, \mathrm{d}s_i + \oint \sigma_{i,j} \, \mathrm{d}s_j \qquad F_x = \oint \left(\mathbf{P}_1 - p \hat{\mathbf{i}} \right) \cdot \mathrm{d}s$$

where we have defined $\mathbf{P}_1 = \sigma_{xx} \hat{\mathbf{i}} + \sigma_{xy} \hat{\mathbf{j}} + \sigma_{xz} \hat{\mathbf{k}}$ such that $\mathbf{P}_1 \cdot \mathrm{d}\mathbf{S}$ is the viscous force exerted in the x direction over a surface $\mathrm{d}\mathbf{S}$, and $\hat{\mathbf{i}}, \hat{\mathbf{j}}$ and $\hat{\mathbf{k}}$ are the unit vectors in the directions x, y and z respectively. But we have already shown that (3.17)

$$\nabla \cdot \left(\mathbf{P}_1 - p \hat{\mathbf{i}} \right) = \left[\mu \nabla^2 \mathbf{v} - \nabla p \right]_x = 0$$

Hence by Gauss's theorem we may calculate the integral over any surface enclosing the sphere. We consider a sphere of very large radius, so that

$$v_x = U + \frac{a}{2r} \left(1 + \frac{x^2}{r^2} \right) \qquad v_y = \frac{a}{2} \frac{xy}{r^3} \qquad v_z = \frac{a}{2} \frac{xz}{r^3}$$

where $a = -\frac{3}{2} U R$. The force in the y and z directions is zero by symmetry. The relevant components of the strain rate tensor are

$$\frac{\partial v_x}{\partial x} = \frac{a}{2} \frac{x}{r^3} \left(1 - \frac{3x^2}{r^2} \right)$$

$$\frac{\partial v_x}{\partial y} + \frac{\partial v_y}{\partial x} = -3a \frac{x^2 y}{r^5}$$

$$\frac{\partial v_x}{\partial z} + \frac{\partial v_z}{\partial x} = -3a \frac{x^2 z}{r^5}$$

Hence substituting for $\mathbf{P}_1$ we obtain

$$\begin{aligned}
F_x &= \left\{ \oint a\mu \left[\left(1 - \frac{3x^2}{r^2} \right) \frac{x}{r^3} \hat{\mathbf{i}} - 3 \frac{x^2 y}{r^5} \hat{\mathbf{j}} - 3 \frac{x^2 z}{r^5} \hat{\mathbf{k}} \right] - \left(p_0 + \frac{a\mu x}{r^3} \right) \hat{\mathbf{i}} \right\} \cdot \mathrm{d}\mathbf{S} \\
&= -3a\mu \oint \frac{x^2}{r^5} \mathbf{r} \cdot \mathrm{d}\mathbf{S} - \oint p_0 \hat{\mathbf{i}} \cdot \mathrm{d}\mathbf{S} \\
&= -3a\mu \int_0^\pi \int_0^{2\pi} \pi \cos^2 \theta \sin \theta \, \mathrm{d}\theta \, \mathrm{d}\phi \\
&= -4\pi a \mu \\
&= 6\pi \mu U R \qquad (3.26)
\end{aligned}$$

Thus the drag in this regime is proportional to the speed U of the fluid flow.

This remarkable result was first derived by Stokes (1851) in the middle of the nineteenth century. It has subsequently been widely used to measure the viscosity of strongly viscous fluids, but more importantly as a correction in Millikan's oil drop experiment to measure the charge on the electron.

Defining the drag coefficient as the dimensionless force,

$$C_D = \frac{F_x}{\frac{1}{2}\rho U^2 S} = \frac{24}{\mathcal{R}} \tag{3.27}$$

where $S = \pi R^2 = (\pi/4)d^2$ is the cross-sectional area and the Reynolds number $\mathcal{R} = \rho U d/\mu$ where $d = R/2$ is the diameter of the sphere.

3.7.1.1 Oseen's correction

Although Stokes' formula is well verified by experiment, there is a serious problem with the solution in the behaviour at large distances from the sphere. Far from the sphere the velocity is not small and in consequence the inertial term becomes important. In the neighbourhood of the sphere, the velocity is well described by Stokes' result and as a consequence the drag is accurately calculated. Oseen (1910) improved Stokes' analysis by retaining an approximation in the form of the additional term $(\mathbf{U} \cdot \nabla)\mathbf{v}$ in the Navier–Stokes equation. Thus the Navier–Stokes equation is reduced to

$$\rho(\mathbf{U} \cdot \nabla)\,\mathbf{v} + \nabla p = \mu\,\nabla^2\mathbf{v} \tag{3.28}$$

As may expected from the above remarks, the solution leads to a marked change in the pattern of the streamlines, but only a small correction to the drag. The drag coefficient becomes

$$C_D = \frac{24}{\mathcal{R}}\left(1 + \frac{3}{16}\mathcal{R}\right) \tag{3.29}$$

Stokes' formula is in good agreement with experiment up to Reynolds number $\mathcal{R} \approx 1$. Oseen's correction extends this up to $\mathcal{R} \approx 5$.

3.7.1.2 Proudman and Pearson's solution

The complete solution valid at both small and large distances from the sphere and the circular cylinder was finally achieved by Proudman and Pearson (1957) using the method of matched asymptotic approximation (see Appendix 6.A). The inner solution is based on an expansion, whose first term is Stokes' formula. The outer solution is based on the Oseen approach. Each solution individually can satisfy only one of the boundary conditions, either no slip at the surface of the sphere (Stokes' solution) or uniform flow at infinity (Oseen). Based on the two expansions with the Reynolds number as the matching parameter, a solution satisfying both boundary conditions is constructed. The drag coefficient taken to next order is

$$C_D = \frac{24}{\mathcal{R}} \left[1 + \frac{3}{16}\mathcal{R} + \frac{9}{160}\mathcal{R}^2 \ln\left(\frac{\mathcal{R}}{2}\right) \right] \tag{3.30}$$

3.7.1.3 Lamb's solution for a cylinder

The problems in the solution arising from Stokes' approximation for the sphere are more severe when applied to a cylinder to the extent that no solution is possible. This effect is known as Stokes' paradox (Milne-Thomson, 1968). Lamb (1932) applied Oseen's method to flow around a cylinder (Lamb, 1932; Batchelor, 1967). The resulting value of the drag coefficient is

$$C_D = \frac{8\pi}{\mathcal{R}\left[1/2 - \gamma - \ln(\mathcal{R}/8)\right]} = \frac{8\pi}{\mathcal{R}\left(2.00225 - \ln\mathcal{R}\right)} \tag{3.31}$$

where $\gamma = 0.57721566$ is Euler's constant.

3.8   Dimensionless Analysis and Similarity

Although a number of approximations can be made, analytic solutions of the Navier–Stokes equation are limited. Dimensional analysis therefore becomes a powerful and widely used tool in fluid mechanics. The technique is particularly valuable when the results can be applied in conjunction with experiments to derive empirical scaling relations of great value for practical application.

Physical quantities are measured in units in which the quantity is referred to a standard value. Whereas the quantities are universal, units are arbitrarily prescribed. In fact units are divided into two types:

- **Primary units**   Certain quantities, normally length, mass, time, temperature and charge, are measured in units which have a primary role and must be set by prescribed standards, normally by international agreement. These define a system of units such as the MKS system (metre, kilogram, second, kelvin, coulomb).

- **Derived units**   The remaining quantities are all measured in units which have some relationship with the primary quantities, generally in their definition, such as

$$Velocity \quad \sim \quad Length/Time$$
$$Force \quad \sim \quad Mass \times Length/Time^2$$
$$Viscosity \quad \sim \quad \{Force/Area\}/\{Velocity/Length\}$$

and are measured in units defined by the pre-set primary units.

The relationship of the derived units to the primary units defines the *dimensions* of the quantity, e.g. $[V] = [L][T]^{-1}$, $[F] = [M][L][T]^{-2}$ and $[\mu] = [M][L]^{-1}[T]^{-2}$. The dimensions clearly represent the changes in the values of quantities as we change from one system of units to another.

The performance of a physical system is expressed in terms of the measured quantities of the variables, which describe its behaviour. Since the fundamental laws of physics cannot vary with a change of units, it must be concluded that any fundamental relationship amongst the variables must be dimensionless, i.e. remain true when the units of measurement are changed. This is expressed as the law of *dimensional homogeneity*. A product of the variables whose value is unchanged by a change of units has no dimension and is known as a *dimensionless product*. The dimensionless functional relationships amongst the physical quantities may be expressed in terms of a *complete set of dimensionless products*. The set is *complete* if any dimensionless product of the variables not included in the set can be expressed as a product of those already included in the set. The mathematical statement of the law of dimensional homogeneity is provided by *Buckingham's Π theorem*, which states that any fundamental law can be expressed as a functional relationship amongst the complete set of independent dimensionless products (Buckingham, 1914). The number of dimensionless products in the complete set is itself determined by Buckingham's theorem. Thus if the set Π_i forms a complete set of n dimensionless products for the system, then there must exist a functional relationship

$$F\left(\Pi_1, \Pi_2, \ldots, \Pi_i, \ldots, \Pi_n\right) = 0 \tag{3.32}$$

The complete set of dimensionless products must be formed from all the physical variables which determine the behaviour of the system. If a dimensionless product is either very small or very large it implies that a corresponding physical quantity is very small and does not normally play a role influencing the relationships amongst the variables. The product may therefore normally be omitted without introducing significant error. In many cases the set of dimensionless products is easily established by inspection without recourse to systematic methods. An outline of the proof of Buckingham's Π theorem and the formal method of finding the complete set of dimensionless products is given in Appendix 3.A. For a more complete account the reader is referred to Durand (1934), Langhaar (1980), Birkhoff (1955) and Barenblatt (1996).

The dimensionless products introduce the concept of *similarity*, where the magnitudes of quantities in two similar systems, a model and a prototype, are related by numerical scaling factors. The scaling factors are established by the requirement that all the dimensionless products in both the model and proto- type must be equal, since they are representative of the same physical system. This technique is very widely applied in fluid mechanics, where experiments on the prototype may be difficult or impractical, and ones on a model relatively straightforward. Typical examples are wind tunnel modelling used for car and aircraft design, and ship tank testing. A range of examples of modelling in applied fluid mechanical problems are to be found in Langhaar (1980) and Sedov (1959).

3.8.1   Similarity and Modelling

Consider a prototype system and a geometrically similar model. We seek to identify conditions under which experiments can be carried out on the model which will replicate measurements made on the prototype in operation. To achieve this requires a knowledge of how experiments on the model relate to those on the prototype, i.e. the scale factors relating the model measurements to the equivalent ones on the prototype. These scale factors are established by identifying the functional relationships amongst the dependent and indepen- dent variables of the overall problem.

To illustrate this method we consider the very simple case of a subsonic wind tunnel. We consider a prototype and model, which are geometrically similar, i.e. every point on the prototype is exactly reproduced on the model, and with all lengths in the same proportion. The flows for both the prototype in its environment and the model in the wind tunnel are described by the Navier–Stokes and the continuity equations, whose characteristic parameters are the density ρ and viscosity μ. The problem is defined by the incoming flow velocity U, and the prototype/model by a single length L. We will assume the problem is steady state, so that time does not enter as a variable. The variables to be measured are quantities such as pressure, P, and velocity, $\mathbf{V}$, as functions of the spatial position, R. From these quantities we form a complete set of dimensionless products

$$\mathcal{P} = \frac{P}{\rho U^2} \qquad \mathcal{V} = \frac{\mathbf{V}}{U} \qquad \mathcal{R} = \frac{\mathbf{R}}{L} \qquad \mathcal{R} = \frac{\rho U L}{\mu}$$

where $\mathcal{R}$ is the Reynolds number of the flow, a quantity depending only on the characteristics of the fluid and the flow. If the prototype and model sys- tems both have the same Reynolds number, all the dimensionless parameters depending only on the independent quantities in the problem are equal, and the systems are said to exhibit *complete similarity*. It follows from the law of dimensional homogeneity that these quantities express the functional relationship between the dependent, the independent and the characteristic parameters. Thus we may write for the dependent variables

$$\mathcal{P} = f_P(\mathcal{R}, \mathcal{R}) \qquad \mathcal{V} = \mathbf{f}_V(\mathcal{R}, \mathcal{R}) \tag{3.33}$$

where f_P and f_V are arbitrary functions determined by experiment on the model. These equations apply to both model and prototype. Hence by casting the experimentally measured results in terms of the dimensionless parameters, we may relate one set to the other, provided the systems exhibit complete similarity.

This condition of complete similarity may be difficult to achieve. For example, if the flow velocity U is near the sound speed c, compressibility effects will become important. In this case the additional dimensionless parameter, the Mach number $\mathcal{M} = U/c$, becomes important. It may be possible to obtain similarity of the Mach number by changing the sound speed, e.g. by a change of gas or by heating. However, achieving the simultaneous equality of $\mathcal{R}$ and $\mathcal{M}$ may not be possible, and an empirical compromise must then be found.

The Reynolds number $\mathcal{R}$ has an important physical significance expressing the importance of viscosity in the flow. Thus

$$\frac{\text{Fluid momentum transfer rate}}{\text{Viscous momentum transfer rate}} \sim \frac{\rho U L}{\mu} = \mathcal{R}$$

In flows with energy transport an additional dimensionless parameter, the Prandtl number, is obtained from

$$\frac{\text{Viscous diffusivity}}{\text{Thermal diffusivity}} = \frac{\mu}{\rho} \bigg/ \frac{\kappa}{\rho c_p} = \frac{\nu}{\chi} = \mathcal{P}$$

where $\nu = \mu\rho$ is the kinematic viscosity determining the diffusion of a velocity disturbance (e.g. vorticity (3.14)) in the fluid, and $\chi = \kappa/\rho c_p$ is the thermal diffusivity expressing the diffusion of a heat perturbation through thermal conduction. This fluid parameter therefore expresses the relative importance of viscous transfer and thermal conduction.

3.8.2   Self-similarity

A useful approach in fluid mechanics is based on the concept of self-similarity, which allows the dimensionality of a problem to be reduced; for example, a problem in one spatial dimension and time may be reduced from a set of partial differential equations in two independent variables to a single ordinary differential equation involving one self-similar parameter. When possible, this reduction finds wide applicability, particularly for a number of time-dependent problems in compressible flow, which we discuss later in Chapter 14. The simplification occurs when it is possible to represent the functional form of the solution in a similar form as time (for example) varies, but with scalings, which are themselves dependent on 'time'. This is only possible when there are only a restricted number of dimensionless parameters.

As an example, consider the following two-dimensional problem of incompressible boundary layer flow over an infinite plate to be treated in Chapter 6. The Navier–Stokes equation can be expressed in terms of the parameters x and y as independent variables, $\mathbf{v}$ and p/ρ as dependent variables, and U, P/ρ and ν as constant parameters, where P and U are the pressure and speed of the incoming flow. Note there is no length amongst the characteristic parameters as the plate is infinite. A complete set of dimensionless products is

$$\frac{\mathbf{v}}{U} \qquad \frac{p/\rho}{P/\rho} \qquad \eta = y\sqrt{\frac{U}{\nu x}}$$

Although the complete set of dimensionless products is not unique, it is impossible to find one containing dimensionless products with either x or y separately, i.e. in one independent variable only. Applying the law of dimensional homogeneity, we have

$$p = P\, f_p(\eta) \qquad \mathbf{v} = U\, \mathbf{f}_v(\eta)$$

The problem is thus reduced from a two-dimensional one to a one-dimensional one with η as the sole independent variable. The solution is self-similar with the profiles of density, pressure, velocity, etc., as functions of the variable η alone. Their spatial profiles are identical in form from point to point, and vary in the spatial scale factor alone. A large number of problems in fluid mechanics can be simplified to self-similar forms, which are often the asymptotic limit of the complete solution.

Appendix 3.A Buckingham's Π Theorem and the Complete Set of Dimensionless Products

Let us suppose we have a problem for which we identify a set of dimensional constants $a_1, \ldots, a_m$, a set of independent variables $x_1, \ldots, x_n$ and a dependent variable y, making a total set of $N = 1 + n + m$ variables $z_1, \ldots, z_N$. The dimensions of the term $[z_i] = [M]^{\alpha_{i,1}} [L]^{\alpha_{i,2}} [T]^{\alpha_{i,3}}$ form the dimensional matrix $\alpha_{i,j}$ where the column i refers to the dimensional power and the row j to the element.

We may form a dimensionless product from these parameters and dimensions as

$$\Pi = \prod_{i=1}^{N} z_i^{k_i} \qquad (3.A.1)$$

Forming the dimensions of the product Π, we find that the powers k_i are given by the solutions of three simultaneous equations

$$\alpha_{ij} \, k_i = 0 \qquad (3.A.2)$$

with $j = 1$ to 3. These form a set of homogeneous equations, whose solutions are governed by the fundamental law of algebra:

> *Disregarding the trivial solution $k_i = 0$, a set of M homogeneous equations in N unknowns possesses exactly $(N - R)$ linearly independent solutions, where R is the rank of the matrix. A set of $(N - R)$ linearly independent solutions is called a fundamental set of solutions.*

The rank of the matrix identifies the number of independent equations, or rows in the dimensional matrix. It is normally, but *not* always, equal to the number of rows M, which in this case is 3.

There is no unique solution. We therefore find any appropriate independent set by assigning appropriate independent values to the powers $k_{(R+1)}$ to k_N and solving for k_1 to k_R. The usual method is to assign the last variable z_N to be the dependent one y, and the next the last to be independent in the inverse order of variation. The most useful set of dimensionless variables is then found by setting initially $k_N = 1$ and subsequently $k_N = 0$. The complete set is found from R independent sets of the variables k_i.

The set of dimensionless products Π_j for $j = 1, R$ thus calculated is a complete set. Any further dimensionless products of the variables can be expressed as products of the existing ones already in the set. This set must therefore represent the complete functional relationship amongst the variables, as required by Buckingham's theorem. Alternative, more convenient forms of the set may obviously be found by multiplication.

Chapter 4

Waves and Instabilities in Fluids

4.1 Introduction

Waves are a very general feature of fluids. Several types of waves, principally associated with surfaces, occur in incompressible fluids. Isotropic fluids cannot support shear motions, which allow waves to develop, consequently transverse waves are forbidden in isotropic fluids.[1] Longitudinal waves involve density changes and are therefore forbidden in incompressible fluids.[2] However, the presence of an interface between two dissimilar fluids allows vorticity to be generated across the surface layer, which can support a wave propagating along the interface. The interface between two different fluids forms a tangential discontinuity across which only the normal component of the velocity is continuous. Although a consequence of the local anisotropy at the surface, the disturbance penetrates deep into the fluid and cannot be simply categorised as either transverse or longitudinal, but embodies elements of both types.

Waves of this type are typically found at surfaces or interfaces, but similar waves may also be generated when the discontinuity is broadened into stratification of density or temperature.

Not all waves are stable in that once generated they propagate with their amplitude remaining nearly constant, gradually decreasing due to damping.

---

[1]Transverse waves are allowed in a plasma with an imposed magnetic field, where the magnetic field destroys the isotropy and allows shear to be transmitted.

[2]Compressible fluids allow density variations and support longitudinal (sound) waves, discussed in Chapter 8.

Introductory Fluid Mechanics for Physicists and Mathematicians, First Edition. Geoffrey J. Pert.
© 2013 John Wiley & Sons, Ltd. Published 2013 by John Wiley & Sons, Ltd.

In contrast in some cases, the wave amplitude grows exponentially as the disturbance feeds on itself converting either potential (from gravity) or kinetic energy (from the background flow) into the wave. Such motions are unstable and play a critical role in many fluid problems. The growth of the wave in these cases is limited by nonlinear effects and ultimately by turbulence (Chapter 5). Surprisingly viscosity acts as a source of instability in shearing flow. We therefore investigate in this chapter the development of instability in both inviscid and viscid flow.

The wave is characterised by a mode usually defined by its wavenumber, k. The frequency, ω, is a function of the wavenumber, determined by the dispersion relation. This relationship is specified by the properties of the fluid and those of the physical environment in which it is found. The frequency is in general a complex quantity, the real part, $\Re(\omega)$, giving the oscillatory frequency of the wave, and the imaginary part, $\Im(\omega)$, the growth or decay rate of the amplitude. If the imaginary part is positive, $\Im(\omega) > 0$, the wave grows exponentially and is linearly unstable. As we shall see, many physical situations give rise to unstable waves, and may ultimately lead to turbulence. Most waves of these types propagate in the plane of the discontinuity as surface waves. In this chapter we examine a number of the more important waves within the incompressible and inviscid approximations. The flows are usually irrotational since we assume that the flow originates in an irrotational fluid. In fact it is found that the introduction of rotation into the flow does not greatly change the resulting dispersion relations. As we shall show in later sections, viscosity and compressibility introduce important new wave systems.

4.2   Small-Amplitude Surface Waves

We consider the behaviour in the neighbourhood of a horizontal contact discontinuity between two dissimilar fluids of different densities in a gravitational field. As a consequence of the gravitational field the fluid will give rise to waves on the surface if the lighter fluid is on top of the heavier, or instability if the heavier is on top.

Let the interface be the surface $z = 0$ and let z be vertically upwards. We suppose each fluid may be described by an irrotational incompressible flow, so that Laplace's equation is valid everywhere except across the interface, where Kelvin's theorem does not hold:

$$\nabla^2 \phi = 0 \qquad (4.1)$$

Consider a small-amplitude sinusoidal wave on the surface varying in the x direction with wavenumber k and frequency ω. The surface displacement therefore has the form

$$\eta = \eta_0 \, \exp[\imath \, (kx - \omega t)] \tag{4.2}$$

The vertical velocity at the surface is

$$v_z = \frac{\mathrm{d}\eta}{\mathrm{d}t} \approx \frac{\partial \eta}{\partial t} \tag{4.3}$$

and the potential

$$\phi(0, t) = \phi_0 \, \exp[\imath \, (kx - \omega t)] \tag{4.4}$$

The flow must have the pressure and normal velocity continuous across the surface. The solution of Laplace's equation matched to the conditions at the interface is

$$\phi_\pm(z, t) = \phi_{\pm 1} \exp\{\imath(kx - \omega t) \mp kz\} + \phi_{\pm 2} \exp\{\imath(kx - \omega t) \pm kz\} \qquad z \gtrless 0 \tag{4.5}$$

the two sets of solutions $\phi_{1,2}$ relating the boundary conditions for large $|z|$ and $\pm$ referring to the upper and lower fluid respectively.

Bernoulli's equation provides the connection with gravity

$$\frac{\partial \phi}{\partial t} + \frac{1}{2}v^2 + \frac{p}{\rho} + gz = \text{const} \tag{4.6}$$

Since the amplitude of the wave is assumed small, the second-order term may be neglected. In the absence of surface tension, the pressure at the interface is the same for each fluid. However, if there is a surface tension σ at the interface there is a force in the z direction normal to the surface, which must be matched by a change of pressure across the surface

$$-\frac{\sigma}{R} \approx -\sigma\frac{\partial^2 \eta}{\partial x^2} = \sigma k^2 \eta \tag{4.7}$$

where $R \approx (\partial^2 \eta / \partial x^2)^{-1}$ is the radius of curvature of the surface. The pressure balance across the surface may be written as

$$p_+ - p_- = -\sigma k^2 \eta \approx \left[-\rho_+\left(\frac{\partial \phi_+}{\partial t} + g\eta\right)\right] - \left[-\rho_-\left(\frac{\partial \phi_-}{\partial t} + g\eta\right)\right] \tag{4.8}$$

and since the normal flow velocity on each side of the interface is the same

$$\frac{\mathrm{d}\eta}{\mathrm{d}t} = \frac{\partial \phi_-(0, t)}{\partial z} = \frac{\partial \phi_+(0, t)}{\partial z} \tag{4.9}$$

Due to the neglect of the second-order term in Bernoulli's equation, equations (4.8) and (4.9) are both linear, and it is easy to see that the principle of superposition may be applied to linear surface waves. Thus two waves generate motions which are independent of one another, the resultant being the simple sum of the two.

4.2.1 Surface Waves at a Free Boundary of a Finite Medium

Consider the case when the upper fluid has zero density $\rho_+ = \phi_+ = p_1 = 0$, and the lower has a depth $z = -h$. At the bottom, the z component of the velocity is zero, therefore

$$\phi_1 \exp(kh) = \phi_2 \exp(-kh) = \tfrac{1}{2}C \tag{4.10}$$

and the potential is

$$\phi = C \cosh\{k(z+h)\} \exp\{\imath (kx - \omega t)\} \tag{4.11}$$

The dispersion relation is obtained from equations (4.8) and (4.9) by adjusting the effective value of gravity to include the surface tension $g' = g + k^2\sigma/\rho$

$$\omega^2 = g'k \tanh(kh) = \left(gk + \frac{\sigma k^3}{\rho}\right) \tanh(kh) \tag{4.12}$$

which gives the phase velocity

$$v_p = \frac{\omega}{k} = \sqrt{\frac{g'}{k}\tanh(kh)} = \sqrt{\left(\frac{g}{k} + \frac{\sigma k}{\rho}\right)\tanh(kh)} \tag{4.13}$$

The waves have a minimum phase velocity when $k = \sqrt{\rho g/\sigma}$ of value

$$v_{p\min} = \sqrt[4]{\frac{4g\sigma}{\rho}} \tag{4.14}$$

4.2.1.1 Capillary waves

It is convenient to separate the short- and long-wavelength motions at the wavelength

$$\lambda^* = 2\pi\sqrt{\frac{\sigma}{\rho g}} \tag{4.15}$$

where the minimum phase velocity occurs. For water and air this corresponds to a wavelength of 1.7 cm, i.e. ripples. Waves with these short wavelengths are dominated by surface tension, and are known as *capillary waves*. Gravity plays little or no part in their behaviour, the restoring force being provided by the tension associated with the surface alone, equation (4.7).

4.2.1.2   Gravity waves

At longer wavelengths the restoring force is due to gravity, capillarity playing little part. Consequently waves of longer wavelength $\lambda > \lambda^*$ are known as *gravity waves* and are more commonly found in nature. We shall therefore restrict our attention to gravity waves henceforward.

The group velocity representing the rate at which energy is transported by the wave is

$$v_g = \frac{d\omega}{dk} = \frac{1}{2} v_p \left(1 + \frac{2kh}{\sinh(2kh)} \right) \tag{4.16}$$

The displacement of the surface is

$$\eta = \eta_0 \exp\{\imath (kx - \omega t)\} \tag{4.17}$$

where the amplitude of the surface displacement is

$$\eta_0 = \imath \frac{\omega C}{g} \cosh(kh) = \imath \frac{kC}{\omega} \sinh(kh) \tag{4.18}$$

If we consider a fluid particle at the point (x, z) and neglect the second-order variation due to the displacement (X, Z), we have that

$$v_x = \frac{dX}{dt} = \frac{\partial \phi}{\partial x} \qquad v_z = \frac{dZ}{dt} = \frac{\partial \phi}{\partial z}$$

and hence the particle displacement

$$\begin{aligned}
X &= \imath \eta_0 \frac{\cosh k(z + h)}{\sinh(kh)} \exp\{\imath (kx - \omega t)\} \\
Z &= \eta_0 \frac{\sinh k(z + h)}{\sinh(kh)} \exp\{\imath (kx - \omega t)\}
\end{aligned} \tag{4.19}$$

At the base of the fluid $z = -h$, the vertical motion is brought to rest. However, near the surface the particle follows an elliptic path, moving forwards at the crest of the wave and backwards in the trough. This behaviour is characteristic of the waves on the surface of the sea.

When the water is deep or the wavelength very short $kh \gg 1$, $\tanh(kh) \approx 1$, the formulae reduce to those for an infinite medium and the phase velocity $v_p = \sqrt{g/k}$. In this case the fluid particle path at the surface is nearly circular. On the other hand if the water is shallow or the wavelength long $kh \ll 1$, $\tanh(kh) \approx kh$, the phase velocity $v_p \approx \sqrt{gh}$, independent of wavelength. This change in behaviour between deep and shallow water waves, which takes place when the wavelength and depth are comparable, is due to the limited penetration of the wave. In deep water, the disturbance of the fluid penetrates as

$\sim \exp(kz)$. Therefore the disturbance decreases by about $1/535$ in a distance of a wavelength, namely $2\pi/k$.

These waves are the familiar waves seen on the surface of the sea and lakes. A number of familiar observations follow:

The forward motion at the crest of the wave and the backward flow at the trough are familiar to swimmers.

Following a heavy storm the first waves reaching the shore have longer wavelengths than those following later–a consequence of the slower phase and group velocities of short wavelengths in deep water.

The phase velocity in deep water $v_p \approx \sqrt{g/k}$. However, as the water shallows ($kh \lesssim 1$) approaching the beach, $v_p \approx \sqrt{gh}$ and the wave slows. The wave will break when the particle velocity equals the phase velocity, and the crest of the wave overtakes the phase of the wave.

Another example of deep water gravity waves is provided by tsunami. These are long-wavelength waves produced by a basal disturbance, usually the seabed, or by a large underwater explosion. The cause of the most powerful tsunamis is usually a movement of the earth's tectonic plates, leading to uplift followed by a rapid subsidence as the plate slips. The wavelength of the resultant wave is very long, typically as much as 200 km, but the amplitude small, generally less than 1 m. Due to the large wavelength the energy stored in the wave is, however, very large, and the group velocity of the wave also large $1/2\sqrt{g/k} \approx 800$ km/h. These characteristics make the tsunami difficult to detect in the open sea. Near to land the sea shallows and the group velocity decreases. The wave therefore 'piles up' on itself, forming a series of large amplitude devastating waves striking the land. Successive waves often increase in strength and may follow after significant intervals. The first intimation of a tsunami on shore is a trough called *drawback* which is easily understood in terms of the rotational motion of the water in the wave, discussed above.

4.2.1.3 Transmission of energy

We may calculate the energy transmitted by the wave by considering the work done per unit width by the pressure on fluid in the direction of the wave

$$\frac{\partial W}{\partial t} = \int_{-h}^{0} p \frac{\partial X}{\partial t}\,dz = \int_{0}^{h} \frac{\rho \omega\, \eta_0\, v_p}{\sinh(kh)} \cosh(kz') \cdot \frac{k\,\eta_0\, v_p}{\sinh(kh)} \cosh kz'\,dz'$$

$$\times \sin^2(kx - \omega t)$$

$$= \frac{\rho \omega\, k\, \eta_0{}^2\, v_p{}^2}{\sinh^2(kh)} \cdot \frac{1}{2}\left(h + \frac{\sinh(2kh)}{2kh}\right) \sin^2(kx - \omega t)$$

The averaged rate at which work is done over a cycle is therefore

$$\frac{\rho \, \omega \, k \, \eta_0{}^2 \, v_p{}^2 \sinh(2kh)}{4 \sinh^2(kh)} \frac{1}{2k}\left(1 + \frac{2kh}{\sinh(2kh)}\right) = \frac{\rho \, \omega \, \eta_0{}^2 \, v_p}{2} \coth(kh) \, v_g = \frac{1}{2} \rho \, g \, \eta_0{}^2 \, v_g$$

(4.20)

The average energy in a column of unit area is the sum of the kinetic and potential energy terms

$$E_{\text{kin}} = \frac{1}{2}\pi \int\limits_{0}^{2\pi} dt \int\limits_{-h}^{0} dz \, \frac{1}{2}\rho \left[\left(\frac{dX}{dt}\right)^2 + \left(\frac{dZ}{dt}\right)^2\right] = \frac{1}{4}\rho \, g \, \eta_0{}^2$$

(4.21)

$$E_{\text{pot}} = \frac{1}{2}\pi \int\limits_{0}^{2\pi} dt \int\limits_{-h}^{\eta} dz \, \rho \, g \, Z = \frac{1}{4}\rho \, g \, \eta_0{}^2$$

The rate of energy flow is therefore equal to the mean energy in a column of unit surface area of the wave multiplied by the group velocity, as might have been expected.

The momentum flow through unit width is easily calculated from the xx component of the momentum flux tensor. Since only the xx term does not average to zero,

$$\frac{1}{2\pi} \int\limits_{0}^{2\pi} dt \sin^2(kx - \omega t) \left\{\int\limits_{-h}^{0} \rho v_x{}^2 dz\right\} = \frac{1}{2}\rho \, g \, \eta_0{}^2$$

(4.22)

Case study 4.I The Wake of a Ship–Wave Drag

A ship travelling through water experiences significant *wave drag* as a result of the surface gravity waves generated by its displacement as it moves through the water. A very simple model can be developed for a ship travelling in a narrow channel, e.g. a canal. It is assumed that the wave is generated at the bow and is normal to the ship's path. The waves are specified by the phase, which is determined by the ship's bow wave. If the waves generated successively do not interfere destructively, the phase velocity $v_p = v_s$, the speed of the ship. The rate at which work is done by the ship is $R v_s$ where R is the resistance to the motion, i.e. the drag. If we consider a fixed plane perpendicular to the path of the ship, the length of the wave train increases by a length v_p per unit time ahead of the plane. The energy increase ahead of the plane, equation (4.21), is therefore $v_p \cdot \frac{1}{2}\rho \, g \, \eta_0{}^2$, from which must be subtracted the energy of the wave groups following the ship crossing the fixed plane, the group velocity v_g being always less than that of the phase v_p

$$\frac{1}{2}\rho \, g \, \eta_0{}^2 \, v_p = \frac{1}{2}\rho \, g \, \eta_0{}^2 \, v_g + R v_s$$

and therefore

$$R = \frac{1}{2}\frac{v_p - v_g}{v_p}\rho \, g \, \eta_0{}^2 = \frac{1}{4}\rho \, g \, \eta_0{}^2 \left(1 - \frac{2kh}{\sinh(2kh)}\right)$$

(4.23)

If the water is shallow the phase velocity cannot exceed $\sqrt{gh}$. Therefore if the speed of the ship exceeds this value, the wave drag vanishes, a fact attested by experiment and familiar to nineteenth-century canal boatmen.

In problem #20 it is argued by dimensional methods that the wave drag generated by a boat, expressed as a drag coefficient, is a function of the Froude number (equation S.58). A typical plot of the drag coefficient versus the Froude number shows a number of maxima and minima. These are due to interference between the waves produced at the stern (a trough) and at the bow (a crest). When the two waves interfere constructively, there is a maximum in the drag, the length of the boat matches an odd number of half wavelengths of the gravity wave.

4.I.i Two-dimensional wake, Kelvin wedge

The simple calculation of the wake following a ship above is unsatisfactory as it assumes that waves propagate longitudinally only in the direction parallel to the ship's motion and takes no account of their transverse propagation. In fact the waves will spread laterally, spreading as a series of linear wavefronts away from the source at the ship's bows or stern. At the bow the wave has a crest and at the stern a trough. The waves are generated with a range of wavelengths, which each satisfy the dispersion relation appropriate to their direction of motion. Consider waves propagating with their wavefronts along a line at an angle θ to the ship's motion. They are generated at (for example) the bow, and the phase velocity must be equal to the component of the ship's velocity normal to the wavefront, $v_p = v_s \sin\theta$ (Figure 4.1). Assuming the waves are deep water waves $h \to \infty$, the wavenumber is given by dispersion relation

$$k = \frac{g}{v_s^2 \sin^2 \theta} \qquad (4.24)$$

The group velocity of gravity waves is less than the phase velocity, $v_g < v_p$. Hence waves cannot propagate ahead of the ship.

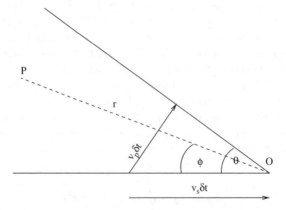

Figure 4.1: Sketch of the wave propagation of one mode of phase velocity v_p generated at the ship O moving with speed v_s at an angle θ. Phase is measured at P at an angle ϕ to the ship's path.

At an arbitrary point on the surface behind the ship, there will be a number of different waves with different phases interacting simultaneously. As a result the disturbance will be cancelled out by destructive interference. However, at certain points groups of waves will interfere constructively and a disturbance of observable amplitude results. At these points the phase of a wave group is stationary. This will occur for a band of waves centred on wavenumber k_0 at inclination θ_0.

Consider a point P relative to the current position of the ship at O (Figure 4.1). The distance OP is r and the angle OP makes with the direction of the ship's motion ϕ. The phase of the wave k_θ at P is

$$\chi_\theta = -k_\theta\, r\, \sin(\theta - \phi) \tag{4.25}$$

Making use of equation (4.24), the derivative of the phase is

$$\frac{\mathrm{d}\chi_\theta}{\mathrm{d}k_\theta} = -r \sin(\theta - \phi) + \frac{1}{2} r \tan\theta \cos(\theta - \phi) \tag{4.26}$$

The stationary phase therefore occurs when

$$\tan(\theta - \phi) = \frac{1}{2} \tan\theta$$

and therefore

$$\tan\phi = \frac{\tan\theta}{\left(2 + \tan^2\theta\right)} \tag{4.27}$$

It is easily shown that ϕ has a maximum value

$$\phi_{\max} = \arctan\left(\frac{1}{2\sqrt{2}}\right) = \arcsin\left(\frac{1}{3}\right) \qquad \text{when} \qquad \theta = \arctan\sqrt{2} \tag{4.28}$$

Consequently all the wave crests lie within a wedge of half angle $\arcsin(1/3) = 19.5°$, known as the *Kelvin wedge*. There are two branches to the profile, one from wavefront angles 0 to $\arctan(\sqrt{2}) = 54.7°$, the other from $54.7°$ to $90°$. At the maximum of the angle ϕ, i.e. at the corner of the wedge, the second derivative of the phase $\mathrm{d}^2\chi/\mathrm{d}k^2$ is also zero. Consequently a broad range of frequencies contributes to the group with stationary phase, and the resultant amplitude of the crest is high.

To be observable the wave must be a crest, i.e. $\chi \approx 2n\pi$, where n is an arbitrary integer. Thus there are a series of waves which form the wake behind the ship, each corresponding to different values of n. The above equation together with equations (4.24) and (4.25) determine the profile of the wave, as follows. Given ϕ, equation (4.27) determines the corresponding angle of the wavefront θ. Equation (4.24) yields the wavenumber k_θ, and finally (4.25) the spatial distance OP. Thus we obtain the polar co-ordinates r, ϕ. There is a useful scaling relation for the waves in that the profiles are similar, $r \propto n v_s^2/g$. A single plot may therefore be used to construct the complete family. The quadratic scaling of wake dimensions with speed was used during the Second World War to measure a ship's speed from aerial photographs.

The profile of the wake is now easily constructed using co-ordinates taken from an origin at the ship along x and perpendicular y to the ship's path. Figure 4.2 shows the typical form. The wake is clearly confined to the Kelvin wedge. The actual wake from a ship is generally more complex as wave structures from other parts of the ship interfere with that from the bows.

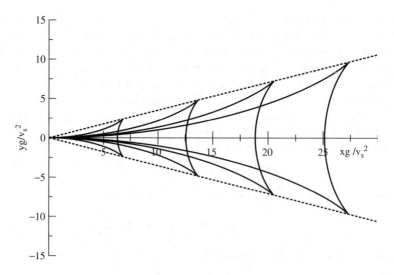

Figure 4.2: Plot of a wake pattern lying within the Kelvin wedge (dashed). The waves correspond to phase shifts with $n = 1$ to 4. The plots are constructed with a scale factor v_s^2/g.

4.3   Surface Waves in Infinite fluids

If the fluid is infinite in both directions, $\phi_{\pm 2} = 0$ since the potential tends to zero at ∞. Then substituting the solution at the surface, we obtain from equations (4.8) and (4.9)

$$\sigma k^2 \eta_0 = \rho_+ (-\imath \omega \phi_{+1} + g\,\eta_0) \quad -\rho_- (-\imath \omega \phi_{-1} + g\,\eta_0)$$
$$-k\phi_{+1} = k\phi_{-1} = -\imath \omega \eta_0$$

Hence we obtain the dispersion relation

$$\omega^2 = \frac{\left[(\rho_- - \rho_+)\,g + \sigma k^2\right]}{(\rho_- + \rho_+)}\,k \tag{4.29}$$

This allows two classes of solutions.

4.3.1   Surface Wave at a Contact Discontinuity

Light fluid above the heavy one, $\rho_+ < \rho_-$: Consequently ω is real. There is therefore a propagating wave along the surface with phase velocity

$$v_p = \frac{\omega}{k} = \sqrt{\frac{(\rho_- - \rho_+)}{(\rho_- + \rho_+)} \cdot \frac{g}{k}} \tag{4.30}$$

4.3.2   Rayleigh–Taylor Instability

Heavy fluid above the light one, $\rho_+ > \rho_-$: Consequently ω is imaginary, and the interface is therefore unstable. A small-amplitude perturbation of the surface grows exponentially, with a small-amplitude growth rate

$$\gamma = \Im(\omega) = \sqrt{\frac{(\rho_+ - \rho_-)}{(\rho_+ + \rho_-)} \cdot gk} \tag{4.31}$$

This growth rate is only maintained for a short time until the instability enters its nonlinear phase, when the growth is approximately proportional to the freefall displacement $\frac{1}{2}gt^2$. In this phase a thin spike of heavy fluid penetrates into the light one, its length increasing approximately as a fraction of $\frac{1}{2}gt^2$. The light fluid forms a bubble which rises through the heavy fluid supported by its buoyancy. As the velocity shear across the contact discontinuity increases, it also becomes unstable to the Kelvin–Helmholtz instability discussed in the next section. As a result vortices develop on the spike, eventually forming a 'mushroom cap'.

During the linear phase, an initial surface perturbation η_0 increases as

$$\eta = \eta_0 \cosh(\gamma t) \qquad \text{with surface velocity} \qquad v = \frac{d\eta}{dt} = \gamma \eta_0 \sinh(\gamma t) \tag{4.32}$$

It follows from Einstein's equivalence principle that gravity and acceleration are equivalent. If gravity is replaced by an equivalent acceleration across the light–heavy fluid boundary, the growth rate for an accelerated interface is given by equation (4.31) with g replaced by the acceleration a.

This instability is very familiar in everyday life. If a tumbler is filled with water and carefully inverted maintaining a plane surface with the heavy water uppermost, the atmospheric pressure on the under surface is sufficient to hold the water in place, but as we know, from experience, the water rapidly falls out. This is due to the rapid growth of the Rayleigh–Taylor instability over the surface. In the language of mechanics, the system is established in a state of unstable equilibrium, in that the potential energy is reduced if the water and air are interchanged. Another way of looking at the problem is that it is a consequence of the buoyancy of the air in the water. Consequently a small bubble initially forms at the surface, rising into the water and displacing it downwards. The displaced water causes an increase in surface pressure forcing more air into the bubble re-reinforcing the motion. Thus the disturbance grows unstably. The Rayleigh–Taylor instability is found in many situations in both terrestrial and stellar environments. Typical examples are found in salt domes and weather inversions, and in the Crab nebula. The instability may also be a limiting factor in the collapse of laser compression targets for inertial fusion.

4.4 Surface Waves with Velocity Shear Across a Contact Discontinuity

We consider the case where the upper and lower fluids have velocities $\mathbf{V}_\pm$ respectively parallel to the surface of discontinuity. The x direction is taken parallel to the velocity difference $\mathbf{V}_+ - \mathbf{V}_-$, with components $U_\pm$. Transforming to the frame in which the common transverse velocity is zero, we may write the total potentials including the perturbation ϕ' as

$$\phi_\pm = U_\pm x + \phi'_\pm \tag{4.33}$$

where x is the direction of the flows $U_\pm$, and y also lies in the surface. The perturbation takes the form of a wave with wavenumber $\mathbf{k} = k_x\hat{\mathbf{i}} + k_y\hat{\mathbf{j}}$.

The displacement of the surface must reflect the motion along the surface. Therefore we must have

$$\frac{\partial \eta}{\partial t} + U_+ \frac{\partial \eta}{\partial x} = \frac{\partial \phi_+}{\partial z} \qquad \frac{\partial \eta}{\partial t} + U_- \frac{\partial \eta}{\partial x} = \frac{\partial \phi_-}{\partial z} \tag{4.34}$$

The approximation to Bernoulli's equation for the pressure must also be adjusted for the applied velocity

$$\frac{p}{\rho} = -\frac{\partial \phi'}{\partial t} - \frac{1}{2}\left\{\left(U + \frac{\partial \phi'}{\partial x}\right)^2 + \left(\frac{\partial \phi'}{\partial y}\right)^2 + \left(\frac{\partial \phi'}{\partial z}\right)^2\right\}$$

$$- gz + \ldots \approx -\frac{\partial \phi'}{\partial t} - U\frac{\partial \phi'}{\partial x} - gz + \ldots$$

Neglecting terms of second order in Bernoulli's equation across the interface and including the surface tension,

$$p_- - p_+ = -\sigma\frac{\partial^2 \eta}{\partial x^2} = -\rho_+\left(\frac{\partial \phi'_+}{\partial t} + U\frac{\partial \phi'_+}{\partial x} + gz\right) - \rho_-\left(\frac{\partial \phi'_-}{\partial t} + U\frac{\partial \phi'_-}{\partial x} + gz\right) \tag{4.35}$$

Considering a travelling wave along the surface as before and noting that only the solutions ϕ'_1 are appropriate since the fluids are infinite, we obtain

$$-\imath\left(\omega - k_x U_+\right)\eta_0 = -k\phi'_{+1} \qquad -\imath\left(\omega - k_x U_-\right)\eta_0 = k\phi'_{-1}$$

$$\rho_+\left[-\imath\left(\omega - k_x U_+\right)\phi'_{+1} + g\,\eta_0\right] - \rho_-\left[-\imath\left(\omega - k_x U_-\right)\phi'_{-1} + g\,\eta_0\right] = -\sigma k^2 \eta_0 \tag{4.36}$$

The dispersion relation is obtained since the equations must be consistent

$$\rho_+\left(\omega - k_x U_+\right)^2 + \rho_-\left(\omega - k_x U_-\right)^2 = g\,k\left(\rho_+ - \rho_-\right) - \sigma k^2 \tag{4.37}$$

Solving for the phase velocity ω/k,

$$
\frac{\omega}{k} = \frac{k_x}{k} \frac{(\rho_- U_- + \rho_+ U_+)}{(\rho_- + \rho_+)} \pm \left\{ \frac{[\sigma\, k^2 + g\,(\rho_- - \rho_+)]}{[k\,(\rho_- + \rho_+)]} \right.
$$

$$
\left. - \frac{k_x^{\,2}}{k^2} \frac{\rho_- \rho_+}{(\rho_- + \rho_+)^2} (U_- - U_+)^2 \right\}^{1/2} \tag{4.38}
$$

Neglecting the surface tension, the solutions represent stable oscillations when ω is real, i.e. $(\rho_-^{\,2} - \rho_+^{\,2})\, g\, k/k_x^{\,2} > \rho_- \rho_+ (U_- - U_+)^2$. Otherwise ω is complex and the wave is unstable with one solution growing and the other decaying. When the heavy fluid is below the light one, gravity provides a stabilising influence, which is stronger for shorter wavelengths. The instability is known as the *Kelvin–Helmholtz instability*. We note that if the velocities of the two fluids are equal $U_+ = U_-$ the flow reduces to the Rayleigh–Taylor instability propagating with the joint flow speed.

When surface tension at the interface is taken into account the system is stabilised for some short wavelengths. With this correction, the results agree well in predicting the onset of wave formation with wind over a surface.

If gravity is negligible, the flow is a vortex sheet with a tangential discontinuity; in this situation the waves are absolutely unstable at all wavelengths. Since $k \geq k_x$ the strongest instability growth occurs for waves travelling parallel to the velocity difference, which therefore dominate for large times. If the shear layer has a finite width, so that the density and velocity spatial variations are continuous, the instability still occurs provided the wavelength is much larger than the width of the shear layer, so that the latter approximates a discontinuity. The appropriate condition, modified for finite width, for instability must still be satisfied. Unstable short-wavelength modes are limited to approximately the width of the shear layer, have the largest growth rate and initially dominate. The effect of surface tension is reflected in the old mariner's method of last resort: that is, oil discharged onto the surface of the sea to reduce the waves.

As with the Rayleigh–Taylor instability, the Kelvin–Helmholtz instability has a simple physical mechanism and is easy to understand. A small upwards perturbation of the surface gives rise to a distortion in the flow of the fluid passing over it. In consequence the velocity of the upper fluid is increased and its pressure on the surface lowered locally. Concomitantly the disturbance in the lower fluid decreases its velocity and increases the pressure. The consequence is a flow of the lower fluid into the bulge increasing its size, and thus the disturbance grows.

The Kelvin–Helmholtz instability is found in many familiar situations giving rise to the growth of waves by the wind over a surface, or the rippling motion of loose flapping sails. A classic example is the development of billow clouds by wind shear in the atmosphere. It is frequently a critical element in the nonlinear, large-amplitude development of many instabilities, e.g. Rayleigh–Taylor, leading to turbulence. The Kelvin–Helmholtz instability also plays a major role in the nonlinear development of the instability itself, introducing progressively longer wavelengths into the motion, as well high-frequency eddies.

As we have seen, vortex sheets are unstable due to Kelvin–Helmholtz instability. If the sheet is finite, the perturbation is strongly developed at the ends and the instability grows rapidly into the nonlinear phase, which causes the sheet to roll up as discussed in Section 11.8.

4.5 Shallow Water Waves

Thus far the wave amplitude has been assumed to be small, so that the flow is treated in the linear approximation, where only first-order perturbation terms are considered. In fact this condition may not be valid, e.g. in the flow of water from a breaking dam or a tidal bore. Many of these flows, which are the finite amplitude, long-wavelength form of those discussed in Section 4.2.1.2, may be treated by a simple method provided the characteristic length of the flow ℓ (e.g. wavelength) is much larger than both the depth of the water h_0 and the amplitude of the disturbance η_0.

Neglecting viscosity and surface tension, the equations of incompressible, ideal flow in two dimensions are

$$
\begin{aligned}
\frac{du}{dt} &= -\frac{1}{\rho}\frac{\partial p}{\partial x} \\
\frac{dw}{dt} &= -\frac{1}{\rho}\frac{\partial p}{\partial z} - g \\
\frac{\partial u}{\partial x} &+ \frac{\partial w}{\partial z} = 0
\end{aligned}
\tag{4.39}
$$

where x is the direction of propagation and z the height measured upwards from the bed, and u and w the corresponding components of the velocity. As for linear waves within the shallow water approximation Section 4.2.1.2, the term dw/dz may be neglected, so that

$$p = p_0 + \rho g \left[h(x,t) - z\right]$$

where p_0 is the atmospheric pressure and $h(x,t)$ is the total height of the surface above the bed. Consequently

$$\frac{du}{dt} = -g\frac{\partial h}{\partial x}$$

The rate of change of the velocity u is therefore independent of the depth z. Thus if u is initially constant with depth it remains so throughout the motion, so that u depends on x and t only,

$$\frac{\partial u}{\partial t} + u\frac{\partial u}{\partial x} = -g\frac{\partial h}{\partial x} \qquad (4.40)$$

Since the bed is solid, $w = 0$ when $z = 0$, and the equation of continuity integrates to give

$$w = -\frac{\partial u}{\partial x} z$$

Assuming the bed is flat, the z component of velocity at the surface is

$$w = \frac{\mathrm{d}h}{\mathrm{d}t} = \frac{\partial h}{\partial t} + u\frac{\partial h}{\partial x}$$

so that

$$\frac{\partial h}{\partial t} + u\frac{\partial h}{\partial x} + h\frac{\partial u}{\partial x} = 0 \qquad (4.41)$$

We introduce the phase velocity of the waves in the linear approximation $c = \sqrt{gh}$ to write

$$\frac{\partial u}{\partial t} + u\frac{\partial u}{\partial x} + 2c\frac{\partial c}{\partial x} = 0$$

$$\frac{\partial (2c)}{\partial t} + u\frac{\partial (2c)}{\partial x} + c\frac{\partial u}{\partial x} = 0$$

Adding and subtracting these two equations we obtain two further equations

$$\left\{\frac{\partial}{\partial t} + (u \pm c)\frac{\partial}{\partial x}\right\}(u \pm 2c) \qquad (4.42)$$

On the lines $C_\pm$, called characteristics $\mathrm{d}x/\mathrm{d}t = v \pm c$, the terms $J_\pm = v \pm 2c$ are constant (characteristic invariants) respectively. The characteristics represent waves moving with the local linear phase speed forwards and backwards in the fluid, carrying information about the state of the flow upstream and downstream in terms of the characteristic invariants. We investigate their properties and the method of calculation based on them in Chapter 9.

Returning to the condition that $\mathrm{d}w/\mathrm{d}t$ may be neglected, the orders of magnitude of the various parameters are easily found as follows:

$$
\begin{array}{lllll}
u\,(\partial u/\partial x) & \sim & g\,(\partial h/\partial x) & \Longrightarrow & u \sim \sqrt{gh} \\
(\partial u/\partial t) & \sim & g\,(\partial h/\partial x) & \Longrightarrow & t \sim \ell/\sqrt{gh} \\
(\partial w/\partial z) & \sim & (\partial u/\partial x) & \Longrightarrow & w \sim \sqrt{gh}\,(h/\ell).
\end{array}
$$

Therefore $\mathrm{d}w/\mathrm{d}t \sim (h/\ell)^2\,g \ll g$ as required.

The shallow water approximation, generalised to include viscosity, Coriolis force and a variable, finds wide application in both oceanographic and atmospheric modelling.

4.6   Waves in a Stratified Fluid

In a gravitational field, an oscillation may be established if the properties of the fluid vary with height. We have already seen such a wave motion at the boundary between two fluids of differing density. However, waves may be established more generally if the fluid is inhomogeneous when the density or temperature varies with height. Since the hydrostatic condition requires that the pressure varies with height, such variation may be due to temperature or compositional variations. There are two areas of physics, where they play important roles:

1. In meteorology layers of differing temperature occur due to the lapse rate in the atmosphere.
2. In oceanography layers of differing salinity, and therefore density, give rise to a stratified fluid.

The force on a fluid particle is associated with the change of buoyancy, resulting from the difference between its density and the local density and the resultant gravitational force. The buoyancy force on a fluid particle given by Archimedes' principle, is due to the net upthrust resulting from the density difference and is therefore simply $\rho' \mathbf{g}$ per unit volume, where ρ' is the density perturbation. We may consider such density variations quite generally as due to a variation of entropy with height in the fluid body. Since the motion is assumed to be adiabatic, the displaced fluid carries with it its entropy, differing from that of the fluid it displaces. Provided the wavelength is small compared with any density variation due to gravity, the fluid behaves as though incompressible. Density changes are therefore considered to be associated with entropy changes only. The wave is treated as a small perturbation to the ambient state of the fluid, density ρ_0, entropy s_0 and pressure p_0, the latter satisfying the hydrostatic pressure condition, equation (1.25), namely $\nabla p_0 = \rho_0 \mathbf{g}$. Introducing the perturbation increments ρ', s' and p', and noting that the change in density is due solely to the change in entropy,

$$\rho' = \left.\frac{\partial \rho}{\partial s}\right|_p s'$$

Since the ambient fluid is at rest, the ambient velocity $\mathbf{v}_0 = 0$ and we have from Euler's equation the perturbation velocity given by

$$\begin{aligned}
\frac{\partial \mathbf{v}'}{\partial t} &= -\frac{\nabla p'}{\rho_0} + \frac{\nabla p_0}{\rho_0{}^2} \rho' \\[2mm]
&= -\frac{\nabla p'}{\rho_0} + \frac{\mathbf{g}}{\rho_0} \left.\frac{\partial \rho}{\partial s}\right|_p s'
\end{aligned} \qquad (4.43)$$

Since the entropy of a fluid particle is constant, the density, depending only on the entropy, must be considered to be constant in the equation of continuity so that

$$\nabla \cdot \mathbf{v}' = 0 \tag{4.44}$$

The equation of entropy conservation (1.16) yields

$$\frac{\partial s'}{\partial t} + \mathbf{v}' \cdot \nabla s_0 = 0$$

Within this approximation, we note that the density perturbation appears only through the product with gravity $\rho' g$; elsewhere it is neglected, e.g. in the equation of continuity. This is a widely used approximation, known as the *Boussinesq approximation*, for dealing with effects resulting from buoyancy. We shall meet it again later in Chapter 7 to treat free convection.

For a wave with wavenumber $\mathbf{k}$ and frequency ω, the perturbation velocity takes the form $\mathbf{v}' = \mathbf{v_0}' \exp[\imath(\mathbf{k} \cdot \mathbf{r} - \omega t)]$, with similar expressions for the entropy and pressure increments, and it follows from the above equations that

$$
\begin{aligned}
\mathbf{k} \cdot \mathbf{v}' &= 0 \\
\imath \omega s' - \mathbf{v}' \cdot \nabla s_0 &= 0 \\
\imath \omega \mathbf{v}' + \frac{1}{\rho_0} \frac{\partial \rho}{\partial s}\Big|_p s' \mathbf{g} - \frac{\imath \mathbf{k}}{\rho_0} p' &= 0
\end{aligned}
$$

Since $\mathbf{k} \cdot \mathbf{v}' = 0$, the wave is transverse, with displacement perpendicular to the direction of propagation. Taking the scalar product of the last equation with $\mathbf{k}$,

$$p' = -\frac{\imath}{k^2} \frac{\partial \rho}{\partial s}\Big|_p s' \mathbf{k} \cdot \mathbf{g}$$

Noting that the entropy gradient is parallel to gravity, and eliminating s', p' and $\mathbf{v}'$, we obtain the dispersion relation[3]

$$\omega^2 = -\frac{1}{\rho} \frac{\partial \rho}{\partial s}\Big|_p g \frac{ds}{dz} \sin^2 \theta \tag{4.45}$$

where θ is the angle the direction of propagation $\mathbf{k}$ makes with the upward vertical $\hat{\mathbf{z}}$, namely $-\mathbf{g}$. If the wave is propagated vertically, $\theta = 0$, the frequency $\omega = 0$, and if horizontally, $\theta = \pi/2$, at the maximum propagation frequency

$$\omega_0 = \sqrt{-\frac{1}{\rho} \frac{\partial \rho}{\partial s}\Big|_p g \frac{ds}{dz}} \tag{4.46}$$

---

[3]Note that the subscript 0 will be omitted henceforward, being unnecessary.

The fluid therefore has oscillations at frequencies in the range $0 < \omega < \omega_0$ at a corresponding angle $\theta = \arcsin(\sqrt{\omega/\omega_0})$. When the direction of propagation is horizontal, the frequency takes its maximum value and the fluid motion consists of vertical columns of rising and falling fluid.

When the fluid is stable, according to the arguments of Section 1.5.3, $ds/dz < 0$ and $\omega_0{}^2 > 0$ is positive. The waves have a real (oscillatory) frequency. On the other hand if $ds/dz > 0$, ω_0 is imaginary, and unstable growth of the disturbance is predicted in agreement with equation (1.37).

We may identify several different characteristic forms of the waves from this general form:

1. *Density gradient stratification* If the stratification is due to density alone

$$\omega_0 = \sqrt{-\frac{g}{\rho}\frac{d\rho}{dz}} \qquad (4.47)$$

   known as the *buoyancy* or *Brunt–Väisälä* frequency. These waves are important in layers of differing salinity in the deep ocean.

2. *Temperature gradient stratification* If the stratification is due to temperature gradient alone, the thermodynamic relations from equations (1.28) and (1.29)

$$-\frac{1}{\rho}\frac{\partial \rho}{\partial s}\bigg|_s = \frac{\alpha T}{c_p} \qquad \text{and} \qquad c_p = T\frac{\partial s}{\partial T}\bigg|_p$$

   we obtain

$$\omega_0 = \sqrt{\alpha g \frac{dT}{dz}} \qquad (4.48)$$

3. *Stratified fluid in both mechanical and thermal equilibrium* If the fluid is in both mechanical and thermal equilibrium, the temperature is constant and therefore, making use of equation (1.29),

$$\frac{ds}{dz} = \frac{\partial s}{\partial p}\bigg|_T \frac{dp}{dz} = -\rho g \frac{\partial s}{\partial p}\bigg|_T = \rho g \frac{\partial V}{\partial T}\bigg|_p = \alpha g$$

   Hence in this case

$$\omega_0 = \sqrt{\frac{T}{c_p}\alpha g}$$

$$\qquad = \frac{g}{\sqrt{c_p T}} \qquad \text{for an ideal gas} \qquad (4.49)$$

4. *Stratified fluid in adiabatic equilibrium* If the fluid has a density and temperature profile given by the adiabatic lapse rate, equation (1.30), the entropy does not vary with height, $ds/dz = 0$, and the frequency $\omega = 0$. The fluid therefore does not support these waves.

5. *Stratified fluid with temperature gradient* The fluid has a temperature gradient dT/dz and is subject to adiabatic convective disturbance. The frequency must be corrected to take this effect into account:

$$\frac{ds}{dz} = \left.\frac{\partial s}{\partial T}\right|_p \frac{dT}{dz} + \left.\frac{\partial s}{\partial p}\right|_T \frac{dp}{dz}$$

$$= \frac{c_p}{T}\frac{dT}{dz} + \alpha\,g \tag{4.50}$$

$$\omega_0 = \sqrt{\alpha\,g\left(\frac{dT}{dz} + \frac{\alpha\,g\,T}{c_p}\right)} \tag{4.51}$$

and is zero if the temperature gradient is equal to the adiabatic lapse rate.

These waves play an important role in the dynamics of atmospheric motions in meteorology, reflecting a difference between the environmental and adiabatic lapse rates.

Waves of this general type have unusual behaviour, because the dispersion relation depends only on the wavenumber $\mathbf{k}$ through the angle, θ, between the wave vector and gravity. Consequently, to calculate the group velocity we must use its three-dimensional expression[4]

$$\mathbf{v}_g = \nabla_k\left(\frac{\omega}{k}\right) = \frac{\partial}{\partial k_i}\left(\frac{\omega}{k}\right) \tag{4.52}$$

where ∇_k is the gradient operator in k space. Hence, since

$$\sin^2\theta = \frac{k^2 - k_z{}^2}{k^2}$$

we obtain the result after differentiation that

$$\mathbf{v}_g = \frac{\omega_0\,k_z{}^2}{k^3\sqrt{k^2 - k_z{}^2}}\left(\mathbf{k} - \frac{k^2}{k_z}\hat{\mathbf{z}}\right) \tag{4.53}$$

It is easy to show that $\mathbf{v}_g \cdot \mathbf{k} = 0$ so that the group velocity is perpendicular to the wave vector, i.e. the phase velocity. Since there must be symmetry about

---

[4]This expression is obtained by extending the usual derivation to three dimensions.

the axis of gravity $\mathbf{g}$, namely $\hat{\mathbf{z}}$, it is clear that the group velocity lies in the plane containing $\mathbf{k}$ and $\mathbf{g}$.[5] The waves therefore have the unusual property that energy is transmitted *along* the wavefront, rather than perpendicular to it in the normal way. The particle motion $\mathbf{v}'$ is, as we have indicated, transverse and therefore parallel to the group velocity. Therefore

$$\frac{v_x{}'}{v_y{}'} = \frac{v_{gx}}{v_{gy}} = \frac{k_y}{k_x}$$

4.7   Stability of Laminar Shear Flow

The steady laminar flow in a pipe is characterised by a velocity parallel to the axis of the pipe, whose value varies across the pipe, but is constant along it. It is therefore characterised by a velocity shear between the wall and the centre. As we saw in Section 4.4 such motion is unstable in inviscid flow due to the Kelvin–Helmholtz instability. As an example, we consider the two-dimensional case where the velocity is constant in the x direction, i.e. $v_x(y) = U(y)$, and the pressure uniform across the duct $p(x) = P(x)$, namely plane Poiseuille flow (Section 3.6.2). To consider the stability, we follow the same approach as that used earlier and apply a sinusoidal perturbation along the duct with amplitude variation across it, namely $A(y) \exp\{i(kx - \omega t)\}$. Considering the perturbations $v_x \to U + v'_x$, $v_y \to v'_y$ and $p \to P + p'$, and assuming the undisturbed flow obeys the Navier–Stokes equations, we obtain

$$\frac{\partial v'_x}{\partial t} + U\frac{\partial v'_x}{\partial x} + v'_y\frac{\partial U}{\partial y} = -\frac{1}{\rho}\frac{\partial p'}{\partial x} + \frac{\mu}{\rho}\nabla^2 v'_x$$

$$\frac{\partial v'_y}{\partial t} + U\frac{\partial v'_y}{\partial x} = -\frac{1}{\rho}\frac{\partial p'}{\partial y} + \frac{\mu}{\rho}\nabla^2 v'_y \qquad (4.54)$$

$$\frac{\partial v'_x}{\partial x} + \frac{\partial v'_y}{\partial y} = 0$$

To reduce this set of equations to a single ordinary differential equation we introduce the streamfunction

$$\psi(x, y, t) = \phi(y) \exp\{i(kx - \omega t)\} \qquad (4.55)$$

with modes defined by the wavenumber k and the complex frequency ω, the growth or decay of the mode being determined by $\Im(\omega)$. The velocity components become

$$v'_x = \frac{\partial \psi}{\partial y} = \dot{\phi}(y) \exp\{i(kx - \omega t)\} \qquad v'_y = -\frac{\partial \psi}{\partial x} = -i k\,\phi(y) \exp\{i(kx - \omega t)\}$$

$$(4.56)$$

---

[5]For if $\mathbf{k}$ lies in the (x, z) plane, then so must v_g (and also the perturbation velocity $\mathbf{v}'$).

The streamfunction form automatically solves the equation of continuity for an incompressible fluid. Hence substituting for v'_x and v'_y and eliminating p', we obtain the *Orr–Sommerfeld equation*

$$(U - c)\left(\ddot{\phi} - k^2\phi\right) - \ddot{U}\phi = -\frac{1}{k\mathcal{R}}\left(\overset{....}{\phi} - k^2\ddot{\phi} + k^4\phi\right) \qquad (4.57)$$

where $c = \omega/k$ and is subject to the boundary conditions for plane Poiseuille flow

$$\phi = \dot{\phi} = 0 \qquad\qquad \text{when } y = \pm h \qquad (4.58)$$

The Reynolds number is defined as $\mathcal{R} = U_{\max} h / \nu$.

The Orr–Sommerfeld equation satisfying the boundary conditions is an eigenvalue problem such that solutions can only be found for particular values of c given the value of the Reynolds number $\mathcal{R}$, i.e. only specific values $c(\mathcal{R})$ generate physically realistic solutions. Of particular interest are the neutral stability solutions $\Im(c) = 0$ which separate the stable and unstable regions of the flow for different values of the wavenumber k. The calculation of this neutral stability curve presents mathematical difficulties. A typical plot of the wavenumber k for varying Reynolds number $\mathcal{R}$ is shown in Figure 4.3.

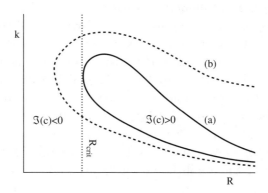

Figure 4.3: Sketch of the neutral stability curve for a representative parallel flow. The region inside the loops is unstable. The full line (a) has flow profile with no point of inflexion and the dashed line (b) includes one.

For Reynolds number below the critical value $\mathcal{R} < \mathcal{R}_{\text{crit}}$ the flow is stable for all values of the wavenumber k. However, once $\mathcal{R} > \mathcal{R}_{\text{crit}}$ a limited group of wavenumbers starts to grow unstably. The value of the critical wavenumber varies from flow to flow; for example, for plane Poiseuille flow, calculations show that $\mathcal{R}_{\text{crit}} = 5772$ (Lin, 1955; Drazin and Reid, 1981).

Since the Reynolds number is generally large for these unstable flows, we may consider the situation if we neglect the terms in $\mathcal{R}$ in the Orr–Sommerfeld equation. When this is done an important result is Rayleigh's inflexion point theorem, which states that: *A necessary condition for linear instability of an inviscid shear flow is that the velocity profile across the duct $U(y)$ must contain a point of inflexion somewhere across the duct.* This implies that if the viscous term is neglected, flow profiles not containing a point of inflexion are stable. Plane Poiseuille flow is such an example, yet investigation shows that the stability plot has the form shown by curve (a) in Figure 4.3 with a region of instability. As $\mathcal{R}$ tends to infinity, the upper and lower branches of the stability curve converge to reduce the region of instability to zero in accord with the inviscid result. When the flow profile has a point of inflexion the behaviour as $\mathcal{R} \to \infty$ is markedly different (Figure 4.3, curve (b)), the two branches diverging, so that the flow remains unstable.

It is interesting that there is a clear difference between viscous and 'non-viscous' flows in this problem. It might be expected that viscosity would tend to damp out any oscillation, but in fact clearly it actually supports the growth. However, Rayleigh's theorem does play a role in that flows with a point of inflexion are more susceptible to instability.

The stability condition derived from the Orr–Sommerfeld equation expresses the behaviour in time of the wave mode k at a fixed point growing exponentially. This is known as an *absolute instability*.

However, we may consider the instability growing in a wave packet moving through the fluid with the group velocity of the wave. The onset of growth will be at wavenumbers near to $k_{\max}$ at the critical Reynolds number $\mathcal{R}_{\mathrm{crit}}$. Waves within the group near onset will grow only if they lie within the narrow region of k space spanned by the stability plot for the given value of $\mathcal{R}$, otherwise they will decay. The group velocity has the usual value $\mathrm{d}\omega/\mathrm{d}k \approx \mathrm{d}\{\Re(\omega_{\mathrm{crit}})\}/\mathrm{d}k$ since $\Im(\omega_{\mathrm{crit}}) = 0$. This instability behaviour is known as *convective instability*. The wave packet may move away from its origin sufficiently rapidly, so that at a fixed point the amplitude decays, although the mode is temporally unstable.

The difference between the two types of instability can be seen from Figure 4.4. The onset condition for each may be distinguished:

$$
\begin{array}{llll}
\text{Absolute instability} & \dfrac{\partial \omega}{\partial k} = 0 & k = k_{\mathrm{crit}} & \Im(\omega) = 0 \\[2ex]
\text{Convective instability} & \dfrac{\partial \omega}{\partial k} \neq 0 & k \approx k_{\mathrm{crit}} & \Im(\omega(k)) > 0
\end{array}
$$

Clearly if the fluid is uniform along the direction of propagation of the wave group, the onset of stability is unchanged by convection. However, in some important applications, the growth rate may change from amplifying to decaying along the trajectory. As a result waves which are absolutely unstable may in fact be stable because they move into a region where $\Im(\omega) < 0$ before strong growth occurs and they subsequently decay. The stability condition can therefore only be established once the full flow is known.

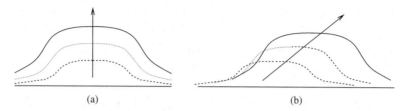

(a)   (b)

Figure 4.4: A sketch to contrast the growth of a wave group subject to absolute (a) and convective (b) instabilities. The arrows show the direction of propagation of the peak of the wave group.

4.8   Nonlinear Instability

Thus far we have considered only the development of the instability when its amplitude is small. In this regime its growth is governed by the linear development exemplified by equations such as the Orr–Sommerfeld, Rayleigh–Taylor or Kelvin–Helmholtz equations. Since the amplitude growth is exponential, this phase is relatively short lived before the waves become sufficiently strong that additional interactions between the waves set in. Near the onset of instability ($\mathcal{R} \gtrsim \mathcal{R}_{\text{crit}}$), the dominant wave is the one which is just unstable (k_{crit}). This grows strongly and is the primary disturbance. Nonlinear terms cause the wave to interact with itself. Landau (Landau and Lifshitz, 1959) described this interaction as one involving the square of the wave amplitude. The primary term is simply the linear growth with rate $\sigma = \Im(\omega)$. Averaging over times which are long compared with the period, terms of odd powers in the amplitude will average out as they contain only sinusoidally varying time dependencies. On the other hand even powers contain a non-zero average. Thus the time average growth is given by

$$\frac{\mathrm{d}|A|^2}{\mathrm{d}t} = 2\sigma |A|^2 - \gamma |A|^4 + \dots \tag{4.59}$$

where γ is a constant, known as the Landau constant. For Reynolds numbers close to the critical value, the linear growth rate $\sigma \approx \text{const}(\mathcal{R} - \mathcal{R}_{\text{crit}})$. This growth equation is easily integrated to give

$$|A|^2 = A_0{}^2 \bigg/ \left\{ \frac{\gamma}{2\sigma} A_0{}^2 + \left(1 - \frac{\gamma}{2\sigma} A_0{}^2\right) \exp(-2\sigma t) \right\} \tag{4.60}$$

where A_0 is the initial amplitude.

If $\gamma > 0$ the instability initially grows exponentially in a similar manner to a linear instability. However, after a time $\tau \sim \ln\left(2\sigma/\gamma - A_0{}^2\right)/2\sigma$, the nonlinearity associated with the fourth-order term sets in, progressively reducing the growth rate. Eventually when $t \gg \tau$ the growth is stabilised and a new configuration of steady flow is generated with a wave of amplitude $|A_{\text{limit}}| = \sqrt{2\sigma/\gamma} \sim \sqrt{\mathcal{R} - \mathcal{R}_{\text{crit}}}$. This condition is known as *supercritical stability*. If the initial amplitude is greater than this limiting value ($A_0 > |A_{\text{limit}}|$) the amplitude will decrease to the limiting value. At the critical Reynolds number the critical mode becomes unstable and will grow to its nonlinear stable amplitude. Consequently, at the critical Reynolds number $\mathcal{R}_{\text{crit}}$ the stability plot *bifurcates* with a new branch depending on the amplitude A_{limit}. As the Reynolds number is increased, new waves become unstable and grow in a similar fashion, but with an oscillation frequency, which is normally incommensurate with the previously generated ones, and with an initial phase which is randomly determined. These in turn saturate to generate a complex pattern of disturbances, which may be Fourier analysed in terms of a set of waves at different frequencies and wavenumbers. Because of the random phases each wave is independent and the ensemble forms a set of rapidly varying, poorly correlated motions. In general we may thus conclude that a series of further bifurcations takes place associated with additional nonlinear terms.

In contrast if $\gamma < 0$ it is evident from equation (4.59) that the nonlinearity increases the growth of the instability. In particular, the solution of equation (4.60) at a time $\tau = \ln\left(2\sigma/|\gamma| + A_0{}^2\right)/2\sigma$ makes the amplitude of the wave infinite. In this region the omitted higher order terms in the Landau equation cannot be neglected. For example, if the sign of the next term is $-\gamma'|A|^6$, the growth is again stabilised.

If the Reynolds number is significantly in excess of the critical value, several modes with differing frequencies become unstable. Consequently once the nonlinear instability is well established, the motions of the fluid contain many different incommensurate frequencies and wavelengths. These waves interact with one another through parametric interactions introduced by the nonlinearities in the governing equations. The waves rapidly become distributed in a 'quasi-random' fashion, and the flow becomes turbulent.

Chapter 5

Turbulent Flow

5.1 Introduction

Turbulence is a very general phenomenon occurring during fluid motion, in which the fluid acquires a rapidly varying random motion superimposed on the steady mean flow. The effect is clearly illustrated in the classical experiment performed by Reynolds in the 1880s (Reynolds, 1883). In this a jet of ink was injected into a tube of water flowing at increasing speeds (Figure 5.1). At low speeds the jet continued along a linear path, as the flow continued along a steady streamline as predicted in laminar flow, typical of Poiseuille flow. As the flow speed increased, the flow no longer remained steady, developing oscillations whose amplitude progressively increased, and eventually became broken, the ink totally filling the flow. At this stage the flow is fully turbulent. Reynolds found that the onset of turbulence is determined by a critical number $\mathcal{R}_{crit}$, whose value varies from one flow configuration to another. For flows with $\mathcal{R} > \mathcal{R}_{crit}$ the flow develops turbulence. Further increase in the speed led to a shortening of the distance over which the transition from laminar to turbulent flow took place. Immediately above the critical value, there is a transition region of unsteady flow before the turbulence is fully developed at Reynolds numbers in excess of about 4000. For pipe flow the critical Reynolds number of $\mathcal{R}_{crit} \approx 2300$ varies from flow to flow depending on the level of turbulence in the incoming flow.

It is found that in turbulent flow, the time-varying turbulent component of the flow consists of a series of superimposed random eddies. These may be visualised either as a set of spatial and temporal fluctuations in velocity with

Introductory Fluid Mechanics for Physicists and Mathematicians, First Edition. Geoffrey J. Pert.
© 2013 John Wiley & Sons, Ltd. Published 2013 by John Wiley & Sons, Ltd.

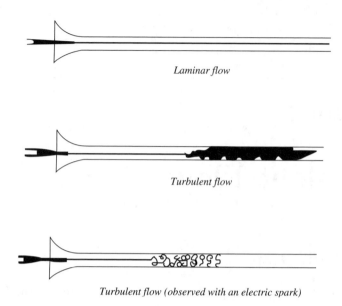

Figure 5.1: Reynolds' sketch of the flow in which ink was injected into water flowing along a pipe (Reynolds, 1883). The flow velocity is increased from below the critical Reynolds number in the upper picture to above in the lower, when the flow becomes unstable and turbulent. The bottom picture shows a short time exposure of the turbulent flow.

scale length λ, or equivalently as a spectrum of waves of spatial frequency k. From the Fourier transform theorem, clearly $\lambda \sim 1/k$. When turbulence is fully developed these eddies have a clear distribution ranging from large ones, whose velocities are of the order of the mean flow speed and dimensions of the order of the characteristic scale of the flow, to small ones whose size is determined by their damping due to viscosity.

An elementary picture of the development of turbulence through this cascade is obtained by considering the breakup of structures in the flow due to nonlinearities. The flow of fluid in a pipe is strongly sheared with the velocity along the centre line corresponding to the mean flow speed and zero at the edge of the pipe. There is a velocity fluctuation of approximately this value moving along the pipe at approximately half the average flow velocity. The nonlinear term in Euler's equation generates spatial harmonics in the usual manner. At medium-eddy Reynolds numbers this gives rise to the nonlinear growth of fluctuations of higher spatial frequency, which progressively increase in number and decrease in size. The eddies are, however, damped by viscosity. Once the Reynolds number of the individual fluctuation becomes small its generation of smaller scale length eddies is damped and no further development occurs. The onset of turbulence is determined by a critical value of the Reynolds number insufficient to damp this behaviour for the largest eddies.

Although turbulence was discovered over a century ago, and has been extensively studied ever since, it has remained an intractable problem for analysis.[1] As will be seen several major difficulties remain:

1. The onset and initial development of turbulence. Several distinct mechanisms have been proposed, each leading to the breakdown of an initially laminar flow into turbulence. The relationship of these different processes for specific problems is not understood.

2. The structure of the velocity spectrum in well-developed turbulence, i.e. the velocities of specific structures, either 'eddies' or 'waves' in the fluid.

3. In several specific turbulent flows, the turbulence is found to contain coherent structures, which may be retained for long times, rather than being homogeneously distributed.

4. Empirical models have been used satisfactorily in engineering applications for many years, many based around dimensional analysis with experimentally determined constants, but still lack a rigorous basis.

Progress towards resolving these issues is slow and difficult. However, advances are being made using computer simulation to address the complexity of turbulence.

5.1.1   The Generation of Turbulence

In general turbulence is generated by a wide variety of effects. Indeed it is difficult, if not impossible, to identify a general mechanism which encompasses all flows. Normally it has its origin in the nonlinear development of the instability of an oscillation in the flow, although, as we shall see, the process by which the flow becomes chaotic may take place by two completely different general routes. As a result, it is not appropriate in this book to give more than a brief general outline of the underlying processes by which turbulence is generated; an excellent survey encompassing both experimental and theoretical ideas is to be found in the book by Tritton (1988). Despite the fact that

---

[1]The difficulty of solving the general problem of turbulence is reflected in several quotations of which the following two are representative:

> When I meet God, I am going to ask him two questions: Why relativity? And why turbulence? I really believe he will have an answer for the first.
>
> (Werner Heisenberg)

and

> I am an old man now, and when I die and go to heaven there are two matters on which I hope for enlightenment. One is quantum electrodynamics, and the other is the turbulent motion of fluids. And about the former I am rather optimistic.
>
> (Horace Lamb)

the specific details of each process may be quite dissimilar, the final state of fully developed turbulence discussed in the next section (Section 5.2) is usually similar in structure.

As we have seen previously in chapter 4, instability is the result of a process by which a perturbation grows essentially without limit. The perturbation itself generates further growth by the disturbance introduced into the flow. In the previous analyses we have examined the linear phases of growth, where the perturbation was treated as a small first-order term alone, and higher order contributions have been neglected. As the perturbation grows, the higher order terms, such as $(\mathbf{v} \cdot \nabla)\mathbf{v}$, increase in relative strength more rapidly and make progressively larger contributions to the flow (Section 4.8).

Higher order terms are of higher power in the amplitude of the wave. Consequently the growth spectrum is not confined to a single harmonic, but higher harmonics, i.e. *frequency doubling*, will be generated. Different incommensurate frequencies develop with significant amplitude and random phase as proposed in the Landau model (Section 4.8). These new waves in turn give rise to further wave mixing introducing sum and difference frequencies through nonlinearity. If these waves are not damped, e.g. by viscosity, the net result is the rapid growth of a distribution of waves over a wide range of frequencies (Landau and Lifshitz, 1959, p.103). This process is essentially random in that the initial phase of the noise, from which growth of the waves of incommensurate frequency occurs, is random. The final turbulent distribution of waves is therefore random. Once the turbulence is fully established, the spectrum of the waves assumes a universal frequency distribution, characteristic (but not universal) of most common forms of fluid dynamical turbulence, discussed in the next section. Turbulence in other systems, e.g. plasmas, may also show a universal distribution, but with different scaling. Nonlinear instability growth of this type leading to turbulence is most often found in shear flows, such as pipes or ducts and boundary layers.

A second route to turbulence is due to *period doubling* and the generation of deterministic chaos (Tritton, 1988). As the strength of the driving source causing some types of oscillation in the fluid increases, the trajectory in the phase space of the oscillation no longer closes on the first circuit, but repeats to close on the second, i.e. the period has doubled. Continuing to strengthen the drive causes the period to continue to increase up to a limit—the *Feigenbaum limit*—when the motion passes through an infinite sequence of loops in phase space without closing, filling a limited region of the phase space. At this point a state of deterministic chaos is reached, and with it the state of fully developed turbulence. Characteristic of this route to turbulence are convective heat flows, typically the Rayleigh–Bénard instability (Section 7.5.4) and its nonlinear development, which have been extensively studied.

In view of the complex and general nature of the processes by which turbulence is established, it is not surprising that its initial development still remains poorly understood. The linear phases of the various instabilities are comparatively simple to analyse, and the general behaviour of their nonlinear development understood through computer simulation. However, the development of the full spectrum of waves and their interactions remains an intractable problem.

Expressed in terms of particle motions rather than their Fourier conjugates (spatial and temporal frequencies), these take the form of a series of eddies of appropriate spatial frequencies as described earlier. The motions give a useful simple physical picture of the fully developed turbulent motion, which is developed in the next section (5.2) on homogeneous turbulence. The eddies lead to extensive mixing of the fluid induced by the turbulence. This may be either beneficial or deleterious depending on the application.

5.2   Fully Developed Turbulence

At large Reynolds number turbulence develops quickly and produces a rapid and irregular variation of velocity with time. The particle paths are irregular and extensive mixing occurs. The particle velocity may be divided into two terms, the mean or flow velocity $\mathbf{u}$ and the fluctuating velocity $\mathbf{v}'$. The velocity of a fluid particle is therefore

$$\mathbf{v} = \mathbf{u} + \mathbf{v}' \qquad (5.1)$$

L.F. Richardson introduced the concept of turbulent motion as a superposition of eddies, or motions, of various sizes[2] and the consequent turbulent cascade. As the Reynolds number increases and the flow becomes unstable, the largest eddies appear first with a size which is characteristic of the dimensions of the flow, due to the fact that, as their gradients are smaller, they are most weakly damped and less rapidly dissipated. An important role is played by the largest eddies, whose size is for example that of the dimensions of the body ℓ. The velocity fluctuation associated with these large eddies is of the same order of magnitude as that of the velocity differential in the mean flow Δu, namely the variation over the distance ℓ. Clearly the length ℓ plays an important role as the distance over which the fluid is mixed by turbulent motions; it is therefore known as the *mixing length*.

---

[2]Although the structure of turbulence is strictly determined by the spatial spectrum of the fluctuations, the concept of simple eddies in configuration space is more easily visualised and therefore appropriate for pedagogic purposes. However, it should be remembered that an eddy of size λ corresponds to a fluctuation of wavenumber $k \sim 1/\lambda$.

The turbulence may be specified in terms of a distribution function for the velocity. Thus consider the relationship between two arbitrary points and times $P_0 = (\mathbf{r_0}, t_0)$ and $P = (\mathbf{r}, t)$. We define the difference in their separation and fluctuation velocities at a particular time $t = t_0$

$$\mathbf{x} = \mathbf{r}(P) - \mathbf{r_0}(P_0) \qquad \text{and} \qquad \mathbf{w} = \mathbf{v}'(P) - \mathbf{v}'(P_0)$$

The distribution law $F(\mathbf{x})$ is the probability of finding a fluctuation $\mathbf{w}$ at P. Clearly, as defined above, the distribution depends on the position P_0. We define (Kolmogorov, 1941b):

1. *Homogeneous turbulence* if the distribution law F is independent of the point P_0 on which it is based.

2. *Isotropic turbulence* if the turbulence is homogeneous and the distribution law is also invariant with respect to rotations and reflections of the co-ordinate system $\mathbf{x}$.

We restrict our study to isotropic turbulence. Considering a Fourier expansion in space of the distribution $F(\mathbf{x})$, we obtain a distribution of structures of wavelengths $\lambda = 2\pi/k$ which correspond to the eddies. The nonlinear term, $(\mathbf{v} \cdot \nabla)\mathbf{v}$, in the Navier–Stokes equation introduces an interaction between the terms: those with wavenumber $\mathbf{k_1}$ interact with those of $\mathbf{k_2}$ to generate a third satisfying $\mathbf{k_3} = \mathbf{k_1} + \mathbf{k_2}$, i.e. eddies of longer wavelengths progressively create shorter wavelength ones. There is a progressive transfer of energy from the longer wavelengths, where eddy motion is created, to shorter ones, where it is dissipated by viscosity due to larger velocity gradients, and their velocities will therefore be less than Δu. Indeed we may expect that there is a spectrum of smaller motions with smaller velocities. In consequence, due to the nonlinearity, fully developed turbulent flow is a cascade, in which energy flows from the largest eddies of size ℓ progressively down to the smallest of size λ_0, which are heavily damped by viscosity.[3]

Using three basic hypotheses Kolmogorov (1941b) was able to analyse Richardson's cascade and thereby identify the basic structure of well-developed turbulence at large Reynolds numbers. These hypotheses are:

1. The small-scale turbulent motions are statistically homogeneous and isotropic. In general the large-scale motions are determined by the boundaries of the motion and are therefore anisotropic. However, during the course of the cascade, isotropy is rapidly introduced by the successive

---

[3]Big whirls have little whirls that feed on their velocity.
   and little whirls have lesser whirls and so on to viscosity.

                                                        (attributed to L.F. Richardson)

nonlinear interactions. Thus when the Reynolds number is high, the statistics of the smaller scale motions have a universal nature, similar for all turbulent flows.

2. At high Reynolds numbers the statistics of the small scales are determined only by the viscosity ν and the rate of energy dissipation ε.

3. The statistics of the intermediate scales are determined only by their size λ and the rate of energy dissipation ε.

Kolmogorov hypothesised that the turbulence exhibits statistical similarity in each of the three distinct separated regions $\ell \gg \lambda \gg \lambda_0$. This implies scale invariance in the inertial range. As a result the characteristic scales of the turbulence are easily derived by dimensional analysis (Landau and Lifshitz, 1959, §32).

We define the eddy Reynolds number

$$\mathcal{R}_\lambda = \frac{\lambda v_\lambda}{\nu} \tag{5.2}$$

where v_λ is the velocity fluctuation associated with an eddy of length λ. For large eddies $\mathcal{R}_\lambda$ is large and viscous damping negligible. On the other hand for the smallest eddies of size λ_0, $\mathcal{R}_{\lambda_0} \sim 1$ and the fluctuation is highly damped.

Since the energy dissipation rate in the flow due to turbulence must be determined by the largest eddies, which are the primary source of the fluctuations, the parameters which characterise these are the density ρ, the size ℓ and the velocity differential Δu. Using dimensional arguments the energy dissipation rate per unit mass must therefore scale as

$$\varepsilon \sim \frac{\Delta u^3}{\ell} \tag{5.3}$$

This energy is dissipated as heat and must be accompanied by work done by an external force or by a loss of kinetic energy. To an external observer, it therefore appears as a viscosity, which must be ascribed to the turbulence, and must be characterised by the parameters of the large eddies (Kolmogorov, 1941a). If we ascribe the energy dissipation in the turbulent flow to a *turbulent viscosity*, the scaling must be given by dimensional analysis as

$$\nu_{\text{turb}} \sim \ell \, \Delta u \tag{5.4}$$

In a purely viscous flow (equation 3.11) we have seen that the energy dissipation per unit mass in a flow with gradient du/dy is given by

$$\varepsilon = \nu \left(\frac{du}{dy} \right)^2 \tag{5.5}$$

In a turbulent flow the scaling of the largest eddies, with velocities Δu and size ℓ, are directly determined by the velocity variation in the fluid, i.e. the velocity gradient, through

$$\frac{du}{dy} \sim \frac{\Delta u}{\ell}$$

The corresponding result for the energy dissipation yields the turbulent viscosity, namely

$$\nu_{\text{turb}} \sim \ell \Delta u \sim \ell^2 \left(\frac{du}{dy}\right) \qquad (5.6)$$

Hence the dissipation rate is

$$\varepsilon \sim \nu_{\text{turb}} \frac{\Delta u^2}{\ell^2} \qquad (5.7)$$

consistent with equation (3.11).

At the critical Reynolds number, the largest eddies fail to be damped by viscosity. Therefore the kinematic viscosity and the turbulence viscosity must be comparable. Hence $\nu_{\text{turb}}/\nu \sim \mathcal{R}/\mathcal{R}_{\text{crit}}$.

The turbulence itself is characterised by the density ρ and by a turbulent flow parameter, e.g. the energy flux from the largest eddies ε, where the turbulence is generated, to the smallest, where it is dissipated. This energy cascades down through the eddies driven by the nonlinearity, before being dissipated by viscosity. The energy transfer rate is a constant passing through the intermediate sizes of eddy ($\lambda_0 \ll \lambda \ll \ell$). The turbulent velocity of an eddy of size λ is therefore

$$v_\lambda \sim (\varepsilon\lambda)^{1/3} \sim \Delta u \left(\frac{\lambda}{\ell}\right)^{1/3} \qquad (5.8)$$

The temporal velocity variation seen at a point is easily obtained, since over time τ the fluid has moved $u\tau$. Therefore over lime τ we see fluctuations associated with an eddy of size $u\tau$ and the fluctuation is

$$v_\tau \sim (\varepsilon\tau u)^{1/3} \qquad (5.9)$$

However, the temporal variation of velocity of a specified fluid particle, i.e. the Lagrangian variation, cannot depend on the mean velocity. It must therefore have the magnitude

$$v'_\tau \sim (\varepsilon\tau)^{1/2}$$

The smallest scale motions are those for which $\mathcal{R}_\lambda \sim 1$. Since

$$\mathcal{R}_\lambda \sim \frac{v_\lambda \lambda}{\nu} \sim \frac{\ell \Delta u}{\nu} \left(\frac{\lambda}{\ell}\right)^{4/3} \sim \mathcal{R} \left(\frac{\lambda}{\ell}\right)^{4/3}$$

the smallest motions have size

$$\lambda_0 \sim \frac{\ell}{\mathcal{R}^{3/4}} \sim \frac{\nu^{3/4}}{\varepsilon^{1/4}} \tag{5.10}$$

known as the *Kolmogorov length scale*. The corresponding velocity

$$v_{\lambda_0} \sim \frac{\Delta u}{\mathcal{R}^{1/4}} \tag{5.11}$$

The simple picture in terms of eddies of size λ, which we have used heretofore, is not suitable for more detailed study. A more formal approach describes the turbulent motion in terms of the Fourier transform of the turbulent velocity, the wavenumber k replacing the scale length $\lambda \sim 1/k$. The energy in the cascade is a continuous distribution and is normally expressed in terms of the energy per unit mass in motions with wavenumber k in the range dk. This quantity is easily obtained by dimensional analysis, since as we have seen the energy distribution depends only on the density, energy flux and the size of the motion $(1/k)$. Thus the Kolmogorov distribution takes the form

$$E(k)\,\mathrm{d}k \sim \varepsilon^{2/3} k^{-5/3}\,\mathrm{d}k \tag{5.12}$$

We may estimate the departure from the power law form of the distribution near the Kolmogorov length by including the viscous dissipation from equation (5.5). When the flux down the cascade is reduced by viscous dissipation

$$\frac{\mathrm{d}\varepsilon}{\mathrm{d}k} \sim -\nu\,k^2\,E$$

Initially, for not too small scale lengths $\lambda \gtrsim \lambda_0$ $(k\lambda_0 \lesssim 1)$ the reduction in the flux from the constant value in the cascade will be small, and the energy density will be approximately given by equation (5.12). Hence substituting for the flux

$$\frac{\mathrm{d}\ln E}{\mathrm{d}\ln k} \sim -\frac{5}{3} - \frac{2}{3}\,\nu\,\frac{k^{4/3}}{\varepsilon^{1/3}} \tag{5.13}$$

which falls off at

$$k \sim \left(\frac{\varepsilon}{\nu^3}\right)^{1/4} = \frac{1}{\lambda_0}$$

as expected.

In recent years Kolmogorov's picture has received considerable further study, and weaknesses have been found. Indeed the structure of turbulence remains one of the major unsolved problems of classical physics. In particular it is found experimentally that the turbulence of the smallest eddies is patchy. There are regions in which the small eddies and their dissipation are strong, and vice versa. Consequently it is found that the required scale invariance is no longer upheld when higher order moments of the velocity are taken. Nonetheless the model works well for low-order moments and therefore gives a useful model of turbulence. Despite these problems, the Kolmogorov $k^{-5/3}$ distribution (5.12) gives reasonable agreement with spectra measured in experiments.

5.3   Turbulent Stress–Reynolds Stresses

The turbulent viscosity may be clearly identified by an argument given by Reynolds (1895) which follows the same lines as that given by Maxwell for the viscosity of gases. We consider the flow of momentum due to the turbulent fluctuations through an element of area δA, whose normal is parallel to the x axis. The velocity components are v_x, v_y and v_z. The momentum in directions x and y flowing through δA in time δt is respectively

$$\delta J_x = \rho\, v_x{}^2\, \delta A\, \delta t$$
$$\delta J_y = \rho\, v_x\, v_y\, \delta A\, \delta t$$

The mean fluxes are therefore

$$\overline{\delta J_x} = \rho\, \overline{v_x{}^2}\, \delta A\, \delta t$$
$$\overline{\delta J_y} = \rho\, \overline{v_x\, v_y}\, \delta A\, \delta t$$

The total velocity $\mathbf{v} = \mathbf{u} + \mathbf{v}'$ including the mean and fluctuation velocities, and since $\overline{\mathbf{v}} = \mathbf{u}$ and $\overline{\mathbf{v}'} = 0$ we obtain

$$\overline{v_x{}^2} = u_x{}^2 + \overline{v_x'{}^2}$$
$$\overline{v_x\, v_y} = u_x\, u_y + \overline{v_x'\, v_y'}$$

Thus the mean momentum transfer in time δt through δA becomes

$$\delta J_x = \rho\left(u_x^2 + \overline{v_x'{}^2} \right) \delta A\, \delta t$$
$$\delta J_y = \rho\left(u_x\, u_y + \overline{v_x'\, v_y'} \right) \delta A\, \delta t$$

which represent terms due to the mean flow and the turbulence. The mean velocity term corresponds to the normal convective flow of momentum, which is included in the momentum flux tensor. The terms resulting from turbulence alone can be considered the x components of a turbulent momentum flux tensor, i.e. as a turbulent stress

$$\sigma_{xx} = -\rho\, \overline{v_x'{}^2} \tag{5.14}$$
$$\sigma_{yx} = \sigma_{xy} = -\rho\, \overline{v_x'\, v_y'} \tag{5.15}$$

with similar terms to complete the tensor in the remaining directions.

We note the signs of the stresses. In fact the cross-correlation terms $\overline{v'_x\, v'_y}$ are generally negative when the velocity gradient is positive. The resultant stresses are therefore frictional and dissipative (compare equation 3.1). This confirms our physical understanding, and is thus consistent with the second law of thermodynamics and the dissipative nature of the turbulence process, as discussed in the previous section (5.2).

5.4   Similarity Model of Shear in a Turbulent Flow−von Karman's Hypothesis

An earlier approach, but with associations to that of Kolmogorov, is due to von Karman. He assumed that the turbulent stresses are local phenomena associated with the distribution of the mean flow velocity in the locality at which the stress is measured. This condition is equivalent to the hypothesis that the turbulent fluctuations are similar from point to point differing only by scale factors in length and time, which themselves depend on the local velocity and its gradients. In this respect, the model has close similarities to that of Kolmogorov, but allows changes in space and time, which appear as variations of the corresponding scale factors. At large Reynolds numbers, such that the turbulence is well developed, the shearing stresses are due to turbulent transport alone, independent of the viscosity. They are therefore functions only of the density ρ and the distribution of the mean velocity, i.e. the derivatives of u with respect to y. Assuming only the lowest derivatives are influential, the shear stress must be a function of the form

$$\sigma = f\left(\rho, \frac{\partial u}{\partial y}, \frac{\partial^2 u}{\partial y^2}\right)$$

The only dimensionless product that can be formed from these variables gives a value for the stress

$$\sigma = \frac{\chi^2 \rho\, (\partial u/\partial y)^4}{(\partial^2 u/\partial y^2)^2} \tag{5.16}$$

where χ is Karman's constant to be determined by experiment.

5.5   Velocity Profile near a Wall in Fully Developed Turbulence−Law of the Wall

Consider the turbulent flow near a smooth wall. The velocity parallel to the wall u will be sheared normal to the wall, y, in the usual fashion. The velocity profile is established by the shear stress, which in this case is due to the momentum transport by turbulence towards the wall. The net flow of momentum to the wall giving rise to the stress clearly depends on the gradient of the mean velocity du/dy, since in the absence of a gradient there would be no net momentum flux. At large Reynolds numbers the turbulent momentum flux far from the wall dominates that due to viscosity, i.e. viscosity is a negligible parameter. The stress therefore depends only on the velocity gradient, density and the distance from the wall. Therefore, by dimensional analysis the shearing stress has the form

$$\tau = \rho \chi^2 y^2 \left(\frac{du}{dy} \right)^2$$

Near the wall the flow velocity is nearly zero and the acceleration small. The shear stress is therefore nearly constant and equal to the wall shear stress τ_0. Prandtl made the further assumption that in this region the shear stress was constant and equal to the wall shear stress.

Introducing the friction velocity $v^* = \sqrt{\tau_0/\rho}$ and integrating, we obtain

$$u = \frac{v^*}{\chi} \ln y + C$$

The constants $\chi \approx 0.4$ and C are established by experiment. In fact it follows from dimensional analysis that the argument of the log must be a dimensionless product, therefore $C = B + \ln(v^*/\nu)$, where $B = 5.5$ is a pure number determined by experiment. Thus we finally obtain

$$\frac{u}{v^*} = 2.5 \ln \left[\frac{y v^*}{\nu} \right] + 5.5 \tag{5.17}$$

which is independent of the Reynolds number.

Although the calculation is strictly speaking only valid near the wall, since the stress has been treated as a constant, it is found that this functional form is well obeyed throughout the entire flow in many cases, e.g. pipe flow.

However, there is a serious problem in the behaviour as $y v^*/\nu \to 0$, where it can be seen that the velocity $u \to -\infty$. This reflects the fact that near the wall, the eddies, which comprise the turbulent fluctuations, are inhibited by the wall. Near the wall, the dominant momentum transfer will be due to viscosity, not turbulence. Thus there is a narrow layer, known as the *viscous sub-layer*,[4] where the shear is given by (3.1) and the velocity has the simple form

$$u = v^* \left(\frac{y v^*}{\nu} \right) \tag{5.18}$$

---

[4]The viscous sub-layer is sometimes called the laminar sub-layer. However, although the momentum transfer is dominated by viscosity, the flow is still turbulent.

An estimate of the thickness of the sub-layer is easily found by plotting the universal distribution law and the viscous sub-layer profiles and noting that the intersection occurs at $yv^*/\nu = 11.8$. The thickness of the laminar sub-layer is therefore approximately $10\,\nu/v^*$.

In fact the situation is slightly more complicated than that discussed above, where the viscous sub-layer is merged directly into the turbulence. In experiments it is found that there is a buffer layer where both viscosity and turbulence are active. The complete velocity profile is well described by

$$\text{viscous sub-layer:} \quad \frac{u}{v^*} = \left(\frac{y v^*}{\nu}\right) \qquad\qquad \left(\frac{y v^*}{\nu}\right) < 5$$

$$\text{buffer layer:} \quad \frac{u}{v^*} = 5.0\ln\left(\frac{y v^*}{\nu}\right) - 3.5 \qquad 5 < \left(\frac{y v^*}{\nu}\right) < 30$$

$$\text{turbulent core:} \quad \frac{u}{v^*} = 2.5\ln\left(\frac{y v^*}{\nu}\right) + 5.5 \qquad 30 < \left(\frac{y v^*}{\nu}\right)$$

$$(5.19)$$

In the preceding analysis we have assumed the wall is smooth. However in many experimental situations the wall is rough with surface irregularities with a mean square root height ε. If the wall is sufficiently rough the viscous sub-layer and the buffer layer are both overwhelmed by the roughness. In this case the viscosity ν is no longer a relevant parameter and is replaced by the roughness height ε. The distribution law again takes the form dictated by dimensional analysis with $\varepsilon \sim \text{const}\,\nu/v^*$, namely

$$\frac{u}{v^*} = 2.5\ln\left(\frac{y}{\varepsilon}\right) + 8.5 \qquad y > \varepsilon \qquad\qquad (5.20)$$

giving good agreement with the classic experiments of Nikuradse (1933) if $\varepsilon\,v^*/\nu > 70$.

5.6   Turbulent Flow Through a Duct

The analysis leading to the law of the wall is only valid near the wall where the shear stress is nearly constant. Away from the wall near to the centre of the duct, the stress will vary in steady flow to balance the pressure gradient. In this core region of the flow, the motion is fully turbulent and viscosity plays a negligible role. The profile of the flow is similar from cross-section to cross-section with the maximum velocity at the duct centre. The characteristic parameters now include the duct (or tube) width h, the distance from the tube axis y' and the velocity on the duct axis $u_{\max}$. Using dimensional analysis, the form of the velocity profile must be

$$\frac{u_{\max} - u}{u_{\max}} = f\left(\frac{y'}{h}\right) \tag{5.21}$$

where the form of the function $f(y'/h)$ is undefined. This important result is known as the *velocity defect law*.

As we shall see, there are a number of different forms of $f(y'/h)$ including a power law form. Despite their differences all give reasonable agreement with each other and with experiment.

The velocity defect law applies to a different region of flow to the law of the wall. We imagine that the two join smoothly in a second buffer zone, which is generally taken to lie at about $0.2h$ from the wall.

We consider a duct of spacing $2h$ with the flow symmetrically distributed about the axis. The pressure gradient is assumed to be constant along the duct $\partial p/\partial x = $ const. Since $-\partial p/\partial x + \partial \tau/\partial y = 0$, the shearing stress is linearly dependent on the distance from the axis

$$\tau = \tau_0 \frac{y'}{h} \tag{5.22}$$

where $y' = h - y$ is the distance from the axis.

5.6.1   Prandtl's Distribution Law

Prandtl assumed that, despite equation (5.22), the turbulent stress across the duct was constant. Thus the law of the wall (5.17) holds across the entire duct. Therefore the velocity on axis

$$u_{\max} = v^*[2.5\ln(h\, v^*/\nu) + 5.5]$$

for a smooth duct.

The velocity defect

$$\frac{(u_{\max} - u(y))}{v^*} = 2.5\ln\left(\frac{y}{h}\right) \tag{5.23}$$

which does not depend on the viscosity and has the correct form.

This result, despite its inconsistency, gives good agreement with experiment. As a result it is often known as the *universal velocity distribution law*.

5.6.2   Von Karman's Distribution Law

In this case we retain the variation of the stress across the duct (5.22), and use expressions from von Karman's similarity theory (Section 5.4), so that

$$\frac{\tau_0}{\rho}\frac{y'}{h} = \chi^2 \frac{(\partial u/\partial y)^4}{(\partial^2 u/\partial y^2)^2}$$

This equation may be integrated twice directly in terms of y', the distance from the axis. In the neighbourhood of the wall, $y' \to h$ and $y \to 0$, the flow becomes that along a plane wall (Section 5.5). Since the distance from the wall $y \ll h$, h is no longer a characteristic parameter of the flow. Therefore the flow in this region is given by the same argument as for the law of the wall, and yields the boundary condition on the turbulent flow

$$\frac{du}{dy} = \frac{1}{\chi}\frac{v^*}{y} = \frac{1}{\chi}\frac{v^*}{(h - y')} \qquad y' \to h$$

The second boundary condition is established by the velocity at the axis

$$u = u_{\max} \qquad y' = 0$$

Performing the first integration we obtain

$$\frac{du}{dy} = \frac{v^*}{\chi\left(2\sqrt{h\,y'} - a\,h\right)} \quad \to \quad \frac{1}{\chi}\frac{v^*}{(h - y')} \qquad \text{as } y' \to h \text{ and iff } a = 1 \quad (5.24)$$

where only the constant of integration $a = 1$ satisfies the boundary condition. Since this equation is also satisfied by the Prandtl law of the wall near the wall, it is clear that the values of $\chi \approx 0.4$ must be the same in both cases. Integrating a second time,

$$u = u_{\max} + \frac{1}{\chi}v^*\left\{\ln\left[1 - \sqrt{\frac{y'}{h}}\right] + \sqrt{\frac{y'}{h}}\right\} \qquad (5.25)$$

At the wall the velocity has a logarithmic singularity, as we found with the universal velocity law (5.17). As in the previous case this is resolved by introducing either the viscous and buffer sub-layers or the surface roughness. In the limit $h \to \infty$, the von Karman distribution (5.25) clearly becomes identical to the universal velocity distribution law (5.17). The velocity defect is

$$\frac{(u_{\max} - u(y))}{v^*} = \frac{1}{\chi}\left\{\ln\left[1 - \sqrt{\frac{y'}{h}}\right] + \sqrt{\frac{y'}{h}}\right\} \qquad (5.26)$$

This velocity defect law thus obeys the result that in turbulent flow the velocity profile across the duct exhibits similarity independent of the Reynolds number.

Generally the two profiles equations (5.23) and (5.26) are nearly equal, although experiment tends to be in slightly better agreement with Prandtl's universal velocity distribution law. Noting that the velocity profile is symmetric about the axis of the duct, we see that both models conflict with expectation.

On the axis of the duct, the second derivative of the mean velocity is infinite, i.e. the curvature of the profile is infinite. This is implausible, and is an example of the more general problem associated with the local models. Similar problems arise in other axially symmetric turbulent flows, such as pipes, jets and wakes, where the symmetry also induces infinite curvature of the profile on axis.

Nonetheless it is found that results generated by this model are generally in good agreement with experimental measurements. The velocity distribution law is therefore used to give the basis for a range of empirical engineering scaling relations, which are found to be widely applicable.

Case study 5.I Turbulent Flow Through a Horizontal Uniform Pipe

Table 5.1: Representative values of surface roughness for various industrial materials

Material	Roughness (mm)
Concrete, coarse	0.25
Concrete, new smooth	0.025
Drawn tubing	0.0025
Glass, plastic, perspex	0.0025
Iron, cast	0.15
Sewers, old	5
Steel, mortar lined	0.1
Steel, rusted	0.5
Steel, structural or forged	0.25
Water mains, old	1.0

In pipe flow we must include an additional parameter to those already considered in our study of the flow along a wall, namely the pipe diameter, D. Dimensional analysis shows that the maximum flow velocity u_{max} on the pipe axis is given by a functional relationship

$$\frac{u_{max}}{v^*} = F\left(\mathcal{R}, \frac{\varepsilon}{D}, 0\right)$$

and that the velocity u a distance y from the wall is

$$\frac{(u_{max} - u)}{v^*} = F\left(\mathcal{R}^*, \frac{\varepsilon}{D}, \frac{(a - y)}{a}\right)$$

where $a = D/2$ is the radius and $\mathcal{R}^* = v^* D/\nu$ the Reynolds number based on the friction velocity and the diameter. The earliest form of the function F was proposed by Darcy based on his experiments:

$$\frac{(u_{max} - u)}{v^*} = 0.08\left(1 - \frac{y}{a}\right)^{3/2}$$

Subsequently it has been assumed that the universal velocity distribution law (5.23) holds for the axis of the pipe, and

$$\frac{(u_{\max} - u)}{v^*} = 2.5 \ln\left(\frac{a}{y}\right)$$

From an engineering point of view the velocity distribution is not a particularly useful quantity. The important engineering parameter is the frictional force at the wall and the corresponding pressure drop required to overcome it and drive the flow along the pipe. Balancing the pressure Δp along the length of the pipe L against the frictional force on the wall,

$$A\Delta p = \tau_0 \, P \, L$$

where P is the perimeter of the pipe and A the cross-sectional area, namely πD and $\pi D^2/4$ for a circular pipe respectively. This relationship is used to define the 'hydraulic diameter' of a non-circular pipe $4A/P$. The preceding relationship is used to define the dimensionless *Fanning friction factor*

$$f = 2\left(\frac{v_*}{\bar{u}}\right)^2 \tag{5.27}$$

such that

$$\tau_0 = \frac{D}{L}\Delta p = \frac{1}{2}\rho\bar{u}^2 f \tag{5.28}$$

where the total mass flow rate through the pipe is $\rho\bar{u}\,A$. The friction coefficient is easily calculated from the velocity distribution law. Assuming the universal velocity distribution law (5.23) holds across the pipe,

$$\bar{u} = \frac{2\pi}{\pi a^2} \int_y^{\prime a} \left[u_{\max} - v^* F\left(1 - \frac{y}{a}\right)\right](a - y)\,dy$$

where y' is the point where $u = 0$. The integrations are easily performed to give for $a \gg y'$

$$\bar{u} = u_{\max} - 3.75\,v^*$$

Experimental measurements (Nikuradse, 1933) slightly change the constant in this result to

$$\bar{u} = u_{\max} - 4.07\,v^*$$

Substituting for the velocity on axis using the universal velocity distribution law (5.23), we obtain

$$\frac{\bar{u}}{v^*} = \begin{cases} 2.5 \ln\left(\dfrac{av^*}{\nu}\right) + 1.75 & \text{smooth} \\[2mm] 2.5 \ln\left(\dfrac{a}{\varepsilon}\right) + 4.75 & \text{rough} \end{cases} \tag{5.29}$$

The friction factor is then obtained in an implicit form after corrections from experimental data

$$\frac{1}{\sqrt{f}} = \begin{cases} 4.0 \log\left(\mathcal{R}\sqrt{f}\right) - 0.4 & \text{smooth} \\ 4.0 \log\left(D/e\right) + 2.28 & \text{rough} \end{cases} \tag{5.30}$$

where $\mathcal{R} = (\bar{u}D/\nu)$ is the Reynolds number based on the mean velocity $\bar{u}$.

This relationship has been cast in a single compact form by Colebrook (1939):

$$\frac{1}{\sqrt{f}} = -4.0 \log \left\{ \frac{\varepsilon/D}{3.7} + \frac{1.25}{\mathcal{R}\sqrt{f}} \right\} \qquad (5.31)$$

The graphical representation of this equation plotted by Moody (Figure 5.2) gives a particularly convenient form for application (Moody, 1944). Table 5.1 shows representative values of the roughness heights for a range of materials with different engineering application, varying from very smooth to old worn pipes of uncertain provenance. The limit at which roughness dominates the viscous layer is given by Rouse's limit

$$\frac{1}{\sqrt{f}} = \frac{\mathcal{R}}{100} \frac{\varepsilon}{d}$$

shown dashed in Figure 5.2.

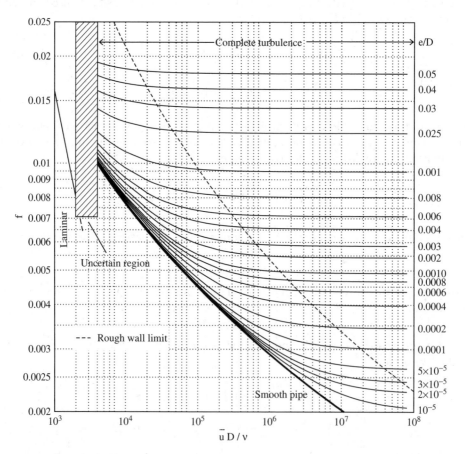

Figure 5.2: Plot of the Fanning friction factor as a function of Reynolds number $\mathcal{R} = \bar{u}D/\nu$ for varying values of the dimensionless roughness ε/D. Also shown is the lower limit of fully developed turbulence where roughness submerges the viscous layer (dashed line).

5.I.i Blasius wall stress correlation

The universal velocity distribution is cumbersome for application, e.g. in the calculation of the wall friction. A much simpler approximation which has a limited range of
validity is the power law distribution

$$\frac{u}{u_{max}} = \left(\frac{y}{a}\right)^{1/n} \tag{5.32}$$

where n is an integer whose value varies depending on the Reynolds number. It is
found from experiments that at the lowest Reynolds numbers $\mathcal{R} \gtrsim 4000$, the value
of n is 6. At $\mathcal{R} \sim 10^5$ the value of n increases to 7. At very large Reynolds numbers
$\mathcal{R} \sim 3 \times 10^6$ the value is 10. Thus for most purposes a value of $n = 7$ is appropriate.
The ratio of the mean velocity to the maximum velocity is easily shown to be

$$\alpha(n) = \frac{\overline{u}}{u_{max}} = 2\int_0^1 (1-z)z^{1/n}dz = \frac{2n^2}{(n+1)(2n+1)}$$

Following a careful analysis of experimental data, Blasius deduced that the Fanning
friction factor for smooth pipes could be written as

$$f = 0.0791\mathcal{R}^{-1/4} = 0.0791\left(\frac{\overline{u}D}{\nu}\right)^{-1/4} \tag{5.33}$$

for $\mathcal{R} < 10^5$. The corresponding wall shear stress is

$$\tau_0 = 0.033\,25\,\rho\,\overline{u}^{7/4}\,\nu^{1/4}\,a^{-1/4} \tag{5.34a}$$

If for the moment we assume that a velocity distribution (5.32) with $n = 7$ is appropriate in corresponding to the upper limit of the Reynolds number, we have that
$\overline{u}/u_{max} \approx 0.817$. Hence

$$\frac{u_{max}}{v^*} = 8.74\left(\frac{v^* a}{\nu}\right)^{1/7} \tag{5.34b}$$

If we make the reasonable assumption that this equation reflects not only the value
on axis, but the general case away from the wall, then

$$\frac{u}{v^*} = 8.74\left(\frac{yv^*}{\nu}\right)^{1/7} \tag{5.35}$$

It is clear that this equation cannot hold adjacent to the wall where $du/dy \to \infty$.
Thus again we see the need to include the viscous and buffer sub-layers near the
wall to account for the reduction of turbulent momentum transfer and the increased
importance of viscosity. It is easy to show that the intersection of the 1/7 power law
velocity profile with that of the linear viscous sub-layer occurs at

$$\frac{u}{v^*} = \frac{yv^*}{\nu} = 12.5$$

which is very similar to the value for the law of the wall, 11.8.

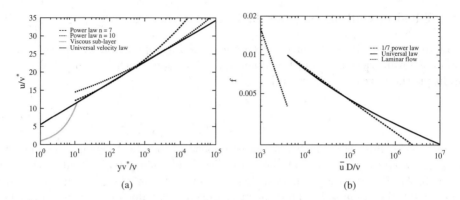

Figure 5.3: Comparison of power law turbulence distributions for velocity and wall shear stress with the universal velocity law forms. (a) Velocity profile and (b) Wall shear stress.

Figure 5.3 shows the values from the power law distributions for velocity (5.35) away from a wall, and for the wall shear stress, i.e. the friction factor (5.33). It can be seen that the $1/7$ power law gives good agreement with the universal law for Reynolds numbers in the range $4000 < \mathcal{R} < 10^5$. For large values of the Reynolds number a power law with a larger value of n is more accurate, as demonstrated by the case $n = 10$ (Figure 5.3(b)), accurate over the range $40\,000 < \mathcal{R} < 10^6$.

Equation (5.34a) or equivalently (5.34b) gives a reasonable estimate of the wall shear stress knowing either the mean mass flow or the centre line velocity, provided a suitable value for n can be identified. If the scaling factor $\mathcal{C}(n)$ in the relationship (equation 5.34b) between the mean velocity and the friction velocity is known from experiment, then

$$\frac{\overline{u}}{v^*} = \mathcal{C}(n) \left(\frac{v^* a}{\nu} \right)^{1/n} \tag{5.36}$$

The parameter $\mathcal{C}(n)$ defines the scalings of all the other quantities, see problem #25. Nikuradse (Schlichting, 1968, p.563) in an extensive experimental study of the velocity profile turbulent flow in a smooth tube, found that as the Reynolds number increased, the flow was well represented by equation (5.36) with increasing n.

Appendix 5.A Prandtl's Mixing Length Model

A simple model introduced by Prandtl allows an estimate of the turbulent parameters to be made using simple physical ideas drawn from the kinetic theory of gases. The process of turbulent mixing is introduced via the concept of a characteristic length ℓ over which the fluid is transported by the turbulent motion, and thereby transferring momentum from one surface of the fluid to another. Although this model is not entirely consistent with scaling methods or with some experimental facts, it is nevertheless a convenient analogy by which a helpful empirical picture of the Reynolds stresses can be visualised. The characteristic scaling length ℓ is regarded purely as a parameter determined by experiment.

Suppose the fluid is flowing steadily in the direction x near a wall with a velocity profile $u(y)$ perpendicular to the wall, but varying with the distance from the wall y. Fluid particles will form 'lumps', which, due to the turbulence, move bodily from one layer of the flow to another carrying the momentum parallel to the wall appropriate to the layer from which they originate. Suppose the distance between the layers is on average ℓ; then the velocity deficit originating from the lower mean velocity layer (below) is

$$\Delta u_1 = u(y) - u(y - \ell) \approx \ell \frac{du}{dy}$$

Similarly, lumps from $y + \ell$ arrive at the layer from above with velocity excess

$$\Delta u_2 = u(y + \ell) - u(y) \approx \ell \frac{du}{dy}$$

These velocity differences give rise to a fluctuation at y, whose absolute value is

$$|\overline{v'_x}| \approx \frac{1}{2}|(\Delta u_1 + \Delta u_2)| \approx \ell \left|\frac{du}{dy}\right| \tag{5.A.1}$$

The characteristic length ℓ is imagined to be the average distance a lump of fluid remains integral, before it disassembles and loses its identity. It is approximately the distance over which the velocity differences match the velocity fluctuations. We can picture the transverse velocity fluctuation arising from the collision of two lumps from $y - \ell$ and $y + \ell$, both moving in the x direction, but with differing speeds and thus with a fluctuation velocity component given by (5.A.1). Due to the incompressible nature of the fluid, they will separate in the y direction with a velocity proportional to the x component of the mean fluctuation speed $|\overline{v'_x}|$:

$$|\overline{v'_y}| \approx \text{const}\, |\overline{v'_x}| \approx \text{const}\, \ell \left|\frac{du}{dy}\right| \tag{5.A.2}$$

To obtain the turbulent stress we need the correlation term $|\overline{v'_x v'_y}|$. Lumps which move upwards with $v'_y > 0$ mostly give rise to $v'_x < 0$ and vice versa. Hence

$$\overline{v'_x v'_y} \approx \text{const}\, |\overline{v'_x}|\, |\overline{v'_y}|$$
$$\approx -\,\text{const}\, \ell^2 \left|\frac{du}{dy}\right| \frac{du}{dy} \tag{5.A.3}$$

If $du/dy > 0$, i.e. the shear velocity gradient is positive and the fluid arrives from larger values of y, the x component of the fluctuation velocity v'_x is positive, but the y component v'_y negative, and vice versa. The cross-correlation term $\overline{v'_x v'_y}$ is therefore negative. On the other hand if the velocity gradient is negative, the correlation term is positive. The form of equation (5.A.3) reflects this sign change. We may include the unknown constant in our definition of the mixing length ℓ, which is not defined. Thus the turbulent shear stress is

$$\sigma = \rho\, \ell^2 \left|\frac{du}{dy}\right| \frac{du}{dy} \tag{5.A.4}$$

where we have taken into account the fact that the sign of the stress must reflect the velocity gradient as noted above. The shear stress thus obtained is frictional, conforming to both expectation and experience. This confirms the validity of the sign of the correlation term (5.A.3) deduced using a non-rigorous argument.

The unsatisfactory nature of the model can be clearly seen in the direct kinetic theory analogy between the motion of the 'lumps' of fluid with molecules between collisions. Between collisions molecules travel in straight lines with no momentum change. In contrast the lumps of fluid are continually interacting with the neighbouring fluid and neither is their path straight nor their velocity constant (due to viscosity). It is possible to fix these problems, but this adds to the complexity of this picture. However, when we compare this result (5.A.4) with that obtained previously based on energy dissipation and dimensional arguments (equation 5.6), the latter gives a more satisfactory derivation, clearly demonstrating that the velocity correlation must be negative. Prandtl's derivation in this section, however, illustrates the importance of mixing in a turbulent fluid.

Chapter 6

Boundary Layer Flow

6.1 Introduction

A crucial step forward in the understanding of fluid behaviour was made at the start of the twentieth century by Prandtl in a remarkable paper, which introduced all the basic concepts of the boundary layer and its separation Prandtl (1904). Previously it had been thought that fluids of vanishingly small viscosity behaved as ideal fluids. As a consequence steady flow exhibited no drag due to d'Alembert's paradox. However, Prandtl argued that even with vanishingly small viscosity a narrow layer existed immediately adjacent to the surface of a body immersed in a flowing fluid. Within this layer there existed a velocity gradient, where viscosity is active due to the fact that fluid immediately adjacent to the wall is at rest (*no-slip condition*). The no-slip condition was introduced by Prandtl (1904) causing considerable controversy. The ideal flow equations, namely Euler's equation and the equation of continuity, lead to an elliptic form, which only allows one boundary condition at the surface (Section 2.6), but the no-slip condition requires two, namely zero components of velocity both normal and parallel to the surface. The inclusion of viscosity increases the order of the governing differential equations within the Navier–Stokes form and therefore admits the additional boundary condition. The mathematical expression of this distinction is provided by the method of matched asymptotics (Appendix 6.A), where the inner solution is the boundary layer approximation near the wall and is based on the full Navier–Stokes form, and the outer asymptotic solution the ideal free stream Euler flow. This difference reflects the asymptotic nature of the ideal flow equations in the limit as viscosity tends to zero. Outside this thin viscous *boundary layer*, the fluid flows with its free stream value as determined

Introductory Fluid Mechanics for Physicists and Mathematicians, First Edition. Geoffrey J. Pert.
© 2013 John Wiley & Sons, Ltd. Published 2013 by John Wiley & Sons, Ltd.

by the ideal flow. The boundary layer thus provides the source of the drag observed in experiments. It is the source of vorticity, locally destroying the condition of irrotationality, and forming the wake. Boundary layer theory is covered in detail in Schlichting (1968) to which reference may made.

We may easily estimate the scale of the boundary layer by a simple argument based on momentum balance. Thus consider the laminar flow over a thin flat plate. The ideal flow is that of a uniform flow of constant velocity equal to that of the incoming flow, U. At the surface of the plate, the fluid is brought to rest. Consequently there exists a velocity difference of U across the boundary layer. If the boundary layer thickness is δ, this corresponds to a velocity gradient $\sim U/\delta$, which will generate a shear stress at the wall of $\mu U/\delta$ opposite to the direction of flow. Thus over a distance of x along the plate, an impulse of $\mu U x/\delta$ per unit width retards the fluid:

$$\text{Momentum removed from the fluid} \sim \mu \frac{U}{\delta} x$$

This momentum loss must be accounted for by slowing down the fluid moving in the boundary layer, where the momentum flux along the wall ρU^2 enters the layer. Since the momentum is lost from a thickness δ over a distance x, we may write

$$\rho U^2 \delta \sim \mu \frac{U}{\delta} x$$

and hence obtain the scaling of the boundary layer thickness

$$\delta \sim \sqrt{\frac{\nu x}{U}} \tag{6.1}$$

We see that the thickness increases as the square root of the distance along the plate. The wall shear stress in consequence is $\tau_0 \sim \sqrt{\rho \mu U^3/x}$ and hence the drag force per unit width $\sqrt{\rho \mu U^3 x}$. The drag coefficient based on the wetted area scales as

$$C_D \sim \mathcal{R}^{-1/2} \tag{6.2}$$

where $\mathcal{R}$ is the Reynolds number based on the length of the plate.

The boundary layer is clearly a region of shear flow. We have seen in Section 4.7 that such flows are inherently unstable, depending on the Reynolds number based on the width of the flow. The corresponding Reynolds number for the boundary layer is established by the width of the layer, which increases along the plate. Hence if the plate is sufficiently long, the flow becomes unstable at the *critical point*. As the instability grows rapidly, turbulence follows quickly at the experimentally measured *point of transition*. Consequently, in many typical situations a major fraction of the boundary layer is turbulent, which increases the drag due to friction, but surprisingly may also lead to a reduction in the total drag.

Physically the boundary layer is a narrow region at whose outer edge viscous forces become asymptotically very small and flow essentially inviscid, thus merging with the irrotational flow around the body. When a no-slip condition is imposed at the surface in inviscid flow, the mathematical problem becomes overdetermined (see Section 2.6). Thus it is possible to satisfy only a limited subset of the boundary conditions by ideal flow. Comparing the Euler and Navier–Stokes equations, it can be seen that the introduction of a small region of viscous perturbation near the surface in the boundary layer increases the order of the governing differential equations, which allows the additional boundary condition and removes the over determination. Physically, with a small coefficient of viscosity, the boundary layer forms a thin *transition layer* within which the higher order Navier–Stokes equation is valid and is continuous with the irrotational free stream, where the viscous perturbation is small. Formally this is accomplished by the method of matching asymptotics in which the asymptotic solution of the boundary layer is matched to that of the free stream at an arbitrary level of approximation (Appendix 6.A).

6.2 The Laminar Boundary Layer in Steady Incompressible Two-Dimensional Flow–Prandtl's Approximation

We consider flow that is incompressible, and everywhere locally planar, laminar and steady. It may be described by the continuity (1.11) and Navier–Stokes (3.13) equations

$$\nabla \cdot \mathbf{v} = 0 \qquad \text{and} \qquad (\mathbf{v} \cdot \nabla)\mathbf{v} = -\frac{1}{\rho}\nabla p + \nu \nabla^2 \mathbf{v} \qquad (6.3)$$

Assuming that the flow is two dimensional only and that the Reynolds number is large, the boundary layer is thin and the flow nearly parallel to the surface of the body. Thus derivatives are large normal to the surface, i.e. across the boundary layer y, compared with those along it x

$$u \gg v \qquad \text{and} \qquad \frac{\partial^2 u}{\partial x^2} \ll \frac{\partial^2 u}{\partial y^2} \qquad (6.4)$$

where u and v are the x and y components of the velocity $\mathbf{v}$ (Prandtl, 1904).

As a consequence the pressure change across the boundary layer necessary to provide any acceleration in the y direction is small and

$$\frac{\partial p}{\partial y} \approx 0 \qquad (6.5)$$

The pressure throughout the boundary layer is therefore equal to that in the free stream, which is determined by Bernoulli's equation

$$\frac{1}{\rho}\frac{dp}{dx} = -U(x)\frac{U(x)}{dx} \tag{6.6}$$

Equations (6.3) thus simplify to

$$u\frac{\partial u}{\partial x} + v\frac{\partial u}{\partial y} - \nu\frac{\partial^2 u}{\partial y^2} = U\frac{dU}{dx} \tag{6.7}$$

$$\frac{\partial u}{\partial x} + \frac{\partial v}{\partial y} = 0 \tag{6.8}$$

Note that we must retain terms containing $v_x \partial/\partial x$ and $v_y \partial/\partial y$, which are of comparable magnitude.

The equations may be generalised to consider the boundary layer flow along the surface of a curved body whose radius of curvature is large compared with the boundary layer thickness. Taking the co-ordinates x along the surface and y perpendicular to it, it can be shown that at the same level of approximation, the only change is to include the transverse pressure gradient in equation (6.5) required to balance the centrifugal term

$$\frac{\partial p}{\partial y} = \kappa\rho u^2$$

where κ is the curvature. However, since $\kappa\delta \ll 1$, the pressure change across the boundary layer is small and the planar form in (6.7) and (6.8) may be used without modification for a curved wall.

If the body has a characteristic scale length ℓ and the flow velocity a characteristic speed U_0, we can cast these equations into dimensionless form by introducing the scaled variables

$$x' = \frac{x}{\ell} \quad y' = \sqrt{\mathcal{R}}\frac{y}{\ell} \quad u' = \frac{u}{U_0} \quad v' = \sqrt{\mathcal{R}}\frac{v}{U_0} \quad U' = \frac{U}{U_0} \tag{6.9}$$

where $\mathcal{R} = \ell U_0/\nu$ is the Reynolds number. On substitution equations (6.7) and (6.8) become

$$u'\frac{\partial u'}{\partial x'} + v'\frac{\partial u'}{\partial y'} - \frac{\partial^2 u'}{\partial y'^2} = U'\frac{dU'}{dx'}$$

$$\frac{\partial u'}{\partial x'} + \frac{\partial v'}{\partial y'} = 0 \tag{6.10}$$

These equations are independent of the fluid properties, and so therefore are the solutions. They reflect an essential similarity of the flows in the boundary layer. Thus if the Reynolds number changes, the thickness of the boundary layer and the transverse velocity change by $1/\sqrt{\mathcal{R}}$.

A useful integral relation can be derived by integrating equation (6.7) through the boundary layer

$$\int_0^h \left(u\frac{\partial u}{\partial x} + v\frac{\partial u}{\partial y} - U\frac{dU}{dx} \right) dy = -\frac{\tau_0}{\rho} \tag{6.11}$$

where h is much larger than the boundary layer thickness, and τ_0 is the wall shear stress.

From the second boundary layer equation (6.8) we may replace

$$v = -\int_0^y \frac{\partial u}{\partial x}\, dy$$

in equation (6.11) and after integrating by parts obtain

$$\int_0^h \left(2u\frac{\partial u}{\partial x} - U\frac{\partial u}{\partial x} - U\frac{dU}{dx} \right) dy = -\frac{\tau_0}{\rho}$$

or, rearranging,

$$\int_0^h \frac{\partial}{\partial x} \left[u\,(U - u) \right] dy + \frac{dU}{dx} \int_0^h (U - u)\, dy = \frac{\tau_0}{\rho} \tag{6.12}$$

Two measures, which are often used, are the displacement thickness δ_1 and the momentum thickness δ_2 defined by the loss of mass and momentum from the free stream into the boundary layer respectively over the distance x:

$$\delta_1 = \frac{1}{U}\int_0^\infty (U - u)\, dy \tag{6.13}$$

$$\delta_2 = \frac{1}{U^2}\int_0^\infty u\,(U - u)\, dy = \frac{\tau_0}{\rho U^2} \tag{6.14}$$

Since the fluid is incompressible, the displacement thickness is the distance the free flow streamlines are shifted due to the boundary layer. The decrease in the volume of flow due to friction $\int_0^\infty (U - u)dy$ is accommodated by a displacement of the free flow of the same volume $U\delta_1$. The momentum thickness is obtained in a similar way by balancing the loss of momentum flow in the boundary layer $\int_0^\infty u(U - u)dy$ with that of the free flow $U^2\delta_2$.

There is a useful relationship amongst these parameters obtained from equation (6.12), namely

$$\frac{\mathrm{d}}{\mathrm{d}x}\left(U^2\,\delta_2\right) + \delta_1\,U\frac{\mathrm{d}U}{\mathrm{d}x} = \frac{\tau_0}{\rho} \tag{6.15}$$

We note that these relations are valid generally for turbulent as well as laminar boundary layers provided the wall shear stress, τ_0, is given the appropriate value.

6.3 Laminar Boundary Layer over an Infinite Flat Plate–Blasius's Solution

In the case of the flow of an incompressible fluid over a flat plate with the incoming flow direction parallel to the plate surface, the ideal flow velocity U is constant over the surface. It is assumed that the fluid in contact with the surface has zero velocity–no-slip condition. Furthermore, if the plate is very long it becomes effectively infinite. In this case the length ℓ is no longer a characteristic parameter of the flow, and equations (6.10) must be transformed in such a way as to remove ℓ from the scaled variables, x' and y', which can therefore only appear in the combination

$$\eta = \frac{y'}{\sqrt{x'}} = y\sqrt{\frac{U}{\nu x}} \tag{6.16}$$

Alternatively, from dimensionless analysis we may argue that the only dimensionless combination involving the independent variables is η, and that in consequence the solutions must be functions of η alone (Section 3.8.2). Thus the problem is cast into a self-similar form with η as the sole independent variable.

Since the flow is incompressible we may solve the equation of continuity by introducing the streamfunction (Blasius, 1908)

$$\psi = \sqrt{\nu\,x\,U}\,f(\eta)$$

so that

$$u \;=\; U\,\dot{f}(\eta)$$

$$v \;=\; \frac{1}{2}\sqrt{\frac{\nu U}{x}}\left[\eta\,\dot{f}(\eta) - f(\eta)\right]$$

and the Navier–Stokes equation becomes

$$\dddot{f} + \frac{1}{2}\,f\,\ddot{f} = 0 \tag{6.17}$$

subject to the boundary conditions

$$f = 0 \quad \dot{f} = 0 \quad \text{at } \eta = 0 \qquad \dot{f} = 1 \quad \text{at } \eta = \infty$$

This third-order ordinary differential equation cannot be evaluated analytically, and the solution must be generated numerically. As the solution extends to $\eta \to \infty$ the boundary layer thickness is conventionally defined as the point at which the streamfunction f is $\frac{1}{2}$ and gives

$$\delta = 1.73 \sqrt{\frac{\nu x}{U}} \tag{6.18}$$

The asymptotic value of the transverse flow velocity approaching the free stream is

$$v \to 0.8604 \sqrt{\frac{U \nu}{x}} \qquad \text{as } y \to \infty$$

This finite velocity component is required since, as the longitudinal velocity is reduced along the plate, the total conservation of the mass flow requires that there must be a weak outward flow. This latter condition, that the velocity is indeed weak, is established by noting that since the Reynolds number is large, $U\nu/x$ is small.

A plot of the normalised longitudinal velocity u/U is shown in Figure 6.1 as a function of the normalised distance from the wall $y\sqrt{U/\nu x}$ and is compared with values used by the momentum integral method to be discussed in the next section.

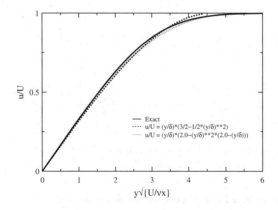

Figure 6.1: The velocity profile in a laminar boundary layer over a flat plate calculated with the exact solution and two approximations. Note the good agreement shown by the approximations.

The wall shear stress is obtained from the velocity gradient at the wall

$$\tau_0 = \sqrt{\frac{\nu U^3}{x}} \, \ddot{f}(0) = 0.332 \sqrt{\frac{\nu U^3}{x}} \tag{6.19}$$

and hence we define the local coefficient of skin friction c_f

$$c_f = 0.664 \, \mathcal{R}_x^{-1/2} \tag{6.20}$$

where $\mathcal{R}_x = Ux/\nu$ is the local Reynolds number, and a factor of 2 has been introduced to take account of the upper and lower wetted surfaces. Hence, integrating along the plate, we obtain the total skin fiction or drag coefficient for a long plate of length ℓ wetted on both sides as

$$C_f = 1.328 \, \mathcal{R}^{-1/2} \tag{6.21}$$

The displacement and momentum thicknesses calculated from this solution are $\delta_1 = 1.721\sqrt{\nu x/U}$ and $\delta_2 = 0.3323\sqrt{\nu x/U}$.

6.4   Laminar Boundary Layer–von Karman's Momentum Integral Method

An alternative simpler and versatile method, limited to two-dimensional flow, is obtained by placing the initial order of magnitude arguments (Section 6.1) on a firmer footing. We consider a section of the boundary layer of length δx a distance x along the plate, extending a height h above the plate into the free stream, Figure 6.2.

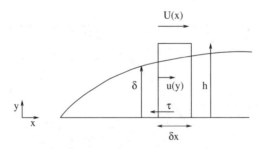

Figure 6.2: The arrangement of the flow cell in a laminar boundary layer over a flat plate used in the momentum integral calculation.

The total momentum flow through the face at x is

$$\text{Momentum flow} = \int_0^h \rho\, u(y)^2 \, dy$$

where $u(y)$ is the velocity along the plate at height y parallel to the plate, the fluid in contact with the plate is at rest, and the component of velocity normal to the plate is zero at the surface of the plate and in the free stream. The momentum flow through the face at $x + \delta x$ is given by a similar expression. Hence the gain in momentum in the interval x to $x + \delta x$ per unit width per unit time must be balanced by the wall shear stress at the plate surface $\tau_0 \delta x$ per unit width and the net pressure force on the faces, $\partial p/\partial x\, \delta x$ per unit area. Thus

$$\frac{\mathrm{d}}{\mathrm{d}x} \int_0^h \rho\, u(y)^2 \mathrm{d}y = -\tau_0 - \frac{\mathrm{d}}{\mathrm{d}x} \int_0^h p(y) \mathrm{d}y \tag{6.22}$$

The mass flow through the face at x is

$$\text{Mass flow} = \int_0^h \rho\, u(y)\, \mathrm{d}y$$

with a similar value at the face at $x + \delta x$. Since no flow enters into the section either by flow through the face parallel to the plate in the free stream, or through the plate, we obtain

$$\frac{\mathrm{d}}{\mathrm{d}x} \int_0^h \rho\, u(y)\, \mathrm{d}y = 0 \tag{6.23}$$

Multiplying by the free stream velocity $U(x)$ we obtain

$$\frac{\mathrm{d}}{\mathrm{d}x} \int_0^h \rho\, U(x)\, u(y) \mathrm{d}y = \rho\, \frac{\mathrm{d}U(x)}{\mathrm{d}x} \int_0^h u(y)\, \mathrm{d}y \tag{6.24}$$

We now make the approximation that the normal velocity is everywhere zero, in view of the fact that the fluid is incompressible, and the values of the boundary conditions at the plate and the free stream (see Section 6.3). Hence $\partial p/\partial y = 0$ and the pressure is constant across the boundary layer. The pressure is given by the value in the free stream, which is determined by Bernoulli's equation:

$$\frac{\mathrm{d}p(x)}{\mathrm{d}x} = -\rho\, U(x)\, \frac{\mathrm{d}U(x)}{\mathrm{d}x}$$

where $U(x)$ is the free stream velocity outside the boundary layer. Subtracting equation (6.22) from equation (6.24) we obtain

$$\frac{\mathrm{d}}{\mathrm{d}x} \int_0^\delta [U - u(y)]\, u(y)\, \mathrm{d}y + \frac{\mathrm{d}U(x)}{\mathrm{d}x} \int_0^\delta [U - u(y)]\, \mathrm{d}y = \frac{\tau_0}{\rho} \tag{6.25}$$

where $\delta(x)$ is the boundary layer height and $u(y) = U$ if $y > \delta$. This equation is easily seen to be equation (6.12). For a flat plate U is constant and in consequence $dp/dx = 0$, and equation (6.25) is simplified by the omission of the term resulting from the velocity gradient. The final equation is that for the wall shear stress, which is due to the viscous stress at the wall,

$$\tau_0 = -\mu \frac{\partial u}{\partial y}\bigg|_0 \tag{6.26}$$

Thus far the analysis has been quite general, with only the assumption regarding constant pressure across the boundary layer, although additional terms may need to be introduced if the surface has curvature. We must now make some assumptions regarding the velocity profile in the boundary layer. The profile must obey a number of boundary conditions:

$$u = 0 \quad \text{at} \quad y = 0 \quad \text{and} \quad u = U \quad \text{at} \quad y = \delta$$
$$\frac{\partial u}{\partial y} = 0 \quad \text{at} \quad y = \delta \quad \text{and} \quad \frac{\partial^2 u}{\partial y^2} = 0 \quad \text{at} \quad y = 0$$

The final condition arises since the flow velocity near the wall is approximately zero, and hence the change in the momentum flow through a small element δy adjacent to the wall is zero. Since this difference is balanced by the change in the shear stress over δy, namely $(\partial/\partial y)(\mu \partial u/\partial y)\delta y = 0$.

The simplest approximation satisfying these conditions is the cubic

$$u(y) = U \left[\frac{3}{2}\left(\frac{y}{\delta}\right) - \frac{1}{2}\left(\frac{y}{\delta}\right)^3 \right] \tag{6.27}$$

Substituting equations (6.27) and (6.26) into equation (6.25) and integrating we obtain

$$\frac{39}{280}\rho U^2 \frac{d\delta}{dx} = \frac{3}{2}\frac{\mu U}{\delta}$$

Hence integrating subject to $\delta = 0$ at $y = 0$, we obtain the boundary layer thickness

$$\delta = 4.64\sqrt{\frac{\mu x}{\rho U}} \tag{6.28}$$

in accordance with our earlier scaling arguments (equation 6.1). The drag is determined by the wall shear stress

$$\tau_0 = 0.323\sqrt{\frac{\rho \mu U^3}{x}} \tag{6.29}$$

Since the plate has two wetted surfaces, top and bottom, the drag coefficient based on the wetted area is therefore

$$C_f = 1.29\, R^{-1/2} \tag{6.30}$$

We may alternatively use a quartic approximation for the velocity profile by introducing the additional constraint

$$\frac{\partial^2 u}{\partial y^2} = 0 \quad \text{as} \quad y \to \infty$$

since there is no acceleration at the outer edge of the boundary layer. The solution is

$$u(y) = U \left(\frac{y}{\delta}\right) \left[2 - 2\left(\frac{y}{\delta}\right)^2 + \left(\frac{y}{\delta}\right)^3\right] \tag{6.31}$$

The value of the thickness $\delta = 5.84\sqrt{\nu x/U}$.

The profiles (Figure 6.1) from these approximate solutions agree well with that from the accurate solution given earlier (Section 6.3). The boundary layer thicknesses δ used in these solutions are essentially scale factors introduced for each approximation, and therefore not comparable from approximation to approximation or with the Blasius solution. However, the displacement and momentum thicknesses do both express physical quantities and reflect the accuracy of the overall approximation and may therefore be compared. Comparisons given by the values of the displacement $(1.732, 1,740, 1.752)\sqrt{\nu x/U}$ and momentum displacement $(0.332, 0.323, 0.343)\sqrt{\nu x/U}$ thicknesses for the exact, cubic and quartic solutions, respectively, are very satisfactory. We note also that the momentum displacement thickness is directly proportional to the wall shear stress, which therefore also shows good agreement amongst the different methods.

6.4.1   Application to Boundary Layers with an Applied Pressure Gradient

The momentum integral method is considerably simpler to apply than direct calculation to two-dimensional flows, but is not suitable in three. It was adapted by Pohlhausen (see Schlichting, 1968, pp. 192–203), to take into account pressure gradients using equation (6.25) and the quartic approximation (6.31) to the velocity profile. Defining the parameter

$$\Lambda = \frac{\delta^2}{\nu} \frac{\mathrm{d}U}{\mathrm{d}x} \tag{6.32}$$

we obtain

$$u = U \left(\frac{y}{\delta}\right) \left\{\left[2 - 2\left(\frac{y}{\delta}\right)^2 + \left(\frac{y}{\delta}\right)^3\right] + \frac{\Lambda}{6}\left[1 - 3\left(\frac{y}{\delta}\right) + 3\left(\frac{y}{\delta}\right)^2 - \left(\frac{y}{\delta}\right)^3\right]\right\} \tag{6.33}$$

It is easily seen that separation, predicted when $du/dy|_0 = 0$ (Section 6.7), occurs if $\Lambda = -12.0$. If the flow has a stagnation point at its leading edge, e.g. an aerofoil, then it is found at the value $\Lambda = 7.052$. However, the simple model leads to inaccuracy as separation is approached, and the line of separation is not accurately predicted.

The displacement and momentum thicknesses are easily obtained by integration (Schlichting, 1968, pp.192–199)

$$\frac{\delta_1}{\delta} = \frac{3}{10} - \frac{\Lambda}{120} \quad \text{and} \quad \frac{\delta_2}{\delta} = \frac{37}{315} - \frac{\Lambda}{945} - \frac{\Lambda^2}{9072} \qquad (6.34)$$

For a laminar boundary layer, the shear stress at the wall follows from $\tau_0 = \mu\, du/dy|_0$

$$\frac{\tau_0\, \delta}{\mu\, U} = 2 + \frac{\Lambda}{6} \qquad (6.35)$$

Unfortunately the parameter Λ is expressed in terms of the ill-defined term δ which is introduced for the purposes of calculation and has no physical significance. Thus to obtain the relation to physical measurements the parameters

$$Z = \frac{\delta_2^2}{\nu}$$

$$K = Z\frac{dU}{dx} = \left(\frac{37}{315} - \frac{1}{945}\Lambda - \frac{1}{9072}\Lambda^2\right)\Lambda \qquad (6.36)$$

are introduced. Using equation (6.15) the differential equation is derived

$$\frac{dZ}{dx} = \frac{F(K)}{U} \qquad (6.37)$$

where

$$F(K) = 2\left(\frac{37}{315} - \frac{1}{945}\Lambda - \frac{1}{9072}\Lambda^2\right)\left(2 - \frac{116}{315}\Lambda + \frac{79}{7560}\Lambda^2 + \frac{1}{4536}\Lambda^3\right) \qquad (6.38)$$

Since the ideal flow profile $U(x)$ around the body is known from earlier measurements or calculations, we may numerically integrate equation (6.37) using equations (6.36) and (6.38), from point to point, starting at the stagnation point at $\Lambda = 7.052$ or $K = 0.0770$ where $F(K) = 0$. By calculating successively Z, K, Λ, $F(K)$ and finally dZ/dx, the integration is advanced along the surface.

6.5   Boundary Layer Instability and the Onset of Turbulence–Tollmein–Schlichting Instability

The laminar boundary layer is a strongly sheared flow with velocity varying from zero at the surface to the free stream value at the boundary layer edge. We therefore expect that if the boundary layer is sufficiently wide that viscosity becomes too weak to damp any oscillatory waves, unstable growth similar to that in a duct will be established. Within the boundary layer the variations in y are much stronger than those along the surface x, since the Reynolds number based on the boundary thickness is $\mathcal{R}_\delta \sim \sqrt{\mathcal{R}_x}$. The local flow in the boundary layer is therefore approximately that between two parallel walls a distance δ_1, the momentum thickness, apart.

Detailed calculations of the solutions of the Orr–Sommerfeld equation using the Blasius velocity profile for the flow along a flat plate with no velocity gradient show the characteristic behaviour (Figure 6.3) for a shear flow with no point of inflexion. The two branches are both asymptotic to zero as $\mathcal{R}_\delta = U\delta_1/\nu \to \infty$. The value of the minimum Reynolds number at which instability occurs is sensitive to the details of the calculation, but $\mathcal{R}_{\text{crit}} \approx 450$ is in good agreement with careful experiments for flat plates.

However, if the ambient flow has a pressure gradient, characteristic of the flow around a finite body, the boundary layer velocity profile has a point of inflexion when $\mathrm{d}p/\mathrm{d}x > 0$ (adverse pressure gradient), and not occurring if $\mathrm{d}p/\mathrm{d}x < 0$. Hence it follows from Rayleigh's point of inflexion theorem that boundary layers with adverse pressure gradients are more susceptible to instability. The neutral stability plot therefore has the characteristic form, Figure 6.3, in which the two branches do not join asymptotically. Consequently, as may be expected, the flows are unstable even at large Reynolds numbers. As may be expected the point of instability is very sensitive to the pressure gradient (Figure 6.3).

Once the boundary layer Reynolds number R_x exceeds the value at the *critical point* where the flow becomes unstable, the instability grows rapidly and soon becomes turbulent. The onset of turbulence is found experimentally a short distance beyond the point of instability at the *point of transition*. For example, for a flat plate the transition point is experimentally found to be at $\mathcal{R}_x \approx 3.5 \times 10^5$ corresponding to $\mathcal{R}_{\delta_1} \approx 1000$ compared with the calculated critical point at $\mathcal{R}_{\delta_1} \approx 450$. The approximate values of the critical point are easily found from the data in Figure 6.3 once the profile of the boundary layer has been calculated by the Pohlhausen method. This approach has been widely used to identify the onset of turbulence theoretically and was used to design *laminar wing sections*, where the profile was designed to allow the airflow to

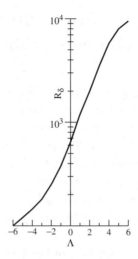

Figure 6.3: Approximate values of the boundary layer Reynolds number $\mathcal{R} = U\delta_1/\nu$ at which the boundary layer becomes unstable as a function of the velocity gradient shape factor Λ (data from Schlichting, 1968).

remain laminar over the wing surface for as long as possible by delaying the region of adverse pressure gradient. This was achieved by making the wing profile with the maximum thickness pushed back along the chord typically to about 50% of the chord compared with 5–20% for a conventional wing, thereby reducing the friction drag and allowing greater speed and longer range. Such aerofoils were used for a number of successful piston-engined aircraft during the Second World War, but their advantages diminished with dirt contamination on the surface. With the advent of jet-powered aircraft the laminar flow concept was adapted as integral to standard wing design. More recently supercritical wings used for transonic flight (Section 12.4.3) have a profile superficially similar to laminar flow wings.

6.6 Turbulent Boundary Layer on a Flat Smooth Plate

The flow along the surface of a flat plate may be visualised as directly comparable with that along the wall of a duct. The width of the duct is replaced by the boundary layer thickness and the maximal velocity on axis by the free stream velocity. Using the universal velocity law, the velocity at the boundary layer edge $y = \delta$ is the free stream value

$$\frac{U}{v^*} = 2.5 \ln \left(\frac{\delta v^*}{\nu} \right) + 5.5 \qquad (6.39)$$

This velocity distribution may be used in the momentum integral formulation of the boundary layer, equation (6.25). However, the final result is complex and difficult to evaluate. Nonetheless we may, at this stage, neglect the contribution of the viscous sub-layer and use the velocity defect relation in the momentum integral (6.25) to obtain

$$v^{*2} = \frac{1}{\chi}\frac{d}{dx}\left\{v^{*2}\delta\left[\frac{U}{v^*} + 5.0\right]\right\} \tag{6.40}$$

Since the local friction coefficient

$$c_f = \frac{\tau_0}{\frac{1}{2}\rho U^2} = 2\frac{v^{*2}}{U^2}$$

c_f will be shown to be very small, $U \gg v^*$. Differentiating equation (6.39) we obtain

$$\frac{d\delta}{dv^*} = -\frac{\delta}{v^*}\left\{0.4\frac{U}{v^*} + 1\right\}$$

and $|v^* \, d\delta/dx| \gg |\delta \, dv^*/dx|$. Hence we obtain

$$\frac{d\delta}{dx} \approx \frac{0.4v^*}{U} \qquad \text{and} \qquad 0.4\,v^*\,x \approx U\,\delta \tag{6.41}$$

This result may be understood as follows. The turbulent fluctuation velocity is given by $|u'| \sim |v'| \sim v^*$ from equation (5.15). Since the turbulence will penetrate into the free stream with a speed approximately equal to the fluctuation velocity, i.e. the friction velocity, the rate of increase in the thickness of the boundary layer will be

$$\frac{d\delta}{dx} \sim \frac{v^*}{U}$$

where $\delta \sim v^*x/U$, as above.

Substituting this result in equation (6.39) we obtain the local friction coefficient

$$0.4\sqrt{\frac{2}{c_f}} = 2.5\ln\left(\mathcal{R}_x\,c_f\right) + 2.875 \tag{6.42}$$

where the numerical constant 2.875 is determined by experiment.

Equation (6.42) is an implicit equation for c_f which must be integrated over the plate to obtain the total drag coefficient $C_f = \int_0^\ell c_f dx$. Since this is a cumbersome result to use, Schlichting obtained a more compact result by fitting an analytic expression to data obtained from the above equation:

$$C_f = 0.455(\log_{10}\mathcal{R}_l)^{-2.58}$$
$$c_f = 0.455(\log_{10}\mathcal{R}_x)^{-2.58} - 0.510(\log_{10}\mathcal{R}_x)^{-3.58} \tag{6.43}$$

The boundary is not entirely turbulent. There is an initial laminar section, until the boundary layer becomes unstable and turbulence is established. We can take this into account by modifying Schlichting's result

$$C_f = 0.455 (\log_{10} \mathcal{R}_l)^{-2.58} - A \mathcal{R}_l^{-1} \tag{6.44}$$

The term A is the difference in the drag force calculated by the appropriate forms of D_{turb} equation (6.43) and D_{lam} equation (6.21) at the onset of turbulence for turbulent and laminar flow in the boundary layer respectively. This correction is illustrated in Figure 6.4(b).

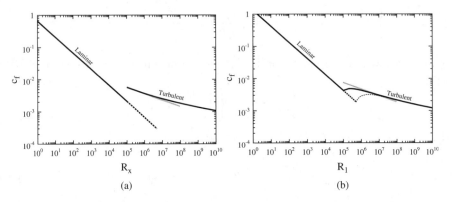

Figure 6.4: Laminar and turbulent boundary layer local and total friction coefficients. Dotted lines show the power law approximation. The onset of turbulence is set at $R_\ell = 10^5$ (full line) and $R_\ell = 5 \times 10^5$ (dashed line). (a) Local friction coefficient and (b) Total friction coefficient.

6.6.1   Turbulent Boundary Layer–Power Law Distribution

For some purposes, the solution given above for the turbulent boundary layer is too complex to be useful. As for pipe flow, a simpler solution based on a power law distribution is appropriate in these cases, e.g. to treat heat transfer in the turbulent boundary layer discussed in Section 7.4.

As the flow near the tube wall is the fully developed form of the boundary layer, we assume that the velocity and stress distributions found in pipe flow (Case Study 5.I.i) may be directly applied to the boundary layer, with the change that the pipe radius is replaced by the boundary layer thickness ($a \rightarrow \delta$) and the velocity on axis replaced by the free stream velocity ($u_{\max} \rightarrow U$). Equation (5.32) in the boundary layer becomes

$$\frac{u}{U} = \left(\frac{y}{\delta}\right)^{1/n} \tag{6.45}$$

The total friction force on one side of the plate per unit width is calculated from the momentum integral (6.25)

$$D(x) = \int_0^\ell \tau_0(x)\,dx = \rho \int_0^{\delta(x)} u\,(U - u)\,dy = \rho U^2\,\delta_2(x) \tag{6.46}$$

where δ_2 is momentum thickness. From the velocity distribution we obtain

$$\delta_2(x) = \frac{n}{(n+1)(n+2)} \delta(x)$$

$$\therefore \quad \frac{\tau_0}{\rho U^2} = \frac{n}{(n+1)(n+2)} \frac{d\delta}{dx}$$

The displacement thickness is also easily calculated to be $\delta_1 = [n/(n+1)] \delta$.
    For a $1/7$ power law, we may apply the Blasius stress formula (5.33)

$$\tau_0 = 0.0225 \rho U^{7/4} \nu^{1/4} \delta^{-1/4} \tag{6.47}$$

Substituting for τ_0 and integrating,

$$\delta(x) = 0.376 x \left(\frac{U x}{\nu}\right)^{-1/5} \tag{6.48}$$

Hence we obtain the drag force per unit width on one side of a plate of length ℓ

$$D(\ell) = 0.036 \rho U^2 \ell \left(\frac{U \ell}{\nu}\right)^{-1/5} \tag{6.49}$$

The local and total friction coefficients for a plate wetted on both sides are

$$c_f = 0.0576 \mathcal{R}_x^{-1/5} \tag{6.50}$$

$$C_f = 0.074 \mathcal{R}_\ell^{-1/5} \tag{6.51}$$

where the factor 0.072 is increased to 0.074 in accordance with experimental measurements. This result is found to give good agreement for the range from the onset of turbulence in the range $5 \times 10^5 < \mathcal{R}_l < 10^7$.
    In this case also the distribution at higher Reynolds numbers is well described by higher values of the power index n. The relevant relations for the boundary layer thickness, drag and drag coefficients are all easily derived in terms of velocity coefficients $\mathcal{C}(n)$ and the friction coefficients $\mathcal{T}(n)$ and $\mathcal{V}(n)$ following the methods used for the flow in a pipe (problem #25):

$$\begin{aligned} U &= \mathcal{C}(n) \{v^* \delta/\nu\}^{1/n} v^* & c_f &= \mathcal{F}(n) (U \delta/\nu)^{-2/(n+1)} \\ \tau_0 &= \mathcal{T}(n) \{U \delta/\nu\}^{-2/(n+1)} \rho U^2 & v^* &= \mathcal{V}(n) \{U \delta/\nu\}^{-1/(n+1)} U \end{aligned} \tag{6.52}$$

The relationships amongst these scaling parameters in boundary layer flow are the same as those in pipe flow (equation P.8), but with the changes to the variables $\overline{u} \rightarrow \alpha(n)^{-1} U$, $a \rightarrow \delta$ and $f \rightarrow c_f$.
    The boundary layer thickness is given by

$$\delta(x) = \mathcal{D}(n) x \left(\frac{U x}{\nu}\right)^{-2/(n+3)} \tag{6.53}$$

where $\mathcal{D}(n) = \{[(n+2)(n+3)/n] \mathcal{T}(n)\}^{(n+1)/(n+3)}$. Values of $\mathcal{D}(n)$ are given in Table 6.1.

Table 6.1: Parameters for power law approximations.

n	7	8	9	10
$\mathcal{C}(n)$	8.74	9.71	10.6	11.5
$\mathcal{T}(n)$	0.022 51	0.01 758	0.01 427	0.01 179
$\mathcal{V}(n)$	0.150	0.1329	0.1195	0.1086
$\mathcal{D}(n)$	0.3707	0.3129	0.2716	0.2386
$\mathcal{F}(n)$	0.036 06	0.027 82	0.022 88	0.018 08

The drag force on one side of the plate is

$$D(\ell) = \mathcal{F}(n)\,\rho\,U^2\,\ell\,\left(\frac{U\,\ell}{\nu}\right)^{-2/(n+3)} \tag{6.54}$$

where $\mathcal{F}(n) = [n/(n+1)(n+2)]\,\mathcal{D}(n)$ is the drag force scaling term.[1]
Hence

$$
\begin{aligned}
C_f &= 2\,\mathcal{F}(n)\,\mathcal{R}_\ell^{-2/(n+3)} \\
c_f &= 2\,\frac{(n+1)}{(n+3)}\,\mathcal{F}(n)\,\mathcal{R}_x^{-2/(n+3)}
\end{aligned}
\tag{6.55}
$$

These larger values of n may be used to construct an approximate model of the boundary layer at larger values of the Reynolds number than applicable to $n = 7$.

6.7 Boundary Layer Separation

The boundary layer introduces a major change to the flow around unstream-lined bodies. This causes the incoming flow to leave the surface of the body and move away into the body of the flow. We have already alluded (Section 2.4) to this behaviour in discussing separation in ideal flow and the consequent non-uniqueness of any ideal flow solution. In fact viscosity through the action of the boundary layer ensures a unique flow pattern in conformity with experiment.

Consider the flow around a cylinder. We have already described the pattern in ideal flow with no separation. In particular we found the pressure was described (2.96) by an expression which had maxima at the stagnation points on the axis where the body streamline joined and left the surface ($\theta = 0$ and π) and minima at the greatest width ($\theta = \pi/2$ and $3\pi/2$), Figure (6.5). The maximum pressure was $p_0 + \frac{1}{2}\rho U^2$ and the minimum $p_0 - \frac{3}{2}\rho U^2$ where U is the incoming flow velocity—a difference of $2\rho U^2$. Near the surface the flow is accelerated from $\theta = 0$ to $\pi/2$ and decelerated from $\theta = \pi/2$ to π, where it is again

---

[1] Note that this term $\mathcal{F}$ (equation 6.54) for the drag coefficient on a flat plate is *not* the same as the alternative expression $\mathcal{F}$ (equation 6.52) for the drag coefficient at the tube wall.

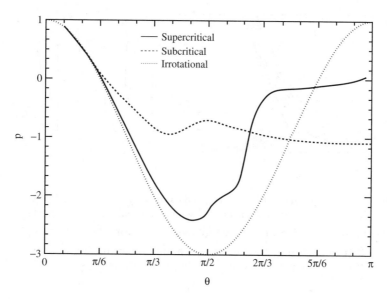

Figure 6.5: The pressure differential $(p - p_0)/\frac{1}{2}\rho U^2$ around a cylinder in laminar (subcritical) and turbulent (supercritical) layers compared with that in irrotational flow.

brought to rest. In the region $\theta = 0$ to $\pi/2$, a fluid particle gains just sufficient kinetic energy to enable it to overcome the adverse pressure gradient from $\theta = \pi/2$ to π.

However, the presence of viscosity in the boundary layer extracts energy from the particle by requiring it to do work against the shear stress, which can only be at the expense of its kinetic energy. Consequently it reaches the minimum pressure with insufficient energy to overcome the pressure barrier and enable it to reach the stagnation point on the downstream axis. In fact, as the fluid is incompressible, there will be a build up of fluid as the speed near to the surface is brought to zero. This clearly is an impossible situation, and the body streamline is forced away from the surface. Separation has occurred. The line of separation is the line on the surface at which separation occurs. Evidently a necessary condition for separation is that the pressure gradient is adverse, i.e. $\partial p/\partial x > 0$ along the surface.

Near the line of separation the streamlines and the corresponding flow velocity profile across the boundary layer take the form shown in Figure 6.6. The line of separation is identified by the condition $\partial u/\partial y\big|_s = 0$.

If the flow in the boundary layer is laminar, this behaviour must be contained in the solution of the boundary layer equations (6.7) and (6.8). Since the approximations made in their derivation fail at the line of separation, where the velocity along the surface is zero, the condition of separation appears as a singularity in the solution. This fact may be used to derive some useful

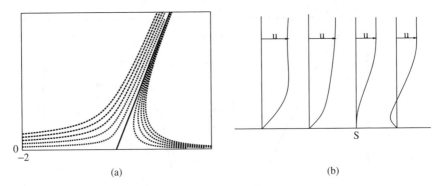

(a)   (b)

Figure 6.6: The streamlines and velocity profiles in the neighbourhood of the line of separation (S). The angle that the surface of separation makes with the surface is shown (α). At the line of separation the velocity gradient away from the surface $\partial u/\partial y\big|_s = 0$. (a) Profiles of streamlines near the line of separation. The arrows indicate the direction of flow and (b) Profiles of flow velocity near the line of separation.

scaling near the line of separation in laminar boundary layers (Figure 6.6a). In particular for geometrically similar flows, the line of separation is independent of the Reynolds number, since it is determined by the scaled parameter $x' = x/\ell$ alone. In the case of flow around a circular cylinder the polar angle at separation is $104.5°$. Similarly the angle at which the surface separating the upstream fluid from that in the flow downstream behind the surface of separation must be given by $\arctan(dy/dx) \sim \arctan(\text{const}/\sqrt{\mathcal{R}})$. In the neighbourhood of the line of separation, the flow may be modified by the backflow and resulting turbulence. Experiment shows that in practice separation always takes place after the line of maximum velocity and is independent of the Reynolds number $\mathcal{R}$. It normally occurs at or slightly later than predicted by the singularity in the boundary layer equations. The surface of separation is found to tend towards the tangent after leaving the surface, provided the latter is smooth. At a salient edge (corner) the flow follows the line of the surface before the corner; the flow inside the corner is generally turbulent and non-steady. The case of two surfaces meeting in a cusp or at a finite angle has already been discussed in connection with the flow around aerofoils, where it was found to be similar to two jets meeting at the junction and to produce a narrow wake.

The area of the wake represented by the separated downstream flow is constant at the body. From the form of the streamlines it is clear that at the line of separation, the velocity gradient away from the surface satisfies $\partial u/\partial y\big|_s = 0$. Behind the line of separation, the flow is back towards the leading edge due to the adverse pressure gradient.

The flow in the downstream separated region has significantly lower pressure (Figure 6.5) than that which would have been generated if the flow had remained attached to the body as in ideal flow. Since the upstream flow approximates to that in ideal flow, where upstream and downstream pressures are balanced, there is a substantial pressure difference across the body. This will contribute a marked drag.

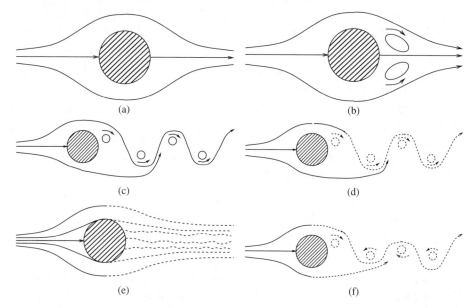

Figure 6.7: Sketches of the flow in the different regimes at various Reynolds number around a cylinder. Turbulent regions are shown dashed. (a) $\mathcal{R} < 5$ Stokes' flow, unseparated flow, (b)$15 \lesssim \mathcal{R} \lesssim 40$ Fixed pair of vortices (c) $40 \lesssim \mathcal{R} \lesssim 150$ Laminar vortex street (d) $150 \lesssim \mathcal{R} \lesssim 3 \times 10^5$ Transition region in which a laminar boundary layer generates a turbulent vortex street (e) $3 \times 10^5 \lesssim \mathcal{R} \lesssim 3.5 \times 10^6$ Laminar boundary layer becomes turbulent (f) $3 \times 10^6 \lesssim \mathcal{R}$ Turbulent boundary layer generating vortex street in a narrow wake.

6.7.1   Viscous Flow Over a Cylinder

The flow of fluid around a cylinder exhibits many of the effects we have already seen in earlier chapters. These are illustrated in Figure 6.7. The different regimes are characterised by a range of Reynolds numbers $\mathcal{R} = UD/\nu$ based on the diameter of the cylinder. Figure 6.8 shows the drag coefficient of a cylinder as a function of the Reynolds number, and shows effects associated with the different flow structures at appropriate Reynolds numbers. Several different regions can be identified:

  a. $\mathcal{R} \lesssim 5$. Stokes' flow regime. The flow is laminar, unseparated and steady. The drag is accurately given by Lamb's formula provided the Reynolds number is small ($\lesssim 0.5$).

  b. $5 \lesssim \mathcal{R} \lesssim 40$. A stationary vortex is formed following separation. The flow remains laminar and steady. Superficially the flow resembles that proposed by

Kirchoff and Rayleigh (Appendix 2.A.1) although that calculation was for inviscid flow, whereas here viscosity plays a dominant role.

c. $40 \lesssim \mathcal{R} \lesssim 150$. The flow is no longer stationary, but remains laminar. Instability of the stationary vortex leads to vortex shedding and the formation of a regular vortex street (Appendix 2.A.2.3). At low Reynolds numbers ($\mathcal{R} \lesssim 90$) the periodicity is governed by instabilities in the wake. At larger $\mathcal{R} \gtrsim 90$ the periodicity is given by the vortex shedding rate determined by the Strouhal number (page 74).

d. $150 \lesssim \mathcal{R} \lesssim 300$. A transition region in which vortices become turbulent although the flow around the cylinder remains laminar.

$300 \lesssim \mathcal{R} \lesssim 3 \times 10^5$. The flow in the wake is fully turbulent. In this regime the flow is doubly unstable, the first instability occurring when the flow separates from the surface and forms the vortex street, the second the instability of the vortex train which becomes irregular and turbulent. Evidence of a turbulent vortex street can be detected. The boundary layer is laminar.

e. $3 \times 10^5 \lesssim \mathcal{R} \lesssim 3 \times 10^6$. The boundary layer becomes turbulent. The wake is narrow and lacks any structure. Drag is strongly reduced.

f. $\mathcal{R} \gtrsim 3 \times 10^6$. The boundary layer and wake are fully turbulent. A turbulent vortex sheet can be identified.

An extensive account of this behaviour, with many photographs illustrating the behaviour shown in Figure 6.7, can be found in the book by Tritton (1988).

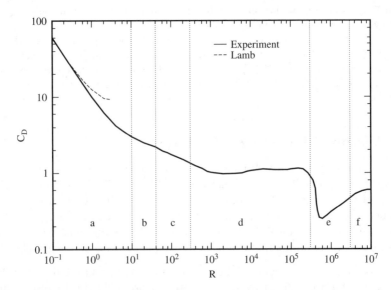

Figure 6.8: The drag coefficient on a cylinder $C_D = F_D / \frac{1}{2} \rho U^2 D$ is shown as a function of the Reynolds number $\mathcal{R} = UD/\nu$ where D is the diameter of the cylinder. The letters refer to the different regimes identified in Figure 6.7. Also shown is Lamb's approximation at low Reynolds number (adapted from Schlichting, 1968).

Figure 6.8 shows the variation of the drag coefficient as the Reynolds number is varied. It can be seen that the scaling of drag is different in the different regimes. Stokes' flow region (a) is characterised by a drag coefficient varying nearly as $1/\mathcal{R}$. The laminar boundary layer separated flow (d) has a nearly constant drag coefficient as discussed in the next section. The laminar/turbulent boundary layer transition leads to a dramatic fall in drag coefficient, known as the drag crisis, followed by a gradual increase in its value as the Reynolds number increases.

6.8   Drag

We may identify three distinct forms of drag:

1. The drag resulting from the pressure difference arising from separation is known as *form* or *pressure drag*. It is the dominant drag force for unstreamlined bodies. The wake formed typically has dimensions comparable with those of the body.

2. The drag directly due to the wall shear stress in the boundary layer known as *friction, skin* or *viscous drag*. This is the dominant drag for streamlined bodies. It gives rise to a narrow wake approximating to the vortex sheet of ideal flow.

3. The drag found in ideal flow in three-dimensional flows, known as *induced drag*. This is associated with streamlined bodies of large aspect ratio, i.e. wings (Section 11.54). In most circumstances (except for wings) it is relatively weak compared with the previous types.

The scaling of form drag in laminar flow is easily established. The pressure drop across the body is a fraction of the pressure drop due to the stagnation pressure on the upstream surface and the much reduced pressure behind the line of separation, a consequence of the disconnected flow, namely the pressure drop $\sim \rho U^2$. The area over which this pressure drop is exerted is a fraction of the cross-section S, as seen by the incoming flow, and determined by the position of the line of separation. In laminar flow the position of the line of separation around the surface is independent of the Reynolds number $\mathcal{R}$ and this fraction of the cross-section of the body is constant. If the boundary layer is turbulent this constancy is lost. The drag force on the body is thus

$$F_D \sim \rho U^2 S$$

and the drag coefficient C_D is consequently constant.[2] In laminar boundary layer flow the form drag coefficient is independent of the Reynolds number.

---

[2] The area parameter S used to define the drag coefficient is not consistently applied. Thus for form drag it is conventionally the cross-section, for friction drag the wetted area, and in

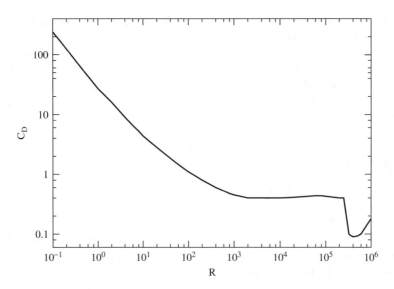

Figure 6.9: Plot of the experimentally measured drag coefficient C_D as a function of the Reynolds number R for a sphere (adapted from Schlichting, 1968). The values from Stokes' and Oseen's theoriesseasured y m at small Reynolds numbers are shown for comparison.

As the Reynolds number is increased a sudden rapid decrease in the drag coefficient is found, a situation known as the drag crisis, a phenomenon exhibited in the flow of both cylinders (Figure 6.8) and spheres (Figure 6.9). Examination of the pressure profiles (Figure 6.5) shows that this is due to a delay of the onset of separation. This is a consequence of the boundary layer becoming turbulent before separation. The mixing associated with the turbulence mixes faster moving fluid from the outer faster moving layers of the boundary layer and the free stream into the slower moving fluid near the surface. This prevents the stagnation in the boundary which gives rise to separation. There is of course also a concomitant increase in the skin drag. Over a limited range of fairly high Reynolds numbers ($3 \times 10^5 \lesssim R \lesssim 3 \times 10^6$ for a cylinder) the boundary layer just separates before becoming turbulent. However, once separated it rapidly becomes turbulent. As a result of entrapment of the flow by the turbulence, the boundary layer reattaches itself after a short gap–a phenomenon known as *reattachment*. At lower Reynolds numbers the boundary layer remains laminar up to separation and becomes turbulent away from the surface. On the other hand, at higher values it becomes turbulent before separation.

---

aerodynamics the plan area of the aerofoil. These differences reflect the different nature of the drag force. Care must be taken to identify which form is being used in any particular application.

Case study 6.I Control of Separation in Aerodynamic Structures

Separation has a severe effect on the performance of aerofoils. At moderate incidence the aerofoil operates in the classical streamline flow mode giving high lift and low drag. However, as the angle of incidence is increased, lift increases together with the pressure differential necessary to generate it. As a result, a situation eventually occurs when it is no longer possible to maintain the streamlined body flow, and the surface of separation on the upper surface moves rapidly towards the point of maximum thickness. This phenomenon is known as the stall, and can be a serious problem as its onset is rapid and occurs at times of high lift, e.g. at take-off. The stall gives rise to high drag, as the region behind the line of separation is at low pressure, in contrast to that upstream. The profiles of lift around both the span and the chord the wing are also modified giving possible control problems.

Clearly it is important to control separation, and to find methods of delaying it to as high an angle an incidence as possible, to improve the lift at slow speeds during take-off and landing. Several methods have been proposed to achieve this:

- Inducing controlled turbulence. This can be achieved by installing small obstructions into the airflow over the wing to initiate the turbulence.

- Introducing a flow of more rapid air from the lower surface into the slower moving upper boundary layer. This is achieved by two types of slotted extensions to the aerofoils. Leading edge slots ('slats') and slotted flaps at the rear. Both work on the principle that faster moving air flows through the slot into the lower parts of the boundary layer. Both are normally used at take-off and landing, and markedly increase both the lift and angle of stall.

- Direct injection of blown gas into the upper wing surface. This has been tried experimentally but is not practical.

- Injection of a different gas into the boundary layer.

- Moving surfaces on the aerofoil. An experimental aircraft with this modification was constructed, but for obvious reasons it was not regarded as a workable option.

6.9   Laminar Wake

Although we have already introduced the idea of the wake in our discussion of irrotational flow (Section 2.3) and the infinitesimally thin vortex sheet trailing away from a wing to be discussed in more detail later (Section 11.8), we have not properly specified the concept of the wake. As we have seen, the flow of a uniform stream around a body at large Reynolds numbers is only modified over a finite locality downstream of the body. This is this region of the wake, where viscosity has played a role in changing the velocity and the flow has become rotational. Thus we may separate the flow into rotational and irrotational regions. Within the latter Bernoulli's equation is expected to remain valid. By definition the wake of a streamlined body is narrow. In contrast the width of the wake of an unstreamlined body is of the order of the cross-section of the body. Depending on the value of Reynolds number and the nature of the flow (streamlined or separated), the wake may be laminar or turbulent. In Stokes' flow at low Reynolds number the disturbance extends over the entire flow, and the concept of the wake loses its meaning.

We consider the body in a steady flow of incompressible fluid at pressure P and velocity U in the x direction. The total force on the body is due to the rate of loss of momentum by the fluid flowing through a closed surface S enclosing the body

$$F_i = - \oiint_S \Pi_{ij}\,\mathrm{d}S_j = - \oiint_S \{\rho v_i v_j + P\,\delta_{ij} - \sigma_{ij}\}\,\mathrm{d}S_j \qquad (6.56)$$

The equation of continuity requires that

$$\oiint_S \rho v_i\,\mathrm{d}S_i = 0 \qquad (6.57)$$

We introduce the velocity perturbation $\mathbf{u} = \mathbf{v} - \mathbf{U}$ and the pressure perturbation p. As in Section 11.8.4 we consider the surface S comprising two planes in (y, z) at x_1 far upstream and x_2 downstream closed by a surface at infinity. Since the perturbation at infinity is zero, the contribution from the closing surface at infinity is zero. The force on the body is therefore

$$F_i = \left(\iint_{x_1} - \iint_{x_2} \right) \{\rho\, u_i\, u_x + p\,\delta_{ix} - \sigma_{ix}\}\,\mathrm{d}y\,\mathrm{d}z \qquad (6.58)$$

As we move downstream away from the body, the gradients in the x direction become progressively smaller. Therefore if the surface S is taken sufficiently far from the body, the gradients in x produce stresses much less than the momentum flux terms $-\rho u_i U_j$, i.e.

$$\mu\,\frac{\partial u_i}{\partial x} \ll \rho U u_i \qquad \text{or} \qquad \frac{\rho U x}{\mu} \gg 1$$

In this case the viscous stress tensor downstream far from the body takes the form

$$\begin{pmatrix} \sigma_{xx} = 0 & \sigma_{xy} = \sigma_{yx} & \sigma_{xz} = \sigma_{zx} \\[2mm] \sigma_{yx} = \mu\,\dfrac{\partial u_x}{\partial u_y} & \sigma_{yy} & \sigma_{yz} = \sigma_{zy} \\[2mm] \sigma_{zx} = \mu\,\dfrac{\partial u_x}{\partial u_z} & \sigma_{zy} = \mu\,\dfrac{\partial u_y}{\partial u_z} & \sigma_{zz} \end{pmatrix} \qquad (6.59)$$

Only stress components associated with the direction x of the incoming flow contribute to the force over planes normal to the flow, as such planes only have an area vector in that direction. The viscous stress components over a plane x_2 far downstream make contributions to the force

$$\iint \sigma_{xj}\,\mathrm{d}S_j \ = \iint_{x_2} \sigma_{xx}\,\mathrm{d}y\,\mathrm{d}z \quad \approx 0$$

$$\iint \sigma_{yj}\,\mathrm{d}S_j \ = \mu \iint_{x_2} \frac{\partial u_x}{\partial y}\,\mathrm{d}y\,\mathrm{d}z \ = \mu \int \left(u_x(y') - u_x(y'')\right)\,\mathrm{d}z$$

$$\iint \sigma_{zj}\,\mathrm{d}S_j \ = \mu \iint_{x_2} \frac{\partial u_x}{\partial z}\,\mathrm{d}y\,\mathrm{d}z \ = \mu \int \left(u_x(z') - u_x(z'')\right)\,\mathrm{d}y$$

where $u_x(y')$ and $u_x(y'')$ and $u_x(z')$ and $u_x(z'')$ are the perturbation velocities at the edge of the wake in the plane x_2. However, since the flow is irrotational outside these limits, the viscous contribution from the remainder of the plane is zero, and we may extend the range of integration to infinity, where the perturbation is zero, $u_y, u_z \to 0$ as $y, z \to \pm\infty$. Hence the total viscous contribution to the momentum loss (6.56) is zero

$$\oiint_S \sigma_{ij}\mathrm{d}S_j = \iint_{x_2} \sigma_x\,\mathrm{d}y\,\mathrm{d}z \approx 0 \qquad (6.60)$$

In the irrotational flow outside the wake, the pressure perturbation $p = -\rho\,\mathbf{U}\cdot\mathbf{u} = -\rho U\,u_x$. Far downstream outside the wake, $u \ll U$, and the pressure perturbation $p \sim \rho u^2 \ll \rho\,\mathbf{U}\cdot\mathbf{u}$ may be neglected. Hence collecting terms together and noting that the perturbation over the upstream surface x_1 is zero, we obtain the final expression for the force

$$F_x = -\rho U \iint_{\text{wake}} u_x\,\mathrm{d}y\,\mathrm{d}z$$

$$F_y = -\rho U \iint_{x_2} u_y\,\mathrm{d}y\,\mathrm{d}z \qquad (6.61)$$

$$F_z = -\rho U \iint_{x_2} u_z\,\mathrm{d}y\,\mathrm{d}z$$

where the integral for the drag F_x is taken over the wake alone, whereas those for the lift F_y and F_z encompass the whole of the downstream plane x_2.

The mass of fluid removed by the body into the wake is

$$-\rho U \iint_{\text{wake}} \mathrm{d}y\,\mathrm{d}z$$

Hence the drag force is just the loss of momentum in the wake resulting from obstruction presented by the body. If the body is symmetric the lift forces are zero, as may be expected from considerations of symmetry. For a body of large aspect ratio, the lift force is that obtained previously in Section 2.4 for a two-dimensional section, confirming the validity of the approximation made when using the ideal flow approximation to calculate lift. However, it should be noted that the effective profile of the aerofoil is modified by the thickness of the boundary layer.

We may estimate the scale of the wake by noting that if the flow is laminar, it must obey the Navier–Stokes equation. Since the wake is narrow, the velocity variations are dominated by the transverse gradients, normal to incoming flow. Thus the viscous and inertial terms for the velocity component parallel to the incoming flow take the forms

$$\nu \frac{\partial^2 u_x}{\partial y^2} \sim \frac{\nu\, u_x}{Y^2} \qquad \text{and} \qquad [(\mathbf{v} \cdot \nabla)\mathbf{v}]_x \sim \frac{U v_x}{x}$$

where Y is the width of the wake. Within the wake these terms are comparable and hence

$$Y \sim \sqrt{\frac{\nu x}{U}} \ll x \tag{6.62}$$

if the wake is narrow.

The area of the wake, i.e. $\sim Y^2$, determines the drag, and consequently $F_x \sim \rho U\, u_x\, Y^2 \sim \rho \nu\, x\, u_x$. Since the drag is independent of the distance downstream from the body, it follows that the x component of the perturbation velocity falls off as

$$u_x \sim \frac{F_x}{\rho \nu\, x} \propto \frac{1}{x} \tag{6.63}$$

A final note of caution: this analysis applies only if the wake is laminar. In fact in most cases the wake is turbulent, and the above results are not valid. A discussion of the turbulent wake follows in Section 6.10.1.

6.10   Separation in the Turbulent Boundary Layer

Consider the steady flow of fluid over the surface of a body at Reynolds number above the critical value for turbulence to develop. The flow is initially laminar, but turbulence will develop in the boundary layer before separation. As we have discussed earlier, if the flow is initially irrotational, then far from the body it will remain so. Whilst close to the body, the vorticity may be non-zero and rotational flow may develop.

For vortices of size greater than the Kolmogorov length, λ_0, equation (5.10), viscous dissipation is insignificant, and the vortices must themselves obey the ideal flow equations, as for example the prototype vortex of Section 2.10.1.2 In particular Kelvin's theorem tells us that over a scale length $\gtrsim \lambda_0$, circulation is conserved along a streamline. Thus the rotational region is bounded by streamlines denoting the onset of rotation (Figure 6.10). In practice this boundary is blurred over a distance $\sim \lambda_0$.

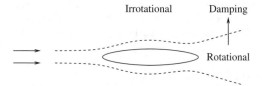

Figure 6.10: Sketch of the flow around a body with turbulent separation.

Due to these blurred edges, fluid may be transported across the boundary by the small eddies and as a result given rotation. Thus the rotational region progressively penetrates into the irrotational (but, of course, not vice versa). There are large vortices which penetrate into the irrotational flow, since vortices can be established in irrotational flow. The resultant flow is both turbulent and irrotational. Since

$$\nabla^2 \phi = \frac{\partial^2 \phi}{\partial x^2} + \frac{\partial^2 \phi}{\partial y^2} + \frac{\partial^2 \phi}{\partial z^2} = 0$$

a rotation in the plane (x, y) is damped in the z direction normal to the plane of their motion since

$$\phi \propto \exp\left[\imath \left(k_x x + k_y y\right) - \sqrt{k_x^2 + k_y^2}\, z\right]$$

Thus small vortices (with large k_x and/or k_y) are rapidly damped normal to their plane.[3] Only large eddies occur in the irrotational flow and these are rapidly damped. The largest eddies are of the same size as the rotational region formed by the wake, and the complete turbulent region therefore about twice that width (Figure 6.11).

Most of the dissipation occurs in the small eddies and therefore in the region of rotational turbulent flow, called the *region of turbulent flow*. Since rotation can only start from the surface of a body, the turbulent region must also start at the body, along a line on the surface, the line of separation. As in laminar flow the position of the line of separation depends on the structure of the boundary layer.

---

[3] The alternative solution with positive sign, $\exp\left(k_z z\right)$, is clearly unphysical.

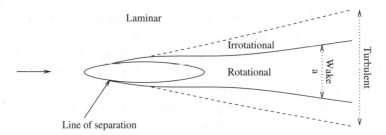

Figure 6.11: Sketch of the flow around a body with turbulent separation identifying the boundary layer, the line of separation, the wake and the regions in which the flow is irrotational or rotational, and laminar or turbulent.

6.10.1   Turbulent Wake

As we noted earlier, at large Reynolds numbers the boundary layer is turbulent and separation occurs to generate a turbulent wake. As before, we consider a flow with incident velocity $\mathbf{U}$, and introduce a mean velocity increment (averaged over the turbulent eddies) $\mathbf{u} = \mathbf{v} - \mathbf{U}$. If the width of the wake is a, the longitudinal velocity is $\sim U$ and the transverse $\sim u$, the streamlines at the edge of the wake make an angle $\sim \mathrm{d}a/\mathrm{d}x \sim u/U$ with the direction of the incoming flow x.

Since the flow outside the wake is irrotational much of the discussion of Section 6.9 for the laminar wake may be applied in this case also, provided the integrals are performed over the same region. Outside the wake, the integrals give zero. Assuming the body is cylindrical, the drag force on the body is

$$F \sim \rho U \, u \, a^2 \tag{6.64}$$

and substituting for u and noting that F is constant,

$$\frac{\mathrm{d}a}{\mathrm{d}x} \sim \frac{F}{\rho U^2 a^2}$$

$$a \sim \left(\frac{F x}{\rho U^2}\right)^{1/3}$$

$$u \sim \left(\frac{F U}{\rho x^2}\right)^{1/3} \tag{6.65}$$

Whether the flow in the wake remains turbulent or the turbulence dies out due to viscosity depends on the Reynolds number

$$\mathcal{R} \sim \frac{u \, a}{\nu} \sim \left(\frac{F^2}{\rho^2 \, U \, x \, \nu^3}\right)^{1/3} \tag{6.66}$$

Thus as $x \to \infty$, $\mathcal{R} \to 0$. Thus at large distances, far from the body, the turbulence gradually decays and the wake eventually becomes laminar.

Appendix 6.A Singular Perturbation Problems and the Method of Matched Asymptotic Expansion

Many problems in physics fall into the category of singular perturbations. In these a small parameter plays a dominant role within a *transition region*. The physicist usually solves such problems by subdividing the domain into two sub-domains: in the first, the *inner region*, it the perturbation plays an active role, and in the second, the *outer region*, it is inactive. The solution in each of the two sub-domains is matched in some appropriate fashion to give the complete solution. Clearly, boundary layer flows fall into this group. The inner region is the boundary layer itself and the outer region the irrotational flow around the body. The small 'perturbation' parameter is the kinematic viscosity ν, and the region of matching is at the boundary layer width δ. Many problems in fluid mechanics, particularly in aeronautics, fall into this category (see Chapter 14). We have already met some: for example, Stokes' flow around a sphere where the perturbation parameter is the Reynolds' number (Section 3.7.1.2). Another important case is shock waves (Chapter 10). The flow around a thin wing, the perturbation parameter, ϵ, being the thickness of the wing (Case study 2.II) is a further typical example. When $\epsilon = 0$ the flow is simply a uniform flow with constant velocity. However, if $\epsilon \to 0+$ the wing section is a branch cut on the real axis across which the tangential velocity changes discontinuously. Problems of this type, when the solutions for $\epsilon = 0$ and the limit as $\epsilon \to 0+$ are different, have been called *asymptotic paradoxes* by Birkhoff (1955). For a full account of perturbation methods applied in fluid mechanics the reader is referred to van Dyke (1975).

Although the method as outlined gives a good qualitative understanding it cannot generate quantitative estimates of the flow. The approach is formalised in the *method of matched asymptotic expansions*, which is used for the class of singularly perturbed problems where the domain may be divided into two (or more) sub-domains, one of which is a transition region of rapid change. Different solutions are constructed in the inner and outer regions, each of which is inaccurate in the other. The solutions in the two regions are generated by a power series expansion in terms of the perturbation parameter. These series are often only semi-convergent, hence asymptotic. In the outer region the problem may be treated as a regular perturbation. However, in the inner region, where the perturbation terms are non-negligible, an alternative small parameter is used. In the limit as the small parameter vanishes, the governing differential equation is reduced in order. Hence only a limited subset of the boundary conditions applies to each solution. A suitable matching between the solutions is required which determines an approximation to the global solution.

The method is best illustrated by an example, which was used by Prandtl (Schlichting, 1968, pp.73–74) to demonstrate the importance of a weak perturbation, which introduces a term of higher order into the governing differential equation and removes an overdeterminacy of the boundary conditions.

Consider a light mass m attached to a spring of constant k and damping d, whose motion satisfies the second-order differential equation

$$m\frac{\mathrm{d}^2 x}{\mathrm{d}t} + d\frac{\mathrm{d}x}{\mathrm{d}t} + kx = 0$$

The mass is accelerated from rest instantaneously to speed U. The boundary conditions are consequently $x = 0$ and $\mathrm{d}x/\mathrm{d}t = U$ at time $t = 0$. If the mass $m = 0$, the second-order term of the equation is lost and a first-order differential equation is generated, which cannot satisfy the two boundary conditions. With finite mass in the limit $m \to 0+$ the problem is well defined and has a simple solution.

The problem is conveniently transformed into dimensionless variables as $(x\,d/m\,U) \to x$ and $(t\,k/d) \to t$, and we introduce the perturbation parameter $\epsilon = (m\,k/d^2)$ to give the dimensionless equation of motion

$$\epsilon\frac{\mathrm{d}^2 x}{\mathrm{d}t} + \frac{\mathrm{d}x}{\mathrm{d}t} + x = 0 \qquad (6.\mathrm{A}.1)$$

with boundary conditions $x = 0$, $\mathrm{d}x/\mathrm{d}t = 1/\epsilon$ at time $t = 0$. The exact solution is

$$x = \frac{1}{\sqrt{1 - 4\,\epsilon}}[\exp(-\lambda\,\lambda - t) - \exp(-\lambda\,\lambda + t)] \qquad (6.\mathrm{A}.2)$$

where

$$\lambda_\pm = \frac{1}{2\epsilon}\left(1 \pm \sqrt{1 - 4\,\epsilon}\right) \qquad (6.\mathrm{A}.3)$$

For ϵ small, $\lambda_+ \approx 1/\epsilon$ and $\lambda_- \approx 1$.

The problem can be visualised from a physical point of view as an initial phase (inner region) in which the dimensionless force due to the spring x is small (as x is small). In this phase, dominated by the initial velocity, the perturbation term plays an important role and the mass moves as

$$\epsilon\frac{\mathrm{d}^2 x}{\mathrm{d}t} + \frac{\mathrm{d}x}{\mathrm{d}t} \approx 0 \qquad \frac{\mathrm{d}x}{\mathrm{d}t} = \exp\left(-\frac{t}{\epsilon}\right) \qquad x = 1 - \epsilon\exp\left(-\frac{t}{\epsilon}\right) \qquad (6.\mathrm{A}.4)$$

This motion is clearly unphysical as $t \to \infty$, and the solution is limited to times $\lesssim \epsilon$ and extensions $x \lesssim 1$ when the mass is nearly brought to rest by the damping.

For long times (outer region) the motion is a damped return from the maximum extension at $x \approx 1$ at time $t \approx 0$ to the rest position in which the mass plays no role. The governing differential equation and solution become

$$\frac{\mathrm{d}x}{\mathrm{d}t} + x = 0 \qquad \text{or} \qquad x = A\exp\left(-t\right) \qquad (6.\mathrm{A}.5)$$

the matching point being taken at the maximum extension, $A = 1$.

Referring back to the exact solution (6.A.2), we can see that the initial motion corresponds to the term in λ_+ and the later to that in λ_-. The matching of the two solutions is, however, poorly defined by this physical picture. We shall see that the solution in this split form is the first terms of the perturbation expansion in the inner and outer regions and will identify a more accurate matching procedure.

We now proceed to tackle the problem by the method of matching asymptotic expansion. We develop a series expansion for the outer solution in terms of ϵ in terms of a set of functions x_n

$$x = \sum_{n=0}^{\infty} \epsilon^n x_n \tag{6.A.6}$$

whose form is found by substituting in equation (6.A.1)

$$\frac{\mathrm{d}x_0}{\mathrm{d}t} + x_0 = 0 \quad \text{and} \quad \frac{\mathrm{d}x_n}{\mathrm{d}t} + x^n = -\frac{\mathrm{d}^2 x_{n-1}}{\mathrm{d}t^2} \tag{6.A.7}$$

and form a recursive set of first-order differential equations subject to, as yet undefined, boundary conditions at the matching point determined by the inner solution.

The inner region is defined by small values of $t \lesssim \epsilon$. We therefore magnify the range of values by introducing a second dimensionless variable $T = t/\epsilon$ in terms of which we again form a set of solutions X_n as suggested by the earlier analysis.[4] Substituting the perturbation expansion

$$X = \sum_{m=0}^{\infty} \epsilon^m X_m \tag{6.A.8}$$

in equation (6.A.1) we get the recursive set of perturbation functions

$$\frac{\mathrm{d}^2 X_0}{\mathrm{d}T^2} + \frac{\mathrm{d}X_0}{\mathrm{d}T} = 0 \quad \text{and} \quad \frac{\mathrm{d}^2 X_m}{\mathrm{d}T^2} + \frac{\mathrm{d}X_m}{\mathrm{d}T} = -X_{m-1} \tag{6.A.9}$$

which are subject to the boundary conditions $X_m = 0$ $(m \geq 0)$ and $[\mathrm{d}X_0/\mathrm{d}T = 1$ and $\mathrm{d}X_m/\mathrm{d}T = 0$ $(m \geq 1)]$ at $T = 0$.

The zero-order solutions are easily obtained

$$X_0(T) = 1 - \exp(-T) \quad \text{and} \quad x_0 = A_0 \exp(-t) \tag{6.A.10}$$

To find the matching condition we make the reasonable hypothesis (*limit matching principle*) that in the limit as the perturbation becomes very small, $\epsilon \to 0$:

The outer limit of the inner region = the inner limit of the outer region

Hence we assume that there exists a value of $t = \delta(\epsilon)$ such as $\delta = \sqrt{\epsilon}$ so that

$$\lim_{\epsilon \to 0} \delta(\epsilon) \to 0 \quad \text{and} \quad \lim_{\epsilon \to 0} (\delta/\epsilon) \to \infty$$

---

[4]This scaling is arbitrarily set by the investigator. Forms may be found to give optimum results. Usually the nature of the problem identifies the appropriate scaling factor. The scaling is introduced so that the 'action' in the inner solution occurs for values of the independent variable $T \sim 1$, and similarly that for the outer $t \sim 1$. Both solutions therefore encompass the essential features of their part of the overall solution.

so that $\lim_{\epsilon \to 0} \{x_0(\delta) = X_0(\delta/\epsilon)\}$, or $x_0(0) = X_0(\infty)$ and $A_0 = 1$. We note that this zero-order solution is identical to that found earlier from physical arguments.

Substituting for x_0 and X_0 we obtain the first-order terms x_1 and X_1

$$x_1 = -t\exp(-t) + A_1\exp(-t) \qquad \text{and} \qquad X_1 = 2\left(1 - \exp(-T)\right) - T\left(1 + \exp(-T)\right) \tag{6.A.11}$$

subject to the boundary conditions. The matching constant A_1 is easily found, since $\epsilon T = t \to 0$ at the limit and $x_1(0) = A_1 = X_1(\infty) = 2$.

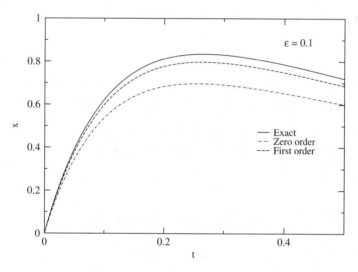

Figure 6.A.1: Comparison of the exact solution with the zero- and first-order perturbation expansions for $\epsilon = 0.1$.

An alternative limit matching principle, which may be used to match perturbation terms of differing orders, is the *asymptotic matching principle* of Kaplun and Lagerstrum, which has the following form (van Dyke, 1975):

> *The mth-order expansion term of the nth-order perturbation term*
> *= The nth-order expansion term of the mth-order perturbation term*

Applying this procedure to the inner and outer terms obtained earlier, we express the inner perturbation term in terms of the outer variable

$$X_1 = (1 - t) + (1 + t)\exp(-t/\epsilon) + 2\epsilon\left(1 - \exp(-t/\epsilon)\right) \approx (1 - t) + 2\epsilon \tag{6.A.12}$$

Correspondingly the expansion of the outer perturbation term is

$$x_1 = \exp(-t) - \epsilon\left(t - A_1\right)\exp(-t) \approx (1 - t) + \epsilon A_1 \tag{6.A.13}$$

for $t \to 0$. Applying the matching condition, we find $A_1 = 2$ as before.

Having obtained the inner and outer solutions we construct a composite solution that is uniformly valid over the full domain by noting that at the matching point δ defined above, $x(\delta) = X(\delta/\epsilon) = x_{\text{match}}$, so that

$$x_{\text{comp}} = x_{\text{inner}} + x_{\text{outer}} - x_{\text{match}} \tag{6.A.14}$$

behaves as the inner solution for small x and the outer for large x, taking the correct value at the matching point.

The two-term (zero and first order) composite solution is therefore

$$x_{comp} = (1 + 2\epsilon) \left[\exp(-t) - \exp\left(-t/\epsilon\right)\right] - t \left[\exp\left(-t/\epsilon\right) + \epsilon \exp\left(-t\right)\right] \quad (6.A.15)$$

Figure 6.A.1 compares the zero- and first-order perturbation terms with the exact solution for $\epsilon = 0.1$. It can be seen that the matched solution gives a good approximation, particularly as higher order terms are included.

Chapter 7

Convective Heat Transfer

7.1 Introduction

Convected heat transfer is associated with the movement of energy by fluid motion. The fluid (liquid or gas) motion may be either self-induced (*natural*) or imposed by external forces (*forced*):

Natural convection Fluid flow arises naturally as a result of buoyancy forces in a gravitational field. Lower density fluid rises and higher sinks giving rise to a rotating circulation pattern, density variation being established by thermal expansion in a temperature gradient. The layered fluid is unstable in the same manner as Rayleigh–Taylor unstable flows (Section 4.3). Such flows are found extensively in nature as in the sun's convective zone, earth's mantle and natural convective cooling.

Forced convection The flow is driven externally by pressures applied to the system. This approach is widely used to transfer heat from a source to a fluid and subsequently to a separated receiver via heat exchangers. In consequence the scalings are extremely important to designers and engineers.

Fluids are able to transfer energy efficiently from one point to another by convection of the heat associated with the internal and kinetic energies of the fluid elements. These two aspects have very different behaviour and it is appropriate to consider them separately.

The basic equation of heat transport is most easily obtained from the entropy transport equation (3.12) and making use of the 'second $T \, \mathrm{d}S$ equation' (1.31)

Introductory Fluid Mechanics for Physicists and Mathematicians, First Edition. Geoffrey J. Pert.
© 2013 John Wiley & Sons, Ltd. Published 2013 by John Wiley & Sons, Ltd.

we obtain

$$\rho\, c_p \frac{\mathrm{d}T}{\mathrm{d}t} = \sigma_{ij}\, \dot{\epsilon}_{ij} + \nabla \cdot (\kappa\, \nabla T) - \frac{T}{\rho} \left.\frac{\partial \rho}{\partial T}\right|_p \frac{\mathrm{d}p}{\mathrm{d}t} \qquad (7.1)$$

In flows in which the temperature varies, the usual incompressible approximation must be used with care due to the temperature variation of the density. A more satisfactory approximation in keeping with the experimental situation is that of isobaric flow, i.e. flow at constant pressure. However, provided the absolute temperature differences are small, and the flow velocity much less than the sound speed, the fluid dynamics are still satisfactorily described by the condition $\nabla \cdot \mathbf{v} = 0$.

Viscous heating is normally small compared with the other heat transfer processes and therefore negligible. If the temperature variation of the density ρ, specific heat at constant pressure c_p and thermal conductivity κ can also be neglected, equation (7.1) reduces to

$$\frac{\mathrm{d}T}{\mathrm{d}t} = \chi \nabla^2 T \qquad (7.2)$$

where $\chi = \kappa / \rho c_p$ is the thermal diffusivity.

7.2   Forced Convection

The simplest, but typical, forced convection problem arises from the flow of fluid over a surface, or through a pipe with a temperature difference between the surface and the bulk fluid external to any boundary or surface layer. As an example consider a pipe of diameter D through which fluid is flowing with mean speed u. If there is a temperature difference Θ between the pipe and the fluid, heat flows at a rate q per unit area per unit time from the fluid to the pipe. The *heat transfer coefficient, H,* is defined as

$$H = \frac{q}{\Theta} \qquad (7.3)$$

For direct thermal conduction through a slab of thickness D and thermal conductivity κ, the heat transfer coefficient is simply shown to be $H = \kappa/D$.

The heat transfer through a composite slab of several different layers, each of thickness D_i and thermal conductivity κ_i, is easily calculated since the heat transfer rate through each is constant. Therefore

$$q = H_i \theta_i \qquad \text{and} \qquad \Theta = \sum_i \theta_i \qquad \text{for each layer}$$

$$\therefore \qquad \frac{1}{H} = \sum_i \frac{1}{H_i} \qquad (7.4)$$

where θ_i is the temperature difference across layer i.

Forced convection is an important engineering discipline. Heat exchangers of various designs are an essential element in power generation, chemical engineering, etc. As a result an extensive set of numerical relations for the design of a variety of complex systems, such as tube banks, has been developed. These systems are principally based on experimental studies and dimensional analysis, and are thus empirical. Underlying them is the physical understanding of heat transfer in flowing media. In these practical systems the flow is normally turbulent, making the system inaccessible to detailed analytic modelling.

Dimensionless parameters

We may identify the characteristic scaling laws using dimensional analysis, noting that the characteristic parameters of the fluid are density ρ, dynamic viscosity η, specific heat at constant pressure c_p and thermal conductivity κ. We also define the kinematic viscosity $\nu = \mu/\rho$ and thermal diffusivity $\chi = \kappa/\rho\, c_p$. To these are added the scale variables of the problem, namely length, temperature difference and heat flow rate (L, Θ, q), which form the complete set of dimensionless products

$$\underbrace{\mathcal{N} = q\,L/\kappa\,\Theta = H\,L/\kappa}_{\text{Nusselt no.}} \qquad \underbrace{\mathcal{P} = c_p\,\eta/\kappa = \nu/\chi}_{\text{Prandtl no.}} \qquad \underbrace{\mathcal{R} = \rho\,u\,L/\eta = u\,L/\nu}_{\text{Reynolds no.}}$$

$$(7.5)$$

For many materials, their characteristic properties, namely viscosity, thermal diffusivity and density, are temperature dependent. In this case the values used to determine the dimensionless parameters are conventionally those at the *film temperature*, which is defined as the mean between the hot and cold surfaces (or fluid) $T_{\text{film}} = \frac{1}{2}(T_1 + T_2)$. This convention will be assumed throughout this chapter unless otherwise stated.

An additional dimensionless parameter is often used, namely the Stanton number

$$\mathcal{S} = H/\rho\,u\,c_p = \mathcal{N}/\mathcal{P}\mathcal{R} \qquad (7.6)$$

The Nusselt number has a simple physical interpretation. Suppose the temperature difference Θ is maintained over the scale length L; then the heat flux due to thermal conduction alone is $q' = \kappa\,\Theta/L$ and the Nusselt number

$$\mathcal{N} = \frac{q}{q'} = \frac{\text{Total heat transfer rate}}{\text{Heat transfer rate due to conduction alone}} \qquad (7.7)$$

In fluid flowing over a surface, the Nusselt number reflects the effect of the boundary region near the surface on the transfer of heat into the body of the fluid, additional to that which would take place due to thermal conduction alone.

The Nusselt number is particularly useful for simply determining the heat flow from one fluid to another across a wall, e.g. in a basic heat exchanger. Suppose the temperature difference between the fluids is Θ and the respective thermal conductivities are κ_1 and κ_2 and thicknesses L_1 and L_2. The wall has conductivity κ and thickness L. The heat transfer coefficient follows immediately from the result for thermal conduction heat flow through a layer of composite materials

$$
\begin{aligned}
H &= \frac{1}{(1/H_1 + 1/H + 1/H_2)} \\
&= \frac{1}{L_1/\kappa_1 \mathcal{N}_1 + L/\kappa + L_2/\kappa_2 \mathcal{N}_2}
\end{aligned}
\tag{7.8}
$$

where $\mathcal{N}_1$ and $\mathcal{N}_2$ are the respective Nusselt numbers.

For small temperature differences Θ, the Nusselt number is approximately constant and the heat transfer rate is approximately directly proportional to the temperature difference

$$
q \propto \Theta
\tag{7.9}
$$

This simple and very general result is known as *Newton's law of cooling*. It is valid for many heat transfer problems, and provides a great simplification in their analysis.

7.2.1 Empirical Heat Transfer Rates from a Flowing Fluid

7.2.1.1 Heat transfer from a fluid flowing along a pipe

Consider fluid flowing along a pipe of inside diameter D. Clearly the Nusselt number fully specifies the heat transfer rate, and furthermore from Buckingham's theorem

$$
\mathcal{N} = f(\mathcal{R}, \mathcal{P})
\tag{7.10}
$$

For the case where the fluid motion is either laminar or turbulent, but no natural convection occurs, we shall find that this relation takes a convenient simplification

$$
\mathcal{N} \approx \mathcal{P}^m f(\mathcal{R})
$$

where m is a number about $1/3$ giving good agreement with a wide range of data. The function $f(\mathcal{R})$ is found by experiment.

As a typical example, consider the heat transfer from the fluid within a pipe in fully developed turbulent flow, where $\mathcal{R} > \mathcal{R}_{\mathrm{crit}} \approx 2100$, and far from the entrance. It is expected that, within a limited range of Reynolds numbers, $f(\mathcal{R}) \sim \mathcal{R}^n$ where n is a constant. Experimentally it is found that $n \approx 0.8$ and hence we have the following empirical formula for the Nusselt number in a smooth pipe in turbulent flow (Fishenden and Saunders, 1950; McAdams, 1973; Welty *et al.*, 1984):

$$\mathcal{N} \approx 0.023\,\mathcal{P}^m\,\mathcal{R}^{0.8} \tag{7.11}$$

valid if $0.6 \leq \mathcal{P} < 160$, $\mathcal{R} \gtrsim 10^4$ and pipe lengths $L \gtrsim 10\,D$, viscosity etc. being measured at the bulk fluid temperature. The best value of the exponent is found experimentally to be $m = 0.4$ if the fluid is heated or 0.3 if it is cooled. The difference is believed to be due to the temperature variation of the viscosity within the viscous sub-layer. This equation, originally due to Dittus and Boelter (1930), is the synthesis of a considerable body of experimental data, and represents a typical example of the empirical engineering approach needed for calculations in this complex field, where analytic solutions are relatively few.[1]

7.2.1.2   Heat transfer from a fluid flowing across a pipe

An important case is provided by an external flow across a cylinder where, as we have seen, separation develops within the boundary layer. In the separated region behind the surface of separation, the flow becomes complex and turbulent, but still contributes to the overall heat transfer between the cylinder and the fluid. This gives rise to a complex distribution of the local Nusselt number over the surface of the cylinder, whose form depends on the Reynolds number. At very low Reynolds number ($\mathcal{R} \lesssim 1$) before the boundary layer is established, the heat flow can be calculated numerically from the governing equations. Once the laminar boundary layer is well established at $10^3 \gtrsim \mathcal{R} \gtrsim 10^5$, separation occurs at about a polar angle of about $85°$. As a result the local heat transfer coefficient exhibits a minimum at this line, due to the thickening of the boundary layer. Following separation, the heat transfer increases in the turbulent wake due to the turbulence. At larger Reynolds number the boundary layer becomes turbulent before separation and a second minimum occurs in the heat transfer profile, one at the laminar/turbulent transition point and one at the point of separation. Empirical studies of experimental data show that the overall Nusselt number is well described by a relatively simple form (McAdams, 1973)

$$\mathcal{N} = C\,\mathcal{R}^n\,\mathcal{P}^{1/3} \tag{7.12}$$

where both C and n are constants based on the range of $\mathcal{R}$, Table 7.1.

---

[1]The original form of the equation, due to Dittus and Boelter (1930), had the correct power law form, but with a different numerical coefficients, which has subsequently been refined (see Winteron (1998)).

Table 7.1: Parameters for heat flow across a cylinder.

$\mathcal{R}$	0.4–4	4–40	40–4000	4000–40 000	40 000–400 000
C	0.999	0.911	0.683	0.193	0.0266
n	0.333	0.385	0.466	0.618	0.805

This expression is often approximated by the values for the regime $1000 < \mathcal{R} < 100\,000$ by the simple expression

$$\mathcal{N} = 0.26\,\mathcal{R}^{0.6}\,\mathcal{P}^{0.3} \tag{7.13}$$

which is valid for most engineering applications.

7.2.1.3 Heat exchanger design

Heat exchangers play an important role in mechanical and chemical engineering enabling heat to be transferred from fluid flowing along one pipe to a second flowing externally. Typically their design is required to satisfy a number of conflicting constraints, which we examine in an elementary fashion. Although this is a typical engineering, rather than physics, problem, heat exchangers are characteristic of those needed to be solved in the practical application of fluid mechanics. Fishenden and Saunders (1950) or McAdams (1973) give the values of typical engineering scaling parameters for the design of pipe banks, and may be consulted for more detail of the design. Typical design constraints to be satisfied are:
- Total temperature difference along the pipe.
- Mass flow rate in the system.
- Weight of pipe.
- Power required to drive flow.
- Total volume.

Using simple scaling arguments it is possible to identify the conflicts between these constraints, and the compromises which have to be made to achieve a satisfactory design.

Normally it is required to estimate the length of pipe necessary to change the temperature of the fluid in the pipe from an inlet temperature T_1 to an outlet temperature T_2. Typically this is achieved in a series of n pipes arranged in parallel, each of length ℓ and diameter D. The mass flow rate is a design parameter

$$M = n\rho u \pi u D^2 / 4$$

Similarly the temperature difference between the fluid entering the pipe and leaving it is a design parameter, so that the total heat transferred per unit time is also a design constant

$$Q = M\,c_p\,(T_2 - T_1) = n\ell \pi u D H (T_2 - T_1)$$

The total heat transfer coefficient is determined as the combination of the terms for the flow from the external flow, through the pipe and into the internal flow by equation (7.4).

If we assume, as an example, that the heat transfer is limited by the flow inside the pipe, then the heat transfer coefficient scales as $H \sim (\rho u)^{0.8}/D^{0.2}$, from which we can obtain the necessary scaling for the length of the pipe $\ell \sim D^{0.8}/n^{0.2}$. Therefore the aspect ratio $D/\sqrt{n}\,\ell$ scales as $n^{0.7}D^{0.2}$.

If their wall thickness is constant, the pipes' total weight scales as $n\ell D \sim n^{0.8}D^{1.8}$. Therefore for minimum weight for a given heat transfer, the pipes should be long with a small diameter through which the fluid moves at high velocity in a system of small aspect ratio.

However, such a system would require a high expenditure of work to drive through the pipes. In turbulent flow the pressure drop scales as $\Delta p \sim \ell u^2/D \sim 1/n^{2.2}D^{4.2}$. The total rate at which work is done is the product of the pressure drop and the volume flow rate, namely

$$W = \Delta p \pi D^2/4u \sim 1/n^{2.2}D^{4.2}$$

Therefore for a minimum work rate both n and D should be made as small as possible. Further constraints such as limited volume are also likely to be imposed, which require ℓD^2 as a design parameter. This requires $nD^{3.5}$ to be set, and the ratio of heat transfer to power dissipation now scales as $\sim 1/D^{3.5}$ or n, implying that a large number of very thin short pipes are required.

7.2.1.4   Logarithmic mean temperature

In many cases the temperature difference along the pipe will not be constant as heat is transferred from one fluid to the other. It is easy to show that provided Newton's law of cooling is maintained, i.e. the Nusselt number is constant, the overall heat transfer is given by the logarithmic mean of the temperature differences at the two ends Θ_1 and Θ_2

$$\overline{\Theta} = \frac{(\Theta_1 - \Theta_2)}{\ln(\Theta_1/\Theta_2)} \tag{7.14}$$

Consider a length δx of the pipe. The heat loss per unit time from the fluid in the pipe is $(\kappa\Theta/D)\,\mathcal{N}\,\pi\,D\,\delta x$ and causes a temperature change $\mathrm{d}\Theta/\mathrm{d}x\,\delta x$ in the fluid mass $\rho\,u\,\pi\,D^2/4$ passing through this cross-section. Hence

$$\rho c_p\,u\,\frac{\pi D^2}{4}\frac{\mathrm{d}\Theta}{\mathrm{d}x}\,\delta x = -\frac{\kappa\Theta}{d}\mathcal{N}\,D\,\delta x$$

and

$$\ln\frac{\Theta_2}{\Theta_1} = -\frac{\kappa}{\rho c_p}\frac{4\mathcal{N}}{u\,D^2}\,\ell \tag{7.15}$$

where ℓ is the length of pipe. The total heat transferred is simply given by the total cooling along the pipe, namely $\rho\,c_p\,u\,(\pi D^2/4)\,(\Theta_1 - \Theta_2)$, which gives an average heat transfer coefficient

$$\overline{h} = \frac{\kappa \Theta}{d} \mathcal{N} = \frac{\rho \, c_p \, u \, (\pi D^2/4) \, (\Theta_1 - \Theta_2)}{\pi \, D \, \ell} = \frac{\kappa}{D} \frac{(\Theta_1 - \Theta_2)}{\ln(\Theta_1/\Theta_2)} \mathcal{N} \qquad (7.16)$$

and equation (7.14) follows. If the temperature difference is small, the logarithmic mean reduces to the arithmetic mean $\frac{1}{2}(\Theta_1 + \Theta_2)$.

7.2.2 Friction and Heat Transfer Analogies in Turbulent Flow

A very useful set of results is obtained by noting the close association between momentum and heat transfer in a flow with fully-developed turbulence. Both transport processes are due to the eddy motion of particles within the fluid.

The analogy allows us to identify the heat flow through the boundary region of turbulent flow in ducts and pipes, if the flow profile has already been empirically determined. Assuming constant viscosity and thermal conductivity, a constant heat flux along the wall, q, establishes a self-similar transverse temperature profile with increasing temperatures along the duct, which is described by a series of increasingly detailed models.

7.2.2.1 Reynolds analogy

In turbulent flow there is a clear relationship between the convective processes which transfer momentum, i.e. friction, and those which transfer energy, i.e. heat. For suppose a fluid particle is moved from the main stream, where its velocity is u to the wall, where it is brought to rest; then the momentum conveyed to the surface is mu, where m is the mass of the particle. At the same time, if there is a temperature difference Θ between the free stream and the wall, and assuming the particle remains in contact long enough to reach the wall temperature, the same mass has transferred heat $m \, c_p \, \Theta$. The ratio of the momentum to heat transfer is thus $u/c_p \Theta$. Since the momentum transfer to the wall per unit area per unit time is the wall shear stress τ, the heat transfer rate per unit area per unit time, q, is given by the *Reynolds analogy*

$$\frac{q}{\tau} = \frac{c_p \Theta}{u} \qquad (7.17)$$

the heat flux q being constant along the wall. This expression is often cast in dimensionless form by introducing the Stanton number $\mathcal{S} = q/\rho \, u \, c_p$ in terms of which the expression is written in terms of the (Fanning) friction factor f, equation (5.27),

$$S = f/2 \tag{7.18}$$

The conditions under which the Reynolds analogy is valid are clearly:
1. The momentum and heat transfer rate coefficients must match, i.e. the turbulent Prandtl number $P_{turb} = 1$ In principle.
2. The friction drag must be directly due to viscosity, i.e. there is no form drag.

The analogy is therefore most appropriate for the flow through pipes and ducts, and for flow over regular surfaces with no separation.

7.2.2.2   Prandtl–Taylor correction

It is implicit in the above derivation that the flow is everywhere turbulent, so that the turbulence carries a fluid particle from the free stream to the wall. Thus the buffer zone is assumed to reach to the wall. In practice, we have seen, the turbulent motion is limited by the viscous sub-layer, Section 5.5, where the turbulent eddies are no longer dominant. Within this zone, momentum transfer is due to viscosity, and energy to thermal conduction.

Let the temperature drop from the wall to the edge of the viscous sub-layer be $b\,\Theta$; then the heat transfer across the layer by thermal conduction is

$$q = \frac{\kappa\, b\, \Theta}{\epsilon}$$

where ϵ is the thickness of the sub-layer. The friction drag across the sub-layer is the wall shear stress

$$\tau = \frac{\mu\, a\, u}{\epsilon}$$

where $a\,u$ is the velocity at the edge of the layer.

Across the region from the free stream to the edge of the sub-layer, the Reynolds relation holds, but with boundary values $b\,\Theta$ and $a\,u$ respectively instead of zero. Hence

$$\frac{q}{\tau} = \frac{(1-b)\,c_p\,\Theta}{(1-a)\,u} = \frac{\kappa\, b\, \Theta}{\mu\, a\, u}$$

Solving,

$$\frac{b}{a} = \frac{P}{[1 + a(P-1)]}$$

where $P = c\,\mu/\kappa$ is the Prandtl number. Hence

$$\frac{q}{\tau} = \frac{c_p\,\Theta}{u} \cdot \frac{1}{[1 + a\,(P-1)]} \tag{7.19}$$

Experiments show $a \sim 0.4 - 0.6$.

If the boundary of the viscous sub-layer is set at $y\,v^*/\nu = 5$, the value of $a = 5\,u\,f/2$, and hence the Stanton number

$$S = \frac{f/2}{1 + 5\,\sqrt{f/2}\,(\mathcal{P} - 1)} \tag{7.20}$$

If $\mathcal{P} = 1$ the relation reverts to the Reynolds form. For gases $\mathcal{P} \sim 0.65 - 1$ and the correction can be neglected. For liquids $\mathcal{P} \gg 1$ most of the temperature drop is across the sub-layer and the Taylor–Prandtl correction significantly reduces the heat flow.

If $a = 0$ the flow is entirely turbulent, and we again recover the Reynolds expression, whereas if $a = b$ the flow is entirely laminar and

$$\frac{q}{\tau} = \frac{\kappa\,\Theta}{\mu\,u}$$

7.2.2.3   Von Karman's correction

Von Karman (1939) modified the Taylor–Prandtl form to use the more complete description of the profile near the wall including the buffer layer, equation (5.19). The analysis is broadly similar to that for the earlier correction. By including the turbulent viscosity and thermal conduction, the basic fluid equations may be written as

$$\tau = \rho\,(\nu + \nu_{\text{turb}})\,\frac{du}{dy}$$
$$h = -\rho c_p\,(\chi + \chi_{\text{turb}})\,\frac{dT}{dy} \tag{7.21}$$

Reynolds' analogy leads to equality between the turbulent thermal diffusivity χ_{turb} with the turbulent kinematic viscosity ν_{turb}. Following Section 5.6.1, the shear stress and heat flux are assumed to be constant through the boundary.

Defining a set of scaled variables,

$$y^+ = \frac{y\,v^*}{\nu} \qquad u^+ = \frac{u}{v^*} \qquad T^+ = \frac{\rho\,c_p\,v^*\,T}{h} \tag{7.22}$$

hence

$$\left(1 + \frac{\nu_{\text{turb}}}{\nu}\right)\frac{du^+}{dy^+} = 1$$
$$\left(\frac{1}{\mathcal{P}} + \frac{\nu_{\text{turb}}}{\nu}\right)\frac{dT^+}{dy^+} = -1 \tag{7.23}$$

In the viscous sub-layer $\nu_{turb} = 0$ and

$$T^+ = T_0^+ - \mathcal{P}y^+$$

where T_{W+} is the normalised wall temperature. In the buffer layer we include diffusion due to both kinematic and turbulence for both the viscosity and thermal conduction. To calculate the turbulent diffusivity, ν_{turb} in the buffer layer we substitute the velocity profile from equation (5.19) into the dynamic equation (7.23) to get

$$\frac{\nu_{turb}}{\nu} = \frac{y^+}{5} - 1 \tag{7.24}$$

Substituting and integrating from the edge of the viscous sub-layer,

$$T^+ = T_{W+} - 5\mathcal{P} - 5\ln\left\{\frac{y^+\mathcal{P}}{5} - (\mathcal{P} - 1)\right\} \tag{7.25}$$

In the turbulent core, the turbulent viscosity dominates and is given by

$$\frac{\nu_{turb}}{\nu} = \frac{y^+}{2.5}$$

Substituting and integrating from the edge of the buffer zone into the core,

$$T^+ = T_{W+} - 5\mathcal{P} - 5\ln\left\{\frac{y^+\mathcal{P}}{5} - (\mathcal{P} - 1)\right\} - 2.5\ln\left(\frac{y^+}{30}\right) \tag{7.26}$$

To proceed we need to calculate the temperature in the free stream. In the turbulent core

$$\frac{dT^+}{dy^+} = -\frac{du^+}{dy^+}$$

in accordance with the Reynolds analogy. It is reasonable to expect that the position where the value of the temperature equals its mean T_m^+ across the duct coincides with that of the velocity u_m^+, so that

$$T_B^+ - T_m^+ = u_m^+ - u_B^+$$

a result consistent with Reynolds' analogy (7.17) applied to the core. The subscript B represents the point at the junction of buffer zone and the core. Hence

$$T_{W+} - T_m^+ = 5\mathcal{P} + 5\ln(1 + 5\mathcal{P}) + u_m^+ - 5\ln 6 - 5 \tag{7.27}$$

Since the friction factor $f = 2/u_m^{+2}$ and the Stanton number $S = 1/u_m^+ (T_0^+ - T_m^+)$, we obtain von Karman's correction

$$S = \frac{f/2}{1 + 5\sqrt{f/2}\left\{\mathcal{P} - 1 + \ln\left[1 + \frac{5}{6}(\mathcal{P} - 1)\right]\right\}} \tag{7.28}$$

7.2.2.4   Martinelli's correction

Von Karman's correction in the previous section makes a notable improvement over the simple Taylor–Prandtl form by including the buffer region in
which both kinematic and turbulent diffusion are active. However, his analysis
assumed that in the fluid core, diffusion is entirely due to turbulence, and the
effective Prandtl number in this region is 1. In practice this is satisfactory for
gases and liquids where $\mathcal{P} \gtrsim 1$.

Martinelli (1947) allowed the Prandtl number in the core to include the
kinematic thermal diffusivity as well as the turbulent. With a uniform heat
flux at the wall, the fluid heats along the duct. In the steady state, a constant
temperature profile is established across the duct separable in (x, y), namely
$C \times T(y)$ (Kays, 1966, pp 104–109). Thus

$$\frac{d}{dy}\left[q(y)\frac{dA}{dy} \right] = -\rho c_p\, Cu(y)T(y)\frac{dA(y)}{dy}$$

where $A(y)$ is the area from axis to the point y. Defining the mean value of the
product $(u(y)T(y))$ namely $(uT)_m$ and integrating across the duct, the wall
heat

$$q_0 = \rho\, c_p\, C\int_0^{A_0} u(y)T(y)dA/\left(\frac{dA}{dy}\right)_0 = \frac{1}{v_{id}}\rho\, c_p\, RC(uT)_m \qquad (7.29)$$

where A_0 is the area of the duct and $v_{id} = 1$ or 2 for planar or cylindrical
systems respectively. $T_M = (uT)_m/u_m$ is known as the *mixed mean fluid temperature*. In the core of the flow both the velocity and temperature are nearly
constant and may be replaced by their mean value. Hence including variations
in both the momentum (shear stress) and heat flux across a duct of width $2h$
using equation (5.22)

$$\begin{aligned}
\tau &= \rho\,(\nu + \nu_{\text{turb}})\frac{du}{dy} &&= \left(1 - \frac{y}{R}\right)\tau_0 \\
q &= -\rho c_p\,(\chi + \chi_{\text{turb}})\frac{dT}{dy} &&= \left(1 - \frac{y}{R}\right)q_0
\end{aligned} \qquad (7.30)$$

where τ_0 and q_0 are the wall shear stress and heat flux respectively. The flow
profile near the wall is given by equation (5.19) including the buffer layer.

In the viscous and buffer layers, $y \ll R$ and the term $(1 - y/R)$ may be
approximated to 1. The temperature across these layers are therefore given by
equations (7.24) and (7.25). The temperature at the boundary between the
buffer layer and the core

$$T_W^+ - T_B^+ = 5\mathcal{P} + 5\ln\{5\mathcal{P} + 1\}$$

Since Reynolds' relation holds $\nu_{\text{turb}} = \chi_{\text{turb}}$ as before and we may use the first of equations (7.30) to calculate the turbulent viscosity in the core

$$1 + \frac{\nu_{\text{turb}}}{\nu} = 0.4 \left(1 - \frac{y^+}{R^+}\right) y^+ \tag{7.31}$$

In the core the kinematic viscosity is small compared with the turbulent. Substituting for turbulent thermal diffusivity in the second equation (7.30)

$$\frac{\mathrm{d}T^+}{\mathrm{d}y^+} = -\frac{\left(1 - y^+/R^+\right)}{\left[1/\mathcal{P} + y^+ \left(1 - y^+/R^+\right)/2.5\right]} \tag{7.32}$$

To obtain the temperature difference between the edge of the buffer layer and a point in the core, we note the following integral

$$\int \frac{(1-x)}{\alpha + x\,(1-x)}\,\mathrm{d}x = \frac{1}{2}\left\{\ln\left[\alpha + x\,(1-x)\right] + \ln\{[z - (1 - 2x)]/[z + (1 - 2x)]\}\right. \tag{7.33}$$

where $z = \sqrt{1 + 4\alpha}$.

Writing $\alpha = 2.5/\left(\mathcal{P}\,R^+\right)$ we obtain

$$\begin{aligned}
T_B^+ - T^+ =& 1.25\ln\left\{\frac{\alpha + (1 - y^+/R^+)\,(y^+/R^+)}{\alpha + (1 - y_B^+/R^+)\,(y_B^+/R^+)}\right\} \\
&+ \frac{1.25}{z}\ln\left\{\frac{[z - (1 - 2y^+/R^+)][z + (1 - 2y_{B^+}/R^+)]}{[z + (1 - 2y^+/R^+)][z - (1 - 2y_{B^+}/R^+)]}\right\}
\end{aligned} \tag{7.34}$$

We need to calculate the mixed mean fluid temperature T_M across the duct. For a circular tube of radius R and taking the temperature at the tube centre T_C

$$\beta = \frac{T_W^+ - T_M^+}{T_W^+ - T_C^+} = \frac{\int\limits^{A_0} u^+ \dfrac{T_W^+ - T^+}{T_W^+ - T_C^+}\,\mathrm{d}A}{\int\limits^{A_0} u^+ \,\mathrm{d}A} \tag{7.35}$$

which is evaluated numerically. $T_M = (u\,T)_m/u_m$ is known as the *mean-mixed fluid temperature*. From equation (7.29) it can be seen that the heat transfer coefficient and therefore the Nusselt and Stanton numbers depend on $(T_W - T_M)$. Values of β for a range of Reynolds' and Prandtl numbers are given by McAdams (1973, p.212)

The temperature difference between the wall and the centre of the duct/tube is given by equations

$$T_W^+ - T_C^+ = 5\mathcal{P} + 5\ln\{5\mathcal{P} + 1\}$$

$$+ 1.25 \ln \left\{ \frac{\alpha}{\alpha + 30/h^+ \ (1 - 30/h^+)} \right\} \quad (7.36)$$

$$+ \frac{1.25}{z} \ln \left\{ \frac{[z-1]}{[z+1]} \ \frac{[z + (1 - 60/h^+)]}{[z - (1 - 60/h^+)]} \right\}$$

where $h^+ = \frac{1}{2} \mathcal{R} \sqrt{f/2}$, $\alpha = 5 \sqrt{2/f} / (\mathcal{PR})$ and $z = \sqrt{1 + 20 \sqrt{2/f} / (\mathcal{PR})}$ where the Reynolds' number is based on the full duct width or diameter.

Since the heat transfer coefficient depends on $(T_W - T_M)$, the Stanton number for the flow

$$S = \frac{1}{u_{m^+} \ (T_W^+ - T_M^+)} = \frac{\sqrt{f/2}}{\beta(T_W^+ - T_C^+)} \quad (7.37)$$

Using equation (7.36) the heat transfer is obtained.

When P > 1 equation (7.37) is equivalent to von Karman's correction (equation 7.28). Martinelli's correction embraces much of the physics of heat transfer in well developed turbulent flow (R > 4000) generalising Reynolds' analogy. It gives reasonable accuracy for liquids and gases (P > 0.6). It overestimates heat transfer in liquid metals (P < 0.1) due to molecular thermal conduction between the eddies (Pturb #1), and to the approximation used to integrate equation (7.30). At large Prandtl numbers (P > 50) errors arise due to the approximate characterisation of the buffer layer (equation 5.19) as thermal conduction is dominated by this region. For a full discussion of these effects see Kays (1966, pp164–173).

As it is not a simple expression, Martinelli's approximation has found little application in engineering design calculations, correlations based on experiment data and more recently computer simulation being preferred.

7.2.2.5 Colburn's modification

The Taylor-Prandtl and von Karman results for heat transfer at surfaces have been based on the 'law of the wall' velocity distribution (§5.5). However turbulent flow in a smooth pipe can also be described by a power law (§5.I.i). If the power law index $n = 8$, the Fanning friction factor

$$f \approx 0.46 \, (\overline{u} \, D / \nu)^{-1/5}$$

Collating the results of several experiments Colburn (1933) deduced that the Stanton number

$$S \, P^{2/3} \approx 0.023 \mathcal{R}^{-1/5} \approx f/2 \quad (7.38)$$

provided that the quantities are measured at the film temperature. Good agreement with experiment is found for the limited range of Prandtl numbers

$0.5 < \mathcal{P} < 50$ and Reynolds' numbers $5000 < (R) < 200000$. Subsequently a number of other power law correlations have been proposed (McAdams, 1973, p.219), Incropera et al. (2007, p.532) mainly based around the Taylor-Prandtl correction. That due to Gnielinski (1976) is probably the most accurate and covers the transition region from laminar to turbulent flow., $0.6 < \mathcal{P} < 2000$ and $3000 < R < 5 \times 10^5$.

7.3   Heat Transfer in a Laminar Boundary Layer

We have earlier examined in some detail the formation and structure of boundary layers along the surface of a body in a flowing fluid (Chapter 6). If the temperatures of the surface and of the fluid are different, we may expect that a growing thermal boundary layer will form as the fluid is progressively cooled along the surface. The heat from the undisturbed free stream diffuses through the boundary layer to the surface in exactly the same way as the momentum.

The close similarity between the formation of the viscous and thermal boundary layer equations is clearly seen by comparing their respective governing equations

$$u\frac{\partial u}{\partial x} + v\frac{\partial u}{\partial y} - \nu\frac{\partial^2 u}{\partial y^2} = 0$$
$$u\frac{\partial \theta}{\partial x} + v\frac{\partial \theta}{\partial y} - \chi\frac{\partial^2 \theta}{\partial y^2} = 0 \qquad (7.39)$$
$$\frac{\partial u}{\partial x} + \frac{\partial v}{\partial y} = 0$$

where θ is the temperature difference between the fluid and the surface. Both sets of equations have identical boundary conditions $u = \theta = 0$ at the surface $y = 0$ and $u = U$, and $\theta = \Theta$, the free stream values outside the boundary layer, which are identical if $\nu = \chi$ or $\mathcal{P} = 1$. This is a further example of the Reynolds analogy and shows that it also holds for laminar flows. More generally we may expect that if $\mathcal{P} > 1$ and $\chi < \nu$, the thermal conduction is weaker and the temperature gradients are larger nearer the wall. The thermal boundary layer is therefore thinner than the viscous one. On the other hand if $\mathcal{P} < 1$, the thermal boundary layer is thicker.

In the particular case where the Prandtl number $\mathcal{P} = 1$, an exact solution for the thermal boundary layer over a flat plate is easily obtained from the corresponding result for the velocity generated by Blasius's solution, Section 6.3.

7.3.1   Boundary Integral Method

More generally we may apply the boundary integral method in Section 6.4 to
the case of an arbitrary Prandtl number. We therefore extend our earlier calculation of the boundary layer on a flat plate to consider the thermal boundary
layer. We note first an important consequence of the similarity of the viscous
and thermal equations (7.28) in this particular case: that is, the structure of
the resultant complete boundary layer will be similar, differing only in scale,
along the plate. Consequently the ratio of the thickness of the thermal bound-
ary layer δ_t to the viscous δ, namely $\xi = \delta_t/\delta$, is constant and depends only on
the Prandtl number.

Following the method as applied to momentum transport of calculating the
change in the momentum flow within a small layer of height h and width δx,
the heat transported through the plane at x per unit time equals

$$\int_0^h \rho\, c_p\, u\, T \mathrm{d}y$$

where T is the temperature of the fluid. The heat conducted from the element
of the wall per unit time is

$$-\kappa \left.\frac{\partial T}{\partial y}\right|_0 \delta x$$

Proceeding as before, we equate the overall change in the heat transported
through the walls of the element to the heat flow to the surface

$$\kappa \left.\frac{\mathrm{d}T}{\mathrm{d}y}\right|_0 \delta x = \left. \int_0^h \rho\, c_p\, u\, T \,\mathrm{d}y \right|_x - \left. \int_0^h \rho\, c_p\, u\, T \,\mathrm{d}y \right|_{x+\delta x} \tag{7.40}$$

Introducing the temperature difference between the wall and the fluid, $\theta = T - T_W$, making use of the mass conservation relation and assuming ρ and c_p
are constant, we obtain the local heat flux to the wall

$$h_x = \kappa \left.\frac{\mathrm{d}\theta}{\mathrm{d}y}\right|_0 = \frac{\mathrm{d}}{\mathrm{d}x} \int_0^{\delta_t} \rho\, c_p\, (\Theta - \theta)\, u \,\mathrm{d}y \tag{7.41}$$

where Θ is the temperature difference between the wall and the free stream.
The boundary conditions are similar to the viscous case, namely

$$\theta = 0 \quad \text{at} \quad y = 0 \quad \text{and} \quad \theta = \Theta \quad \text{at} \quad y = \delta$$

$$\frac{\partial \theta}{\partial y} = 0 \quad \text{at} \quad y = \delta \quad \text{and} \quad \frac{\partial^2 \theta}{\partial y^2} = 0 \quad \text{at} \quad y = 0$$

The last condition arises because the fluid adjacent to the wall is stationary and therefore not heated.

For the solution we introduce a suitable approximation for the temperature distribution. As before, a cubic in y is appropriate satisfying the above boundary conditions

$$\theta(y) = \Theta \left[\frac{3}{2} \left(\frac{y}{\delta_t} \right) - \frac{1}{2} \left(\frac{y}{\delta_t} \right)^3 \right] \tag{7.42}$$

which gives

$$\Theta U \frac{d}{dx} \int_0^{\delta_t} \left\{ 1 - \frac{3}{2} \left(\frac{y}{\delta_t} \right) + \frac{1}{2} \left(\frac{y}{\delta_t} \right)^3 \right\} \left\{ \begin{array}{ll} \frac{3}{2} \left(\frac{y}{\delta} \right) - \frac{1}{2} \left(\frac{y}{\delta} \right)^3 & \text{if } \left(\frac{y}{\delta} \right) < 1 \\ 1 & \text{otherwise} \end{array} \right\}$$

$$dy = \frac{3}{2} \frac{\chi \Theta}{\delta_t} \tag{7.43}$$

Since the integral is clearly a function of the ratio of the boundary layer thicknesses, $\xi = \delta_t/\delta$, we obtain

$$\delta_t f \left(\frac{\delta}{\delta_t} \right) \frac{d\delta_t}{dx} = \frac{3}{2} \frac{\chi}{U}$$

$$\frac{39}{280} \delta \frac{d\delta}{dx} = \frac{3}{2} \frac{\nu}{U} \tag{7.44}$$

where

$$f(\xi) = \left\{ \begin{array}{ll} \frac{3}{20} \xi - \frac{3}{280} \xi^3 & \text{if } \xi < 1 \\ \frac{3}{8} - \frac{3}{8} \xi^{-1} + \frac{3}{20} \xi^{-2} - \frac{3}{280} \xi^{-4} & \text{otherwise} \end{array} \right. \tag{7.45}$$

If, as we argued earlier, the ratio of the viscous and thermal boundary thicknesses, ξ, is constant along the surface, we may solve these equations subject to the condition that the boundary layer thickness is zero at the start of the plate, $\delta = \delta_t = 0$ at $x = 0$:

$$\delta = \sqrt{\frac{280}{13}} \sqrt{\frac{\nu x}{U}} \quad \text{and} \quad \delta_t = \sqrt{3f(\xi) \frac{\chi x}{U}} \tag{7.46}$$

from which it is clear that the condition $f(\xi) = \text{const}$ is valid. The value of ξ is found from the ratio of the two thicknesses (Figure 7.1)

$$\xi^2 f(\xi) = \frac{39}{280} P^{-1} \tag{7.47}$$

The heat transfer coefficient varies along the plate as the boundary layer thickens. To account for this we introduce a local heat transfer coefficient due

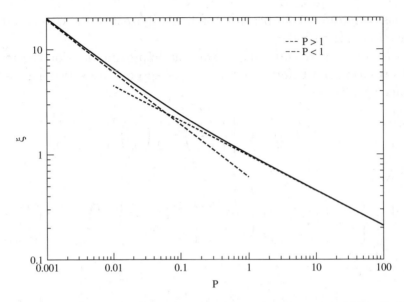

Figure 7.1: Plot of the ratio of the thickness of the laminar thermal boundary layer to that of the viscous ξ as a function of the Prandtl number n. Also shown are the approximations for large and small Prandtl number.

to the surface heat flux

$$h_x = -\kappa \frac{\partial T}{\partial y}\bigg|_0 = \frac{3}{2}\frac{\kappa}{\delta_t} = 0.323 \frac{\kappa}{\xi}\sqrt{\frac{U}{\nu x}}$$

The local Nusselt number is

$$\mathcal{N}_x = 0.323\,\xi^{-1}\sqrt{\mathcal{R}_x}$$

The mean Nusselt number is obtained by integrating along the length L of the plate over both surfaces

$$\overline{\mathcal{N}} = 0.646\xi^{-1}\sqrt{\mathcal{R}_L} \tag{7.48}$$

Two limiting cases are readily found:

1. Thermal conduction weak: $\xi < 1$, $f(\xi) \approx 3/20\,\xi$ and $\delta/\delta_t \approx 1.026\sqrt[3]{\mathcal{P}}$.

   For the range $0.6 < \mathcal{P} < 10$, the mean Nusselt number, $\overline{\mathcal{N}} \approx 0.664 \sqrt[3]{\mathcal{P}}\sqrt[2]{\mathcal{R}_L}$, forms the basis for the Colburn correction.

2. Thermal conduction strong: $\xi \gg 1$, $f(\xi) \approx 3/8$ and $\delta/\delta_t \approx 1.64\sqrt{\mathcal{P}}$.

   The mean Nusselt number $\overline{\mathcal{N}} \approx 1.128 \sqrt[2]{\mathcal{P}}\sqrt[2]{\mathcal{R}_L}$.

7.4   Heat Transfer in a Turbulent Boundary Layer on a Smooth Flat Plate

As we have seen, the velocity distribution across the turbulent boundary layer is in general described by either the law of the wall, equation (5.17), or the more complete form incorporating the buffer layer, equation (5.19). Hence for a smooth plate we may apply von Karman's correction of the Reynolds analogy to the friction factor given by equation (6.42), and enable the local heat transfer coefficient to be calculated. However, it is clear that the resulting expression involves an implicit form for the friction factor, and is not therefore convenient for calculation.

Fortunately a simpler approach is possible. As we have seen, Blasius's power law distribution, Section 6.42, gives a good approximation over the Reynolds number range $5 \times 10^5 < R < 10^7$ where the velocity distribution is

$$u = U \, (y/\delta)^{1/7}$$

with thickness

$$\delta = 0.376 \, x \, (U \, x/\nu)^{-1/5}$$

Applying the Reynolds analogy it follows that the temperature distribution will have the same form as the velocity, i.e.

$$\theta = \Theta \, (y/\delta_t)^{1/7}$$

where θ is the temperature difference between the fluid and the wall. If the Prandtl number $\mathcal{P} = 1$, the thermal boundary layer thickness δ_t is equal to that of the velocity δ. Clearly equation (7.30) is valid in this case also, but with turbulent velocity distribution. It may be simply integrated to give the local heat flux to the wall

$$h_x = 0.0292 \, \rho \, c_p \, U \, \left(\frac{\nu}{U x} \right)^{1/5} \tag{7.49}$$

a result in accordance with the Reynolds analogy.

However, this analysis has assumed that the thickness of the thermal boundary layer is the same as that of the velocity. As we have argued, this is not the case unless the Prandtl number $\mathcal{P} = 1$. The correction for $\mathcal{P} \neq 1$ is equivalent to the Prandtl–Taylor correction and related forms discussed earlier. For this problem, the simple empirical form due to Colburn, which introduces a factor of $\sqrt[3]{\mathcal{P}}$, is convenient, and is valid over the range of Prandtl numbers $0.6 < \mathcal{P} < 50$. The local and complete Nusselt numbers are consequently

$$\mathcal{N}_x = 0.0292 \, \mathcal{P}^{1/3} \mathcal{R}_x^{\,4/5}$$
$$\mathcal{N}_m = 0.0365 \, \mathcal{P}^{1/3} \mathcal{R}_m^{\,4/5} \tag{7.50}$$

More generally it may be assumed that the correction to the boundary layer thickness is a power n of the Prandtl number

$$\frac{\delta_t}{\delta} = \frac{1}{\mathcal{P}n}$$

In this case the Nusselt numbers become

$$\mathcal{N}_x = 0.0292\, \mathcal{P}^{(1-8n/7)}\, \mathcal{R}_x^{4/5}$$

$$\mathcal{N}_m = 0.0365\, \mathcal{P}^{(1-8n/7)}\, \mathcal{R}_m^{4/5}$$

$$(7.51)$$

Colburn's approximation varying as $\mathcal{P}^{1/3}$ corresponds to $n = 0.583$. However, limited experimental data suggests that a better approximation is $n \approx 0.5$ and that the dependence of the Nusselt number on the Prandtl number is as $\mathcal{P}^{0.43}$ Bennett and Myers, 1982, p.374. However, since the range of allowed values of the Prandtl number is relatively small, the differences between these values are consequently also not excessive.

It must be remembered that this equation for the overall Nusselt number is based on the assumption that the boundary layer is turbulent at the start of the plate. In practice this is unlikely to be the case until an initial section of flow, which is laminar, has been established. This extends until the Reynolds number along the plate $\mathcal{R}_x \approx 500\,000$.

7.5   Free or Natural Convection

In free convection the movement of the fluid is due to buoyancy induced by thermal expansion as it is heated. As we saw in Section 1.5.3 the fluid in a gravitational field will be unstable if the temperature gradient exceeds the adiabatic lapse rate. In fact the onset of substantive fluid motion may be inhibited by viscosity. The resultant flow sets up currents in the fluid tending to restore equilibrium, and may be either laminar or turbulent depending on the physical conditions. The moving fluid transports heat at a rate determined by the flow velocity generated, which is itself dependent on the temperature gradient.

In a typical situation heat is applied to a hot plate vertically below a cold surface. The hot fluid at the bottom expands, becomes buoyant and rises to the top where it transfers its heat to the cold surface. On cooling its density increases and the fluid sinks forming a cycle of heating and cooling. A stable rotational system may be established if the temperature gradient is not too large.

As before we may use dimensional analysis to identify the characteristic parameters of this behaviour. The heat transfer is described by the Nusselt number

$$\mathcal{N} = \frac{HL}{\kappa} \tag{7.52}$$

where H is the heat transfer coefficient and L the length scale as before.

The rate at which heat is transferred by free convection depends the buoyancy terms due to thermal expansion β and the gravitational acceleration g and the temperature difference Θ. Clearly the heat transfer rate will also depend on the length scale L, the viscosity ν and the thermal diffusivity (or conductivity) χ. From dimensional analysis it follows that the Nusselt number is a function of

$$\underbrace{\mathcal{G} = \beta\,\Theta\,g\,L^3/\nu^2}_{\text{Grashof no.}} \qquad \underbrace{\mathcal{P} = \nu/\chi}_{\text{Prandtl no.}} \tag{7.53}$$

The physical significance of the Grashof number lies in determining the relative importance of viscous, inertia and buoyancy forces, and plays a similar role to the Reynolds number in forced convection flows. Thus the relative scale of the forces per unit mass due to inertia, viscosity and buoyancy are

$$\left|(\mathbf{v}\cdot\nabla)\mathbf{v}\right| \sim u^2/L \qquad \left|\nu\nabla^2\mathbf{v}\right| \sim \nu\,u/L^2 \qquad \left|\mathbf{g}\,\Delta\rho/\rho\right| \sim \beta\,g\,\Theta$$

respectively.

If the inertia term is comparable with the buoyancy the scaled velocity $u \sim \sqrt{\beta\,g\,L\,\Theta}$, and the viscous term is small if

$$\frac{u^2/L}{\nu u/L^2} \sim \sqrt{\beta\,g\,L^3\,\Theta/\nu^2} \sim \sqrt{\mathcal{G}} \gg 1$$

On the other hand if the viscosity dominates the scaled velocity $u \sim \beta\,g\,L^2\,\Theta/\nu$, and the inertia terms are small if

$$\frac{u^2/L}{\nu u/L^2} \sim \beta\,g\,L^3\,\Theta/\nu^2 \sim \mathcal{G} \ll 1$$

Thus if $\mathcal{G} \gg 1$, viscous force is negligible and inertia balances buoyancy; on the other hand if $\mathcal{G} \ll 1$ the viscous force balances the buoyancy. In particular the ratio of the rates of heat transfer due convection and conduction are

$$\frac{\text{Convection}}{\text{Conduction}} \quad \sim \quad \mathcal{G}^{1/2}\,\mathcal{P} \qquad\qquad \mathcal{G} \gg 1$$

$$\frac{\text{Convection}}{\text{Conduction}} \quad \sim \quad \mathcal{G}\,\mathcal{P} \qquad\qquad \mathcal{G} \ll 1$$

An additional dimensionless product

$$\mathcal{G}\mathcal{P} = \mathcal{R}_a = \beta\,\Theta\,g\,L^3/\nu\,\chi \tag{7.54}$$

is known as the Rayleigh number. If $\mathcal{G}$ is sufficiently small that the viscous force balances the buoyancy and $\mathcal{P}$ is not too small, the heat transfer is expected to depend on $\mathcal{R}_a$ alone. Similarly when $\mathcal{G}$ is large and the convection is turbulent, the Nusselt number is expected to depend on $\mathcal{G}\mathcal{P}^2$.

7.5.1   Boussinesq Approximation

As we have seen in Section 4.6 stratified fluid may be treated by the Boussinesq approximation to take account of the buoyancy terms, whilst remaining essentially incompressible. Free convection may be treated in like manner, thereby reducing the overall complexity of the problem. In this case the density change giving rise to the buoyancy is due to thermal expansion.

Thus if the temperature is written as $T = T_0 + T'$ where T_0 is the background mean temperature and T' the temperature increment, the density increment is

$$\rho' = -\rho_0\,\beta\,T' \tag{7.55}$$

where ρ_0 is the mean density, which is treated as a constant. The pressure $p = p_0 + p'$ where p_0 is the background pressure corresponding to the thermal state (ρ_0, T_0) at the position $\mathbf{r}$ in mechanical equilibrium with the gravitational field $\mathbf{g}$

$$p_0 = \rho_0\,\mathbf{g}\cdot\mathbf{r} + \text{const}$$

Neglecting the second-order terms in the increments, the pressure term in the Navier–Stokes equation (3.13) takes the form

$$\frac{\nabla p}{\rho} \approx \mathbf{g} + \frac{\nabla p'}{\rho_0} + \beta T'\,\mathbf{g}$$

the last term being the buoyancy. Substituting in equation (3.13), the Navier–Stokes equation becomes

$$\frac{\partial\mathbf{v}}{\partial t} + (\mathbf{v}\cdot\nabla)\mathbf{v} = -\frac{\nabla p'}{\rho_0} - \beta T'\,\mathbf{g} + \nu\,\nabla^2\mathbf{v} \tag{7.56}$$

The transport of heat is governed by the entropy equation (3.12). Since the flow is strongly subsonic, the pressure variation due to density changes is extremely small. The appropriate specific heat is that at constant pressure. Therefore

$$T ds \approx \left.\frac{\partial s}{\partial T}\right|_p dT = c_p\,dT$$

In addition the viscous work term is of second order and therefore neglected. Hence the temperature variation is given by

$$\frac{\partial T}{\partial t} + (\mathbf{v} \cdot \nabla)T \approx \chi \nabla^2 T \tag{7.57}$$

Since the flow in the continuity equation is taken to be steady, we finally obtain the *Boussinesq approximation*

$$(\mathbf{v} \cdot \nabla)\,\mathbf{v} = -\frac{1}{\rho}p' - \beta\,T'\,\mathbf{g} + \nu\,\nabla^2\mathbf{v}$$
$$(\mathbf{v} \cdot \nabla)\,T' = \chi\,\nabla^2 T' \tag{7.58}$$
$$\nabla \cdot \mathbf{v} = 0$$

where we have dropped the subscript 0.

Since the Boussinesq approximation is widely used for problems involving buoyancy in both geophysics and meteorology, it is important to clearly identify the approximations that have been made in this derivation:[2]

1. Density changes from the mean value are small due to temperature differences from the volume expansivity, β and scale temperature difference Θ, namely $\beta\,\Theta \ll 1$.
2. Volume changes are small due to compression associated with the bulk compressibility, α and the pressure change due to change in height, namely $\alpha\,L\,g \ll 1$
3. Volume change is small due to heating by energy released by the change in gravitational potential, namely $\beta\,L\,g/c_p \ll 1$.
4. Temperature difference Θ is small compared with the absolute temperature T in the form $\beta\,L\,g\,T/c_p\,\Theta \gg 1$

Equations (7.47) are easily put into dimensionless form by introducing the dimensionless variables

$$\tilde{\mathbf{v}} = \mathbf{v}L/\nu \qquad \tilde{p}' = p'L^2/\rho\nu \qquad \tilde{\theta}' = \theta'/\Theta$$

to give the forms

$$(\tilde{\mathbf{v}} \cdot \tilde{\nabla})\,\tilde{\mathbf{v}} = -\tilde{\nabla}\tilde{p}' - \mathcal{G}\,\tilde{\theta}'\hat{\mathbf{g}} + \tilde{\nabla}^2\tilde{\mathbf{v}}$$
$$(\tilde{\mathbf{v}} \cdot \tilde{\nabla})\,\tilde{\theta}' = \frac{1}{\mathcal{P}}\,\tilde{\nabla}^2\tilde{\theta}' \tag{7.59}$$
$$\tilde{\nabla} \cdot \tilde{\mathbf{v}} = 0$$

where $\tilde{\nabla}$ is the gradient operator related to the scaled distance $\mathbf{r}/L$ and $\hat{\mathbf{g}}$ the unit vector in the direction of gravity. The similarity of flows governed by these equations referred to earlier is clearly seen. If the Boussinesq approximation is not valid additional dimensionless terms play a role.

---

[2]These approximations are discussed in detail by Tritton (1988).

7.5.2 Free Convection from a Vertical Plate

As noted earlier, analytic solutions to the Boussinesq equations for free convection are few. One of the most useful is provided by the application of the
boundary layer methods to free convection of the fluid rising from a heated
plate mounted vertically. Provided the Grashof number is not too large the
flow is laminar and a progressively widening layer of heated fluid forms moving
up and along the plate. Experiments have shown that the onset of turbulence
occurs at a Grashof number $\mathcal{G} \approx 4 \times 10^8$. The flow is similar to that in the normal boundary layer, limited by viscosity, except that the flow is self-generated
rather than driven. The thicknesses of the viscous and thermal structures are
therefore equal in this case.

We will address this problem in two ways directly equivalent to those used
in Chapter 6, namely the direct approach using similarity and the momentum
integral method.

7.5.2.1 Similarity analysis

Applying the approximations made in Section 6.2 to the Boussinesq equations (7.47) we obtain

$$v_x \frac{\partial v_x}{\partial x} + v_y \frac{\partial v_x}{\partial y} = \nu \frac{\partial^2 v_x}{\partial y^2} + \beta g \theta$$

$$v_x \frac{\partial \theta}{\partial x} + v_y \frac{\partial \theta}{\partial y} = \chi \frac{\partial^2 \theta}{\partial y^2} \qquad (7.60)$$

$$\frac{\partial v_x}{\partial x} + \frac{\partial v_y}{\partial y} = 0$$

where x and y are co-ordinates taken parallel and normal to the plate respectively with the origin at the lower end. θ is the temperature difference between
the local fluid and the free stream.

We define the similarity variable

$$\xi = Cy/x^{1/4} \qquad\qquad C = \left[\beta g \Theta / 4 \nu^2 \right]^{1/4} \qquad (7.61)$$

where Θ is the temperature difference between the wall and the free stream.
Introducing the functions $f(\xi)$ and $g(\xi)$, we may write

$$v_x = 4 \nu C \sqrt{x} \dot{f}(\xi) \qquad v_y = \nu C \left[\dot{f}(\xi) - 3 f(\xi) \right] / \sqrt{x} \qquad \theta = \Theta g(\xi) \quad (7.62)$$

which yield on substitution in equations (7.60)

$$\dddot{f} + 3 f \ddot{f} - 2 \dot{f}^2 + g = 0 \qquad \text{and} \qquad \ddot{g} + 3 \mathcal{P} f \dot{g} = 0 \qquad (7.63)$$

The boundary conditions are $f(0) = \dot{f}(0) = 0$, and $g(0) = 1, \dot{f}(\infty) = g(\infty) = 0$. It follows from (7.50) that the thickness of the boundary layer scales as $\delta \sim \sqrt{x}/C$. Since the Prandtl approximation is only valid if the boundary layer is thin, $\delta \ll L$, the length of the plate, it follows that this condition requires that $\mathcal{G}^{1/4} \gg 1$.

Equations (7.52) may be integrated by a standard numerical method. However, the two point boundary conditions introduce complications, which may (for example) be solved by a shooting method to obtain the eigenvalues, which depend on the Prandtl number $\mathcal{P}$. The local Nusselt number is obtained from the thermal conduction heat transfer rate at the surface of the plate

$$\mathcal{N} = -\frac{1}{\kappa\,\Theta} \int_0^L \kappa \left.\frac{\partial\theta}{\partial y}\right|_0 dx = -\frac{4}{3} C\, L^{3/4} \dot{f}(0, \mathcal{P}) = F(\mathcal{P})\, \mathcal{G}^{1/4} \tag{7.64}$$

7.5.2.2 Boundary layer integral approximation

The preceding calculation is accurate provided the Grashof number is large and the flow laminar. However, it is a cumbersome process to calculate the numerical values for arbitrary values of the Prandtl number. Fortunately the boundary layer integral method provides a good approximation to the above method and yields an analytic solution in a closed form. The method is based on von Karman's momentum integral method for the viscous boundary layer, Section 6.4. It therefore follows a simple physical picture in which conservation of momentum and conservation of energy are directly applied across the layer.

The velocity is zero both at the wall and external to the layer (i.e. the free stream); the viscous force at the edge is also zero $u(0) = u(\delta) = du/dy|_\delta = 0$. The temperature difference between the fluid in the boundary layer and external to the layer θ is Θ at the wall, $\theta(0) = \Theta$. At the boundary layer edge, the temperature difference and the heat flow are both zero, $\theta(\delta) = d\theta/dy|_\delta = 0$. There is a further boundary condition which follows from the first equation of the set (7.49), namely $\partial^2 u/\partial y^2 = -\beta\,g\,\Theta$. Since the fluid flow is directly caused by the temperature difference, the thickness of the thermal and velocity layers must be equal, δ.

The momentum equation is directly obtained from equation (6.25) with the pressure gradient term replaced by the buoyancy and noting that the free stream velocity is zero

$$\frac{d}{dx}\int_0^\delta \rho\,u(y)^2\,dy = \beta\,g\int_0^\delta \theta\,dy - \nu\left.\frac{du}{dy}\right|_0 \tag{7.65}$$

where δ is the thickness of the boundary layer and the integrated heat flow

$$\frac{\mathrm{d}}{\mathrm{d}x} \int_0^\delta u\,\theta\,\mathrm{d}y = -\chi \left.\frac{\mathrm{d}\theta}{\mathrm{d}y}\right|_0 \tag{7.66}$$

The simplest approximate profiles for the temperature and velocity across the layer are simple polynomials satisfying the boundary conditions

$$\theta = \Theta \left(1 - \frac{y}{\delta}\right)^2 \qquad \text{and} \qquad u = U\frac{y}{\delta}\left(1 - \frac{y}{\delta}\right)^2 \tag{7.67}$$

where $U(x)$ is an unknown scaling velocity to be evaluated. The maximum velocity is easily shown to be $4\,U/27$ at $y = \delta/3$.

The integrals across the layer are easily evaluated to give

$$\frac{1}{105}\frac{\mathrm{d}}{\mathrm{d}x}\left(U^2\,\theta\right) = \frac{1}{3}\beta g\Theta\delta - \nu\frac{U}{\delta} \qquad \text{and} \qquad \frac{1}{30}\Theta\frac{\mathrm{d}}{\mathrm{d}x}(U\,\delta) = 2\chi\frac{\Theta}{\delta} \tag{7.68}$$

To solve these equations we try power law variations along the plate

$$U = C_1 x^m \qquad \text{and} \qquad \delta = C_2 x^n \tag{7.69}$$

which when evaluated yield the following results

$$
\begin{aligned}
m &= \frac{1}{2} \qquad \text{and} \qquad C_1 = 5.17\nu\left(\frac{20}{21} + \frac{\nu}{\chi}\right)^{-1/2}\left(\frac{\beta g\Theta}{\nu^2}\right)^{1/2} \\
n &= \frac{1}{4} \qquad \text{and} \qquad C_2 = 3.93\left(\frac{20}{21} + \frac{\nu}{\chi}\right)^{1/4}\left(\frac{\beta g\Theta}{\nu^2}\right)^{-1/4}\left(\frac{\nu}{\chi}\right)^{-1/2}
\end{aligned}
\tag{7.70}
$$

Substituting these values into equations (7.56) and introducing the Prandtl and Grashof numbers, the maximum velocity at distance x is

$$\frac{x\,u_{\max}(x)}{\nu} = 0.766\,(0.952 + \mathcal{P})^{-1/2}\,\mathcal{G}_x^{1/2} \tag{7.71}$$

and the layer thickness

$$\frac{\delta}{x} = 3.93\,\mathcal{P}^{-1/2}\,(0.952 + \mathcal{P})^{1/4}\,\mathcal{G}_x^{-1/4} \tag{7.72}$$

where $\mathcal{G}_x$ is the Grashof number based on the length x.

The local Nusselt number is

$$\mathcal{N}_x = \frac{x}{\kappa\Theta}\left(-\kappa\left.\frac{\partial\theta}{\partial y}\right|_0\right) = 2\frac{x}{\delta} = 0.508\,\mathcal{P}^{1/2}\,(0.952 + \mathcal{P})^{-1/4}\,\mathcal{G}_x^{1/4} \tag{7.73}$$

   Assuming the temperature is constant over the plate, we may integrate along
the plate to obtain the overall Nusselt number

$$\mathcal{N}_L = 0.677\,\mathcal{P}^{1/2}\,(0.952 + \mathcal{P})^{-1/4}\,\mathcal{G}^{1/4} \tag{7.74}$$

Figure (7.2) shows a comparison of the overall Nusselt number calculated
from the exact integration of the Boussinesq approximation quoted by
Schlichting (1968) with values from the boundary integral approximation
equation (7.63). The agreement is within 10% over the complete range of the
Prandtl number, which is satisfatory for most purposes.

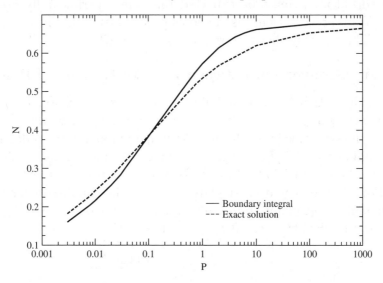

Figure 7.2: Comparison of the values of the overall Nusselt number for free convection
from a vertical plate calculated from equation (7.63) and by integration of the exact
equations (7.52) from Schlichting (1968) as a function of the Prandtl number $\mathcal{P}$.

7.5.3   Free Convection from a Heated Horizontal Plate

It may be expected that there is a marked difference between the cases where
the hot surface is uppermost or lower. If the hot surface faces upwards the
convection currents develop freely above the surface. On the other hand if
the hot surface faces downwards convection is inhibited. In fact the Nusselt number for plates facing upwards is about twice that for those facing
down. Experimentally it is found that if the Rayleigh number is not too high
($10^5 < \mathcal{R}_a < 2 \times 10^7$) the Nusselt number is well approximated by

$$\mathcal{N} \approx C\mathcal{R}_a^{1/4} \tag{7.75}$$

where $C \approx 0.54$ for plates facing upwards, and $C \approx 0.27$ for those facing down.[3]

---

[3]The scale length of the plate is taken as the ratio of the plate surface to the perimeter.

At larger Rayleigh numbers $(2 \times 10^7 < \mathcal{R}_a < 3 \times 10^{10})$ the flow above the top heated plate becomes turbulent and the Nusselt number becomes

$$\mathcal{N} \approx C' \mathcal{R}_a^{1/3} \tag{7.76}$$

where $C' \approx 0.14$. For downward-facing plates equation (7.64) is still valid.

The behaviour exhibited in the change of power from $1/4$ to $1/3$ as the Rayleigh number increases is characteristic of free convection. The latter behaviour is relatively easy to understand in that once the flow becomes strongly turbulent, it must be expected that the experimentally observable, the heat transfer coefficient, is no longer dependent on the characteristic length L. This is easily shown to require that $\mathcal{N} \sim \mathcal{G}^{1/3}$.

7.5.4 Free Convection between Parallel Horizontal Plates

If we consider the behaviour of fluid between two horizontal plates whose temperatures T_1 and T_2 differ, the lower plate being hotter than the upper $(T_1 > T_2)$, the system is in equilibrium with no fluid motion provided the temperature difference $T_1 - T_2$ between the plates is small. This equilibrium is stable with the viscosity balancing the buoyancy induced by the expansion of the fluid due to the heated lower plate $T_1 > T_2$. The stability depends on the Rayleigh number calculated from the temperature difference $(T_1 - T_2)$ and the plate separation d; that is, the fluid is stable if

$$\mathcal{R}_a < \mathcal{R}_{a \text{ crit}} \approx 1708 \tag{7.77}$$

see Figure 7.4 which shows the variation of the Nusselt number as convection is initiated, following the onset of the instability. In this limit, when equilibrium is disturbed, regions of rising and falling fluid occur (Figure 7.3). Fluid heated by the lower plate rises up to the top plate where it is cooled and falls again. A continuous cycle of circular flow is established with horizontal motion at the top and bottom. For Rayleigh numbers not too much above critical $(\mathcal{R}_a \lesssim 5 \times 10^3)$ the cells are well established and take the form of long rolls of rising and falling fluid as sketched in Figure 7.3.

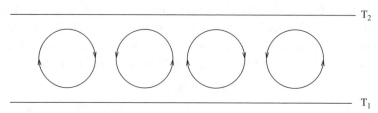

Figure 7.3: Sketch of the convection currents established at Rayleigh numbers just greater than critical $\mathcal{R}_a > \mathcal{R}_{a \text{ crit}}$. The flow forms in Bénard cells–long cylindrical rolls.

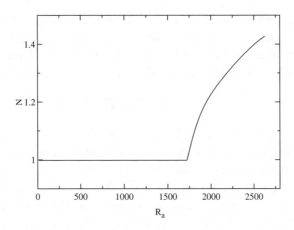

Figure 7.4: Sketch of the variation of the Nusselt number near the onset of free convection between parallel plates.

As the Rayleigh number increases ($\mathcal{R}_a \lesssim 2 \times 10^4$) the cells become less well-ordered three-dimensional structures whilst still remaining steady. Further increases of $\mathcal{R}_a$ lead to the development of periodic variations, which give way to less regular variations. Convection cells can still be observed, although the flow within is turbulent, at Rayleigh numbers well above critical ($\mathcal{R}_a \lesssim 10^6$) (Figure 7.5). Tritton (1988) contains a full account of the formation and development of Bénard cells illustrated with excellent photographs, and should be consulted for further information.

At Rayleigh numbers well above threshold ($\mathcal{R}_a \gtrsim 10^5$) only fluid passing close to the lower boundary is heated. The resulting hot rising regions are quite narrow, and so are the cold falling ones (figure 7.5). Much of the fluid circulates isothermally as turbulence mixes it. Turbulence is generated within the narrow hot and cold boundary regions.

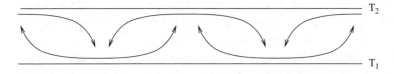

Figure 7.5: Sketch of the convection currents established at Rayleigh numbers much greater than critical $\mathcal{R}_a \sim 10^6$. The flow forms in turbulent cells of rising and falling fluid.

At still larger Rayleigh numbers ($\mathcal{R}_a \gtrsim 10^6$) the cellular structure disappears and we have the region known as turbulent convection. The temperature gradient (averaged in time) is large at the boundaries, but the fluid is nearly

isothermal at the centre. In the centre the fluid efficiently transports heat, but
the boundaries constrain the eddy motion, and large temperature gradients
are necessary to support the heat flux by conduction. In this regime each plate
acts independently as a heated plate, the scaling changes to the $1/3$ power and

$$\mathcal{N} \approx 0.069 \mathcal{R}_a^{1/3} \mathcal{P}^{0.074} \qquad (7.78)$$

has been proposed for the range $3 \times 10^5 < \mathcal{R}_a < 7 \times 10^9$.

If the Rayleigh number is less than critical or the hot plate is the top surface,
heat transfer is by thermal conduction alone and $\mathcal{N} = 1$.

An important practical situation arises from the scaling of the critical
Rayleigh number. Given a fixed separation of the surfaces and a fixed
temperature difference, the heat flow between the surfaces can be minimised
by reducing the plate separation. Indeed if the spacing is limited by inserting
a series of intermediate plates, the separation of each gap and the temperature
difference are both diminished, reducing the Rayleigh number. If this is
reduced below the critical value, heat transfer is due to conduction alone and
thereby minimised by eliminating convection. This simple technique is the
basis of insulating fibrous and porous material as well as metal foil insulation.

7.5.4.1   Rayleigh–Bénard instability

The instability between parallel horizontal plates of an incompressible fluid in a
temperature gradient can be investigated within the Boussinesq approximation
by considering small sinusoidal perturbations in the (x, y) directions with temporal variation $\exp(-\imath \omega t)$ where ω is complex, so that $\Im(\omega)$ represents growth
or decay of the wave. Just below the threshold of the instability, the fluid is at
rest and therefore $\omega = 0$. Thus as we approach and finally achieve the condition
of marginal stability at threshold the fluid is still just stationary. Under the
condition of marginal stability, a stationary pattern of Bénard cells is established. At the onset of the instability we may therefore assume that the flow
is stationary and $\omega = 0$, which greatly simplifies the analysis. This condition,
known as the *principle of exchange of stabilities*, can be formally justified from
the approximations and the boundary conditions (see Chandrasekhar, 1981).

The boundary conditions on the flow are that, at the solid surface, the perturbations associated with the instability in temperature and normal velocity
are zero. In addition since the flow has a no-slip condition at the walls, it follows from the continuity condition that $\partial v_z / \partial z = 0$ at the walls, where z is
taken across the duct.

The initial conditions across the duct are that the velocity is zero, and the
temperature varies linearly as $-Az$. The perturbation is described by the perturbation terms of velocity $\mathbf{v}'$, pressure p' and temperature T'. Substituting

into the Boussinesq equations (7.47) and neglecting second-order terms as usual, we obtain

$$\nu \nabla^2 \mathbf{v}' = \left(\frac{\nabla p'}{\rho}\right) + \beta g T'$$

$$\chi \nabla^2 T' = -A v'_z \qquad \nabla \cdot \mathbf{v}' = 0 \tag{7.79}$$

Making use of the second equation, the boundary conditions on the velocity component v'_z can be reduced to ones on higher order derivatives of T'.

We eliminate the term in p' from the first equation by making use of the vector operation $\nabla \wedge \nabla \wedge \equiv \nabla \nabla \cdot - \nabla^2$ to obtain

$$\nu \nabla^4 \mathbf{v}' = -\beta \left(\nabla^2 T' - \frac{\partial^2 T'}{\partial z^2}\right) \mathbf{g} \tag{7.80}$$

or for the z component

$$\nabla^4 v_z = \frac{\beta g}{\nu}\left(\frac{\partial^2 T'}{\partial x^2} + \frac{\partial^2 T'}{\partial y^2}\right) \tag{7.81}$$

Assuming the perturbation has a sinusoidal variation in (x, y),

$$T'(x, y, z) = T(z)\exp\left(\imath \mathbf{k} \cdot \mathbf{r}\right)$$
$$v'_z(x, y, z) = W(z)\exp\left(\imath \mathbf{k} \cdot \mathbf{r}\right)$$

where $\mathbf{k}$ is a vector in the (x, y) plane. Introducing the dimensionless distance $\zeta = z/d$ with boundaries at $\zeta = \pm 1/2$, and $a = kd$, we obtain

$$\left(\frac{d^2}{d\zeta^2} - a^2\right) T = -\frac{A d^2}{\chi} W$$

$$\left(\frac{d^2}{d\zeta^2} - a^2\right)^2 W = \frac{a^2 d^2 \beta g}{\nu} T \tag{7.82}$$

Hence eliminating either W or T, we obtain identical differential equations for T and W

$$\left(\frac{d^2}{d\zeta^2} - a^2\right)^3 T + \mathcal{R}_a a^2 T = 0$$

$$\left(\frac{d^2}{d\zeta^2} - a^2\right)^3 W + \mathcal{R}_a a^2 W = 0 \tag{7.83}$$

whose solutions subject to the boundary conditions determine the value of the wavenumber corresponding to marginal stability for a given Rayleigh number. The boundary conditions can be written as

$$T = \left(\frac{d^2}{d\zeta^2} - a^2\right) T = \frac{d}{d\zeta}\left(\frac{d^2}{d\zeta^2} - a^2\right) T = 0 \qquad \text{at} \quad \zeta = \pm\frac{1}{2}$$

or

$$W = \frac{\mathrm{d}W}{\mathrm{d}\zeta} = \left(\frac{\mathrm{d}^2}{\mathrm{d}\zeta^2} - k^2\right)^2 W = 0 \qquad \text{at} \quad \zeta = \pm\frac{1}{2}$$

We thus have two ordinary differential equations in the two linked variables T and W either of which can be used to find the result. Since the equations are homogeneous of sixth order and we have six boundary conditions, the problem resolves to the determination of the eigenvalue for a particular value of the Rayleigh number $\mathcal{R}_a$. To find this we note that the solution may be either even or odd in ζ and that solutions have the form $\exp(\alpha\,\zeta)$. Substituting we find the six roots of α

$$\alpha^2 = -a^2(\tau - 1) \qquad \text{and} \qquad \alpha^2 = a^2[1 + 1/2\,\tau\,(1 \pm \imath\sqrt{3})] \qquad (7.84)$$

where $\tau = \sqrt[3]{\mathcal{R}_a/a^4}$. Linear combinations of these solutions generate the general solutions, the eigenfunction being the one matching the boundary conditions for the given value of $\mathcal{R}_a$, and found by solving a transcendental equation for a in terms of τ. Hence $\mathcal{R}_a$ is obtained. The marginal solution corresponds to the lowest value of $\mathcal{R}_a$ for which a solution exists, i.e. the first possibility to arise as the instability is approached. This problem is treated in detail by Chandrasekhar (1981) and by Drazin and Reid (1981).

7.5.5 Free Convection around a Heated Horizontal Cylinder

The convective heat flow from a long cylinder of diameter D is basically similar to that from a flat plate. Simple power law correlations are available to give good empirical values for the heat loss, namely

$$\mathcal{N} = C\,\mathcal{R}_a^n \qquad (7.85)$$

$\mathcal{N}$ and $\mathcal{R}_a$ being calculated using the cylinder diameter D. The best fit constants C and n are found empirically to be given by the values in Table 7.2.

Table 7.2: Parameters for heat flow across a cylinder.

$\mathcal{R}_a$	$1.0^{-10}-10^{-2}$	$10^{-2}-10^2$	10^2-10^4	10^4-10^7	10^7-10^{12}
C	0.675	1.02	0.850	0.480	0.125
n	0.058	0.148	0.188	0.250	0.333

Case study 7.I Positive Column of an Arc

A somewhat surprising application of these approximations allows the development of simple scaling relations to describe the positive column of a high-pressure, horizontal, unconfined arc (Suits and Poritsky, 1939; Cobine, 1985). An arc is a plasma column normally carrying a current of several amps with a low potential drop of a few tens of volts in gas at about atmospheric pressure. The regions associated with the electrodes are relatively small. Most of the heat is generated in the column, which may be several centimetres long between the electrode regions known as the *positive column*. A long positive column is stretched upwards by strong convection currents arising from the strong heating. The positive column is formed from plasma at a temperature of about 5000°C, which is highly conducting and self-sustaining. It approximately forms a cylinder about 1 centimetre in diameter.

It can be argued that the major heat loss from the positive column is due to strong convection currents generated in the surrounding gas by the large temperature difference between it and the plasma core. Typically less than 10% of the heat is estimated to be lost by direct convection of core gases. A simple model of the convection is therefore to consider the hot core as equivalent to a solid cylinder. Assuming that the results of the previous section may be extrapolated to the high temperatures involved, the convective heat loss is given by equation (7.74) with values of n given in Table 7.2.

The heat loss per unit length of the cylindrical core must balance the heat generated by ohmic heating, namely Ei where E is the electric field and i the current. Thus we may write

$$Ei = \pi\,D\,h = \pi\,\kappa\,\Theta\,C\,\mathcal{R}_a{}^n$$

$$= \pi\,\kappa\,\Theta\,C\left(\frac{D^3 M^2 p^2 g\,\Theta}{\mathcal{P}\,\mu^2\,R_g{}^2\,T^3}\right)^n \tag{7.86}$$

where D is the diameter of the plasma core, Θ is the temperature between the core and the background gas, M is the molecular mass, R_g the gas constant, and we have used the ideal gas equation for the density $\rho = p\,M/R_g\,T$ and the thermal expansivity $\beta = 1/T$ appropriate to an ideal gas. We also make use of the result that the Prandtl number for a gas is a constant (≈ 0.73 for air).

One further result from the theory of low-temperature plasmas is required: that is, the drift velocity of the current carriers is proportional to the field. Since the electrons are the principal current carriers and $\bar{v}_e = \mu_e E$, where μ_e is the electron mobility, the current is therefore

$$i = \frac{\pi D^2}{4} n_e e \mu_e E \tag{7.87}$$

The electron density n_e in the core depends only on the temperature and can be treated as constant. Hence

$$\frac{i}{E} = \text{const}\,(E\,i)^{2/3\,n}$$

Eliminating the diameter D we obtain the current/voltage relation for the arc, namely

$$E = \text{const}\,i^{(2-3n)/(2+3n)} \tag{7.88}$$

which is the well-known negative resistance characteristic of the arc. For an arc burning in air at a temperature ~ 5000°C, the film temperature is approximately 2500°C and the kinematic viscosity $\nu \sim 5 \times 10^4\,\mathrm{m^2/s}$. Thus for an arc diameter of 10^{-2}, the Grashof number $\mathcal{G} \sim 75$, and the Prandtl number $\mathcal{P} \approx 0.73$, it is estimated that the parameter $n \approx 0.1$ for the arc in air, in which case the power of the characteristic is -0.74, compared with a value of -0.60 experimentally. The agreement is satisfactory given the crudeness of the approximations.

The model can also be applied to the variation of diameter D and field E with external pressure p. The agreement is acceptable. For further details see Suits and Poritsky (1939) or Cobine (1985).

Confirmation of the importance of convection has been provided by the behaviour of arcs in a gravity-free environment. As a result the voltage drop and the current are reduced by the absence of strong cooling.

Chapter 8

Compressible Flow and Sound Waves

8.1   Introduction

The condition for the fluid to be treated as incompressible is evidently that the density changes are small, $\Delta\rho$, i.e.

$$\frac{\Delta\rho}{\rho} \ll 1$$

However, in supersonic flow, it will be shown that the flow is approximately adiabatic and therefore

$$\Delta p \approx \left.\frac{\partial p}{\partial \rho}\right|_S \Delta p = c^2 \Delta\rho$$

where c is the adiabatic sound speed. We note that since $\partial p/\partial \rho\,|_s$ is positive in all materials, the sound speed is a real quantity. In steady flow it follows from Bernoulli's equation that the pressure changes satisfy the condition

$$\Delta p < \rho v^2$$

where v is the flow speed. Hence we obtain the condition for steady incompressible flow

$$\frac{\Delta\rho}{\rho} < v^2 \left.\frac{\partial \rho}{\partial p}\right|_S = \frac{v^2}{c^2} = M^2$$

Thus in subsonic flow, Mach number $M = v/c \ll 1$, the flow may be treated as incompressible. In fact, as we shall see (Section 12.2), we may treat flows

Introductory Fluid Mechanics for Physicists and Mathematicians, First Edition. Geoffrey J. Pert.
© 2013 John Wiley & Sons, Ltd. Published 2013 by John Wiley & Sons, Ltd.

with $M \lesssim 1$ by a perturbation of the incompressible solution. Compressibility effects are therefore predominantly associated with flows, where $v \gtrsim c$.

If the flow is non-steady, we may extend the preceding discussion to include flows which change over a time τ and scale length ℓ. If the time-dependent terms dominate in Euler's equation, we have

$$\frac{\partial v}{\partial t} \sim -\frac{1}{\rho}|\nabla p| \quad \text{or} \quad \frac{v}{\tau} \sim \frac{1}{\rho}\frac{\Delta p}{\ell} \quad \text{and} \quad \frac{\Delta \rho}{\rho} \sim \frac{\ell v}{\tau c^2}$$

Hence the condition for incompressibility is $\tau \gg \ell/c$, i.e. the time taken for pressure waves moving at the sound speed to pass through the characteristic region of the flow. Taking these results together we find that the condition for incompressible flow to be a good approximation is

$$\text{Max}\{v, \ell/\tau\} \ll c \tag{8.1}$$

We now show that sonic and supersonic flows are generally, although not entirely, dissipationless. Dissipation is due to two effects:

1. Viscosity. The importance of viscosity within the bulk of the flow is determined by the Reynolds number $\mathcal{R} \gtrsim \ell c/\nu$, where we have used the sound speed instead of the flow velocity since $v \gtrsim c$ in compressible flow. Using results from kinetic theory, the kinematic viscosity $\nu \approx \frac{1}{3}\bar{c}\lambda$ where $\bar{c}$ is the mean thermal speed of the molecules and λ their mean free path. Hence

$$\mathcal{R} \sim \frac{c\,\ell}{\bar{c}\,\lambda} \sim \frac{\ell}{\lambda} \gg 1 \tag{8.2}$$

since the fluid approximation is only valid for scale lengths $\ell \gg \lambda$. Thus viscosity is only important for supersonic flows $M > 1$ over distances of the order of the mean free path or less.

2. Thermal conduction. The importance of heat transport in the fluid by thermal conduction is determined by the ratio of the heat flux due to thermal conduction $\kappa T/\ell$ to the flux of kinetic energy ρv^3

$$\frac{\kappa T}{\ell \rho v^3} = \frac{1}{\mathcal{PR}}\frac{c^2}{v^2} \sim \frac{1}{\mathcal{PR}} \ll 1 \tag{8.3}$$

where κ is the thermal conductivity, $c_p T \sim c^2 \lesssim v^2$ and the Prandtl number for compressible fluids, namely gases, $\mathcal{P} = \mu c_p/\kappa \sim 1$. Hence we conclude that thermal conduction is only significant when viscosity is also.

Consequently sonic and supersonic flows are essentially dissipationless. Viscosity and thermal conduction normally only play a role in some minor situations, such as the damping of sound waves, apart from the shock layer (Section 10.5).

8.2   Propagation of Small Disturbances

Let us consider a uniform fluid flowing with constant velocity $\mathbf{v_0}$ and with specified pressure p_0, density ρ_0. A small disturbance is introduced at some point in the fluid, and propagates through the flow. Consider small increments to the velocity, pressure and density respectively

$$\mathbf{v} = \mathbf{v_0} + \mathbf{v'} \qquad \rho = \rho_0 + \rho' \qquad p = p_0 + p'$$

Since the flow is dissipationless (ideal) Euler's equation holds and

$$\frac{\partial \mathbf{v}}{\partial t} + (\mathbf{v} \cdot \nabla)\mathbf{v} = -\frac{1}{\rho}\nabla p$$

and is complemented by the equation of continuity

$$\frac{\partial \rho}{\partial t} + \nabla \cdot (\rho \mathbf{v}) = 0$$

We transform to the frame moving with the fluid. Since the disturbance is a small perturbation, we may linearise the equations to give

$$\frac{\partial \rho'}{\partial t} + \rho_0 \nabla \cdot \mathbf{v'} = 0 \tag{8.4}$$

$$\frac{\partial \mathbf{v'}}{\partial t} + \frac{1}{\rho_0}\nabla p' = 0 \tag{8.5}$$

where $\mathbf{v'}$, p' and ρ' are the increments of velocity, pressure and density respectively.

Introducing the adiabatic sound speed since the flow is dissipationless

$$p' = \left.\frac{\partial p}{\partial \rho}\right|_s \rho'$$

equation (8.4) becomes

$$\frac{\partial p'}{\partial t} + \rho_0 \left.\frac{\partial p}{\partial \rho}\right|_s \nabla \cdot \mathbf{v'} = 0 \tag{8.6}$$

Introducing the velocity potential $\mathbf{v'} = \nabla\phi'$ we obtain from equation (8.5) by integration

$$p' = -\rho_0 \frac{\partial \phi'}{\partial t}$$

since $p' = 0$ if $\phi' = 0$, i.e. there is no disturbance. Thus

$$\frac{\partial^2 \phi'}{\partial t^2} - c^2 \nabla^2 \phi' = 0 \tag{8.7}$$

where the adiabatic sound speed

$$c = \sqrt{\left.\frac{\partial p}{\partial \rho}\right|_S} = \sqrt{\frac{\gamma p}{\rho}}$$

for a polytropic gas. As noted earlier this is always a real quantity, so that wave solutions exist.

The disturbance propagates through the fluid with speed c relative to the fluid, and the velocity of the wave is transformed back in the laboratory frame (see Section 9.1). These wave perturbations are the familiar sound waves.

8.2.1   Plane Waves

Suppose the disturbance is uniform in the plane (yz); then the governing differential equation becomes

$$\frac{\partial^2 \phi'}{\partial x^2} - c^2 \frac{\partial^2 \phi'}{\partial t^2} = 0 \tag{8.8}$$

whose general solution is

$$\phi' = f(x - ct) + g(x + ct) \tag{8.9}$$

representing forward $(\rightarrow +x)$ and backward $(\rightarrow -x)$ propagating waves in the x direction respectively. Taking the gradient we see that the only non-zero component is $\partial \phi'/\partial x$ representing a velocity perturbation in the x direction, i.e. parallel to the direction of propagation. Sound waves are therefore longitudinal, as is well known.

In a forward-travelling wave $v_x = \dot{f}(x - ct)$, the velocity perturbation may be simply expressed in terms of the pressure and density fluctuations

$$v' = \frac{p'}{\rho_0 c} = \frac{c \rho'}{\rho_0} \tag{8.10}$$

The ratio $z = p'/v' = \rho_0 c$ is known as the specific acoustic impedance. The temperature perturbation

$$T' = \left.\frac{\partial T}{\partial p}\right|_s p' = \frac{T}{c_p} \left.\frac{\partial V}{\partial T}\right|_p \rho_0 c v' = \frac{c \alpha T v'}{c_p} \tag{8.11}$$

where α is the volumetric coefficient of expansion $(1/V)\, \partial V/\partial T$.

8.2.2   Energy of Sound Waves

The energy density of the fluid is given by the sum of the internal and kinetic energies per unit volume, namely $\rho\left(\epsilon + \frac{1}{2}v^2\right)$. Thus retaining only the lowest order terms in the perturbation, the energy density is

$$\rho_0\,\epsilon_0 + \rho'\,\frac{\partial\left(\rho\,\epsilon\right)}{\partial\rho} + \frac{1}{2}\rho'^2\,\frac{\partial^2\left(\rho\,\epsilon\right)}{\partial\rho^2} + \frac{1}{2}\rho_0\,v'^2$$

Since a sound wave is adiabatic, all derivatives are taken at constant entropy. The second term may be rewritten using the first law of thermodynamics, $d\epsilon = T\,ds - p\,dV$, in terms of the enthalpy as

$$\left.\frac{\partial\left(\rho\,\epsilon\right)}{\partial\rho}\right|_s = \epsilon + \frac{p}{\rho} = h$$

$$\left.\frac{\partial^2\left(\rho\,\epsilon\right)}{\partial\rho^2}\right|_s = \left.\frac{\partial h}{\partial\rho}\right|_s = \left.\frac{\partial h}{\partial p}\right|_s\left.\frac{\partial p}{\partial\rho}\right|_s = \frac{c^2}{\rho}$$

and hence the energy density becomes

$$\rho_0\,\epsilon_0 + h_0\,\rho' + \frac{1}{2}\frac{c^2\,\rho'^2}{\rho_0} + \frac{1}{2}\rho_0\,v'^2$$

The first term is the constant energy density of the ambient fluid, and the second term, the change of energy as the density changes, averages to zero as there is no net mass variation in the ambient fluid. Neither of these terms therefore contribute to the average energy density, which is given by the integral over the volume. We may therefore regard the remaining terms as the energy density of the sound wave, namely

$$E = \frac{1}{2}\rho_0\,v'^2 + \frac{1}{2}\frac{c^2\,\rho'^2}{\rho_0} \tag{8.12}$$

In a travelling wave these two terms are equal so that $E = \rho_0\,v'^2 = c^2\,\rho'^2/\rho_0$, but this result is not true in general. However, for a general set of disturbances, this result holds for the averaged energy density. This is an expression of the general theorem that the mean potential and mean kinetic energies of an oscillating system are equal, provided the amplitude is small.

To calculate the energy flux $\mathbf{q}$ we make use of energy conservation in the form

$$\frac{\partial E}{\partial t} = \rho_0\,\mathbf{v}'\cdot\frac{\partial\mathbf{v}'}{\partial t} + \frac{c^2}{\rho_0}\rho'\frac{\partial\rho'}{\partial t} = -\mathbf{v}'\cdot\nabla p' - p'\,\nabla\mathbf{v}' = -\nabla\cdot\left(p'\,\mathbf{v}'\right) = -\nabla\cdot\mathbf{q} \tag{8.13}$$

and hence the energy flux

$$\mathbf{q} = p' \, \mathbf{v}' \qquad (8.14)$$

In a travelling wave, the pressure increment $p' = c \, \rho_0 \, v'$ and the energy flux is parallel to the velocity perturbation, whose direction is given by unit vector $\hat{\mathbf{n}}$, so that

$$\mathbf{q} = c \, E \, \hat{\mathbf{n}} \qquad (8.15)$$

thus yielding the familiar result that the energy flux is the product of the energy density and the group velocity of the wave.

8.3   Reflection and Transmission of a Sound Wave at an Interface

A sound wave is incident at an angle (to the normal) θ_1 on a plane interface between fluid at density ρ_1 and sound speed c_1, and fluid at ρ_2 and c_2. Along the interface the pressure and normal component of velocity must be unchanged in the two media. Since the media are assumed to be isotropic, it is evident from the symmetry of the system that all waves generated must lie in the plane of incidence. At the interface the incoming wave may generate two outgoing waves, the backward-going reflected wave and the forward-going transmitted wave. Since the boundary conditions do not depend on time or the tangential components, it follows that the frequency and the tangential projection of the wave vector parallel to the interface must be unchanged. The angle that the reflected wave makes with the normal is $\pi - \theta_1$ and that of the forward-going refracted transmitted wave θ_2. Hence since $k_\parallel = (\omega/c) \sin \theta$ we have that

$$\sin \theta_1 / \sin \theta_2 = c_1/c_2 \qquad (8.16)$$

Taking the plane (x, y) to be that of the incident wave, the incident, reflected and transmitted waves take the form

$$
\begin{aligned}
\phi_i &= A_i \exp \{\imath \omega \, [\ (x/c_1) \cos \theta_1 + (y/c_1) \sin \theta_1 - t]\} \\
\phi_r &= A_r \exp \{\imath \omega \, [-(x/c_1) \cos \theta_1 + (y/c_1) \sin \theta_1 - t]\} \\
\phi_t &= A_t \exp \{\imath \omega \, [\ (x/c_2) \cos \theta_2 + (y/c_2) \sin \theta_2 - t]\}
\end{aligned}
\qquad (8.17)
$$

The total velocity potential on the upstream side is $\phi_i + \phi_r$ and on the downstream side ϕ_t. Therefore noting that the pressure perturbation $p = -\rho \, \partial\phi/\partial t$ and that the normal velocity perturbation $v_x = \partial\phi/\partial x$, we obtain, from the pressure and normal velocity perturbations on each side of the interface, the amplitudes of the reflected and transmitted waves in terms of the incident wave

$$\rho_1 \, (A_i + A_r) = \rho_2 \, A_t \qquad \text{and} \qquad (A_i - A_r) \cos \theta_1/c_1 = A_t \cos \theta_2/c_2 \qquad (8.18)$$

Solving for A_t/A_i and A_r/A_i gives the amplitude transmission and reflection coefficients

$$\frac{A_t}{A_i} = \frac{2\,\rho_1\,c_2\,\cos\theta_1}{\rho_2\,c_2\,\cos\theta_1 + \rho_1\,c_1\,\cos\theta_2} \quad \text{and} \quad \frac{A_r}{A_i} = \frac{(\rho_2\,c_2\,\cos\theta_1 - \rho_1\,c_1\,\cos\theta_2)}{(\rho_2\,c_2\,\cos\theta_1 + \rho_1\,c_1\,\cos\theta_2)}$$

(8.19)

If $\rho_1\,c_1\,\cos\theta_2 > \rho_2\,c_2\,\cos\theta_1$, $A_r < 0$, and there is a change of phase on reflection, the compression part of the cycle reflecting as an expansion.

The normal energy flux at the surface for each wave is given by $\left(\rho\omega^2\,A/c\right)\cos\theta$, hence the intensity coefficients for reflection R and transmission T are

$$R = \left\{\frac{(\rho_2\,c_2\,\cos\theta_1 - \rho_1\,c_1\,\cos\theta_2)}{(\rho_2\,c_2\,\cos\theta_1 + \rho_1\,c_1\,\cos\theta_2)}\right\}^2 = \left\{\frac{(\rho_2\,\tan\theta_2 - \rho_1\,\tan\theta_1)}{(\rho_2\,\tan\theta_2 + \rho_1\,\tan\theta_2)}\right\}^2$$

$$T = \frac{4\,\rho_1\,\rho_2\,c_1\,c_2\,\cos\theta_1\,\cos\theta_2}{(\rho_2\,c_2\,\cos\theta_1 + \rho_1\,c_1\,\cos\theta_2)^2} = \frac{4\,\rho_1\,\rho_2\,\tan\theta_1\,\tan\theta_2}{(\rho_2\,\tan\theta_2 + \rho_1\,\tan\theta_2)^2}$$

(8.20)

It follows immediately that $R + T = 1$ as required by the conservation of energy.

8.4   Spherical Sound Waves

If the sound wave diverges/converges symmetrically from/to a point, the wave equation takes the simple form

$$\frac{\partial^2\phi'}{\partial t^2} - c^2\frac{1}{r^2}\frac{\partial}{\partial r}\left(r^2\frac{\partial\phi'}{\partial r}\right) = 0$$

(8.21)

Substituting

$$\phi' = \frac{\psi'}{r}$$

(8.22)

reduces the equation to

$$\frac{\partial^2\psi'}{\partial t^2} - c^2\frac{\partial^2\psi'}{\partial r^2} = 0$$

(8.23)

which, as we have seen, has the general solution

$$\psi' = f(r - ct) + g(r + ct)$$

(8.24)

representing outgoing and incoming waves respectively. Thus the general solution is

$$\phi' = \frac{f(r - ct)}{r} + \frac{g(r + ct)}{r}$$

(8.25)

The arguments of the arbitrary functions f and g are retarded potentials, familiar from electromagnetism, and the $1/r$ term arises from the inverse square law as the energy flux decreases/increases with the area to maintain a constant total energy flow.

We consider now the sound wave generated by a source at the origin. Clearly only the outgoing wave is present, so that $\phi' = f(r - ct)/r$. The time dependence of the retarded potential only becomes important as the phase shift becomes large, i.e. when the distance r becomes of the order of the wavelength λ of the wave.

Provided that the size of the source ℓ is small compared with the wavelength λ, we may neglect the temporal term in the wave equation for short distances, i.e. for $\ell \ll r \ll \lambda$. The wave equation reduces to Laplace's equation

$$\nabla^2 \phi' = 0 \tag{8.26}$$

which has the solutions

$$\phi' = -a\,\frac{1}{r} + \mathbf{b} \cdot \nabla\left(\frac{1}{r}\right) + \dots \tag{8.27}$$

corresponding to monopole, dipole and multipole terms of the electrostatic field. The strengths a, $\mathbf{b}$, ... are determined by the details of the source itself in the neighbourhood of the origin. This solution represents the asymptotic form of the exact solution ($r \gg \ell$) taking into account the detailed form of the source. In the context of the method of matched asymptotics of Appendix 6.A, this solution is the outer solution. The values of the constants a, $\mathbf{b}$... are generated by matching to the inner solution in the neighbourhood of the source.

If we consider distances much greater than the size of the source, ℓ, namely $r \gg \ell$, only the lowest order terms contribute. Consider the term $-a/r$. As we showed in Section 2.8, a potential of this form represents a source or sink with a volume flow of $4\pi a$ out through a surface surrounding the origin.

In incompressible flow (as in Section 2.8) this represents a change in the total volume of fluid inside a surface surrounding the body generating the sound wave at the origin. It therefore represents a change in the volume of the body, whose surface is moving, pushing or pulling fluid through the fixed surface.. Thus if $V(t)$ is the volume of the source and $\dot{V}$ the rate of change of volume, the source term $a = \dot{V}/4\pi$.

Matching the solution for $\ell \ll r \ll \lambda$ to the general form,

$$\phi' = -\frac{\dot{V}(t - r/c)}{4\pi r} \tag{8.28}$$

The velocity perturbation due to the wave follows immediately

$$\mathbf{v}' = \frac{\ddot{V}(t - r/c)}{4\pi c r}\,\hat{\mathbf{n}} \tag{8.29}$$

at sufficiently large distances such that the term in r^{-2} may be neglected.

The intensity I of the wave, namely the average total energy emitted per unit time, is obtained by integrating the energy flux over the surface of a large sphere of radius r

$$I = \int_S \rho \, c \, \overline{v^2} \, \mathrm{d}s = \frac{\rho}{4 \pi c} \, \overline{\ddot{V}}^{\,2} \tag{8.30}$$

and depends on the time average of the square of the second time derivative of the body volume.

8.5   Cylindrical Sound Waves

We follow the same procedure as for spherical waves, but the analysis is more complicated due to the two-dimensional nature of the waves. We consider a wave homogeneous in the z direction which specifies the axis of cylindrical symmetry. The wave depends on time and the distance from this axis R. The general axisymmetric solution is obtained by the summation of many spherical waves originating at/converging to the axis from the point z. Thus the spherical radius $r = \sqrt{R^2 + z^2}$. The lowest order wave (monopole solution) is therefore obtained by integrating equation (8.25) from $z = -\infty$ to $z = \infty$, or equivalently r from $r = R$ to $r = \infty$, namely

$$
\begin{aligned}
\phi'(R, t) &= \int\limits_R^\infty \frac{f(ct - r)}{\sqrt{r^2 - R^2}} \, \mathrm{d}r + \int\limits_R^\infty \frac{g(ct + r)}{\sqrt{r^2 - R^2}} \, \mathrm{d}r \\[2mm]
&= \int\limits_{-\infty}^{(ct-R)} \frac{f(\zeta)}{\sqrt{(ct - \zeta)^2 - R^2}} \, \mathrm{d}\zeta + \int\limits_{(ct+R)}^\infty \frac{g(\zeta)}{\sqrt{(\zeta - ct)^2 - R^2}} \, \mathrm{d}\zeta
\end{aligned}
\tag{8.31}
$$

representing the outgoing and incoming waves respectively.

We consider again the case of an outgoing wave generated by the pulsating body, where the characteristic scale of the perturbing body ℓ is small compared with the wavelength λ and when the distance R satisfies $\ell \ll R \ll \lambda$ so that the phase terms may be neglected. The relevant solution of Laplace's equation for a line source in cylindrical geometry is

$$\phi' = a \ln(\chi R) = \frac{\dot{S}(t)}{2\pi} \ln\left(\chi(t) R\right) \tag{8.32}$$

where $\chi(t)$ is an undetermined function of time. Comparing this with the result for a two-dimensional source from Section 2.10.1 and using the same argument as before, we see that the source term a = cross-sectional area.

We may relate this expression to the general cylindrical wave solution for outward-going waves as follows. For $R \to 0$ we may write

$$\phi' = \lim_{R_0 \to \infty} \int_{-R_0/c}^{(t-R/c)} \frac{f(t')}{\sqrt{c^2(t-t')^2 - R^2}} \, dt'$$

$$\approx f(t) \left[\operatorname{arccosh}\left(\frac{x}{R}\right) \right]_R^{R_0} \approx -f(t) \ln\left(\frac{R}{R_0}\right) \tag{8.33}$$

where R_0 imposes a cut-off necessary to avoid the logarithmic singularity as $R \to -\infty$. In reality this is avoided by the behaviour of the source term as $t \to -\infty$.

Comparing equation (8.32) with (8.33) we see that $f(t) = c\dot{S}(t)/2\pi$ and that the term $\chi(t)$ reflects the fall-off of the source $\dot{S}(t)$ implicit in the requirement that $\dot{S}(t)$ tends to zero sufficiently rapidly as $t \to -\infty$ to allow the integral to converge.[1]

Thus finally we obtain a general expression for the perturbation potential due to a pulsating line source

$$\phi' = -\frac{c}{2\pi} \int_{-\infty}^{(t-R/c)} \frac{\dot{S}(t')}{\sqrt{c^2(t-t')^2 - R^2}} \, dt' \tag{8.34}$$

---

[1] The equivalence of these two forms (8.32) and (8.33) is easily seen from electrostatics, in the calculation of the potential due to a line charge: (8.32) by Gauss's theorem and (8.33) by an integral over the individual charges along the line.

Chapter 9

Characteristics and Rarefactions

9.1 Mach Lines and Characteristics

Consider a disturbance initiated at point O in fluid moving steadily and uniformly at velocity v. In the fluid frame the head of the disturbance will propagate into the field as a sphere of radius $c\tau$ at a time τ after its initiation. In the laboratory frame we have the situation depicted in Figure 9.1. At time τ the disturbance initiated at O has reached points P on a sphere of radius $c\tau$ whose centre C is displaced from O by the flow by $v\tau$. If the flow is subsonic, $v < c$, the perturbation moves backwards as well as forwards and in time the entire space is reached by the disturbance (Figure 9.1(a)). In contrast if the flow is supersonic, $v > c$, the perturbation only moves forwards swept along by the flow. Only a restricted space enclosed by the cone, formed by the tangents to the sphere, is reached by disturbance (Figure 9.1(b)). This region contained by the cone, known as the *Mach cone*, whose half angle μ is given by

$$\mu = \arcsin\left(\frac{c}{v}\right) \tag{9.1}$$

clearly defines the *range of influence* of the point O. The surfaces of the cone are known as *Mach lines* in two dimensions, or *Mach surfaces* more generally. The angle μ that the Mach lines make with the flow is known as the *Mach angle*.[1]

---

[1]**Historical note**   The role of pressure waves travelling at the sound speed defining the nature of supersonic flow was identified by Mach in a series of papers published between 1887 and 1890 in the *Reports of the Meetings of the Scientific Academy of Vienna*, which are

Introductory Fluid Mechanics for Physicists and Mathematicians, First Edition. Geoffrey J. Pert.
© 2013 John Wiley & Sons, Ltd. Published 2013 by John Wiley & Sons, Ltd.

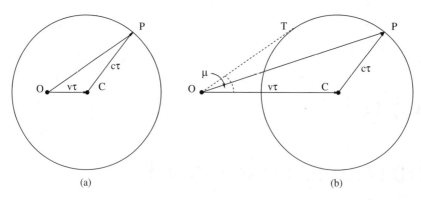

(a)   (b)

Figure 9.1: The development of a disturbance in uniformly flowing fluid as seen in the laboratory frame: (a) shows the situation when the flow velocity is less than the sound speed and (b) when it is greater. The disturbance originates at the centre O, which moves to C in the time τ. The disturbance is thus represented by a point P. If the flow is supersonic all points lying outside the cone formed by the tangent to the sphere T do not see the disturbance.

The two Mach lines leaving the point clearly contain information, the *characteristic invariant*, about the nature of the flow perturbation introduced at O in an appropriate form. The characteristic invariant is a quantity whose value remains constant on a particular characteristic. In general the form of the characteristic invariant is not easily expressed in analytic form, but we shall examine two particular cases where it can be identified. Clearly two Mach lines can only intersect at the point of origin, otherwise the flow disturbance would be multi-valued.

Although the concept of Mach lines has been developed in the context of steady uniform flow, it is clear that this concept applies more generally when the flow is non-uniform and both the velocity and sound speed are local quantities. Indeed we shall show subsequently (Section 9.4) that the Mach lines are a particular example of the general system of characteristics applicable to the particular case of steady two-dimensional irrotational flow. In general the Mach lines are defined by the angle μ between the Mach line and the streamline by the condition

$$\sin \mu = \frac{c}{v} \tag{9.2}$$

Thus the component of the velocity normal to the Mach line is the sound speed c. In the case of two-dimensional steady flow in Cartesian geometry, we shall find in Section 9.4 that an analytic form of the characteristic invariant may be found, but of a rather complex nature.

summarised in (B, 1890; Reichenbach, 1983). The terms 'Mach line', 'Mach cone' and 'Mach angle' are due to Meyer (1908), and 'Mach number' to Ackeret (1925).

Similar arguments may be applied in the case of one-dimensional time-varying flow where the characteristics are pressure waves running ahead and behind the flow, i.e. with speeds

$$\frac{\mathrm{d}r_{\pm}}{\mathrm{d}t} = v \pm c \tag{9.3}$$

In the case of a uniform one-dimensional flow the characteristic invariants are particularly simple, as shown in Section 9.3.[2]

9.2   Characteristics

We now present some general results concerning the nature of characteristics. Many of the adiabatic equations of fluid motion are of hyperbolic form. A feature of the solutions of this equation is that they may be represented by a set of waves, along which an invariant quantity is propagated. The Mach lines discussed above are the characteristics for the particular case of steady two-dimensional irrotational flow, but in all cases of supersonic, steady flow and time-dependent flow of compressible fluid, it can be shown that characteristics can be found. However, the subsonic steady flow equations are of the elliptic type, and real characteristics do not exist. This difference can be traced to the behaviour of the waves originated by small disturbances as discussed above.

The nature of characteristics may be illustrated by the following discussion. Suppose we have a set of partial differential equations in two dependent variables (u, v) and independent variables (x, y), namely

$$\mathcal{L} = 0 \qquad \text{and} \qquad \mathcal{M} = 0 \tag{9.4}$$

where $\mathcal{L}$ and $\mathcal{M}$ are differential operators in u and v and their derivatives. If the equations are hyperbolic, there exist linear combinations of $\lambda \mathcal{L} + \mu \mathcal{M}$ such that there exist lines $y(x)$ along each of which a solution of the combination $\phi(u, v)$ remains constant. These lines are the characteristics and the corresponding functions ϕ are their characteristic invariants. This procedure is illustrated by the examples in Sections 9.3 and 9.4. An alternative, but equivalent, approach to characteristics is found in Section 10.A.1.

---

[2]**Historical note**   The general theory of characteristics in one-dimensional time-dependent flow was developed by Riemann (1860) in a remarkable paper. The role of the characteristics in determining the flow through the invariants is clearly defined. Using an isentropic equation of state the analysis considered both waves of expansion and compression, the latter leading to discontinuity.

9.2.1   Uniqueness Theorem

An important consequence of this concept of characteristics is the establishment of the uniqueness of the flow at any point, given the state of the flow along some initial line $\mathcal{I}$, provided certain basic conditions are met. The following uniqueness theorem can be proved (Courant *et al.*, 1928).

Consider a solution with continuous second derivatives in the region ABP bounded by two characteristics C_+ and C_- through P and the interval AB bounded by the initial line $\mathcal{I}$ between them (Figure 9.2). Suppose another solution also with continuous second derivatives is given in ABP with the same values on AB; then the second solution is identical to the first in region ABP.

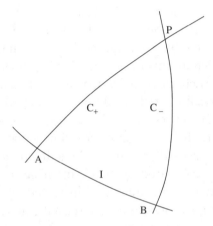

Figure 9.2: The domain of dependence of P. On the characteristic C_+ the invariant J_+ is constant, and on C_-, J_- is constant.

Note this does not establish that the two solutions are identical outside ABP. The initial values of the problem are only specified on the sector AB of the line $\mathcal{I}$ intercepted by the two outermost characteristics through P. The *domain of dependence* of P is the region of space enclosed by the characteristics through P. This is the region of space whose values can influence the flow at P.

Similarly the state of the flow at P can influence the flow in the space enclosed by the outgoing characteristics through P–the *range of influence*.

The uniqueness theorem has the important consequence that solutions of inviscid compressible fluid dynamics provide a unique solution to the Cauchy (or initial value) problem provided they possess continuous second derivatives. The problem is defined such that: given a set of initial values on a boundary $\mathcal{I}$, the solution at a specified point is uniquely determined within the appropriate space, namely its domain of dependence.

9.2.2 Weak Discontinuities

Weak discontinuities are defined as solutions where the variables are continuous, but the first derivative is discontinuous. Since small disturbances are propagated along characteristics, it is clear that weak discontinuities must also follow a characteristic. More specifically the uniqueness theorem excludes solutions with discontinuous derivatives. Since the values of the dependent variables are continuous across a weak discontinuity, the characteristics on each side of the discontinuity must be parallel. Points on either side of the discontinuity, whose domain of dependence is formed by the characteristics through them, will have continuous derivatives. Thus the weak discontinuity must propagate along a characteristic.

Weak discontinuities are typically found at a transition from a uniform state, where the upstream gradients are zero and change to ones reflecting the motion of the fluid. Similarly the transition from non-steady to steady flow must also occur along a characteristic. Therefore the head of a rarefaction or expansion wave must travel along a characteristic into the ambient steady state fluid.

9.2.3 The Hodograph Plane

Many problems involve two independent variables and two dependent ones, from which the flow can be calculated. Examples are one-dimensional time-dependent flow, where the independent variables are (x, t) and the dependent ones (v, c), and steady two-dimensional irrotational flow, where the variables are (x, y) and (u, v). In such cases we define a transformation between the space of the independent variables (x, y) and that of the dependent ones (u, v), the latter being known as the hodograph plane. This transformation is well established provided the Jacobian $\partial(u, v)/\partial(x, y)$ and its inverse exist. The hodograph plane is spanned by the characteristics in the same way as configuration space.

9.2.4 Simple Waves

A simple wave is defined as a flow in which one family of characteristics all lie along a single line in the hodograph plane. In this case the Jacobian for the transformation between the hodograph and the configuration space clearly does not exist.

Let the family of the characteristics containing a single line be C_-, for example. On this line, in the hodograph plane, the values of the dependent variables will be fixed by the characteristic invariant. Any characteristic of the other family C_+ will intersect this line at only one point, and will therefore have the values of the dependent variables associated with the C_- invariant fixed everywhere on that characteristic. The consequence is that the gradient of an individual member of the family of the second characteristic C_+ in configuration space will be constant, as the values of both flow variables are then set by their own invariant. Thus the characteristics C_+ form a family of straight lines of different gradient in configuration space.

Simple waves form an important class of flows, being those adjacent to a steady flow. Consider the characteristics C_-. The values of the dependent variables are everywhere constant. Thus the value of the characteristic invariant on each C_- line is the same. The characteristics lie on a single line in the hodograph plane. The flow is therefore a simple wave.

Consequently we obtain the important result that all flows adjacent to a steady state are simple waves. The weak discontinuity separating the steady and non-steady flows propagates along one of the two characteristics from the start of the disturbance, C_+ in the above example, and is a straight line path in configuration space. No characteristic of this family from the originating point can penetrate into the steady flow region. Thus the domain of dependence of any point in the steady flow excludes the originating point, and the range of influence of the latter excludes the former.

9.3 One-Dimensional Time-Dependent Expansion

Consider a cylindrical tube containing gas at pressure p_0 and density ρ_0 at rest, $v_0 = 0$. The tube is closed by a piston which is withdrawn resulting in an expansion into the space vacated by the piston. The characteristic variables of the problem are therefore the independent variables: distance along the cylinder from the initial position of the piston x and time from the instant the piston starts to withdraw t. The dependent variables are the flow velocity along the cylinder v and one of the thermodynamic state variables, typically the sound speed c. Since the flow is adiabatic, entropy is constant and only one state variable is independent, because the others are determined by the equation of state. The flow is described by the equation of continuity

$$\frac{\partial \rho}{\partial t} + v\frac{\partial \rho}{\partial x} + \rho\frac{\partial v}{\partial x} = 0 \tag{9.5}$$

and by Euler's equation

$$\frac{\partial v}{\partial t} + v\frac{\partial v}{\partial x} + \frac{1}{\rho}\frac{\partial p}{\partial x} = 0 \tag{9.6}$$

Introducing the adiabatic sound speed

$$c^2 = \frac{\partial p}{\partial \rho}\bigg|_s$$

and since the flow is adiabatic

$$dp = \frac{\partial p}{\partial \rho}\bigg|_S d\rho = c^2 d\rho \tag{9.7}$$

we obtain by substitution the equations of the characteristics

$$\left[\frac{\partial}{\partial t} + (v \pm c)\frac{\partial}{\partial x}\right]\left[v \pm \int \frac{c\,d\rho}{\rho}\right] \tag{9.8}$$

Thus on the characteristic lines

$$
\begin{aligned}
C_+ && \frac{dx}{dt} = v + c && J_+ = v + \int \frac{c\,d\rho}{\rho} \\
C_- && \frac{dx}{dt} = v - c && J_- = v - \int \frac{c\,d\rho}{\rho}
\end{aligned}
\tag{9.9}
$$

where J_+ and J_- are the characteristic invariants, known as Riemann invariants in this case. From their governing equations it is evident that the characteristics propagate at the sound speed relative to the flow, a result in conformity with the physical picture described in Section 9.1. The complete set of invariants is completed by the entropy

$$C_0 \quad \frac{dx}{dt} = v \quad J_0 = S$$

A polytropic gas has the adiabatic equation of state $p/\rho^\gamma = \text{const}$, so that $c^2 = \text{const}\,\gamma\,\rho^{(\gamma-1)}$ and hence

$$\int \frac{c\,d\rho}{\rho} = \frac{2c}{(\gamma - 1)}$$

Thus the Riemann invariants can be written as

$$J_\pm = v \pm \frac{2c}{(\gamma - 1)} \tag{9.10}$$

Since the ambient fluid into which the expansion propagates has a uniform state with zero initial velocity, the rarefaction as described above forms a simple wave. The characteristics, C_+, start from the initial state and then propagate into the expansion. Their invariant is therefore determined by the initial conditions of the ambient gas, and once set remains constant throughout the flow. As the initial conditions are uniform, the value of the Riemann invariant is

$$J_+ = v + \frac{2c}{(\gamma - 1)} = \frac{2c_0}{(\gamma - 1)} \tag{9.11}$$

everywhere throughout the flow, where c_0 is the initial sound speed. The set of characteristics C_+ thus falls on a single line in the hodograph plane (9.11).

The gradients of the characteristics C_- are given by (9.9)

$$\frac{\mathrm{d}x}{\mathrm{d}t} = v - c$$

Along an individual characteristic both v and c are constant and the characteristics themselves are straight lines, each of different gradient. If the piston path is known we may successively calculate the flow throughout the expansion. Thus let the piston trajectory be given by $x_p(t)$. The gas flow velocity at the piston must be equal to the piston velocity at time τ, namely

$$v = \frac{\mathrm{d}x}{\mathrm{d}t}\bigg|_\tau$$

Hence, since the Riemann invariant (equation 9.11) is known, the gradient of the corresponding C_- characteristic leaving the piston at this time is

$$\frac{\mathrm{d}x}{\mathrm{d}t}\bigg|_{C_-} = \frac{(\gamma + 1)}{2} \frac{\mathrm{d}x}{\mathrm{d}t}\bigg|_\tau - c_0 \tag{9.12}$$

The trajectory of each C_- characteristic may be plotted starting from the appropriate point on the piston path. The velocity and sound speed are the values corresponding to those at the piston $v = v_p$ and $c = c_0 - [(\gamma - 1)/2]v_p$. Note that as the piston speed increases, the sound speed decreases. If the piston stops being accelerated, the succeeding C_- characteristics are all parallel and the flow is uniform with the corresponding values of v and c appropriate to the piston speed. Finally if the piston velocity $v_p > [2/(\gamma - 1)]c_0$ the sound speed $c < 0$, indicating a void has formed behind the piston as the flow cannot maintain contact with the piston.

At any instant, the *head* of the rarefaction is the point reached by the disturbance leading to expansion of the ambient uniform gas. Clearly it is determined by the first C_- characteristic reaching that point. The head therefore propagates back into the undisturbed gas with speed c_0, i.e. with the sound speed relative to the flow. The *tail* of the rarefaction is correspondingly the C_- characteristic at which the expansion is complete, either at the piston or at a region of steady flow. In this case also the characteristic moves at the sound speed relative to the flow.

9.3.1   The Centred Rarefaction

The case where the piston is instantaneously accelerated to a steady velocity at time $t = 0$ is a particularly simple case of the preceding one. The C_- characteristics must reflect the fact that for times $t < 0$ their gradient is $-c_0$ and for

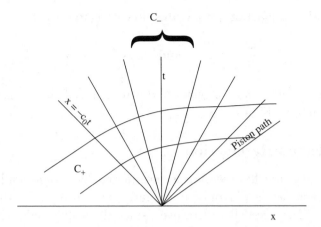

Figure 9.3: Sketch of the characteristics of a centred rarefaction behind an instantaneously accelerated piston.

times $t > 0$ it is given by equation (9.12). Hence the family of C_- characteristics start with one at a gradient $-c_0$ propagating into the undisturbed fluid, followed by a series of straight line characteristics of increasing gradient each starting at the origin, bounded by the one at which the flow velocity equals the piston speed (Figure 9.3). This has gradient given by equation (9.12). There then follows a further region of uniform flow adjacent to the piston. The characteristics associated with the expansion thus form a fan centred on the origin. The equation of the characteristics C_- is

$$\frac{x}{t} = v - c$$

and hence, making use of the values associated with the C_+ invariant,

$$
\begin{aligned}
c &= \frac{2}{(\gamma + 1)} c_0 - \frac{(\gamma - 1)}{(\gamma + 1)} \frac{x}{t} \\
v &= \frac{2}{(\gamma + 1)} \left[c_0 + \frac{x}{t} \right]
\end{aligned}
\tag{9.13}
$$

Along the line

$$x/t = [2/(\gamma - 1)] c_0$$

the sound speed $c = 0$ and consequently the density $\rho = 0$. Thus if the piston expands rapidly such that $v_p > [2/(\gamma - 1)] c_0$, the gas density falls to zero behind the piston. The gas cannot be sufficiently accelerated to keep up with the piston, and a cavity is formed. Alternatively if the gas expands into a void, the leading edge of the expansion moves outwards with speed $[2/(\gamma - 1)] c_0$.

The tail of the rarefaction propagates downstream only if

$$\frac{dx}{dt}\bigg|_{\text{tail}} < 0 \quad \text{and} \quad v_p < \frac{2}{\gamma + 1} c_0 \tag{9.14}$$

otherwise the characteristics propagate downstream of the initial piston position as illustrated in Figure 9.3.

9.3.2   Reflected Rarefaction

Suppose the upstream flow section is terminated by a rigid wall. We assume that the characteristics C_- only propagate back into the undisturbed gas, i.e. condition (9.14) is satisfied. However, since the wall is rigid, the flow velocity at the wall must be zero $v = 0$, a condition which is not satisfied by any but the first C_- simple wave characteristic (head). The boundary condition is satisfied by the establishment of a set of C_+ characteristics leaving the wall, which modify the flow such that the boundary condition is satisfied, and which represent the reflected C_- characteristic. Since the velocity normal to the wall is zero, the Riemann invariant of the C_+ characteristic is unchanged from that of the C_- one, and its direction is that of a reflection at the wall. The immediate flow involving this set of characteristics C_+ and C_- is *not* simple and their trajectories are not straight.

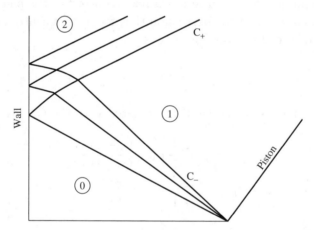

Figure 9.4: Reflection of a centred rarefaction propagating into undisturbed gas ⓪; downstream of the rarefaction behind the piston is region ① and behind the developed reflected wave, region ②.

The flow therefore contains three regions of steady flow, Figure 9.4:

- Region ⓪   The initial state before the expansion starts spanned by $C_\pm$ characteristics with $J_\pm = \pm 2c_0/(\gamma - 1)$.

- Region ① The region following the expansion fan, bounded by the constant motion of the piston, spanned by characteristics C_+ and C_- with invariants

$$J_+ = 2c_0/(\gamma - 1) \quad \text{and} \quad J_- = 2v_p - 2c_0/(\gamma - 1)$$

The velocity and the sound speed in region ① follow directly

$$v_1 = \frac{1}{2}\{J_+ + J_-\} = v_p$$

$$c_1 = \frac{1}{2}\frac{\gamma - 1}{2}\{J_+ - J_-\} = c_0 - \frac{1}{4}(\gamma - 1)v_p$$

- Region ② Following the reflection, this region is spanned by C_- characteristics from region ① and C_+ characteristics reflected at the wall. In this region the flow is brought to rest by the rigid wall:

$$J_- = 2v_p - 2c_0/(\gamma - 1) \quad \text{and} \quad J_+ = 2c_0/(\gamma - 1) - 2v_p$$

The velocity and the sound speed in region ② are easily calculated

$$v_2 = \frac{1}{2}\{J_+ + J_-\} = 0$$

$$c_2 = \frac{1}{2}\frac{\gamma - 1}{2}\{J_+ - J_-\} = c_0 - \frac{1}{2}(\gamma - 1)v_p = 2c_1 - c_0$$

from which the pressure and temperature ratios across the reflected rarefaction may be obtained:

$$\frac{p_2}{p_0} = \left[2\left(\frac{p_1}{p_0}\right)^{(1-\gamma)/2\gamma} - 1\right]^{2\gamma/(\gamma-1)}$$

$$\frac{T_2}{T_0} = \left[2\left(\frac{p_1}{p_0}\right)^{(1-\gamma)/2\gamma} - 1\right]^{2}$$

(9.15)

Since $c_2 - c_0 = 2(c_1 - c_0)$ it follows that for a weak rarefaction the pressure change $p_2 - p_0 \approx 2(p_1 - p_0)$ as for a sound wave.

The flow between regions ① and ② is a simple wave spanned by straight line C_+ characteristics from the individual reflections, forming a second expansion fan approximately centred at the mid-point of the reflection at the wall. In the immediate neighbourhood of the reflection, where the two expansion fans intersect, the flow is no longer simple, and is more difficult to calculate.

If the condition (9.14) is not satisfied, the interaction is of infinite duration and simple analytic results are not achievable.

9.3.3 Isothermal Rarefaction

If thermal conduction in the fluid is very strong (e.g. in plasma), the fluid may be maintained at approximately the same temperature throughout the expansion. The required condition is that the thermal heat flux balances the work done accelerating the gas, but with very small temperature gradient. In this case the equation of state becomes one of constant temperature, $T = \mathrm{const.}$ Although the flow is not dissipationless, the method of characteristics may be used, but temperature, instead of entropy, is constant on the streamlines. The appropriate sound speed is therefore the isothermal sound speed $\bar{c} = \sqrt{\partial p/\partial \rho}\big|_T$ ($= \sqrt{p/\rho}$ in an ideal gas). It is easy to see that an isothermal fluid corresponds to the limit of a polytropic gas with $\gamma \to 1$, i.e. infinitely large internal energy.

The velocity of the flow in a centred isothermal rarefaction follows directly from equation (9.13) with $\gamma = 1$. Thus $v = \bar{c}_0 + x/t$ and $\bar{c} = \bar{c}_0$, where $\bar{c}_0$ is the isothermal sound speed at the head. Inserting this velocity profile into the equation of continuity and assuming a density profile of the form $\rho = \rho_0 \exp(-\alpha x/t)$, we obtain the equation of continuity

$$\frac{\partial \rho}{\partial t} + v\frac{\partial \rho}{\partial x} + \rho\frac{\partial v}{\partial x} = \left[\frac{\alpha x}{t^2} - \left(\bar{c}_0 + \frac{x}{t}\right)\frac{\alpha}{t} + \frac{1}{t}\right]\rho = 0$$

and hence conclude that the density profile $\rho(x,t) = \rho_0 \exp(-x/\bar{c}_0 t)$ is consistent, where ρ_0 is the density at the head.

The total energy in the rarefaction at time t is

$$\int_0^\infty \rho(x,t)\left[\epsilon_0 + \frac{1}{2}v^2\right]\,\mathrm{d}x = \rho_0 \int_0^\infty \exp\left(-\frac{x}{\bar{c}_0 t}\right)\left[\epsilon_0 + \frac{1}{2}\left(\bar{c}_0 + \frac{x}{t}\right)^2\right]\,\mathrm{d}x$$

$$= \rho_0\,\bar{c}_0\,t\left[\epsilon_0 + \frac{5}{2}\bar{c}_0^{\,2}\right]$$

where ϵ_0 is the specific internal energy at the head of the rarefaction. In the rest frame of head of the rarefaction the flow speed $v_0 = \bar{c}_0$ The convected flux of energy into the rarefaction per unit time is

$$\rho_0\,v_0\left(h_0 + \frac{1}{2}v_0^{\,2}\right) = \rho_0\,\bar{c}_0\left(\epsilon_0 + \frac{3}{2}\bar{c}_0^{\,2}\right)$$

The downstream isothermal flow is maintained by the excess flow of heat from the head of the rarefaction. Hence the additional energy flux required to maintain the isothermal rarefaction in an ideal gas is

$$\rho_0\,\bar{c}_0^{\,3} = p_0\,v_0 \tag{9.16}$$

9.4   Steady Two-Dimensional Irrotational Expansion

The equation of continuity

$$\rho \frac{\partial v_x}{\partial x} + \rho \frac{\partial v_y}{\partial y} + v_x \frac{\partial \rho}{\partial x} + v_y \frac{\partial \rho}{\partial y} = 0 \tag{9.17}$$

and Bernoulli's equation

$$v_x \, dv_x + v_y \, dv_y + \frac{c^2}{\rho} \, d\rho = 0 \tag{9.18}$$

which may be combined to give

$$\left(v_x^2 - c^2\right) \frac{\partial x_x}{\partial x} + v_x \, v_y \left(\frac{\partial v_x}{\partial y} + \frac{\partial v_y}{\partial x}\right) + \left(v_y^2 - c^2\right) \frac{\partial v_y}{\partial y} = 0 \tag{9.19}$$

together with the condition for irrotationality

$$\frac{\partial v_x}{\partial y} - \frac{\partial v_y}{\partial x} = 0 \tag{9.20}$$

form the governing equations. To find the characteristics we form linear combinations of these equations using undetermined multipliers λ_1 and λ_2, namely

$$\lambda_2 \left(v_x^2 - c^2\right) \frac{\partial v_x}{\partial x} + (\lambda_1 + \lambda_2 \, v_x \, v_y) \frac{\partial v_x}{\partial y} - (\lambda_1 - \lambda_2 \, v_x \, v_y) \frac{\partial v_y}{\partial x}$$

$$+ \lambda_2 \left(v_y^2 - c^2\right) \frac{\partial v_y}{\partial y} = 0 \tag{9.21}$$

We now look for a reduction of this equation to characteristic form, i.e. along the line $dy/dx = \zeta$, some function $\phi(x, y)$ is constant. Hence

$$\left[\frac{\partial}{\partial x} + \zeta \frac{\partial}{\partial y}\right] \phi(x, y) = \frac{\partial \phi}{\partial v_x} \frac{\partial v_x}{\partial x} + \frac{\partial \phi}{\partial v_y} \frac{\partial v_y}{\partial x} + \zeta \left(\frac{\partial \phi}{\partial v_x} \frac{\partial v_x}{\partial y} + \frac{\partial \phi}{\partial v_y} \frac{\partial v_y}{\partial y}\right) = 0 \tag{9.22}$$

Hence, noting that these two forms must be the same apart from an undetermined factor $f(v_x, v_y)$,

$$\frac{\partial \phi}{\partial v_x} = f(v_x, v_y) \lambda_2 \left(v_x^2 - c^2\right) = f(v_x, v_y) \zeta^{-1} (\lambda_1 + \lambda_2 \, v_x \, v_y)$$

$$\frac{\partial \phi}{\partial v_y} = f(v_x, v_y)(-\lambda_1 - \lambda_2 \, v_x \, v_y) = f(v_x, v_y) \zeta^{-1} \lambda_2 \left(v_y^2 - c^2\right) \tag{9.23}$$

which have consistent solutions if

$$\zeta^{-2} \left(v_y^2 - c^2\right) - 2\,\zeta^{-1} v_x\, v_y + \left(v_x^2 - c^2\right) = 0 \tag{9.24}$$

Hence the gradients of the characteristics are

$$\zeta_\pm = \frac{v_x\, v_y \pm c\sqrt{v_x^2 + v_y^2 - c^2}}{v_x^2 - c^2} \tag{9.25}$$

and only have real solutions if $v_x^2 + v_y^2 > c^2$, i.e. the flow is supersonic.

In the (x, y) plane the characteristics are the Mach lines, i.e. lines making an angle μ with the streamline. The streamline angle $\theta = \arctan(v_y/v_x)$, and the Mach angle $\mu = \arctan(c/\sqrt{v^2 - c^2})$ where $v = \sqrt{v_x^2 + v_y^2}$ is the total speed. The Mach lines are at an angle $\theta \pm \mu$ with respect to the x axis. Thus

$$
\begin{aligned}
\tan(\theta \pm \mu) &= \frac{\tan\theta \pm \tan\mu}{1 \mp \tan\theta \tan\mu} \\[2mm]
&= \frac{v_y/v_x \pm c/\sqrt{v^2 - c^2}}{1 \mp v_y\, c/(v_x\sqrt{v^2 - c^2})} \\[2mm]
&= \frac{\left[v_y\sqrt{v^2 - c^2} \pm v_x\, c\right]\left[v_x\sqrt{v^2 - c^2} \pm v_y\, c\right]}{v_x^2\left(v^2 + c^2\right) - v_y^2\, c^2} \\[2mm]
&= \frac{v_x\, v_y \pm c\sqrt{v^2 - c^2}}{v_x^2 - c^2} = \zeta_\pm
\end{aligned} \tag{9.26}
$$

and the Mach lines and characteristics coincide in accordance with our earlier physical arguments (Section 9.1).

9.4.1   Characteristic Invariants

The characteristic invariants for this problem are much more difficult to generate than in the previous case. In principle they are contained in the functions $\phi(x, y)$ given by equation (9.22). However, it is easier to obtain their values by direct analysis of the form of the streamlines and the characteristics. To do this we will need to identify derivatives along and normal to the streamlines, which will differ from the laboratory values as this is a curvilinear co-ordinate system. As we will need to calculate the curl and div of vector quantities we define a local set of Cartesian co-ordinates (x, y) to which we can refer the curvilinear set. Let θ be the angle that the streamline makes with the

x direction. Then $v_x = v\cos\theta$ and $v_y = v\sin\theta$ where v is the total velocity. Taking the derivatives and allowing $\theta \to 0$ we obtain

$$\frac{\partial v_x}{\partial x} = \frac{\partial v}{\partial s} \quad \frac{\partial v_x}{\partial y} = v\frac{\partial\theta}{\partial s} \quad \frac{\partial v_y}{\partial x} = v\frac{\partial\theta}{\partial n} \quad \frac{\partial v_y}{\partial y} = \frac{\partial v}{\partial n} \qquad (9.27)$$

where s is the length of arc along the streamline and n along its normal.

Thus the equation of continuity and the condition of zero vorticity become

$$\frac{\partial(\rho v)}{\partial s} + \rho v\frac{\partial\theta}{\partial n} = 0$$

$$\frac{\partial v}{\partial n} - v\frac{\partial\theta}{\partial s} = 0$$

Further, it follows from the strong form of Bernoulli's equation for a polytropic gas that

$$v\,dv = -\frac{1}{\rho}\,dp = -c^2\,\frac{d\rho}{\rho} \qquad (9.28)$$

holds for variation in any direction. Thus we obtain

$$v\frac{\partial\theta}{\partial n} = -\frac{\partial v}{\partial s} - \frac{v}{\rho}\frac{\partial\rho}{\partial s} = -\frac{\partial v}{\partial s}\left(1 - \frac{v^2}{c^2}\right) = \cot^2\mu\,\frac{\partial v}{\partial s}$$

since $v = c\operatorname{cosec}\mu$. Hence along a characteristic

$$
\begin{aligned}
dv &= \frac{\partial v}{\partial s}\,ds + \frac{\partial v}{\partial n}\,dn \\
&= v\left(\tan^2\mu\,\frac{\partial\theta}{\partial n}\,ds + \frac{\partial\theta}{\partial s}\,dn\right) \\
&= v\tan\mu\left(\frac{\partial\theta}{\partial n}\,dn + \frac{\partial\theta}{\partial s}\,ds\right) \\
&= v\tan\mu\,d\theta \qquad\qquad (9.29)
\end{aligned}
$$

since $dn = \tan\mu\,ds$. Therefore

$$v\frac{d\theta}{dv} = \cot\mu \qquad (9.30)$$

The component of the velocity along the characteristic is $v_t = v\cos\mu$ and hence, since $c = v\sin\mu$,

$$
\begin{aligned}
dv_t &= \cos\mu\,dv - v\sin\mu\,d\mu = v\sin\mu\,(d\theta - d\mu) \\
&= c\,(d\theta - d\mu)
\end{aligned}
$$

To proceed we introduce Bernoulli's equation remembering that it holds throughout the flow as the latter is irrotational. Therefore

$$d\theta - d\mu = \frac{dv_t}{k\sqrt{v_{max}^2 - v_t^2}}$$

$$\theta - \mu = \frac{1}{k}\arcsin\left(\frac{v_t}{v_{max}}\right) - \frac{\pi}{2} + A_\pm$$

where $k = \sqrt{(\gamma - 1)/(\gamma + 1)} = c_*/v_{max}$. When $v = c_*$, the critical velocity, $\mu = \pi/2$ and $\theta = A$.

Finally

$$\arcsin\left\{\frac{v_t}{v_{max}}\right\} = \arctan\left\{\frac{k\,v_t}{c}\right\} = \arctan\{k\cot\mu\}$$

and we obtain along the characteristic

$$\theta = \mu + \frac{1}{k}\arctan\{k\cot\mu\} - \frac{\pi}{2} + A \tag{9.31}$$

which gives the characteristic invariant. The angle μ defines the value of the limit speed from Bernoulli's equation

$$v^2 = \frac{v_{max}^2}{1 + [2/(\gamma - 1]\sin^2\mu} \tag{9.32}$$

The second characteristic is inclined at an equal angle of opposite sign to the streamline, and thus has μ replaced by $-\mu$. Therefore the two invariants are

$$J_\pm = \theta \mp f(\mu) \tag{9.33}$$

$$f(\mu) = \mu + \frac{1}{k}\arctan\{k\cot\mu\} - \frac{\pi}{2} \tag{9.34a}$$

Since $\cot\mu = \tan(\pi/2 - \mu) = \sqrt{M^2 - 1}$ we may write the invariant as a function of the local Mach number M in terms of the Prandtl–Meyer function

$$f(\mu) = \nu(M) = \sqrt{\frac{(\gamma + 1)}{(\gamma - 1)}}\arctan\left\{\sqrt{\frac{(\gamma + 1)}{(\gamma - 1)}(M_1{}^2 - 1)}\right\}$$

$$- \arctan\left\{\sqrt{(M_1{}^2 - 1)}\right\} \tag{9.34b}$$

The Prandtl–Meyer function is defined to set $\nu(1) = 0$. Equation (9.32) can also be written in terms of the Mach number

$$v^2 = \frac{v_{max}^2}{1 + [2/(\gamma - 1)]M^{-2}} \tag{9.35}$$

The characteristic form in terms of the Mach number is often easier to use than that in terms of Mach angle. For convenience the Prandtl–Meyer function $\nu(M)$ is often tabulated, together with the Mach angle μ. An extensive tabulation of the Prandtl–Meyer function can be found in Anderson (2007).

Using equations (9.34a) and (9.34b) together with equation (9.32) for the two characteristics passing through any point, we may calculate the angle that the flow makes with a prescribed direction $\theta = J_+ + J_- - \pi$ and the total velocity from $f(\mu)$ or $\nu(M) = J_- - J_+$, and thus from the latter the sound speed, pressure and density using Bernoulli's formula. Hence the flow is uniquely calculated. This leads to a useful method of calculation in which the characteristics and hence the flow are progressively traced through the flow field–*the method of characteristics*–which is used to calculate the flow through ducts and nozzles.

9.4.2 Expanding Supersonic Flow around a Corner

This problem is very similar to the rarefaction treated earlier. The flow is a simple wave since the incoming flow along the surface is uniform until the start of the corner is reached. One set of characteristics, in this case C_+, fills the entire space with a uniform value of the invariant J_+.[3] Consequently since J_- is constant on any C_- characteristic, the angle θ that the streamline makes with the initial line, defined parallel to the surface, is constant on the characteristic. Since the streamline is parallel to the surface at the point where the characteristic C_+ meets the surface, the angle of the tangent to the surface defines the value of the flow velocity vector, θ, on the characteristic C_- leaving the surface at the point. The value of the Mach angle is given by the value of J_+ and θ, and hence the angle of the characteristic to the initial line of the line is determined, and thus the entire flow on that characteristic.

9.4.3 Flow around a Sharp Corner–Centred Rarefaction

If the corner is reduced to an angle the characteristics expressing the rotation of the flow become a fan originating in the corner, similar to the earlier case in time-dependent flow. The characteristics, which rotate the flow, are straight and bounded by the initial and final ones where the flow velocity is parallel to the surface. In each case the characteristic must originate in the corner and therefore so must this family of C_- characteristics (Figure 9.5). Thus suppose the angle of the corner is χ; then the flow is rotated by the expansion through χ to become a uniform flow parallel to the surface.

---

[3]The sense of the angles Ψ, θ and μ is taken from the upstream flow into the expansion and dictates the role of the individual characteristics.

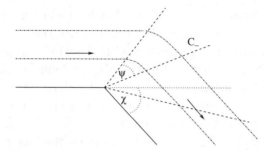

Figure 9.5: The flow around a sharp corner showing the streamlines (dot–dash lines) and C_+ characteristics (dashed lines).

Consider a supersonic flow incoming at Mach number M_1, velocity v_1 and sound speed c_1. A particular C_- characteristic makes an angle ψ with the C_- characteristics in the incoming flow, θ is the angle the streamline makes with the surface of the corner of entry and μ the Mach angle. On this Mach line the velocity and sound speed are constant. The C_+ characteristics all originate in the upstream initial flow and terminate on the surface of the angled plates. The invariant J_+ is therefore constant throughout the flow and equal to its value in the upstream flow

$$
\begin{aligned}
J_+ &= \theta - f(\mu) = -f(\mu_1) \\
&= \theta - \nu(M) = -\nu(M_1)
\end{aligned} \tag{9.36}
$$

Since the invariant J_-

$$
J_- = \theta + f(\mu) = \theta + \nu(M) \tag{9.37}
$$

is constant on an individual C_- characteristic, it follows that the streamline angle θ, the Mach angle μ, the Mach number and the flow variables v_r, v_ψ and c are all constant on the characteristic C_-, where v_r and v_ψ are the components of the velocity parallel and perpendicular to the characteristic. Therefore the characteristic C_- is a straight line at an angle ψ, as expected since the wave is a simple wave.

Defining the zero of ψ when θ is also zero, i.e. at the first characteristic, the angles ψ and θ obey the geometric relation

$$
\psi + \mu = \theta + \mu_1 \tag{9.38}
$$

Eliminating the streamline angle θ from equations (9.36) and (9.38) we obtain

$$
\psi = [f(\mu) - \mu] - [f(\mu_1) - \mu_1]
$$

and substituting for $f(\mu)$

$$\arctan(k\cot\mu) = k\psi + \arctan(k\cot\mu_1) = k\psi + \varepsilon \qquad (9.39)$$

where $\varepsilon = \arctan(k\cot\mu_1) = k\sqrt{M_1^2 - 1}$. The ratio $k\,v_r/v_\psi = k\cot\mu = \tan(k\psi + \varepsilon)$ defines the Mach angle.

Since $v_\psi = c$, Bernoulli's equation may be expressed as

$$v_r^2 + v_\psi^2 + \frac{2}{(\gamma - 1)}v_\psi^2 = v_r^2 + \frac{1}{k^2}v_\psi^2 = v_{\max}^2 = \frac{1}{k^2}c_*^2 \qquad (9.40)$$

Noting that $k = \sqrt{(\gamma - 1)/(\gamma + 1)} = c_*/v_{\max}$, equation (9.40) takes the form

$$(k\,v_r)^2 + v_\psi^2 = c_*^2 \qquad (9.41)$$

which is consistent with

$$v_r = k\,c_* \sin(k\psi + \varepsilon) \quad \text{and} \quad v_\psi = c_* \cos(k\psi + \varepsilon) \qquad (9.42)$$

At a particular azimuthal angle ψ of the C_- characteristic, the Mach angle is given by $\cot\mu = [\tan(k\psi + \varepsilon)]/k = v_r/v_\psi$. The spatial co-ordinates of the flow variables are then found from the stream angle θ, which is given by the invariant $\theta = f(\mu) - f(\mu_1)$ using either (9.36) or (9.38).

The sound speed, pressure and density in terms of their stagnation values (c_0, p_0, ρ_0), equations (1.48 and 1.49), are calculated from the sound speed $c = v_\psi$. Since the flow is adiabatic, $(p/p_0) = (\rho/\rho_0)^\gamma$, $c^2 = \gamma p/\rho = c_*^2 \cos^2(k\psi + \varepsilon)$ and $c_*^2 = [2/(\gamma + 1)]c_0^2$, we obtain the conditions in the downstream flow:[4]

$$\frac{p}{p_0} = \left\{\frac{2}{(\gamma + 1)}\cos^2(k\psi + \varepsilon)\right\}^{\gamma/(\gamma - 1)} \qquad \frac{\rho}{\rho_0} = \left\{\frac{2}{(\gamma + 1)}\cos^2(k\psi + \varepsilon)\right\}^{1/(\gamma - 1)}$$

$$(9.43)$$

As the streamlines turn through an angle θ, the characteristic is rotated through to an angle given by equation (9.38), $\psi = \theta + \mu_1 - \mu$. The angle of the final characteristic leaving the surface when $\theta = \chi$ is therefore given by

$$f(\mu_2) = \chi + f(\mu_1) \quad \psi_2 = \chi + \mu_2 - \mu_1 \qquad (9.44)$$

---

[4]**Historical note** The full solution to this problem of supersonic flow around a sharp corner is due to Prandtl and Meyer, using a method not explicitly involving characteristics. Meyer was a graduate student and Prandtl his supervisor. The full solution, along with results for oblique shocks, is found in Meyer's PhD thesis, (Meyer, 1908).

Unfortunately this equation requires the inversion of the function $f(\mu)$, which is not elementary. It is therefore simpler at this stage to use the tabulated values of the Prandtl–Meyer function to find the Mach number from $\nu(M_2) = \chi + \nu(M_1)$, and hence the streamline angle and flow variables.

As we saw earlier for the centred rarefaction, there is a limiting condition beyond which the initial state of the gas cannot sustain the expansion. In this case there is a maximum angle beyond which the sound speed becomes negative, i.e. the flow cannot follow the surface and cavitation occurs. This occurs at the corner angle χ_{lim} where $c_2 = 0$ and $\mu_2 = 0$, i.e.

$$\chi_{\text{lim}} = \frac{1}{k} \left(\frac{\pi}{2} - \varepsilon \right) - \mu_1 \tag{9.45}$$

The flow just described is reversible. In that we inject a flow opposite to that exiting from the angle into the corner from the exit, the characteristics would also be reversed and the flow would also reverse and therefore compress. However, an alternative expanding flow is possible with the streamlines following the pattern described above. In practice the latter expanding flow is found.

9.4.3.1   The complete Prandtl–Meyer flow

If the initial flow is sonic, $M_1 = 1$, we obtain the extreme case of the above flow. The initial C_- characteristic is identified by the condition that $v_r = 0$ since $v_1 = c_1$ and $\mu_1 = \pi/2$. Hence $\varepsilon = 0$. The limiting angle of the corner thus takes its maximum value

$$\chi_{\text{lim}} = \left(\frac{1}{k} - 1 \right) \frac{\pi}{2} = \left(\sqrt{\frac{(\gamma + 1)}{(\gamma - 1)}} - 1 \right) \frac{\pi}{2} \tag{9.46}$$

The flow through the rarefaction is easily calculated in terms of the critical velocity, and pressure and density in terms of the angle ψ. Figure 9.6 shows the values through the expansion for gas of index $\gamma = 7/5$ plotted as functions of the local Mach number $M = v/c$.

If we consider an incident flow at a Mach number $M_1 \neq 1$, the flow may be directly related to the points on the complete Prandtl–Meyer plot, where the local Mach number equals that of the incoming flow, since the characteristic invariants of each flow will be the same (apart from an additive constant in the values of θ and ψ). Consequently the values of the angles μ_1, ψ_1, θ_1 and the flow variables $c/c_*, p/p_*$ and ρ/ρ_* from the graph may be used to set the initial values to the flow, and the remainder of the complete expansion used to give the values for an incomplete expansion. More accurately we may

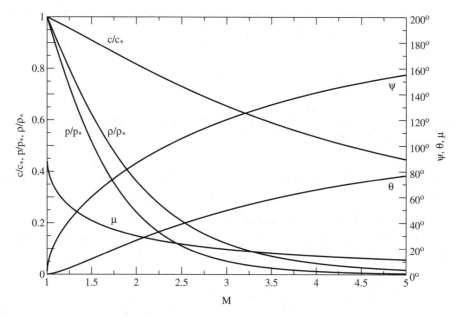

Figure 9.6: The flow variables calculated through a complete Prandtl–Meyer expansion for a polytropic gas of $\gamma = 7/5$. The plots shows the sound speed c/c_*, pressure p/p_*, density ρ/ρ_* in terms of the critical values, and the characteristic angle ψ, Mach angle μ and streamline deflection θ in terms of the local Mach number M.

introduce the Prandtl–Meyer function, $\nu(M)$, equation (9.34b), and use its tabulated values. It follows from equations (9.44) and (9.38) that the rotation of the flow is given by

$$\chi = \nu(M_2) - \nu(M_1) \qquad (9.47)$$

Then, given the incoming flow Mach number M_1, we obtain $\nu(M_1)$ from the table. From the value of χ, $\nu(M_2)$ and thus M_2 are obtained. The pressure and density ratios p_2/p_1 and ρ_2/ρ_1 follow directly from the stagnation ratio pressure and density ratios given by equation (1.49) since the upstream and downstream Mach numbers are known.

9.4.3.2   Weak rarefaction

If the expansion is weak so that the deflection of the flow is small, the rarefaction becomes equivalent to a weak discontinuity lying along a single characteristic C_- at the Mach angle. The properties are very easily evaluated in terms of the velocity increment, from which the sound speed, pressure and density changes are easily calculated. Since the C_+ characteristics are everywhere constant, we may use equation (9.29) to calculate the change in the velocity across the C_- characteristic, i.e. along the C_+ one. Thus for a small deflection θ

$$
\begin{aligned}
v_2 - v_1 &= v_1 \theta \tan \mu \\
&= v_1 \theta \, \frac{1}{\sqrt{M_1^2 - 1}}
\end{aligned}
\tag{9.48}
$$

since $\tan \mu = 1/\sqrt{M_1^2 - 1}$. From Bernoulli's equation the increment in the sound speed is therefore

$$
c_2^2 - c_1^2 \approx -(\gamma - 1) v_1 (v_2 - v_1) = -(\gamma - 1) v_1^2 \theta \, \frac{1}{\sqrt{M_1^2 - 1}}
\tag{9.49}
$$

where M_1 is the Mach number of the flow.

Since the flow is isentropic, we may use the adiabatic equation of state to calculate the pressure and density changes:

$$
\begin{aligned}
\frac{p_2 - p_1}{p_1} &= -\frac{\gamma}{(\gamma - 1)} \frac{c_2^2 - c_1^2}{c_1^2} = -\gamma \frac{v_1^2}{c_1^2} \theta \, \frac{1}{\sqrt{M_1^2 - 1}} \\
\frac{\rho_2 - \rho_1}{\rho_1} &= -\frac{1}{(\gamma - 1)} \frac{c_2^2 - c_1^2}{c_1^2} = -\frac{v_1^2}{c_1^2} \theta \, \frac{1}{\sqrt{M_1^2 - 1}}
\end{aligned}
\tag{9.50}
$$

Chapter 10

Shock Waves

10.1 Introduction

Shock waves are associated with compression, whereas the flows we have so far investigated have all involved expansion. Returning to the piston problem, we reverse the direction of travel of the piston into the gas, rather than out. The gradients of the characteristics C_- are still given by equation (9.12). Thus as $v_p < 0$, the slope of the characteristic progressively increases. Physically this is due to the fact that the sound speed in adiabatic change varies as

$$c \propto \rho^{(\gamma-1)/2}$$

and the sound speed increases in compression.

The characteristics C_- thus converge and eventually intersect. This, however, cannot occur as the flow would become multi-valued at the intersection, which is physically impossible.[1] The model we have used must therefore have

[1]**Historical note**   The argument that the characteristics intersect leading to discontinuity is due to Riemann (1860) The difficulty appeared earlier in Poisson (1808) in studies of compressive finite amplitude sound waves (simple waves) and was identified more definitively later by Stokes (1848) and Earnshaw (1860), when the compressive part of the sound wave was shown to progressively steepen. Riemann (1860) proposed that at this point the flow underwent a discontinuous transition. However, the concept of discontinuities in solutions of the (differential) hyperbolic equations caused mathematical difficulties, which have only been recently resolved, Appendix 10.A.3 (see Salas, 2007). Furthermore Rayleigh (1945, vol. 2, §250–253) showed that because the flow was assumed to be adiabatic, the condition of energy conservation could not be satisfied. However, this latter problem had already been resolved by Rankine (1870) and subsequently by Hugoniot (1887, 1889) who independently realised that the flow was non-isentropic. The lack of continuity was accounted for by the introduction of dissipational factors within a narrow shock layer accounting for the discontinuity in the context of asymptotics.

Introductory Fluid Mechanics for Physicists and Mathematicians, First Edition. Geoffrey J. Pert.
© 2013 John Wiley & Sons, Ltd. Published 2013 by John Wiley & Sons, Ltd.

become invalid. The reason for this is easily seen, for as the intersection is approached, the gradients in the physical quantities, velocity and temperature become large. Consequently we can no longer neglect the effects of viscosity and thermal conduction, and they contribute strongly and ensure the single-valued nature of the flow. This transition is known as a *shock*, and takes place over a narrow region of thickness of the order of the mean free path, the *shock layer*. Since both viscosity and thermal conduction give rise to dissipation, the shock transition must lead to an increase in the entropy of the flow. This increase in entropy requires that the velocity of the flow on entry to the shock be super-sonic, i.e. the Mach number $M_1 = v_1/c_1 > 1$ where v_1 and c_1 are respectively the flow velocity relative to the shock and the sound speed on entry. The shock layer is therefore a narrow zone in which the full dissipational equations must be solved, sandwiched between two asymptotic flows of isentropic fluid. In principle the complete flow could be treated by the method of matched asymptotic expansion (Appendix 6.A), but this is normally unnecessary.

As the intersection is approached the envelope of the characteristics C_- takes the form of a cusp known as the *shock cusp*.

10.2 The Shock Transition and the Rankine–Hugoniot Equations

As the shock is typically only a few mean paths in thickness, within the fluid description, it has zero width. Thus we may regard the shock as a discontinuity in which the flow variables change their values from their upstream to their downstream values as they pass through the shock. Since the shock is very narrow and contains no external sources of mass, momentum or energy, the flux of each of the three quantities leaving (downstream) the shock must equal that entering (upstream). This condition is most easily established in the rest frame of the shock, where if the upstream values are (v_1, ρ_1, p_1, h_1) and the downstream ones (v_2, ρ_2, p_2, h_2) we obtain for the case where the shock is normal to the flow:[2]

Mass conservation $\qquad\qquad\qquad\quad \rho_1 v_1 = \rho_2 v_2$

Momentum conservation $\qquad\quad \rho_1 v_1^2 + p_1 = \rho_2 v_2^2 + p_2 \qquad\qquad$ (10.1)

Energy conservation $\qquad \left(h_1 + \tfrac{1}{2}v_1^2\right)\rho_1 v_1 = \left(h_2 + \tfrac{1}{2}v_2^2\right)\rho_2 v_2$

---

[2]**Historical note**   These conservation equations were independently derived by Rankine (1870) and Hugoniot (1887, 1889). The existence of discontinuities in the solutions of the inviscid hyperbolic equations proved to be a major stumbling block to the acceptance of the theory of shock waves. Their existence was not finally accepted until the work of Becker (1922) (Section 10.5.1) established that stable solutions were possible. Nowadays it is known

The third of these equations may be simplified to obtain

$$h_1 + \frac{1}{2}{v_1}^2 = h_2 + \frac{1}{2}{v_2}^2 \tag{10.2}$$

Thus Bernoulli's equation is applicable across a shock in addition to its role in ideal flow.

These equations are completed by the equation of state in the form $h(p, \rho)$, so that there are then three equations for three unknown quantities. For example, if the upstream (v_1, p_1, ρ_1) values in a polytropic gas are known, a unique solution is easily obtained.

In many cases the situation in the laboratory system involves the shock moving into an ambient gas at rest. In that case the transition from the shock frame to the laboratory frame is readily established by the transformation $v = U - u$, where u and v are the velocities in the laboratory and shock frames respectively, and U the shock speed. Since the velocity is taken as positive for flow *into* the shock, the velocity of the shock into the undisturbed gas is U, and $u_1 = 0$ so that the ambient gas is stationary. In such experimental situations the boundary conditions are usually specified either by the Mach number of the shock moving into gas at rest (M_1), or by the downstream pressure (p_2) driving the shock.

10.2.1 Rankine–Hugoniot Equations for a Polytropic Gas

The Rankine–Hugoniot jump relations across a shock of given Mach number M_1 propagating into a polytropic gas are relatively easy to obtain (problem #32). The results are

$$y = \frac{\rho_2}{\rho_1} = \frac{v_1}{v_2} = \frac{(\gamma + 1){M_1}^2}{(\gamma - 1){M_1}^2 + 2} \tag{10.3a}$$

$$\Pi = \frac{p_2}{p_1} = \frac{2\gamma {M_1}^2 - (\gamma - 1)}{(\gamma + 1)} \tag{10.3b}$$

$$\frac{T_2}{T_1} = \frac{c_2^2}{c_1^2} = \frac{\left[2\gamma {M_1}^2 - (\gamma - 1)\right]\left[(\gamma - 1){M_1}^2 + 2\right]}{(\gamma + 1)^2 {M_1}^2} \tag{10.3c}$$

$$\frac{v_2}{c_1} = \frac{(\gamma - 1){M_1}^2 + 2}{(\gamma + 1)M_1} \tag{10.3d}$$

$$\frac{v_2}{c_2} = \sqrt{\left\{ \frac{(\gamma - 1){M_1}^2 + 2}{2\gamma {M_1}^2 - (\gamma - 1)} \right\}} \tag{10.3e}$$

that generalised solutions called weak solutions of the hyperbolic equations exist satisfying the general integral relations, expressions of the conservation laws, equivalent to the Rankine–Hugoniot relations. Experimentally shock waves were first observed by Mach and Salcher (1887), although their attribution was unclear at that time. An interesting review of the controversy surrounding 'waves of permanent type' is given by Salas, (2007).

We note the important result that $v_2 < c_2$ since $\gamma > 1$, a condition which can be extended to gases with a general equation of state. This ensures that downstream characteristics can reach the shock, and enables energy from the driving source to be supplied to the shock to account for the increasing energy in the flow in the laboratory frame, where the ambient gas is at rest.

The density and pressure ratios can conveniently be written in terms of each other

$$\Pi = \frac{p_2}{p_1} = \frac{(\gamma+1)\,\rho_2 - (\gamma-1)\,\rho_1}{(\gamma+1)\,\rho_1 - (\gamma-1)\,\rho_2} = \frac{(\gamma+1)\,y - (\gamma-1)}{(\gamma+1) - (\gamma-1)\,y} \quad (10.4a)$$

$$y = \frac{p_2}{\rho_1} = \frac{(\gamma+1)\,p_2 + (\gamma-1)\,p_1}{(\gamma+1)\,p_1 + (\gamma-1)\,p_2} = \frac{(\gamma+1)\,\Pi + (\gamma-1)}{(\gamma+1) + (\gamma-1)\,\Pi} \quad (10.4b)$$

It is clear from equation (10.4a) that the compression ratio varies between the limits $(\gamma-1)/(\gamma+1) \leq y \leq (\gamma+1)/(\gamma-1)$ corresponding to shocks with zero and infinite upstream pressure respectively.

A useful relationship, due to Prandtl, is easily obtained from the preceding analysis between the product of the upstream and downstream velocities and the maximum velocity $v_{\max}$ (1.43), and thus the critical velocity c_* (1.47). Since

$$v_{\max}^2 = \frac{2\gamma}{(\gamma-1)}\frac{p_1}{\rho_1} + v_1{}^2$$

we may substitute for $\gamma\, p_1/\rho_1 = c_1{}^2$ in (10.3a) to obtain

$$\frac{v_2}{v_1} = \frac{(\gamma-1)}{(\gamma+1)}\left(1 + \frac{(v_{\max}{}^2 - v_1{}^2)}{v_1{}^2}\right)$$

$$v_1 v_2 = \frac{(\gamma-1)}{(\gamma+1)} v_{\max}{}^2 = c_*^2 \quad (10.5)$$

10.2.1.1   Strong shocks

If the Mach number is very large, $M_1 \gg 1$, the shock relations take a particularly simple form

$$y = \frac{\rho_2}{\rho_1} = \frac{(\gamma+1)}{(\gamma-1)}$$

$$\Pi = \frac{p_2}{p_1} = \frac{2}{(\gamma+1)} M_1{}^2 \quad (10.6)$$

$$\Theta = \frac{T_2}{T_1} = \frac{2(\gamma-1)}{(\gamma+1)^2} M_1{}^2$$

The remaining terms are easily found. We note that the compression ratio tends to a constant value and the downstream pressure scales as the square of the Mach number.

10.3   The Shock Adiabat

The flux $j = \rho_1 v_1 = \rho_2 v_2$ through the shock is given by the second Rankine–Hugoniot relation, which can be written $p_1 + j^2 V_1 = p_2 + j^2 V_2$ in terms of the specific volume $V = 1/\rho$. Hence

$$j^2 = \frac{p_2 - p_1}{V_1 - V_2} \tag{10.7}$$

This requires that if $p_2 > p_1$ then $V_1 > V_2$ and vice versa. This relationship allows a further useful expression

$$
\begin{aligned}
v_1 - v_2 &= j\left(\frac{1}{\rho_1} - \frac{1}{\rho_2}\right) = j\,(V_1 - V_2) \\
&= \sqrt{[(p_2 - p_1)(V_1 - V_2)]} \tag{10.8}
\end{aligned}
$$

Using the third Rankine–Hugoniot relation

$$h_2 - h_1 - \frac{1}{2}(V_2 + V_1)(p_2 - p_1) = 0$$

and substituting the internal energy ϵ for the enthalpy h,

$$\epsilon_2 - \epsilon_1 + \frac{1}{2}(V_2 - V_1)(p_2 + p_1) = 0 \tag{10.9}$$

This relationship is known as the shock adiabat, or, when plotted as a function of the final state $p_2(V_2)$, as the Hugoniot curve. For general materials it is a function of the equation of state, and, if known experimentally, it may be used to deduce useful information about the form of the equation of state. Clearly the adiabat is a function of the initial conditions. Each different upstream state (p_1, V_1) will generate a different Hugoniot.

The Hugoniot for a polytropic gas is particularly simple, namely

$$\frac{p_2 - p_1}{p_2 + p_1} = -\gamma \frac{V_2 - V_1}{V_2 + V_1} = \gamma \frac{\rho_2 - \rho_1}{\rho_2 + \rho_1} \tag{10.10}$$

This equation may be combined with the shock momentum conservation equation (10.1) in the form

$$p_2 - p_1 + \frac{\rho_1 \rho_2}{\rho_2 - \rho_1}(v_2 - v_1)^2$$

to give the useful relation for a polytropic gas

$$(p_2 - p_1)^2 = \frac{\rho_1}{2} (v_2 - v_1)^2 \left[(\gamma + 1) p_2 + (\gamma - 1) p_1 \right]$$
$$= \frac{\rho_2}{2} (v_2 - v_1)^2 \left[(\gamma + 1) p_1 + (\gamma - 1) p_2 \right] \qquad (10.11)$$

The Hugoniot for this case may be simply expressed in a universal form in terms of the pressure ratio $\Pi = p_2/p_1$ and the density ratio $y = \rho_2/\rho_1$ shown in Figure 10.1 for the case $\gamma = 5/3$.

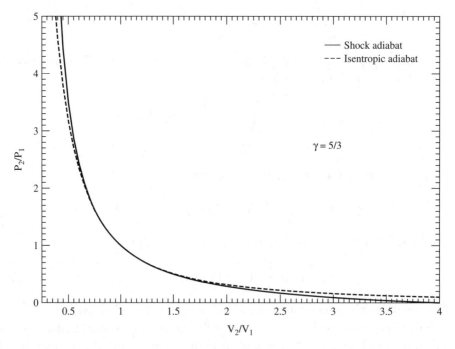

Figure 10.1: The Hugoniot curve for a polytropic gas with adiabatic index $\gamma = 5/3$ in terms of the specific volume V. Note that the pressure axis is plotted on a logarithmic scale.

Figure 10.1 exemplifies the typical form of the adiabat for many materials. The curve is concave towards larger final density, having a hyperbolic shape. The curve extends from a limiting compression at high upstream pressure through the initial condition to the limiting rarefaction at low pressure. There are two possible families of transitions satisfying the Rankine–Hugoniot equations. In general the compressive transition involves dissipation and gives rise to an entropy increase. Since discontinuous solutions formed in expansion decrease entropy, we conclude that there is an additional constraint to the Rankine–Hugoniot relations imposed by the second law of thermodynamics, which ensures the uniqueness of a shock discontinuity, i.e.

$$s_2 > s_1 \qquad (10.12)$$

The gradient of the chord (p_1, V_1) to (p_2, V_2) is $-j^2 = -\rho_1{}^2 v_1{}^2$ and therefore expresses the shock velocity. Hence allowing the final state to be nearly the same as that of the initial, we see that the gradient at the initial state yields the upstream sound speed through

$$\left.\frac{\mathrm{d}p}{\mathrm{d}\rho}\right|_1 = -(\rho_1 c_1)^2$$

In fact, as we show, the adiabatic curve through (p_1, V_1) has a third-order contact with the shock adiabat for weak shocks (Section 10.3.1). This is illustrated in Figure 10.2.

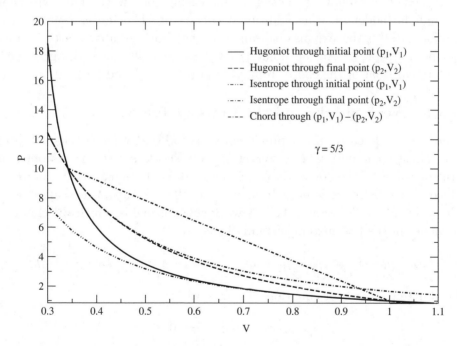

Figure 10.2: Upstream and downstream Hugoniot curves for a polytropic gas with adiabatic index $\gamma = 5/3$ plotted together, so that the upstream curve starts at $(p_1 = 1, V_1 = 1)$ and the downstream one at $(p_2 = 10, V_2 = 18.5)$. Also plotted are the adiabatic curves through each point, and the chord between the initial and final points.

Figure 10.2 allows the identification of many of the characteristic properties of shocks directly. Since the form is representative of nearly all materials, the results apply very generally. The shock adiabat lies above the isentrope through the initial point for compression, but below for expansion, thus confirming that shocks are only allowed in compression. The shock adiabat touches and intersects the isentrope at the initial point, in conformity with our later conclusion that the entropy change is of third order (Section 10.3.1). The magnitude of the gradient of the adiabat is greater than that of the chord from (p_1, V_1) to (p_2, V_2), and therefore the shock velocity is larger than the initial sound speed $v_1 > c_1$. We show later that the entropy gain is of third order in the pressure jump, and it follows that the entropy increases along the adiabat as the pressure increases. A result consistent with problem #33, which can be established quite generally (Landau and Lifshitz, 1959, §84).

In expansion, the shock adiabat would lie below the isentrope, reflecting a decrease in entropy for a transition obeying the Rankine–Hugoniot relations. This would be forbidden by the second law of thermodynamics. However, the Hugoniot plot allows extension to increasing specific volume (expansion) although forbidden by this additional constraint. The initial point (p_1, V_1) therefore lies on the expansion branch of the shock adiabat starting at the final point (p_2, V_2).[3] At the point (p_2, V_2) the magnitude of the gradient is greater than that of the chord from (p_1, V_1) to (p_2, V_2) and hence $v_2 < c_2$.

10.3.1   Weak Shocks and the Entropy Jump

It may be expected from our previous discussion that weak shocks will propagate along a characteristic. However, since a shock is a strong discontinuity with the values of the flow variables changing discontinuously, this is not guaranteed. Nonetheless it is easy to show that the velocity of a weak shock is indeed the adiabatic sound speed. A weak shock is one in which the change in the values of the parameters is small. Thus

$$p_2 = p_1 + \delta p \qquad \rho_2 = \rho_1 + \delta \rho \qquad v_2 = v_1 + \delta v \qquad h_2 = h_1 + \delta h$$

The Rankine–Hugoniot equations linearise to

$$\rho_1 \, \delta v + v_1 \, \delta \rho = 0$$
$$\rho_1 v_1 \, \delta v + \delta p = 0$$
$$\delta h + v_1 \, \delta v = 0$$

---

[3]Note that the shock adiabat from (p_2, V_2) is not the same as that from (p_1, V_1), even though both have the initial and final points in common.

Introducing the entropy change δs and making use of the thermodynamic relation $dh = T ds + (1/\rho) dp$ we reduce the equations to

$$\frac{\partial p}{\partial s}\bigg|_\rho \delta s + \left[\frac{\partial p}{\partial \rho}\bigg|_S - v_1{}^2\right]\delta\rho = 0$$

$$\left[\rho T + \frac{\partial p}{\partial s}\bigg|_\rho\right]\delta s + \left[\frac{\partial p}{\partial \rho}\bigg|_S - v_1{}^2\right]\delta\rho = 0 \tag{10.13}$$

Since the temperature T is non-zero, these relations are only consistent if $\delta s = 0$. Thus we conclude that to first order the shock is adiabatic. It is also clear that the shock velocity is

$$v_1 \approx \sqrt{\frac{\partial p}{\partial \rho}\bigg|_S} \approx c_1$$

The velocity change across the shock is clearly given by

$$v_1 - v_2 = \frac{\partial p}{\partial \rho}\bigg|_S \frac{(\rho_2 - \rho_1)}{\rho_1 v_1}$$

and since $v \approx v_1 \approx v_2 \approx c = \sqrt{\partial p/\partial\rho|_S}$

$$\frac{(v_1 - v_2)}{v} \approx \frac{(\rho_2 - \rho_1)}{\rho} \approx \frac{(p_2 - p_1)}{\rho c^2} \tag{10.14}$$

To identify the nature of the entropy change we must take the expansion to higher order. Expanding the enthalpy in Taylor's series in terms of entropy and pressure,

$$\delta h = T\,\delta s + V\,\delta p + \frac{1}{2}\frac{\partial V}{\partial p}\bigg|_S \delta p^2 + \frac{1}{6}\frac{\partial^2 V}{\partial p^2}\bigg|_S \delta p^3 \tag{10.15}$$

where $V = 1/\rho$ is the specific volume and we have made use of the thermodynamic relation used above to treat the derivatives of the enthalpy. From the third Rankine-Hugoniot equation we obtain

$$\delta h = \frac{1}{2}(v_1 + v_2)\,\delta v = \frac{1}{2}(V_2 + V_1)\,\delta p$$

$$= \frac{1}{2}\left\{2V_1\,\delta p + \frac{\partial V}{\partial p}\bigg|_S \delta p^2 + \frac{1}{2}\frac{\partial^2 V}{\partial p^2}\bigg|_S \delta p^3\right\}$$

Hence we obtain the result that

$$\delta s = \frac{1}{12\,T_1}\frac{\partial^2 V}{\partial p^2}\bigg|_S \delta p^3 \tag{10.16}$$

and we infer that the entropy change is third order in the pressure and density.

Since for nearly all materials $\partial^2 V/\partial p^2\big|_S > 0$, we conclude that entropy increases across the shock, and that a weak shock is thermodynamically irreversible provided the pressure increases across the shock, i.e. the shock is compressing. This entropy increase is the result of viscosity and thermal conduction within the shock front transition region as discussed earlier. It is also clear that shocks cannot form in expansion, an entropy decrease being forbidden by the second law of thermodynamics. There is therefore an entropy condition to be added to the Rankine–Hugoniot equations defining the shock:

$$s_2 \geq s_1 \tag{10.17}$$

This important result has only been derived for weak shocks, although applying to general materials. The generalisation to consider arbitrary shock strength and any material was achieved independently by Bethe (1942) and Weyl (1944), who identified additional criteria to ensure an entropy gain in compressive discontinuities. Fortunately for most materials these conditions are satisfied, and it may be assumed quite generally that shocks form in compression subject to the Rankine–Hugoniot relations. The result is proved for a polytropic gas in problem #33.

This result is important in the context of the Cauchy problem. As we have seen, solutions are well defined provided they are continuous. In compression we have shown that the flow is discontinuous when a shock is formed, the shock relations being expressions of the integral (conservation law) forms of the inviscid compressible fluid dynamical equations. In fact the Rankine–Hugoniot equations are not unique as a solution of the Cauchy problem; that is, if the upstream flow parameters are given, there are multiple solutions downstream. Once the entropy condition (10.17) is added, the uniqueness of the solution is obtained, as can be seen from the discussion of the Hugoniot plot in Section 10.3. A more formal discussion is given in Appendix 10.A.

For weak shocks equation (10.14) takes the form

$$\frac{(v_1 - v_2)}{v} = \frac{(\rho_2 - \rho_1)}{\rho} = \frac{1}{\gamma}\frac{(p_2 - p_1)}{p} \tag{10.18}$$

This result is identical to that for an adiabatic change, as is expected since the entropy increase in weak shocks is only third order in the shock strength.

10.4   Shocks in Real Gases

Thus far we have considered shock waves in ideal polytropic gases only, i.e. ones for which the enthalpy is

$$h = \frac{\gamma}{\gamma - 1}\frac{p}{\rho}$$

However a shock wave may involve considerable heating of the gas, which may result in changes in the vibrational and electronic structure of the molecules. These may take a short time to achieve, but the gas behind the shock will achieve a state of thermal equilibrium in which the various modes are balanced. The value of the polytropic constant γ is determined by the number of degrees of freedom f of the gas molecules. For a monatomic gas only motional degrees are allowed and $f = 3$, and for diatomic molecules two additional rotational degrees are allowed, $f = 5$. The polytropic constant is

$$\gamma = \frac{f+2}{f}$$

and for monatomic gas $\gamma = 5/3$ and for diatomic gas $\gamma = 7/5$.

As the temperature is increased the internal energy of the molecules becomes comparable with the vibrational and ultimately the electronic excitation energies. As a result progressively more degrees of freedom become involved and γ is decreased. The value of γ behind the shock wave may be estimated by considering the total specific enthalpy

$$\frac{(\gamma' - 1)}{\gamma'} = \varepsilon + pV = \varepsilon_t + \varepsilon_r + \varepsilon_v + \varepsilon_e + pV \tag{10.19}$$

where the terms ε are the specific internal energies associated with the various different modes and V the specific volume. However, we note that the vibrational and electronic modes do not become fully developed as classical degrees of freedom until

$$\bar{e} \sim kT \gg \Delta E$$

where ΔE is an appropriate energy interval.

In addition to these excitation effects the gas will also be dissociated and ionised if the temperature is sufficiently high. Generally ionisation becomes important before electronic excitation contributes significantly to the enthalpy (Figure 10.3).

Suppose the cold gas, into which the shock is driven, is composed of molecules M, which dissociate into two monatomic molecules A and B, which are themselves subsequently ionised. The total enthalpy per molecule of the gas, M, is made up of components from the dissociated molecules and their products, the ions and electrons:

$$\begin{aligned}
h' &= (1 - \alpha)\,\varepsilon_M + \varepsilon_V + \alpha\,D + \alpha\,x_A\,I_A + \alpha\,x_B\,I_B \\
&\quad + \alpha\,(1 - x_A)\,\varepsilon_{A^\circ} + \alpha\,x_A\,\varepsilon_{A^+} + \alpha\,(1 - x_B)\,\varepsilon_{B^\circ} + \alpha\,x_B\,\varepsilon_{B^+} \\
&\quad + \alpha\,(x_A + x_B)\,\varepsilon_{e^-} + pV
\end{aligned}$$

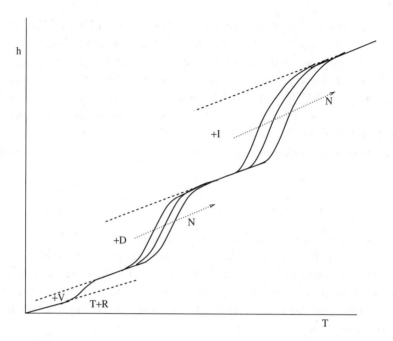

Figure 10.3: Sketch of the variation of specific enthalpy (h) with temperature (T) for a diatomic molecular gas whose constituent atoms are singly ionised. The regions where vibrational excitation (V), dissociation (D) and ionisation (I) play a role are indicated. The variations of dissociation and ionisation with initial number density of the molecule (N) are shown.

where α is the dissociation fraction and x_A and x_B the fractional ionisation of A and B respectively. D is the dissociation energy per mole and I_A and I_B the ionisation energies of A and B per mole of M. For a diatomic molecule the internal energy of a molecule comprises thermal, rotational and vibrational energies

$$\varepsilon_M = \frac{3}{2}kT + kT + \varepsilon_V$$

For a diatomic molecule the total vibrational energy of the molecule is determined by a Boltzmann distribution amongst the set of vibrational energy levels. The mean energy of a single mode of oscillation of frequency ν is

$$\varepsilon_V \approx \frac{h\nu}{[\exp(n\,h\nu/kT) - 1]}$$

where h is Planck's constant.

The dissociation fraction is determined by the law of mass action

$$\frac{\alpha^2}{1-\alpha} = \frac{K}{N_M} \qquad\qquad \text{heterogeneous molecules}$$

$$= \frac{K}{4N_M} \qquad\qquad \text{homogeneous molecules}$$

where N_M is the density of molecules M and K the dissociation constant, which varies as

$$K = \frac{\mathcal{Z}_{\text{atoms}}}{\mathcal{Z}_{\text{molecule}}} \exp(-D/kT)$$

where D is the dissociation energy and $\mathcal{Z}$ are the partition functions.

The ionisation fractions are described by Saha's equation, which is similar to the law of mass action

$$\frac{xZ}{1-x} \approx \frac{\hbar^3}{N_A (2\pi kT)^{3/2}} \frac{g_0}{2g_1} \exp(-I/kT)$$

where I is the ionisation energy, g_0 and g_1 the statistical weights of the neutral and ion respectively, Z the fractional number of electrons and N_A the number of atoms of species A. Only one stage of ionisation is assumed. If the density is low and the temperature high, Saha's equation may need to be replaced by one based on a coronal equilibrium in which total detailed balance is not assumed.

The internal energies of the ions are mainly their kinetic energies only, excitation playing a relatively small part in the overall energy balance, because the bound electrons are more readily ionised than excited. The internal energies of the atoms, ions and electrons per particle are therefore simply their kinetic energy

$$\varepsilon_{A^\circ} = \varepsilon_{A^+} = \varepsilon_{B^\circ} = \varepsilon_{B^+} = \varepsilon_{e^-} = \frac{3}{2} kT$$

The total pressure follows in a similar manner using Dalton's law of partial pressures. Thus the pressure per molecule of M is

$$p' = (1-\alpha)p_M + \alpha(1-x_A)p_{A^\circ} + \alpha x_A (p_{A^+} + p_{e^-})$$
$$+ \alpha(1-x_B)p_{B^\circ} + \alpha x_B (p_{B^+} + p_{e^-})$$

But in thermal equilibrium the pressures are all due to the thermal energy

$$p_M = p_{A^\circ} = p_{A^+} = p_{B^\circ} = p_{B^+} = p_{e^-} = kT$$

and the pressure takes the simpler form

$$\begin{aligned} p &= [1 - \alpha + \alpha(1 + x_A + x_B)] NkT \\ &= [1 + \alpha(x_A + x_B)] NkT \end{aligned} \qquad (10.20)$$

where N is the number density of the molecules M. The specific enthalpy is

$$h = \frac{1}{\mathcal{M}} \left\{ \frac{5}{2} kT [1 + \alpha + \alpha(x_A + x_B)] + kT(1-\alpha) \right.$$
$$\left. + \varepsilon_V + \alpha D + \alpha x_A I_A + \alpha x_B I_B \right\} \qquad (10.21)$$

where $\mathcal{M}$ is the molecular mass of M.

10.5   The Hydrodynamic Structure of the Shock Front

As we have discussed, the actual shock cannot be a mathematical discontinuity in which the jumps in velocity, density and pressure occur abruptly. Rather the behaviour of the fluid in the discontinuity must be governed by the viscosity and thermal conduction in the fluid, which generate the entropy increase necessary to support the shock. Such an analytic model was developed for polytropic gases by Rayleigh (1910), Taylor (1910) and Becker (1922). In general a purely hydrodynamic picture in terms of viscosity and thermal conduction is not satisfactory. The continuum approximation cannot be maintained as the shock thickness is of the same order of magnitude as the mean free path. In addition the use of bulk averaged transport coefficients, which imply that relaxation amongst the different energy modes is instantaneous, is invalid. A full discussion of the structure of the shock front is given in Zel'dovich and Raizer (1967, chap. 7). Becker's model, which we follow here, allows a description of the essential properties of the shock transition in a gas, particularly if the shock is weak, when we may express these changes using fluid dynamical principles of conservation. In the shock frame, the flow is steady: the conservation laws of mass (1.18), momentum (3.9) and energy (3.11) require that the corresponding total fluxes are constant through the shock

$$
\begin{aligned}
\rho v &= C_1 = \rho_1 M_1 c_1 \\
\rho v^2 - \tau &= C_2 = \left(\frac{1}{\gamma} + M_1^2 \right) \rho_1 c_1{}^2 \\
\rho h\, v + \frac{1}{2} \rho v^2 - \tau v - q &= C_3 = \left(\frac{1}{(\gamma-1)} + \frac{1}{2} M_1{}^2 \right) \rho_1 c_1{}^3
\end{aligned}
\tag{10.22}
$$

where $\tau = \frac{4}{3} \mu'\, \mathrm{d}v/\mathrm{d}x - p$ is the total stress and $q = \kappa\, \mathrm{d}T/\mathrm{d}x$ the heat flux due to thermal conduction. The effective ('longitudinal') viscosity is associated with the longitudinal stress for a one-dimensional flow and includes the second coefficient of viscosity $\frac{4}{3} \mu' = \frac{4}{3} \mu + \zeta$. To investigate this flow we assume a perfect gas whose pressure $p = n k T$ where $n = \rho/M$ is the number density of particles, k Boltzmann's constant and M the molecular mass. The internal energy may be written in terms of the temperature and the number of degrees of freedom f of the molecule as $\frac{1}{2} f n k T$.

The growth of entropy through the shock layer is determined by the viscous forces and thermal conduction from equation (3.12)

$$
\rho v \frac{\mathrm{d}s}{\mathrm{d}x} = \frac{1}{T} \left\{ \mu' \left(\frac{\mathrm{d}v}{\mathrm{d}x} \right)^2 + \frac{\mathrm{d}}{\mathrm{d}x} \left(\kappa \frac{\mathrm{d}T}{\mathrm{d}x} \right) \right\}
\tag{10.23}
$$

The entropy jump is obtained by integrating this equation across the shock

$$\rho v \left(s_2 - s_1 \right) = \int\limits_{-\infty}^{\infty} \frac{1}{T} \left\{ \mu' \left(\frac{dv}{dx} \right)^2 + \frac{d}{dx} \left(\kappa \frac{dT}{dx} \right) \right\} dx$$

$$= \int\limits_{-\infty}^{\infty} \left\{ \mu' \frac{1}{T} \left(\frac{dv}{dx} \right)^2 + \kappa \left(\frac{1}{T} \frac{dT}{dx} \right)^2 \right\} dx \geq 0 \qquad (10.24)$$

after an integration by parts. The total entropy generation across the shock is therefore positive as required. However, the entropy generated has been shown elsewhere (Sections 10.3.1 and 10.2.1) to be independent of the values of the viscosity and thermal conductivity. The role of viscosity and thermal conductivity is to determine the structure and thickness of the shock layer but not its limits.

This is a remarkable conclusion, in that a steady state flow between two constant states of flow can be established with values independent of the effects responsible for it. It is not obvious that such flows are possible. However, this can be rigorously established for a shock of arbitrary strength in a general fluid. To demonstrate this conclusion we consider in detail two cases:

1. Arbitrary shocks in polytropic gas with Prandtl number = 0.75 (Becker's solution).

2. Weak shocks in a general material.

The generalisation of both these cases involves the formal treatment of the governing differential equation without calculating the detailed structure. We note in passing that these are the two cases for which we proved that shocks satisfying the entropy condition existed. Extensive experimental experience confirms these conclusions and shows that shocks exist in this form.

In general the entropy has a maximum within the transition layer. Clearly the term due to viscosity is positive throughout the layer. However, the thermal conduction heat flux flows from the high temperature downstream towards the low temperature upstream. Since the gradient is zero at each limit, the temperature profile has a point of inflexion. Heat is removed from the fluid downstream of the point of inflexion, flows upstream and is added upstream. The entropy generation due to thermal conduction alone is therefore positive upstream of the temperature point of inflexion and negative downstream. This may give rise to the maximum in the entropy.

The thermal conductivity may be written in terms of the longitudinal Prandtl number $\mathcal{P}$

$$\kappa = \frac{c_p \, \mu'}{\mathcal{P}} = \frac{\gamma}{(\gamma - 1)} \frac{\mathcal{R}}{\mathcal{P}} \mu'$$

where $\mathcal{R} = k/\mathcal{M}$ is the gas constant. Neglecting the second coefficient of viscosity, kinetic theory yields for the first coefficient of viscosity

$$\mu' \approx \mu = \sqrt{\frac{2\mathcal{R}T}{\pi}}\, \nu\rho\ell \qquad (10.25)$$

for hard-sphere elastic molecules. The constant $\nu \approx 0.998$ and ℓ is the mean free path.

10.5.1 Polytropic Gas Shocks

We next consider the general case of arbitrary Mach number, but for a gas with restricted properties, namely ideal with a Prandtl number $\mathcal{P} = 3/4$.[4] The conservation equations (10.22) may be reduced to

$$
\begin{aligned}
C_1\, v + p - \mu' \frac{\mathrm{d}v}{\mathrm{d}x} &= C_2 \\
\frac{1}{(\gamma - 1)} C_1\, \mathcal{R}\, T\, v + C_2\, v - \frac{1}{2} C_1\, v^2 - \kappa \frac{\mathrm{d}T}{\mathrm{d}x} &= C_3
\end{aligned}
\qquad (10.26)
$$

The problem may be more conveniently treated by introducing a set of dimensionless variables to avoid unnecessary constants. We also introduce the dimensionless length in terms of mean free paths from equation (10.25) by neglecting the second viscosity

$$
w = \frac{C_1}{C_2}\, v \qquad \theta = \left(\frac{C_1}{C_2}\right)^2 \mathcal{R}\, T \qquad \text{and} \qquad s = \frac{x}{\ell}
$$

Equations (10.26) take the simpler form

$$
\begin{aligned}
\frac{4}{3} A\, \frac{\sqrt{\theta}}{w}\, \frac{\mathrm{d}w}{\mathrm{d}s} &= w + \frac{\theta}{w} - 1 \\
\gamma \mathcal{P}^{-1} A\, \frac{\sqrt{\theta}}{w}\, \frac{\mathrm{d}\theta}{\mathrm{d}s} &= \theta - \frac{(\gamma - 1)}{2}\left[(1 - w)^2 + \alpha\right]
\end{aligned}
\qquad (10.27)
$$

where $A = \sqrt{2/\pi}\,\nu$, and $\alpha = 2\, C_1\, C_3/C_2{}^2 - 1$.

As the distance tends to infinity (the boundaries of the shock) the gradients reduce to zero. Therefore

$$
w_1 + w_2 = \frac{2\gamma}{(\gamma + 1)} \qquad \text{and} \qquad w_1\, w_2 = \frac{(\gamma - 1)}{(\gamma + 1)} \cdot \frac{2\, C_1\, C_3}{C_2{}^2}
$$

whose values are consistent with those obtained directly (equations 10.3).

---

[4]**Historical note**   Solutions when either the viscosity ($\mathcal{P} = 0$) or the thermal conductivity ($\mathcal{P} = \infty$) are individually zero were obtained independently by Rayleigh (1910) and Taylor (1910). The solution for the case $\mathcal{P} = 3/4$ was obtained by Becker (1922). The formal behaviour of the general problem of arbitrary Prandtl number is due to von Mises (2004). No general analytic solution is possible.

In general these equations must be solved numerically. However, there is an analytic solution due to Becker (1922), which is valid for the particular value of the Prandtl number $\mathcal{P} = 3/4$. This value is typical of many gases, e.g. for air $\mathcal{P} = 0.88$, although not applicable to any particular gas. Substituting a quadratic in w for θ it is easily shown that both equations (10.27) are identical if

$$\theta = \frac{(\gamma - 1)}{2\gamma}\left(\frac{2\,C_1\,C_3}{C_2{}^2} - w^2\right) = \frac{\theta_2\,w_1{}^2 - \theta_1\,w_2{}^2}{w_1{}^2 - w_2{}^2} - \frac{\theta_2 - \theta_1}{w_1{}^2 - w_2{}^2}\,w^2 \tag{10.28}$$

This solution, when $\mathcal{P} = 3/4$, has the remarkable property, which is given either directly from equation (10.22) or from equation (10.28), that the sum of the enthalpy and kinetic energy flows is constant through the shock, namely

$$c_p\,T + \frac{1}{2}\,v^2 = c_p\,T_1 + \frac{1}{2}\,v_1{}^2 \tag{10.29}$$

which is a form of Bernoulli's equation. The equation may be expressed in terms of the dimensionless variables as

$$\frac{\gamma}{(\gamma - 1)}\,\theta + \frac{1}{2}\,w^2 = \frac{C_1 C_3}{C_2{}^2} \tag{10.30}$$

Since the gradients at both boundaries w_1 and w_2 are zero, equation (10.27) must take the form

$$\frac{4}{3}\,A\,\frac{\sqrt{\theta}}{w}\,\frac{dw}{ds} = -(w_1 - w)(w - w_2) \tag{10.31}$$

If the viscosity is constant through the shock, then there is an analytic solution to this differential equation

$$\frac{(w_1 - w)^{w_1}}{(w - w_2)^{w_2}} = \frac{(w_1 - \sqrt{w_1 w_2})^{w_1}}{(\sqrt{w_1 w_2} - w_2)^{w_2}}\exp\left\{\frac{3(w_1 - w_2)}{4(w_1 + w_2)}\,\frac{w_1}{A\sqrt{\theta_1}}\,s\right\} \tag{10.32}$$

where the distance $s = x/\ell$ is measured in mean free paths from $s = 0$ at the point of inflexion in velocity $w = \sqrt{w_1 w_2}$.

This equation has a convenient form in terms of the velocity ratio $\eta = v/v_1 = w/w_1$ in the shock and the Mach number M_1

$$\frac{(\eta - \eta_2)^{\eta_2}}{(1 - \eta)} = \frac{(\sqrt{\eta_2} - \eta_2)^{\eta_2}}{(1 - \sqrt{\eta_2})}\exp\left\{-\frac{3}{4\nu}\sqrt{\frac{\pi}{2\gamma}}\,\frac{(M_1{}^2 - 1)}{M_1}\,s\right\} \tag{10.33}$$

This solution clearly approaches the initial and final conditions asymptotically as the flow moves from $-\infty$ to $+\infty$. The thickness of the shock is estimated from

$$s' = \frac{(w_1 - w_2)}{|dw/ds|_{\max}} \approx \frac{4}{3}\,A\,\frac{\sqrt{\theta_1}}{w_1}\,\frac{w_1{}^2 - w_2{}^2}{(w_1 + w_2) - 2\sqrt{w_1 w_2}} \tag{10.34}$$

Figure 10.4 shows the variation of the shock thickness with Mach number for a gas with polytropic index $\gamma = 7/5$ and constant viscosity. Since $(w_1 - w_2)/(w_1 + w_2) = (M_1{}^2 - 1)/M_1{}^2$ and $w_1/\sqrt{\theta_1} = \sqrt{\gamma}M_1$, the shock front thickness decreases approximately as $\sim M_1/(M_1{}^2 - 1)$. There is clearly a serious problem

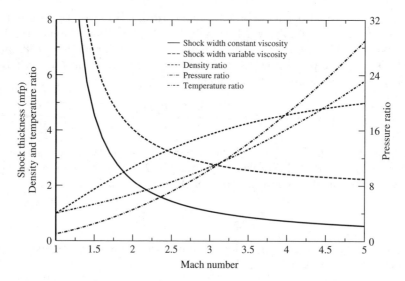

Figure 10.4: Plots of the shock layer thickness calculated in terms of the mean free path for gas of constant viscosity as a function of the Mach number and polytropic constant $\gamma = 7/5$. The constant and corrected temperature widths are plotted. Also shown are the density, pressure and temperature ratios across the shock.

with this solution as the thickness for Mach numbers greater than approximately 3 is less than the mean free path. In part this may be ascribed to the neglect of the variation of the viscosity through the shock. However, for variations over these short lengths, the hydrodynamic approximations fail and the shock should be described in terms of a kinetic theory model.

Figure 10.5 shows the variation of the velocity, pressure and temperature though a Mach 2 shock in a gas with index $\gamma = 1.4$. It can be seen that the thickness over which the variation takes place is about two mean free paths, consistent with Figure 10.4. The entropy difference shows a maximum at the point of inflexion of the velocity profile. This is found quite generally. It does not, however, reflect any violation of the second law of thermodynamics, as the entropy is only required to increase over the full transition. Using the first law of thermodynamics in the form

$$\rho T\,ds = dh - \frac{1}{\rho}\,dp$$

and making use of equation (10.29) and the second equation of the set (10.22), the entropy generation rate may be written as

$$\rho v T \frac{ds}{dx} = \rho v\left[-v\frac{dv}{dx} - \frac{1}{\rho}\frac{d}{dx}\left(\frac{4}{3}\mu'\frac{dv}{dx}\right) + v\frac{dv}{dx}\right] = -v\frac{d}{dx}\left(\frac{4}{3}\mu'\frac{dv}{dx}\right) = -\frac{4}{3}\mu' v\frac{d^2v}{dx^2}$$

$$(10.35)$$

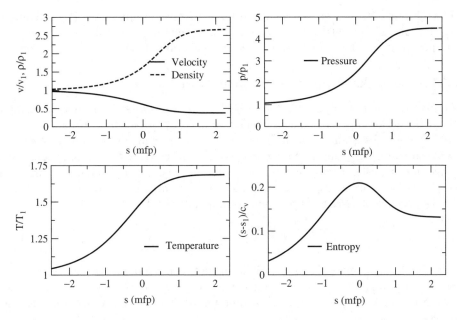

Figure 10.5: Plots of velocity ratio v/v_1, density ratio ρ/ρ_1, temperature ratio T/T_1, pressure ratio p/p_1 and entropy gain $(s - s_1)/c_v$ through a Mach 2 shock in a polytropic gas with $\gamma = 1.4$.

for constant viscosity and Prandtl number $\mathcal{P} = 3/4$. The entropy generation rate therefore increases up to the point of inflexion of the velocity profile, and falls thereafter. This occurs as a result of the different behaviours of the two terms comprising the entropy generation rate. Due to viscosity, the rate from equation (10.23) increases throughout the shock. Thermal conduction, on the other hand, varies as $\kappa \, \mathrm{d}^2 T/\mathrm{d}x^2$ and therefore changes sign at the point of inflexion.

Following Thomas (1944), we may approximately account for the variation of the viscosity and thermal conduction in the shock by modifying equation (10.32) to include the temperature change by using the approximate form from equations (10.25) and (10.28):

$$s' = \frac{(w_1 - w_2)}{|\mathrm{d}w/\mathrm{d}s|_{\max}} \approx \frac{4}{3} A \left\{ \frac{(w_1 - w_2) \left[(w_1 + w_2)(w_1 \, w_2 - (w_1 + w_2 - 1) \, w^2) \right]^{1/2}}{(w_1 - w)(w - w_2)} \right\}_{\min} \tag{10.36}$$

Figure 10.4 shows these corrected values for a range of Mach numbers. It can be seen that the shock front thickness is now typically a few mean free paths wide provided the shock is not too strong. For weak shocks the viscous model is not too inaccurate, but we will discuss real gas effects in the next sections, which modify the structure of the shock front.

The solution of the governing equations for an arbitrary value of the Prandtl number is due to von Mises (2004) and is expressed in a general form that does not allow an analytic expression. The results show the same general behaviour as Becker's solution.

10.5.1.1 Shocks supported by heat transfer

We noted earlier that is possible for a shock to be supported by heat transfer alone without the action of viscosity, a case first examined by Rayleigh (1910) and by Taylor (1910). It might be argued that both viscosity and thermal conduction are controlled by viscosity and that one is consequent on the other. However, it is also possible to have heat transfer totally independent of collisions, e.g. by radiative transfer. In general this heat flux is always anti-parallel to the temperature gradient $-\mathrm{d}T/\mathrm{d}x$.

In the absence of viscosity the momentum equation takes the form

$$p + \rho v^2 = C_2 = \frac{1}{\gamma}\left(1 + \gamma {M_1}^2\right) \tag{10.37}$$

The flow on the Hugoniot plot is a straight line between the points (p_1, V_1) and (p_2, V_2).

Since the left hand side of the first equation (10.27) is zero, we obtain the temperature through the shock layer as a function of the velocity in the dimensionless variables

$$\theta = w\left(1 - w\right) \tag{10.38}$$

which has a maximum at $w = 1/2$. The behaviour of the flow depends on whether this maximum lies within the shock layer flow. Noting that

$$w_1 = \frac{\gamma {M_1}^2}{1 + \gamma {M_1}^2} \qquad \text{and} \qquad w_2 = \frac{\gamma\left[(\gamma - 1){M_1}^2 + 2\right]}{(\gamma + 1)\left(1 + \gamma {M_1}^2\right)} \tag{10.39}$$

clearly w_1 and w_2 have the limits $\gamma/(1 + \gamma) \leq w_1 \leq 1$ and $\gamma/(1 + \gamma) \geq w_2 \geq (\gamma - 1)/(\gamma + 1)$ for small and large Mach numbers M_1 respectively:

- For weak shocks, $M_1 < \sqrt{(3\gamma - 1)/[\gamma\,(3 - \gamma)]}$, when $\gamma/(\gamma + 1) > w_2 > 1/2$, the maximum lies outside the flow. The temperature gradient is therefore $\mathrm{d}T/\mathrm{d}x > 0$ everywhere. The heat flow is counter to the stream as required to generate the necessary pressure gradient. This is consistent with the preceding weak shock layer calculation.

- For strong shocks, $M_1 > \sqrt{(3\gamma - 1)/[\gamma\,(3 - \gamma)]}$, and $w_1 > 1/2 > w_2$. The temperature maximum therefore lies within the shock layer. The heat flow below the point $w = 1/2$ is therefore downstream taking the heat away from the head of the layer where it is required. Writing the second equation of (10.26) in terms of the heat flux and substituting for the temperature,

$$\frac{C_1}{{C_2}^2}\, q = -\mathcal{K}\,\frac{\mathrm{d}\theta}{\mathrm{d}x} = -\frac{(\gamma + 1)}{2}\left(w_1 - w\right)\left(w - w_2\right) \tag{10.40}$$

where $\mathcal{K}(w, \theta)$ is a rate coefficient of appropriate form for the type of heat transfer process. This equation becomes inconsistent when the sign of the temperature gradient changes at the maximum, provided the velocity is monotonically decreasing. The latter can easily be shown to be the case from the above equation if $\mathcal{K}$ is constant, but also more generally The flow requires an *isothermal discontinuity*, a discontinuous isothermal jump, from the point where $\theta = \theta_2$ to the final downstream flow at $w = w_2$. This can only be accomplished by the introduction of viscosity, which is contrary to our original assumption.

We may therefore conclude that it is possible to have shocks supported solely by heat transfer provided the shock is sufficiently weak, the necessary dissipation and entropy growth being provided by heat transfer alone. For strong shocks this is no longer possible without the isothermal discontinuity.

10.5.2   Weak Shocks

For weak shocks the pressure and entropy differences are small across the flow. We therefore seek to expand the flow variables in terms of powers of the pressure and entropy differences using the conservation equations and Taylor's expansions (Landau and Lifshitz, 1959, §87). This follows closely the procedure used in Section 10.3.1. We shall find that the reciprocal of the shock thickness δ is first order in pressure difference, but that the entropy difference is second order. We need therefore to retain terms up to (and including) second order in pressure and only first order in entropy.[5] Thus the momentum and energy conservation equations take the form

$$(p - p_1) + C_1{}^2 (V - V_1) - \frac{4}{3} C_1 \mu' \frac{\mathrm{d}V}{\mathrm{d}x} = 0 \tag{10.41}$$

and

$$(h - h_1) + \frac{1}{2} C_1{}^2 (V^2 - V_1{}^2) - \frac{4}{3} C_1 \mu' \frac{\mathrm{d}V}{\mathrm{d}x} - \frac{\kappa}{C_1} \frac{\mathrm{d}T}{\mathrm{d}x} = 0 \tag{10.42}$$

where, as before, we have used the specific volume V rather than density. Expanding the change in specific volume by Taylor's series,

$$V - V_1 = \left.\frac{\partial V}{\partial p}\right|_s (p - p_1) + \frac{1}{2} \left.\frac{\partial^2 V}{\partial p^2}\right|_s (p - p_1)^2 + \left.\frac{\partial V}{\partial s}\right|_p (s - s_1) \tag{10.43}$$

It will transpire that the reciprocal of the thickness of the front is of order $(p - p_1)$ and that in consequence spatial derivatives of the form $\mathrm{d}/\mathrm{d}x$ increase the order by one. Hence the term $\mathrm{d}s/\mathrm{d}x$ is of third order and therefore negligible, whilst $\mathrm{d}p/\mathrm{d}x$ is second order. Thus

$$\frac{\mathrm{d}V}{\mathrm{d}x} = \left.\frac{\partial V}{\partial p}\right|_s \frac{\mathrm{d}p}{\mathrm{d}x} + \left.\frac{\partial V}{\partial s}\right|_p \frac{\mathrm{d}s}{\mathrm{d}x} \approx \left.\frac{\partial V}{\partial p}\right|_s \frac{\mathrm{d}p}{\mathrm{d}x} \tag{10.44}$$

Substituting for $\mathrm{d}V/\mathrm{d}x$ and $V - V_1$ in equation (10.41) we obtain

$$\left[1 + C_1{}^2 \left.\frac{\partial V}{\partial p}\right|_s\right] (p - p_1) + \frac{1}{2} C_1{}^2 \left.\frac{\partial^2 V}{\partial p^2}\right|_s (p - p_1)^2 + \left.\frac{\partial V}{\partial s}\right|_p (s - s_1) = \frac{4}{3} \mu' C_1 \left.\frac{\partial V}{\partial p}\right|_s \frac{\mathrm{d}p}{\mathrm{d}x} \tag{10.45}$$

---

[5] This expansion is in line with our earlier conclusion that the total entropy jump was of third order in pressure.

Multiplying this equation by $(V_1 + V_2)/2$ and subtracting from equation (10.42) we obtain

$$(h - h_1) - \frac{1}{2}(p - p_1)(V_1 + V_2) - \frac{2}{3}\mu' C_1 (V - V_1) \frac{dV}{dx} - \frac{\kappa}{C_1} \frac{dT}{dx} = 0 \qquad (10.46)$$

The third term is of third order and therefore negligible, and from the definition of the enthalpy and retaining only terms up to second order,

$$h - h_1 = \int_1 T\, ds + V\, dp \approx T(s - s_1) + \frac{1}{2}(V + V_1)(p - p_1) \qquad (10.47)$$

Expanding dT/dx and neglecting the term in ds/dx as before in equation (10.43),

$$T(s - s_1) = \frac{\kappa}{C_1} \left.\frac{\partial T}{\partial p}\right|_s \frac{dp}{dx} \qquad (10.48)$$

Eliminating the entropy change in equation (10.45) we finally obtain an equation for pressure across the shock front

$$\frac{1}{2}C_1^2 \left.\frac{\partial^2 V}{\partial p^2}\right|_s (p - p_1)^2 + \left[1 + C_1^2 \left.\frac{\partial V}{\partial p}\right|_s\right](p - p_1) = C_1 \left\{-\frac{\kappa}{T} \left.\frac{\partial V}{\partial s}\right|_p \left.\frac{\partial T}{\partial p}\right|_s + \frac{4}{3}\mu' \left.\frac{\partial V}{\partial p}\right|_s\right\} \frac{dp}{dx} \qquad (10.49)$$

This equation may be simplified by noting that as $x \to \pm\infty$, $dp/dx \to 0$ and $p \to p_1$ or $p \to p_2$, the final upstream and downstream states of the shock. Hence the quadratic on the left hand side of the equation may be expressed as $(p - p_1)(p - p_2)$ and we obtain the differential equation

$$\begin{aligned}
\frac{dp}{dx} &= \frac{1}{2}C_1 \left\{\left.\frac{\partial^2 V}{\partial p^2}\right|_s \middle/ \left[-\frac{\kappa}{T} \left.\frac{\partial V}{\partial s}\right|_p \left.\frac{\partial T}{\partial p}\right|_s + \frac{4}{3}\mu' \left.\frac{\partial V}{\partial p}\right|_s\right]\right\}(p - p_1)(p - p_2) \\
&= \frac{1}{2}C_1 \frac{c^2}{V^2}\left\{\left.\frac{\partial^2 V}{\partial p^2}\right|_s \middle/ \left[\frac{4}{3}\mu' + \frac{\kappa}{T}\left(\left.\frac{\partial T}{\partial p}\right|_s\right)^2 \frac{c^2}{V^2}\right]\right\}(p - p_1)(p_2 - p) \quad (10.50)
\end{aligned}$$

since $\left.\partial V/\partial s\right|_p = \left.\partial T/\partial p\right|_s$.

Since the shock is weak the viscosity and thermal conduction are nearly constant. Setting the origin of the co-ordinate $x = 0$ at the point of inflexion of the pressure, $\frac{1}{2}(p_1 + p_2)$ (at the mean pressure point), and integrating we obtain

$$\begin{aligned}
x &= \frac{4aV^2}{(p_2 - p_1)\left.\partial^2 V/\partial p^2\right|_s} \ln\left\{\frac{(p - p_1)}{(p_2 - p)}\right\} \\
&= \frac{4aV^2}{\frac{1}{2}(p_2 - p_1)\left.\partial^2 V/\partial p^2\right|_s} \operatorname{arctanh}\left\{\frac{p - \frac{1}{2}(p_1 + p_2)}{\frac{1}{2}(p_2 - p_1)}\right\} \qquad (10.51)
\end{aligned}$$

where approximating $C_1 \approx c_1/V_1$ since the shock is weak, we obtain

$$a = \frac{V}{2c^3}\left[\frac{4}{3}\mu' + \frac{\kappa}{T}\left(\left.\frac{\partial T}{\partial p}\right|_s\right)^2 \frac{c^2}{V^2}\right]$$

which is a quantity associated with the decay of sound waves.

Hence we obtain the layer thickness

$$\delta = 8aV^2/(p_2 - p_1)\, \partial^2 V/\partial p^2\big|_s \tag{10.52}$$

and

$$p = \frac{1}{2}(p_2 + p_1) + \frac{1}{2}(p_2 - p_1)\tanh\left(\frac{x}{\delta}\right) \tag{10.53}$$

The density and hence the velocity variations are easily obtained to first order

$$\rho - \rho_1 \approx \frac{1}{c^2}(p - p_1) \qquad \text{and} \qquad \frac{v}{v_1} = \frac{\rho_1}{\rho} \tag{10.54}$$

The entropy follows from equation (10.48)

$$s - s_1 = \frac{\kappa}{16\,a\,c\,V\,T}\,\frac{\partial T}{\partial p}\bigg|_s\,\frac{\partial^2 V}{\partial p^2}\bigg|_s (p_2 - p_1)\,\mathrm{sech}^2\left(\frac{x}{\delta}\right) \tag{10.55}$$

The entropy change has a maximum at $x = 0$, i.e. at the point of inflexion associated with thermal conduction. If the shock is a result of viscosity alone this maximum should vanish. This result is in accord with our earlier observations. The entropy change tends to zero on the downstream side of the shock, which is clearly in error. This discrepancy reflects the fact that the calculation has been carried out to second-order accuracy only, but we have previously (Section 10.3.1) shown that the entropy jump is a third-order term.

Weyl (1944) considered the generalisation of this result for an arbitrary material, and showed that stable shock layers representing a matched solution between two ideal flows existed. As may be expected this general result does not allow a solution in a closed analytic form.

When the fluid is an ideal polytropic gas the adiabatic equation of state is $pV^\gamma = \text{const.}$ Thus

$$\frac{\partial T}{\partial p}\bigg|_s = \frac{(\gamma - 1)\,T}{\gamma}\,\frac{1}{p} \qquad \frac{\partial p}{\partial V}\bigg|_s = -\frac{\gamma p}{V} = -\frac{\gamma RT}{V^2} = -\frac{c^2}{V^2} \quad \text{and} \quad \frac{\partial^2 V}{\partial p^2}\bigg|_s = \frac{(\gamma + 1)}{\gamma^2}\,\frac{V}{p^2}$$

and making use of Maxwell's relations we obtain

$$a = \frac{2V}{c^3}\left[\frac{4}{3}\mu' + \kappa\left(\frac{1}{c_v} - \frac{1}{c_p}\right)\right] = \frac{2V}{c^3}\left[\frac{4}{3}\mu' + \frac{(\gamma - 1)\kappa}{c_p}\right] = \frac{2V}{c^3}\,\mu'\left(\frac{4}{3} + (\gamma - 1)\mathcal{P}^{-1}\right) \tag{10.56}$$

where $\mathcal{P} = c_p\,\mu'/\kappa$ is the Prandtl number for the longitudinal viscosity. Taylor and Maccoll (1935) obtained this result directly from equations (10.26) for a polytropic gas.

It follows from equation (10.56) that the shock layer is formed for all values of the Prandtl number, i.e. independently of whether the dissipation is due to viscosity or thermal conduction or a combination of the two. However, the two processes act differently to decelerate the flow and increase the entropy:

- **Viscosity** is a friction force which acts to convert the directed motion of the flow into random thermal motion. The heat is generated by collisional scattering from the conversion of the kinetic energy of the one-dimensional flow into randomly thermal modes.

- **Thermal conduction** transfers heat from the hot regions upstream to the cold gas downstream. The consequent increase in pressure at the higher density causes the flow to decelerate. The heating of the gas is a consequence of the compression resulting from the increased pressure. This mechanism can only occur in continuous flow if the shock is not too strong.

10.6   The Shock Front in Real Gases

In Section 10.5 we examined the structure of the shock front as generated by viscosity and thermal conduction. The very important result was found that dissipation allowed the generation of a shock transition independent of the details of the dissipation, the latter only determining the thickness of the front, which was typically only a few mean free paths. In real gases, this approach correctly identifies the underlying physics, but is not accurate. Fortunately, exact values of shock wave thickness are rarely needed, and a qualitative estimate is normally sufficient. The problems are:

- The scale length of a few mean free paths is not sufficient to allow the approximations of continuum theory.

- The time scale of the flow through the front is insufficient to allow thermal equilibrium to be established.

To overcome these problems, kinetic theory has been used to calculate the shock thickness (Zel'dovich and Raizer, 1967, chap. 7). The thickness thereby obtained was significantly greater than that due to Thomas's continuum theory. Hornig and co-workers obtained experimental measurements of front thickness in approximate agreement with the kinetic theory results (Bradley, 1962, chap. 7), using an optical technique based on optical reflection within the shock front.

When the fluid is a monatomic gas and the shock not too strong, the role of the shock front is simply to randomise a fraction of the directed motion of the gas in front of the shock into the thermal energy behind it. This is accomplished by a series of collisions amongst the gas molecules, which lead to a thermal equilibrium (Maxwell–Boltzmann) molecular velocity distribution. Typically only two or three collisions are required to substantially reduce the flow velocity, although slightly more are needed to achieve full equilibrium. Within the fluid picture this is due to the viscosity, and the thickness estimated thereby is not inconsistent with this kinetic picture.

When the gas is polyatomic, the situation is more complicated in that additional degrees of freedom are introduced. Suppose the shock is so strong that molecular vibration or electronic excitation is not established. To illustrate this behaviour suppose the only additional mode introduced is rotation. This typically relaxes into an equilibrium distribution more slowly than the motional modes–usually 10 to 15 collisions. In hydrogen it takes about 300 collisions, due to the low moment of inertia of the molecule and the consequent large separation of the rotational levels. The rotational relaxation occurring more slowly is therefore nearly independent of the translational one. The gas therefore rapidly relaxes to conditions behind a shock with three degrees of freedom, i.e. $\gamma = 5/3$ associated with translational motion, within the model discussed earlier. The slower relaxation of the rotational modes increases the number of degrees of freedom and decreases the temperature correspondingly. A diatomic gas with two rotational modes relaxes to the Rankine–Hugoniot final state of a gas with $\gamma = 7/5$. The increase in entropy from the initial condition behind the shock front to the final one is due to the dissipation associated with the mode relaxation. The time taken for the energy distribution of a mode to relax to the thermal equilibrium value is known as the *relaxation time*.

In fact rotational relaxation is usually sufficiently fast that it can be accommodated within the shock front and its relaxation process is not identifiable in the manner described above. Such modes which are fully populated in the shock front are known as *active modes*. The energy separation of the states is generally small compared with the thermal energy available making excitation easy. However, vibrational and electronic excitation levels generally have large energy gaps exceeding the thermal energy. Their relaxation is generally slow compared with the active modes, the relaxation playing a distinctive role in the temporal development of the flow behind the shock (Figure 10.6). They are known as *inert modes*. The relaxation time for the active (motional and rotational) modes is typically a few collision intervals, due to the small energy gaps. For the inert modes relaxation times are generally of the order of microseconds, although these are small compared with the characteristic times of the flow and therefore are effectively a discontinuity. However, they are readily measured experimentally.

The slow relaxation of the inert modes causes viscosity to play very little part in the changes in the pressure and density during this phase of relaxation. As a result the pressure and specific volume vary linearly during this phase. Figure 10.7 shows the effects of this for a gas with one inert mode. It can be seen that the initial compression from (p_1, V_1) to (p_2', V_2') associated with the active modes takes places along a curved path, whereas the final stage to (p_2, V_2) is along a straight line on the Hugoniot plot. Note that due to the additional modes, the final state Hugoniot, including the inert modes, lies below (in pressure) that for the active modes alone. If the shock is sufficiently weak, it may happen that the shock chord lies below the tangent to the active mode Hugoniot (but above the full one). In this case no intermediate shock

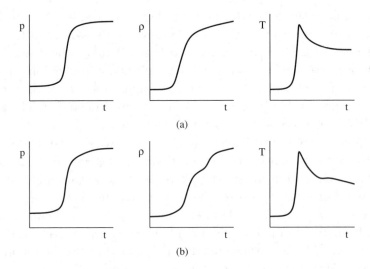

Figure 10.6: Sketch of the variations of the pressure, density and temperature with time behind the shock front for a gas with one inert mode (a) and one with two (b). Note the change in relaxation rate when thermal equilibrium is nearly achieved for the faster mode.

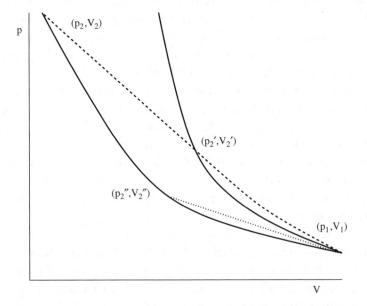

Figure 10.7: Sketch of the Hugoniot curves for a gas with one inert mode. The upper curve is the Hugoniot plot for the active modes of a shock (dashed), initial state (p_1, V_1) and passing through the intermediate state (p_2', V_2'). The relaxation to the lower final Hugoniot is along the straight line to the final state (p_2, V_2). If the shock is weak the shock transition to the active modes alone may not be allowed and the shock moves directly along a straight line (dotted) to the final state (p_2'', V_2'').

is able to form,[6] but one may with full relaxation. In this case the active modes thermalise almost synchronously with the relaxation of inert modes. This latter situation is rare as the excitation of vibrational modes requires reasonably strong shocks. The most likely situation where this may occur is a weak shock in hydrogen.

The second viscosity ζ is clearly associated with volume change. Relaxation processes such as this may therefore be described within the fluid picture by an appropriate value of this coefficient. However, it will be appreciated that in this case, the value of this viscosity will depend on the characteristic time in the problem and the relaxation rate, i.e. it cannot be treated as a tabulated constant, unlike the first coefficient or thermal conductivity. Fortunately its use is very infrequent.

As the shock strength increases, additional modes are brought into play: firstly vibration and dissociation, subsequently electronic excitation and ionisation. All these processes are slow to reach equilibrium. Relaxation times are measured in nanoseconds and microseconds instead of collision intervals. The slow rate of relaxation means the region behind the shock front is thick, and clearly structured as the temperature falls to accommodate the energy required to excite the vibrational and electronic energy levels. Since the translational relaxation rates are very much faster than those of excitation, the motional modes have a Maxwellian distribution and a well-defined temperature, although total equilibrium is not yet achieved. Within the relaxation zone there is a clear structure as the temperature decreases, the density increases and the pressure falls, which can be observed experimentally.

10.7   Shock Tubes

A shock tube is a relatively simple device for generating shock waves in the laboratory under controlled conditions (Figure 10.8). It is constructed from a circular or rectangular tube usually of steel. High-pressure gas in the *driver section* is separated from a low-pressure gas in the *driven section*. The separating diaphragm is rapidly removed allowing the high-pressure gas to expand into the low-pressure driven gas. The expansion into the low-pressure gas produces a wave, which eventually settles down into a stable shock. A centred rarefaction propagates back into the driver section. The driver and driven gases are separated by a large discontinuity in density and temperature, known as the contact surface, but pressure and velocity are constant across it. Although the contact surface is generally stable, strong mixing between driver and driven gases occurs due to the turbulence resulting from the rupture of the diaphragm.

---

[6]Since the shock is subsonic with respect to the active modes.

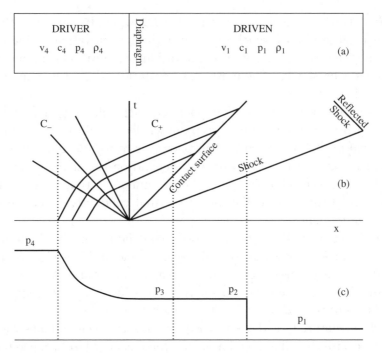

Figure 10.8: Schematic of the basic shock tube configuration (a). The high-pressure driver gas is separated from the low-pressure driven gas by a diaphragm. Removal of the diaphragm drives a shock wave into the driven gas, separated from the driver gas by a contact surface discontinuity (b). A centred rarefaction propagates back into the driver. At the end of the shock tube the shocked gas is brought to rest in a reflected shock. The resulting pressure distribution is shown in (c).

Measurements are therefore always made after the shock has travelled some distance and is well separated from the contact surface. Different gases are normally used as the driver and driven.

If the tube is closed by a rigid end, a reflected shock propagates back through the shocked gas. This is often prevented by a large evacuated expansion tank, called a *dump chamber*, separated from the shock tube by a second diaphragm, where the moving shocked gas is brought to rest. A reflected rarefaction is also formed from the end of the driver section, which normally plays no role in the shock-forming process.

The sudden rupture of the diaphragm at a controlled pressure is critical to the reproducibility of the generated shocks, and can by accomplished in a number of ways:

- A mechanically driven knife is used to puncture the diaphragm.
- Annealed diaphragms of plastic or metal with well-defined bursting pressure are used, often scored to rupture evenly.

The driver pressure is raised slowly until the diaphragm bursts. Plastic is used for low pressure and metal, typically copper or aluminium, for high.

Shock tubes have a variety of applications which make use of both the high temperatures generated and the rapid rise in temperature:

- Chemical kinetics measuring molecular reaction rates and dissociation times.
- Aerodynamics as a hypersonic wind tunnel allowing high temperatures and pressures, e.g. to simulate atmospheric re-entry of spacecraft.
- Testing conditions in the turbine section of jet engines.
- Generating plasma flows under controlled conditions.

A major limitation for these applications is the limited flow time of a few milliseconds between the transit of the shock and the contact surface or the reflected shock.

10.7.1   Shock Tube Theory

A simple model of the shock tube for polytropic gases is easily developed using the theory of the centred rarefaction (9.3.1) and the Rankine–Hugoniot equations for a polytropic gas, equations (10.3).

If M_s is the Mach number of the shock propagating into the undisturbed gas[7], the pressure ratio across the shock is (equation 10.3b)

$$\frac{p_2}{p_1} = \frac{2\gamma_1 M_s{}^2 - (\gamma_1 - 1)}{\gamma_1 + 1} \tag{10.57}$$

where γ_1 is the polytropic index of the driven gas in which the shock forms.

The velocity of the gas in the laboratory frame flowing behind the shock is given by

$$u_2 = U - v_2 = \left[1 - \frac{(\gamma_1 - 1)M_s{}^2 + 2}{(\gamma_1 + 1)M_s{}^2}\right]U = \frac{2}{(\gamma_1 + 1)}\left(\frac{M_s{}^2 - 1}{M_s}\right)U \tag{10.58}$$

where v_2 is the velocity of the flow behind the shock in the shock frame.

The pressure and velocity are constant through the shocked gas, and therefore the pressure is p_2 at the contact surface. Since the pressure and velocity are constant across the contact surface, the pressure and velocity at the tail of the rarefaction are

$$p_3 = p_2 \qquad \text{and} \qquad u_3 = u_2 \tag{10.59}$$

The pressure ratio across the rarefaction is due to an adiabatic change

$$\frac{p_4}{p_3} = \left\{\frac{c_4}{c_3}\right\}^{2\gamma_4/(\gamma_4 - 1)} = \left\{\frac{c_4}{c_4 - \frac{1}{2}(\gamma_4 - 1)u_3}\right\}^{2\gamma_4/(\gamma_4 - 1)} \tag{10.60}$$

---

[7]The velocity of the shock U propagating into the undisturbed gas in the laboratory frame is equal to the velocity of the gas entering the shock in the shock frame v_1, but in the opposite direction.

Collecting all these results together we obtain the pressure ratio between the initial driver and driven gases

$$\frac{p_4}{p_1} = \left\{ \frac{2\gamma_1 M_s^2 - (\gamma_1 - 1)}{(\gamma_1 + 1)} \right\} \left\{ 1 - \frac{(\gamma_4 - 1)}{(\gamma_1 + 1)} \frac{c_1}{c_4} \frac{(M_s^2 - 1)}{M_s} \right\}^{-2\gamma_4/(\gamma_4 - 1)} \qquad (10.61)$$

It is clear that there is a maximum shock strength which can be generated for the largest pressure ratio

$$M_{s\,(\mathrm{lim})} = \frac{1}{2} \frac{(\gamma_1 + 1)}{(\gamma_4 - 1)} \frac{c_4}{c_1} + \sqrt{1 + \left[\frac{1}{2} \frac{(\gamma_1 + 1)}{(\gamma_4 - 1)} \frac{c_4}{c_1} \right]^2} \approx \frac{(\gamma_1 + 1)}{(\gamma_4 - 1)} \frac{c_4}{c_1} \qquad (10.62)$$

Equation (10.61) is rather inconvenient in not allowing closed solutions for the Mach number as a function of the pressure ratio. As one normally wants to know the Mach number as a function of the pressure ratio, solutions are usually obtained graphically, Figure 10.9. To obtain a particular Mach number for a minimum pressure ratio, it is clear from equation (10.62) that the term containing the ratio of sound speeds should be maximised, i.e. the ratio c_1/c_4 should be minimised. The driver gas should therefore be a light gas such as hydrogen or helium, the latter being generally preferred for safety reasons.

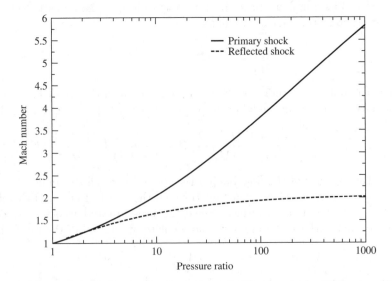

Figure 10.9: Variation of the pressure ratio p_4/p_1 for an argon shock driven by helium, all gases at room temperature. Also shown is the Mach number of the reflected shock when the primary shock is brought to rest.

To obtain very high-shock Mach numbers $M_s \sim 20$ it is necessary to use very high driver pressures with hydrogen. This has led to the combustion-driven shock tube where the driver gas is a mixture of hydrogen and oxygen, which is uniformly ignited by a thin wire heating element. A thick scored metal diaphragm is used to prevent the expansion starting before the maximum pressure is reached, but allows the rupture to take place smoothly and (reasonably) reproducibly. The driver gas is both light (hydrogen) and hot, both of which increase the sound speed and therefore the shock strength. The optimum composition oxygen:hydrogen ratio is approximately 8%:92%. To increase the Mach number further, the flow is constricted by reducing the cross-section at the junction of the combustion chamber with the flow tube, typically by a factor of about 2. By these means shock Mach numbers of about 18 are generated in argon.

10.8   Shock Interaction

In this section we will consider the behaviour of a planar shock when it interacts with either a rigid wall or a second shock, both of which are parallel to the initial shock. When the shock is incident normally on a rigid wall or a second shock, the situation is reasonably simple. The case of two shock waves interacting obliquely is much more complicated and we will leave a discussion of that situation to more advanced standard texts such as Courant and Friedrichs (1948, chap. 3 §D) and Landau and Lifshitz (1959, §93).

10.8.1   Planar Shock Reflection at a Rigid Wall

Of particular practical importance is the case when the shock tube is closed by a rigid plate. The flow is brought to rest by a reflected shock wave which propagates back into the already shocked gas. As a result the density and temperature in the working zone are both increased. The primary shock moves into undisturbed gas at shock speed U with parameters ρ_1, p_1, c_1 and $u_1 = U - v_1$ $= 0$ in the laboratory frame. Behind the shock, the gas has parameters ρ_2, p_2, c_2 and $u_2 = U - v_2$, also in the laboratory frame. The relationship of the downstream parameters to the upstream ones is given by the Rankine–Hugoniot equations (10.1).

The reflected shock moves back away from the wall with a speed U' in the laboratory frame into the downstream gas behind the primary shock (Figure 10.10). We recall that the positive direction is that of flow into the moving shock, i.e. the opposite of that for the primary shock. The conditions downstream of the reflected shock ρ_3', p_3', c_3' and $u_3' = v_3' - U' = 0$ are similarly determined by the Rankine–Hugoniot equations, from those behind the primary shock, where the upstream flow velocity is $u_2' = v_2' - U' = -u_2$ in the frame of the reflected shock. These equations may be solved quite generally for the parameters behind the reflected shock given the Mach number of the primary shock M_s, but not in analytic form for a general equation of state. However, for a polytropic gas the solution is relatively simple.

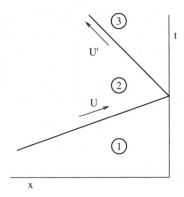

Figure 10.10: Sketch of trajectories of the incident and reflected shocks at a rigid wall. The incident shock moves with speed U and the reflected one, U'.

The Mach number of the flow into the reflected shock, $M_r = (u'_2 - U')/c_2$, determines the density, pressure and temperature state behind the reflected shock following relations (10.3).

From the equation for the velocity behind the primary shock (10.3d) we obtain in the laboratory frame

$$u_2 = v_2 - U = -\frac{2}{(\gamma+1)}\left(\frac{M_s{}^2 - 1}{M_s}\right)U \tag{10.63}$$

which is negative since the flow is in the direction of the moving shock.

In the frame of the reflected shock, the incoming velocity is $v'_2 = U' + u'_2$ and the outgoing one $v'_3 = U'$ since $u'_3 = 0$. Hence from the velocity ratio across a shock (10.3a)

$$U' = \frac{[(\gamma-1)M_r{}^2 + 2]}{(\gamma+1)}[U' + u'_2] = \frac{[(\gamma-1)M_r{}^2 + 2]}{2(M_r{}^2 - 1)}u'_2$$

where $M_r = (U' + u'_2)/c_2$. Since the reflected and incident shocks move in opposite directions, $u'_2 = -u_2$, and we obtain

$$\frac{u_2}{c_2} = -\frac{2}{(\gamma+1)}\left(\frac{M_r{}^2 - 1}{M_r}\right)$$

$$= -\frac{2\left(M_s{}^2 - 1\right)}{\sqrt{\left\{\left[2\gamma M_s{}^2 - (\gamma-1)\right]\left[(\gamma-1)M_s{}^2 + 2\right]\right\}}} \tag{10.64}$$

making use of the ratio of sound speeds (10.3c) across the initial shock. Hence we may calculate the reflected shock Mach number from that of the primary shock in the shock tube, Figure 10.9.

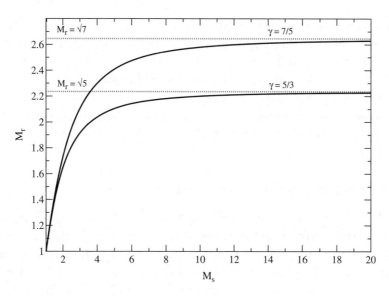

Figure 10.11: Plots of the reflected shock Mach number M_r as a function of the incident Mach number M_s for polytropic gases of index $\gamma = 5/3$ and $\gamma = 7/5$.

Figure 10.11 shows the Mach number of the reflected shock M_r as a function of the incident Mach number M_s. For weak shocks $M_r \approx M_s \sim 1$ and the reflected wave is simply the direct reflection of the incoming 'sound' wave. The overall pressure jump behind the reflected shock is readily shown to be $(p_3 - p_1) = 2(p_2 - p_1)$ as in a sound wave. For strong incident shocks, on the other hand, it is easily seen that the reflected shock Mach number has a limiting value

$$M_{r \lim} = \sqrt{\frac{2\gamma}{(\gamma - 1)}} = \sqrt{f + 2}$$

where $f = 2/(\gamma - 1)$ is the number of degrees of freedom of the driven gas. The corresponding density (y_r), pressure (Π_r) and temperature (Θ_r) ratios across the reflected shock are

$$y_r = \frac{\gamma}{(\gamma - 1)} = \frac{f + 2}{2} \qquad (10.65a)$$

$$\Pi_r = \frac{(3\gamma - 1)}{(\gamma - 1)} = f + 3 \qquad (10.65b)$$

$$\Theta_r = \frac{(3\gamma - 1)}{\gamma} = \frac{2(f + 3)}{(f + 2)} \qquad (10.65c)$$

There is clearly only a limited maximum density rise that can be achieved across a shock and reflected shock, which is subsequently brought to rest, however large the driver pressure. This limiting value is $\gamma(\gamma + 1)/(\gamma - 1)^2$, which is

10 for a monatomic gas with $\gamma = 5/3$, ($f = 3$). The limit values for the overall pressure and temperature ratios are easily found from equations (10.6). The overall temperature jump is $2(\gamma - 1)(3\gamma - 1)/\gamma(\gamma + 1)^2 M_s^2$, a value of $0.45 M_s^2$ for $\gamma = 5/3$, the bulk of the incident kinetic energy being converted to heat as a consequence of the entropy increase.

Reflected shocks are useful for several applications enabling the pressure and temperature to be easily increased with little increase in experimental effort in shock tube experiments.

10.8.1.1   Collision between two planar shocks

Consider the case where two shocks of Mach numbers M_s and M'_s are travelling in opposite directions along the x direction, Figure 10.12. They collide normally, forming two reflected shocks M_r and M'_r travelling in opposite directions away from the collision. Behind the reflected shocks both the downstream pressure $p_3 = p'_3 = p$ and the downstream flow velocity in the laboratory frame $u_3 = -u'_3 = u$ must be equal.[8]

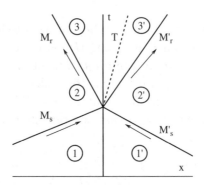

Figure 10.12: Sketch of trajectories of the two incident and reflected shocks. The incident shocks move with Mach numbers M_s and M'_s and the reflected one with M_r and M'_r. The contact surface is T.

If the two shocks are not identical, we have a set of equations based on the polytropic Rankine–Hugoniot equations (10.3) which allow us to express the two compatibility equations for velocity and pressure. We proceed by making use of the relations (10.3) to derive the pressure and flow velocities behind each incident shock derived earlier. The flow velocity behind each primary shock is given by equation (10.63) for each shock. Making use of equations (10.11) we

---

[8]Since the signs follow the convention on page 243, the downstream flow velocities are opposite for the two shocks.

may write

$$(p - p_2)^2 = \frac{\rho_2}{2} (u_2 - u)^2 [(\gamma + 1) p + (\gamma - 1) p_2]$$

$$(p - p_2')^2 = \frac{\rho_2'}{2} (u_2' + u)^2 [(\gamma + 1) p + (\gamma - 1) p_2']$$

(10.66)

Thus we have a pair of nonlinear simultaneous equations for the common pressure $p = p_3 = p_3'$ and flow velocity $u = u_3 = -u_3'$, which may be solved given the values of p_2, p_2', u_2 and u_2' calculated from the Mach numbers of the incoming shocks M_s and M_s'. A numerical solution using the Newton–Raphson method is satisfactory, and efficiently gives the values of the pressure p and velocity u behind the reflected shocks. The Mach numbers M_r and M_r' (equations 10.3), and the density ratios y_r and y_r' (equation 10.4a), across the reflected shocks are obtained from the corresponding pressure ratios $\Pi_r = p/p_2$ and $\Pi_r' = p/p_2'$, and thence the temperature and sound speed ratios are obtained.

When the two shocks have different strengths, the density behind each reflected shock is different. There is therefore a contact surface discontinuity across which the pressure and flow velocity are constant, and between gas which has passed through the first and the second shock. The velocity of the contact surface itself is equal to that of the gas, namely u, and is towards the weaker shock due to the higher pressure behind the stronger primary shock. To illustrate this behaviour the collision between two shocks of Mach numbers $M_s = 8$ and $M_s' = 4$ in a gas with $\gamma = 5/3$ generates reflected shocks having Mach numbers $M_r = 1.633$ and $M_r' = 3.188$ respectively. The overall compressions from the initial condition are $y = 7.194$ and $y' = 10.4$. The common flow velocity behind the shocks is $u = -3.108$.

If the two shocks are of equal strength, it is easily seen that symmetry demands that the flow velocity behind the shocks must be zero, $u = 0$. Hence the problem may be considered as an extension of the flow described in the previous section, where the shock is reflected from a solid wall, the contact surface taking the place of the wall.

10.8.2   Overtaking Interactions

Thus far we have considered only interactions in which the flows collide. Since a shock moves supersonically with respect to the gas ahead of it and subsonically with respect to that behind, we conclude that a trailing shock will always catch and interact with one ahead of it. Furthermore a rarefaction moves sonically with respect to both the upstream and downstream fluid (9.3). Therefore in a similar fashion a shock will always catch a rarefaction and vice versa a rarefaction a shock. On the other hand the separation of two rarefactions will remain constant. A succinct general account of these interactions is given by Bradley (1962, chap. 3 §5).

10.8.2.1   Shock overtaking a shock

The flow in this case is similar to that of the colliding shocks and may be analysed by a similar method based on equation (10.11). The result is a forward-going shock and a backward rarefaction. Between the two is a contact surface

$$ S_> \, S_> \; \longrightarrow \; \mathcal{R}_< \, \mathcal{T} \, S_> $$

where $S_>$ is a forward-going shock, $\mathcal{R}_<$ a backward-going rarefaction and $\mathcal{T}$ a contact surface.

10.8.2.2   Shock–rarefaction overtaking

The interaction of a shock with a rarefaction is a complex interaction as a consequence of the extended physical form of the rarefaction. As a result the interaction may take a long time and lead to the generation of a complex series of waves. However, if the overtaking wave is much stronger than the overtaken one, the interaction may be completed in a finite time and yields

$$ S_> \, \mathcal{R}_> \; \longrightarrow \; S_< \, \mathcal{T} \ldots \mathcal{T} \, S_> $$
$$ \mathcal{R}_> \, S_> \; \longrightarrow \; \mathcal{R}_< \, \mathcal{T} \ldots \mathcal{T} \, S_> $$

where the backward-going shocks $S_<$ are a series of weak compression waves which will in due course coalesce into a shock. The set of contact surfaces $\mathcal{T} \ldots \mathcal{T}$ is formed from different parts of the interaction.

10.8.2.3   Shock interaction with a contact surface

Two different modes of interaction are possible depending on a general acoustic impedance. In Section 8.3 we found that a sound wave was reflected with or without phase reversal if the acoustic impedance ρc was increased or decreased across the surface respectively. Generalising this result to consider shock waves crossing the boundary such that the total velocity and pressure are continuous across the density discontinuity, we find that the acoustic impedance is replaced by

$$ A = \frac{1}{c} \left[\gamma \, (\gamma + 1) + (\gamma - 1) \, \frac{1}{\Pi} \right]^{1/2} \tag{10.67} $$

where Π is the pressure ratio generated by the applied pressure of the incident shock in each gas.

1. **Generalised acoustic impedance decreases across the interface.**
   Both forward-going and backward shocks result:

$$\mathcal{S}_> \, \mathcal{T} \longrightarrow \mathcal{S}_< \mathcal{T} \, \mathcal{S}_>$$

2. **Generalised acoustic impedance increases across the interface.**
   A forward-going shock and backward rarefaction result:

$$\mathcal{S}_> \, \mathcal{T} \longrightarrow \mathcal{R}_< \, \mathcal{T} \, \mathcal{S}_>$$

Simpler sufficient conditions may be derived (Courant and Friedrichs, 1948; Bradley, 1962), but are not general.

The latter case has an interesting application. Rock and concrete are strong in compression, but weak in tension. An explosive charge placed against the front surface drives a shock through a block. At the rear, the wave is reflected back as a rarefaction which fractures the material. The 'dam busting' bouncing bomb during the Second World War was based on this effect.

10.9 Oblique Shocks

Thus far we have only considered planar shocks in which the flow is normal to the plane of the shock. In fact shocks can occur where the flow is inclined at an arbitrary angle to the shock. This situation is easily treated by a simple extension to the theory of normal shocks.[9]

We resolve the incoming flow speed $\mathbf{v}_1$ into its normal v_{1n} and tangential component v_{1t}. Consider the situation as seen in the frame moving with velocity v_{1t} along the shock front. In this frame only the normal component of velocity is seen, and the shock is therefore developed as a normal shock front. The Rankine–Hugoniot equations (10.1) therefore hold, but with the total velocities replaced by their normal components. Thus the downstream values (p_2, ρ_2, v_{2n}) are given by the Rankine–Hugoniot relations from the set (p_1, ρ_1, v_{1n}). Returning to the laboratory frame, it is clear that only the normal velocity is changed, and that therefore $v_{2t} = v_{1t}$.

Since the transverse component of the velocity is constant across the shock, the difference in the velocity vectors across the shock $(\mathbf{v}_1 - \mathbf{v}_2)$ is normal to the shock, a result that will be useful later.

---

[9]**Historical note** The original analysis of oblique shocks is due to Prandtl and Meyer (1908) in the latter's PhD thesis (Meyer, 1908), which also contains experimental observations of shocks in supersonic flow. The shock polar (§10.14) was introduced by Busemann (1930).

Using the third Rankine–Hugoniot equation,

$$h_1 + \frac{1}{2}\left(v_{1n}{}^2 + v_{1t}{}^2\right) = h_2 + \frac{1}{2}\left(v_{2n}{}^2 + v_{2t}{}^2\right)$$

$$h_1 + \frac{1}{2}v_1{}^2 = h_2 + \frac{1}{2}v_2{}^2 \tag{10.68}$$

Prandtl's relation (10.5) holds for the normal components of the velocity. It must therefore be modified in a similar fashion by the inclusion of the tangential velocity component

$$v_{1n}\,v_{2n} = \frac{(\gamma - 1)}{(\gamma + 1)}\left(v_{\text{max}}{}^2 - v_t{}^2\right) \tag{10.69}$$

The effect of the shock is to refract the streamlines towards the shock. The shock angle β is determined by the following conditions:

$$v_{1n} = v_1 \sin\beta$$

$$v_{2n} = \frac{1}{y}v_{1n} = \frac{1}{y}v_1 \sin\beta \tag{10.70}$$

$$v_{2t} = v_{1t} = v_1 \cos\beta \tag{10.71}$$

The downstream flow makes an angle θ with the initial flow, i.e. an angle $(\beta - \theta)$ with the shock front. Therefore

$$v_2 \sin(\beta - \theta) = \frac{1}{y}v_1 \sin\beta$$

$$v_2 \cos(\beta - \theta) = v_1 \cos\beta \tag{10.72}$$

We now particularise the flow to that of a polytropic gas for which the compression ratio is given by (10.3a) in terms of the Mach number of the normal flow $M = v_1 \sin\beta/c_1$. Thus

$$\tan(\beta - \theta) = \frac{1}{y}\tan\beta = \frac{[(\gamma - 1)M^2 + 2]}{(\gamma + 1)M^2}\tan\beta \tag{10.73}$$

defines the angle of the shock.

The angle through which the flow is deflected is easily found to be given by

$$\tan\theta = 2\cot\beta\frac{M_1{}^2 \sin^2\beta - 1}{M_1{}^2[\gamma + \cos(2\beta)] + 2} \tag{10.74}$$

where M_1 is the Mach number of the incoming flow, so that $M = M_1 \sin\beta$.

This is an important relation for the application of oblique shocks in aeronautics. Unfortunately the form is not very convenient as one generally has the angle of rotation of the flow given, and needs to know the angle of the shock to the incoming flow. The inversion of the above equation is not simple, although reasonably rapid using some method for solving nonlinear equations, e.g. Newton–Raphson. To avoid this complication a plot of θ versus β for differing Mach numbers M_1 is convenient. This is known as the θ–β–M plot and is shown in Figure 10.13. Once the shock angle is known together with the Mach number of the incoming flow, the flow downstream of the shock is readily calculated, since $M = M_1 \sin \beta$.

We may note a number of important points about this plot:

1. There are two possible solutions to the oblique shock: one is always subsonic and the other normally supersonic in the flow behind the shock. The subsonic and supersonic regions are separated by the sonic line, where the downstream flow velocity equals the sound speed. The strong shock region occurs at larger shock angles β, so that the shock is more nearly normal, and therefore stronger. The weak region at smaller angles represents a more oblique, and therefore weaker, shock.

2. There is a maximum angle through which the flow can be turned. The flow behind the shock at this limit angle is always just supersonic. The limit line separates the weak and strong shock regions. Shock angles greater than the limit line are strong and vice versa.

3. If the angle of rotation of the flow $\theta = 0$, the two possible solutions correspond to a normal shock, $\beta = 90°$, the strong case with the downstream Mach number $M_2 < 1$; or in the weak case to a weak shock wave at the Mach angle μ leaving the flow unchanged.

4. As the deflection angle is increased the shock becomes stronger.

5. Experimentally it is found that the weak branch is found, the strong branch not naturally occurring.

Weak oblique shock

When the shock lies on the weak branch and the deflection of the flow through the shock is small, $\theta \sim 0$, it follows from equation (10.74) that $M_1 \sin \beta \approx 1$, i.e. the shock angle β equals the Mach angle μ. This is expected as a weak shock is equivalent to a weak discontinuity, which as we have seen propagates along the Mach line. In this case the normal component of the velocity $v_{1n} \approx c_1$ or

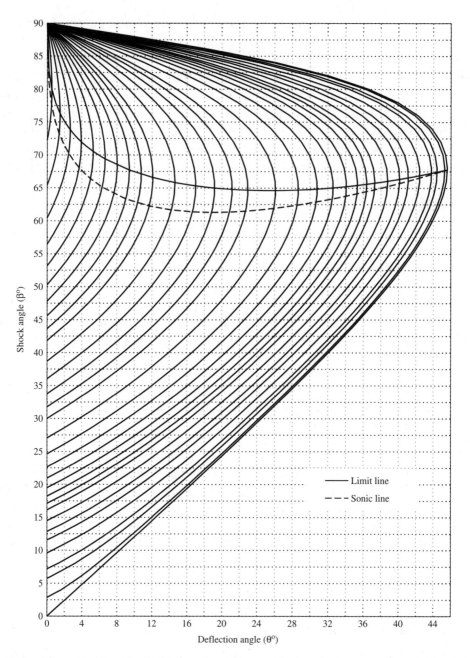

Figure 10.13: The $(\theta-\beta-M)$ plot for a gas of $\gamma = 1.4$ appropriate to air. The different curves apply for increasing Mach numbers 1.05, 1.1, 1.15, 1.2, 1.25, 1.3, 1.35, 1.4, 1.45, 1.5, 1.6, 1.7, 1.8, 1.9, 2.0, 2.2, 2.4, 2.6, 2.8, 3.0, 3.2, 3.4, 3.6, 3.8, 4.0, 4.5, 5.0, 6.0, 8.0, 10.0, 20.0, ∞ moving outwards. The limit line separates the weak and strong shock regions, the weak region corresponds to small shock angle β and the strong to large. The sonic line separates subsonic and supersonic flow behind the shock.

$M \approx 1$. The downstream properties of the shock are determined by $(M - 1)$ or equivalently the deflection angle of the velocity θ. Thus we have

$$
\begin{aligned}
v_{1n} &= v_{1t} \tan \beta \\
v_{2n} &= v_{2t} \tan (\beta - \theta) \\
&\approx v_{1n} \frac{(\tan \beta - \theta \sec^2 \beta)}{\tan \beta}
\end{aligned}
$$

Hence the velocity increment through the shock is

$$
\mathbf{v_2 - v_1} = v_{2n} - v_{1n} \approx -v_1 \theta \sec \beta \tag{10.75}
$$

As may be predicted, this result is identical to that for a small deflection in a rarefaction except, of course, that the sign of the increment is reversed.

The increments in the sound speed, pressure and density are easily obtained by the use of Bernoulli's formula across the shock:

$$
\begin{aligned}
v_2{}^2 - v_1{}^2 &= 2v_{1n}(v_{2n} - v_{1n}) + (v_{2n} - v_{1n})^2 \approx -2v_1{}^2 \theta \tan \beta \\
\therefore \quad c_2{}^2 - c_1{}^2 &\approx (\gamma - 1) v_1{}^2 \theta \frac{1}{\sqrt{M_1{}^2 - 1}}
\end{aligned}
$$

Since the shock is weak, the flow is nearly adiabatic. Therefore we may with little error use the adiabatic equation of state to calculate the pressure and density changes, or alternatively use the results for a weak normal shock (10.18):

$$
\begin{aligned}
\frac{p_2 - p_1}{p_1} &= \gamma \frac{v_1{}^2}{c_1{}^2} \theta \frac{1}{\sqrt{M_1{}^2 - 1}} \\
\frac{\rho_2 - \rho_1}{\rho_1} &= \frac{v_1{}^2}{c_1{}^2} \theta \frac{1}{\sqrt{M_1{}^2 - 1}}
\end{aligned} \tag{10.76}
$$

We note that these values are identical to those obtained for a weak rarefaction (9.50), with the sign of the deflection reversed, as may be expected since both represent the change resulting from the flow through a single Mach line.

10.9.1 Large Mach Number

When the Mach number is large, the compression ratio is independent of the Mach number (equation 10.6). Writing $c = \cot \beta$ and $c_0 = \cot \theta$ we obtain the quadratic equation

$$
y\, c^2 - (y - 1)\, c_0\, c + 1 = 0
$$

whose solution yields two possible values for the shock angle β, one supersonic (weak shock) and one subsonic (strong shock) on the downstream side.

Which one occurs in practice depends on the physical situation, but usually the supersonic one occurs. However, the above equation only has real solutions if

$$c_0^2 (y-1)^2 > 4y \quad \text{or} \quad \theta < \arcsin\left\{\frac{(y-1)}{(y+1)}\right\} = \arcsin\left(\frac{1}{\gamma}\right) \quad (10.77)$$

The shock angle β is easily calculated from the value of c at the limit, namely

$$\beta = \arcsin\sqrt{\frac{(\gamma+1)}{2\gamma}} \quad (10.78)$$

For larger angles of flow or shock, no shock attached at the apex is possible.

10.9.2   The Shock Polar

Consider the flow in the hodograph plane using Cartesian co-ordinates with the x direction parallel to the incoming flow. The normal and transverse velocity components are

$$v_{1n} = v_1 \sin \beta \quad v_{2n} = v_{2x} \sin \beta - v_{2y} \cos \beta$$
$$v_{1t} = v_{2t} = v_1 \cos \theta = v_{2x} \cos \theta + v_{2y} \sin \theta \quad (10.79)$$

Hence
$$\tan \beta = \frac{v_1 - v_{2x}}{v_{2y}} \quad (10.80)$$

Using Prandtl's relation for oblique flow (10.69), the substitutions (10.79) and $v_{max} = \sqrt{(\gamma+1)/(\gamma-1)}\, c_*$,

$$\frac{v_{2x} \sin \beta - v_{2y} \cos \beta}{v_1 \sin \beta} = \frac{(\gamma-1)}{(\gamma+1)} + \frac{2c_*^2}{(\gamma+1)\, v_1^2 \sin^2 \beta} \quad (10.81)$$

Substituting for β from equation (10.80) we obtain the equation for the shock polar

$$v_{2y}^2 \left\{\frac{2}{(\gamma+1)} v_1^2 - v_1 v_{2x} + c_*^2\right\} = (v_1 - v_{2x})^2 \left(v_1 v_{2x} - c_*^2\right) \quad (10.82)$$

where the definitions of the limit speed (1.43) and the critical velocity (1.47) have been used to simplify the expression.

Figure 10.14 shows a typical plot of the shock polar at Mach number $M_1 = 1.96$ where $M_1 = v_1/c_1$ is appropriate to the incoming flow, rather than the normal component only as used previously. The graph is a strophoid and related to the folium of Descartes. The curve is a plot of the function (10.82). The flow velocity behind the shock is given by the vector OP in both magnitude and direction. The range of the downstream flow is clearly seen between the

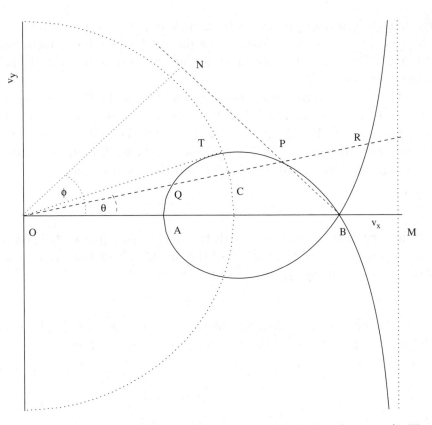

Figure 10.14: The shock polar for a polytropic gas with adiabatic index $\gamma = 5/3$. The critical speed $c_* = 1$ and the incoming flow speed $V_1 = 1.5$, Mach number $M = 1.96$. The line OP represents the flow behind the shock in magnitude and direction. The line OT is the tangent to the polar representing the limiting angle. Note there are three possible intersections of a flow with a given direction with the polar.

limiting values, A, $u_0 = c_*^2/v_1$ and B, v_1, both limits corresponding to flow normal to the shock. There are three intersections of the flow at a specified angle with the polar. Two represent possible final states P and Q, the third, R, corresponds to a final flow speed greater than v_1 and is therefore forbidden. However, the Rankine–Hugoniot equations are reversible, irreversibility only being introduced by the entropy condition. The intersection R therefore represents a flow starting at R and terminating at B. The polar has an asymptote for the branch $v_2 > v_1$ at M where $u_2 = u_0 + 2v_1/(\gamma + 1)$.

The sonic limit C, a circle of radius c_*, identifies the transition between subsonic AC (strong shock) and supersonic BC (weak shock) flow in the downstream flow. The sonic point at which the downstream flow is sonic is easily calculated by equating $v_2{}^2 = c_*{}^2$. When possible the supersonic (weak shock) flow downstream is normally established.

The line corresponding to the limiting angle $\theta_{\lim}$ is shown on Figure 10.14 as the tangent OT to the polar from the origin. In this case the limiting angle is about $20.2°$. The limiting condition is established close to the sonic transition, in supersonic flow.

Since the vector OB represents the incoming flow and OP the outgoing flow, it follows that BP is the normal to the shock,[10] a result which can also be seen from equation (10.80). In consequence ON is the tangential velocity and BN the normal component in front of the shock. Furthermore, since BP is perpendicular to ON, $\angle OBP = \pi/2 - \beta$, and the shock angle

$$\beta = \arctan \frac{(v_1 - v_{2x})}{v_{2y}} \tag{10.83}$$

The limiting angle is not too difficult to calculate by equating the gradient of the shock polar at T to the gradient of the line OT. This leads to a quadratic equation in the value of v_{2x} at the limit, namely

$$a\, v_{2x}{}^2 + b\, v_{2x} + c = 0$$

whose coefficients are complicated but may be expressed as simple functions of the incoming flow speed and the two bounds $u_0 = c_*^2/v_1$ and $u_2 = u_0 + 2v_1/(\gamma + 1)$:

$$a = v_1 - \frac{1}{2}(u_2 - u_0)$$

$$b = -\frac{1}{2}v_1(3u_0 + u_2)$$

$$c = u_0\, v_1\, u_2$$

The velocity v_{2y} is obtained from the shock polar (10.82) and hence the limiting angle. Figure 10.15 shows plots of the limiting angle for shock formation as a function of the Mach number of the incoming flow M_1 for two different polytropic indices. Also shown is the approximation (10.85), which can be seen to overestimate the angle for Mach number $M_1 \gtrsim 1.2$. As noted earlier the downstream flow at the limiting angle is just subsonic (strong shock).

When the Mach number of the incoming flow is large, $M \gg 1$, the shock polar reduces to a circle passing through the abscissa points $\sqrt{(\gamma - 1)/(\gamma + 1)}\,c_*$ and $\sqrt{(\gamma + 1)/(\gamma - 1)}\,c_*$; it is then a trivial problem to show that

$$\sin(\theta_{\lim}) = \frac{\left[\sqrt{(\gamma + 1)/(\gamma - 1)} - \sqrt{(\gamma - 1)/(\gamma + 1)}\right]}{\left[\sqrt{(\gamma + 1)/(\gamma - 1)} + \sqrt{(\gamma - 1)/(\gamma + 1)}\right]} = \frac{1}{\gamma} \tag{10.84}$$

in accord with our earlier calculation.

---

[10] Since $\mathbf{OB} - \mathbf{OP} = \mathbf{v}_1 - \mathbf{v}_2$ which is normal to the shock.

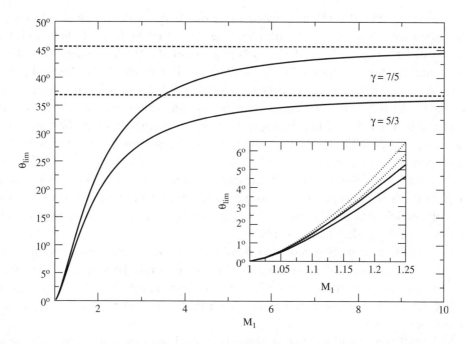

Figure 10.15: The limiting angle for the formation of an attached shock for polytropic gases with adiabatic index $\gamma = 5/3$ and $7/5$ as a function of Mach number M_1 of the incoming flow. The inset plot shows the values for weak shocks (full line) compared with the approximation (10.85) (dotted line), the upper curves referring to $\gamma = 7/5$ and the lower ones to $\gamma = 5/3$.

When the shock is weak, $M \approx 1$, v_1 and v_2 are both nearly equal to c_*, thus $v_{2x} \approx v_2$, $v_{2y} \approx c_* \theta$ and θ is small, The shock polar takes the simple form

$$\theta^2 \approx \frac{(\gamma + 1)\,(v_1 - v_2)^2\,(v_1 + v_2 - 2c_*)}{2c_*{}^3}$$

Hence it is easy to show that

$$\theta_{\lim} = \frac{4\sqrt{\gamma + 1}}{3\sqrt{3}}\left(\frac{v_1}{c_*} - 1\right)^{3/2} = \frac{8\sqrt{2}}{3\sqrt{3}(\gamma + 1)}\{M_1 - 1\}^{3/2} \tag{10.85}$$

The values generated by this approximation are shown in Figure 10.15 compared with those obtained from the accurate solution. It can be seen that the approximation is reasonable for Mach numbers $M_1 \lesssim 1.2$, where the angles of attachment are quite small.

10.9.3   Supersonic Flow Incident on a Body

Consider a wedge of half angle θ with a supersonic flow symmetrically incident upon it. A shock will form on each side starting from the apex at an angle β to the incoming flow direction, provided a shock is allowed. If the shock rotates the flow through an angle θ, the resulting flow downstream of the shock will be parallel to the surface of the wedge and satisfy the necessary conditions for a uniform flow along the wedge surface. This requires that the wedge half angle $\theta < \theta_{\text{lim}}$. If this condition is satisfied, an oblique shock will form, whose strength will be correspondingly reduced to that appropriate for the normal component of the incoming flow speed $M = v_{1n}/c_1$ rather than the flow speed itself $M_1 = v_1/c_1$. The pressure generated on the surface of the wedge will be correspondingly reduced.

However, if the wedge exceeds the limiting angle $\theta > \theta_{\text{lim}}$ the shock can no longer attach to the apex of the wedge. The shock therefore detaches itself to provide sufficient space between the shock and the wedge surface to allow the flow to rotate parallel to the surface.

The problem now becomes a difficult one to solve and recourse may be made to computational methods. As the wedge angle becomes progressively larger, the shock is further detached from the surface. The limiting case at which the wedge has increased to $90°$ is a blunt body, when the flow stagnates at the leading edge.

Although the complete flow is difficult to evaluate we may deduce the pressure on the axis of the body where a stagnation point is established. Thus consider a symmetric blunt body with the flow incident normally along the line of symmetry with velocity v_1 (Figure 10.16). After passage through the shock, normal to its surface, the velocity is reduced to v_2. The streamline along the axis will touch the surface of the body and continue to follow it symmetrically above and below the line of the incoming flow. Therefore by symmetry the flow is brought to rest at the point where the axial streamline meets the surface $v_3 = 0$. Since the flow is adiabatic behind the shock, Bernoulli's equation (1.41) may be used along the streamline between the two points ② and ③ with the adiabatic equation of state to obtain

$$\frac{p_3}{p_2} = \left\{ 1.0 + \frac{(\gamma - 1)}{2} \frac{v_2{}^2}{c_2{}^2} \right\}^{\gamma/(\gamma-1)} \tag{10.86}$$

The ratios of the pressure and the Mach numbers across the shock are easily obtained from the Rankine–Hugoniot set (10.3). Hence we obtain the pressure ratio between the stagnation point on axis and the incoming flow

$$\frac{p_3}{p_1} = \frac{p_3}{p_2} \frac{p_2}{p_1} = \frac{2\gamma M^2 - (\gamma - 1)}{(\gamma + 1)} \left\{ 1.0 + \frac{(\gamma - 1)}{2} \frac{(\gamma - 1)M^2 + 2}{2\gamma M^2 - (\gamma - 1)} \right\}^{\gamma/(\gamma-1)}$$

$$= \left(\frac{(\gamma + 1)}{2} \right)^{(\gamma+1)/(\gamma-1)} \frac{M_1{}^2}{\left\{ \gamma - (\gamma - 1)/2M_1{}^2 \right\}^{1/(\gamma-1)}} \tag{10.87}$$

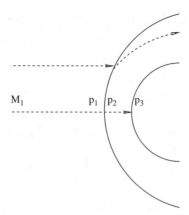

M_1 $\qquad\qquad\qquad$ p_1 p_2 p_3

Figure 10.16: Sketch of the shock wave and streamlines (dashed) around a sphere with a detached shock. The pressures p_1, p_2 and p_3, at the corresponding points ①, ② and ③ refer to the upstream flow, the downstream flow behind the shock and the stagnation point on the surface of the sphere.

For strong shocks the pressure ratio is $[(\gamma+1)/2]^{(\gamma+1)/(\gamma-1)} M_1{}^2$ at the surface of the body compared with $2\gamma/(\gamma+1)M_1{}^2$ behind the shock due to the adiabatic flow behind the shock, i.e. a factor of 2.53 for $\gamma = 5/3$ and 2.56 for $\gamma = 7/5$, which is a substantial increase.

10.10   Adiabatic Compression

It is possible to achieve compression without generating a shock provided the characteristics of the flow do not intersect. In the simplest case of the very slow compression of a finite volume of gas by a piston at one end, when the piston velocity v always remains slow compared with the sound speed c throughout the compression, the characteristics leaving the piston always remain separated provided their transit time across the volume ℓ/c is small compared with any change due to the piston velocity or the sound speed. This is the important case considered in classical thermodynamics as a reversible adiabatic change.

The limiting case of such compression is one where all the characteristic meet at some point at a particular time as the piston is pushed (in the x direction) into a uniform ambient gas. Let all the characteristics C_+ converge on the point $x = 0$ at time $t = 0$. Since the C_- characteristics start in the undisturbed flow, the wave is a simple wave and the velocity and sound speed are constant on the C_+ characteristics, which are therefore straight lines starting at the piston and in this limiting case pass through the point $(0, 0)$. Let the start of the piston motion be from the point $x' = \ell$ at time $t' = \tau$. Since the first C_+ characteristic propagates from $x = \ell$ to $x = 0$ at speed c_0, the collapse time must be $\tau = \ell/c_0$, where c_0 is the sound speed in the ambient gas. The equation of an arbitrary C_+ characteristic is

$$\left.\frac{\mathrm{d}x}{\mathrm{d}t}\right|_+ = v(t') + c(t') = \frac{x}{t} \tag{10.88}$$

where $v(t')$ and $c(t')$ are the flow velocity and sound speed on the C_+ characteristic leaving the piston at t'. Since the Riemann invariant J_- is everywhere constant, we proceed as for the centred rarefaction to give

$$v(t') = \frac{2}{(\gamma + 1)} \left\{ \frac{x'}{t'} + c_0 \right\}$$
$$c(t') = \frac{(\gamma - 1)}{(\gamma + 1)} \frac{x'}{t'} + \frac{2}{(\gamma + 1)} c_0 \tag{10.89}$$

where $x(t')$ is the piston position. Noting that the flow velocity at the piston must equal the piston speed, we obtain the differential equation

$$\frac{\mathrm{d}x'}{\mathrm{d}t'} = \frac{2}{(\gamma + 1)} \frac{x'}{t'} - \frac{2}{(\gamma + 1)} c_0 \tag{10.90}$$

which may be directly integrated with the boundary condition $x = \ell$, $t = \tau$ to give

$$x' = -\frac{2}{(\gamma - 1)} c_0 \tau \frac{t'}{\tau} + \frac{(\gamma + 1)}{(\gamma - 1)} c_0 \tau \left(\frac{t'}{\tau}\right)^{2/(\gamma+1)}$$
$$v(t') = -\frac{2}{(\gamma - 1)} c_0 + \frac{2}{(\gamma - 1)} c_0 \left(\frac{t'}{\tau}\right)^{-(\gamma-1)/(\gamma+1)} \tag{10.91}$$
$$c(t') = c_0 \left(\frac{t'}{\tau}\right)^{-(\gamma-1)/(\gamma+1)}$$

From the sound speed we may immediately calculate the pressure and density on each C_+ characteristic, namely

$$p(t') = p_0 \left(\frac{t'}{\tau}\right)^{-2\gamma/(\gamma-1)}$$
$$\rho(t') = \rho_0 \left(\frac{t'}{\tau}\right)^{-2/(\gamma-1)} \tag{10.92}$$

where p_0 and ρ_0 are the pressure and density of the ambient gas respectively.

Since the wave is a simple wave, the density and sound speed are constant along the characteristic. The total mass through which the characteristic C_+ passes must equal that between the piston and the centre, i.e. the total mass of gas along each characteristic is constant. Since the speed of the characteristic through the gas is c, the length of path from its start at the piston to convergence is $c(t') \, t'$, and the constant mass is therefore $\rho(t') \, c(t') \, t'$. Substituting the adiabatic equation of state in this result we obtain equations (10.92) and hence (10.91).

The state of the gas at any particular point (x, t) during the compression is easily found by identifying the particular characteristic C_+ on which the point lies. The characteristic leaving the piston at (x', t') passes through the points (x, t) such that

$$
\begin{aligned}
\xi &= \frac{x}{c_0 \, t} = \frac{x'}{c_0 \, t'} \\
&= \frac{(\gamma + 1)}{(\gamma - 1)} \left(\frac{t'}{\tau} \right)^{-(\gamma - 1)/(\gamma + 1)} - \frac{2}{(\gamma - 1)}
\end{aligned}
\tag{10.93}
$$

Hence the sets of equations (10.92) and (10.91) can all be expressed in terms of the single variable $\xi \in (1, \infty)$ expressing the complete solution

$$
\begin{aligned}
\frac{v}{c_0} &= \frac{2}{(\gamma + 1)} \, (\xi - 1) \\
\frac{c}{c_0} &= \frac{(\gamma - 1)}{(\gamma + 1)} \xi + \frac{2}{(\gamma + 1)} \\
\frac{p}{p_0} &= \left[\frac{(\gamma - 1)}{(\gamma + 1)} \xi + \frac{2}{(\gamma + 1)} \right]^{2\gamma/(\gamma - 1)} \\
\frac{\rho}{\rho_0} &= \left[\frac{(\gamma - 1)}{(\gamma + 1)} \xi + \frac{2}{(\gamma + 1)} \right]^{2/(\gamma - 1)}
\end{aligned}
\tag{10.94}
$$

The solution exhibits self-similarity. All points with the same value of ξ have the same state. More generally we can see that the spatial profiles of v, c, p and ρ have the same form at different times but with scale factors which depend on the time alone. The problem is therefore reduced to one in a single dimensionless variable ξ alone. We shall return to a fuller discussion of this behaviour in a later chapter.

The motion is independent of the initial starting point ℓ. If ℓ is increased the corresponding collapse time τ is increased proportionately, so that the variable ξ is unchanged. The functional form of the solution in terms of the similarity parameter ξ remains the same, and only the length scale is changed by an appropriate scale factor.

As the final stages of compression are reached, $t \to \tau$, the piston reaches the convergence point, and both the sound speed and the fluid velocity c, $v \to \infty$, hence the pressure and density also. This is clearly an unachievable final state.

However, this behaviour underlies the essential compression required by designs of fuel capsules for inertial confinement fusion.

Appendix 10.A An Alternative Approach to the General Conservation Law Form of the Fluid Equations

In the laboratory (or Eulerian) frame, the general equations of fluid mechanics can be written in several forms, of which the general conservation form is appropriate for a general discussion:

$$
\begin{aligned}
\frac{\partial \rho}{\partial t} + \nabla \cdot (\rho \mathbf{v}) &= 0 \\
\frac{\partial}{\partial t}(\rho \mathbf{v}) + \nabla \cdot \mathbf{\Gamma} &= 0 \\
\frac{\partial}{\partial t}[\rho(\epsilon + \tfrac{1}{2} v^2)] + \nabla \cdot \left[\rho \mathbf{v}(\epsilon + \tfrac{1}{2} v^2) - \mathbf{v} \cdot \boldsymbol{\tau} + \mathbf{q}\right] &= 0
\end{aligned}
\tag{10.A.1}
$$

where ϵ is the specific internal energy of the fluid, $h = \epsilon + p/\rho$ the specific enthalpy, and $\mathbf{q}$ the heat flux vector. $\mathbf{\Gamma}$ is the momentum flux tensor

$$
\Gamma_{i,j} = \rho\, v_i\, v_j - \tau_{i,j}
\tag{10.A.2}
$$

and $\tau_{i,j}$ is the total stress tensor

$$
\tau_{i,j} = \sigma_{i,j} - p\, \delta_{i,j}
\tag{10.A.3}
$$

where $\sigma_{i,j}$ is the viscous stress tensor and $\delta_{i,j}$ the Kronecker delta (1 if $i = j$ and 0 otherwise).

The set of equations is of the general conservative form

$$
\frac{\partial u}{\partial t} + \nabla \cdot \mathbf{f} = 0
\tag{10.A.4}
$$

where u is a general conserved quantity and $\mathbf{f}$ the flux associated with it.

Clearly for any volume V enclosed by surface S

$$
\frac{\mathrm{d}}{\mathrm{d}t} \int_V u\, \mathrm{d}V + \int_S \mathbf{f} \cdot \mathrm{d}\mathbf{s} = 0
\tag{10.A.5}
$$

expresses the same result in an integral form expressing global balance within the system.

10.A.1 Hyperbolic Equations

Generally, in many fluids viscosity and thermal conduction are weak, i.e. the fluid behaves as an ideal dissipation less continuum, and to a first approximation dissipation can usually be neglected. We therefore seek to develop approximations in an ideal inviscid fluid, and introduce corrections to the flow where necessary to take account of dissipative effects. In these circumstances the equations of fluid dynamics (10.A.1) form a set of nonlinear hyperbolic equations.

Let us consider the simple one-dimensional case of the time-dependent Eulerian non-dissipative system, which can be expressed in the general form

$$\frac{\partial u}{\partial t} + A\frac{\partial u}{\partial x} = 0 \qquad (10.A.6)$$

where u and f are general 'vector sets' (column matrix) of the fluid variables

$$u = \{\rho,\ \rho v,\ \rho(\epsilon + \tfrac{1}{2}v^2)\}$$
$$f = \{\rho v,\ (\rho v^2 + p),\ \rho v(h + \tfrac{1}{2}v^2)\}$$

and A is the matrix $\partial f/\partial u$. This set of three equations represents the general set of conservation equations in one dimension, and takes the form of three first-order quasi-linear differential equations in two variables x and t.

The equations are hyperbolic if the eigenvalues of A are real. In the case of an ideal fluid, they can be shown to be $v + c$, v, $v - c$ where c is the isentropic sound speed: $c^2 = \partial p/\partial \rho\,|_s$.

Since the eigenvalues are distinct we may diagonalise A by a similarity transformation $A' = PAP^{-1}$ so that

$$P\frac{\partial u}{\partial t} + A'P\frac{\partial u}{\partial x} = 0 \qquad (10.A.7)$$

where A' is diagonal, and P is a function of the variables u only. Hence we obtain the set

$$\left[\frac{\partial}{\partial t} + \lambda_i\frac{\partial}{\partial x}\right]\Gamma_i = 0 \qquad (10.A.8)$$

for the eigenvalues λ_i. The functions Γ_i, obtained from $\int P\,du$, are constant on the lines $dx/dt = \lambda_i$ respectively, and are the Riemann invariants. For this case of the ideal fluid equations, they take the form

$$[v + \int c\,d\rho/\rho], \qquad s, \qquad [v - \int c\,d\rho/\rho]$$

respectively, where s is the specific entropy.

In principle a knowledge of the initial values of these quantities allows us to integrate along the characteristics in space–time and determine the flow.

10.A.2 Formal Solution

At any point (x, t) we may determine the three values of the variables u if the Riemann invariants are known, provided we know the equation of state, i.e. $p(\rho, s)$, or more succinctly $c(\rho, s)$, by tracing the development of the fluid variables in space and time along the characteristics. Thus in principle we integrate along the field of characteristics to determine the flow. Clearly the value of u at the point (x_0, t_0) is determined only by points which lie within the outermost characteristics through (x_0, t_0)–its domain of dependence (Figure 10.A.1). Similarly it can only influence future events within its outermost outgoing characteristics–the range of influence. The existence of such solutions has been formally demonstrated (Courant and Hilbert, 1962).

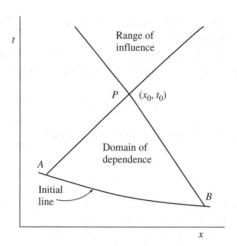

For this system the following uniqueness theorem is valid (Courant and Hilbert, 1962; Courant and Friedrichs, 1948):

> Consider a solution with continuous second-order derivatives in the region ABP bounded by two characteristics through P, and the domain of dependence AB cut off by them on the initial line I. Suppose another solution with continuous second derivatives is given in ABP which assumes the same values on AB as the first. Then the second solution is identical to the first within the domain ABP.

10.A.3 Discontinuities

The uniqueness theorem quoted above requires that the solution be continuous up to the second order everywhere within the domain of dependence. However, discontinuities arise naturally within fluid mechanics by virtue of the nonlinearity of the governing equations (Courant and Friedrichs, 1948; Landau and Lifshitz, 1959). These occur in two forms, namely strong and weak discontinuities. Strong discontinuities involve a jump in value of the flow variables and include shocks and contact discontinuities. Shocks involve flow through the discontinuity, whereas tangential discontinuities do not. Weak discontinuities are continuous, but with a change of gradient. As we have seen, weak discontinuities, which include the head of rarefaction waves, propagate along the characteristics (Landau and Lifshitz, 1959).

If two characteristics were to intersect, the Riemann invariant would become indeterminate. Such intersections can occur in compression, where they lead to shocks. In this neighbourhood, the ideal flow condition breaks down. When the characteristics approach over distances of the order of the mean free path, large gradients in the flow are established, and dissipation becomes important. This leads to an entropy increase in accordance with the second law of thermodynamics across the shock. The thickness of the 'discontinuity' is of the order of the mean free path, which is vanishingly small in the continuum theory. In general fluid theory, the collisional behaviour at a molecular level gives rise to dissipational processes, namely viscosity and thermal conduction. Within the fluid description, the shock 'discontinuity' is controlled by dissipation, principally viscosity. The shocks therefore appear as narrow zones within which viscous effects are dominant (10.5), which in molecular terms are typically a few mean paths thick (Zel'dovich and Raizer, 1967).

Considering the case of a general fluid with dissipation, the general conservation law in integral form (10.A.5) requires that across a normal shock

$$S = \frac{f(u_2) - f(u_1)}{(u_2 - u_1)} \tag{10.A.9}$$

where S is the shock speed, the shock moving from left to right, f the fluxes normal to the shock, and u_1 and u_2 the values of u on the upstream (left) and downstream (right) sides of the shock respectively. Note that this relation must be obeyed for all the elements of u with the same value of S. In one dimension these are the familiar Rankine–Hugoniot relations for a normal shock (equation 10.1).

10.A.4 Weak Solutions

The question arises as to whether strong discontinuities in the form of shock waves can be accommodated within inviscid theory in the absence of dissipation, i.e. as solutions of the hyperbolic equations. In recent years the extension of the description of functions to include functions which are non-differentiable has made this possible, thereby formally resolving many of the problems which troubled nineteenth-century fluid dynamics.

Discontinuities arise naturally in the general solution of nonlinear hyperbolic equations. At such points the differential equations break down, but these are expressions of more general integral relations, e.g. the integral forms of the conservation laws. Using a test function, namely a smooth function of compact support, we may construct generalised solutions by integration of the product of the solution and test functions. In this way the derivatives are transferred from the solution to the test function. The solution, called a *weak solution*, satisfies this integral relation for all allowable test functions. The Rankine–Hugoniot equations (10.A.9) form such a set of weak solutions for the fluid equations (Lax, 1954). Weak solutions of the dissipationless fluid equations therefore satisfy the differential equations where the derivatives exist, and the Rankine–Hugoniot relations across discontinuities. However, in contrast to the actual physical situation, rarefaction as well as compression discontinuities can occur, i.e. the solution is not unique.

Two questions arise if we wish to use the dissipationless equations alone for calculation:

1. Is the solution unique? Clearly the previous uniqueness theorem fails when discontinuities occur.

2. Is the solution without dissipation the limit of vanishing viscosity to the complete fluid dynamical problem?

In fact both questions are related. The discontinuities in ideal flow can be shown not to be uniquely specified by the Rankine–Hugoniot condition. In the real fluid such uniqueness is imposed by the additional requirement for an entropy increase. In the inviscid case, this is established by the following results, which have been proved for one-dimensional systems (Oleinik, 1963a):

1. Solutions of dissipative equations converge to weak solutions of the hyperbolic system in the limit of vanishing viscosity (Oleinik, 1963a).

2. Such solutions of the viscous system satisfy

$$S[u, u_1] \geq S \geq S[u_2, u] \qquad (10.A.10)$$

known as the entropy condition, where S is the shock speed (10.A.9), and

$$S[a, b] = \frac{f(b) - f(a)}{(b - a)} \qquad (10.A.11)$$

where u lies between u_1 and u_2 (Hopf, 1969/70; Quinn, 1971).

3. Weak solutions of the hyperbolic system satisfying the entropy condition are unique (Quinn, 1971; Oleinik, 1963b).

The entropy condition has a clear relation with many of the familiar properties of shocks (see equation 10.17). Since the limit of a weak shock is a sound wave, which propagates at the local sound speed $S[u, u] \rightarrow c(u)$, the shock speed is supersonic with respect to the upstream flow $S \geq c(u_1)$ and subsonic with respect to the downstream one $S \leq c(u_2)$. Indeed the entropy condition reflects the facts that the shock adiabat is generally convex and that physical shocks are compressive as required by the second law of thermodynamics (Courant and Friedrichs, 1948). The entropy condition thus ensures that the shock in the ideal fluid satisfies the second law of thermodynamics, and that there is an entropy increase across the shock.

Chapter 11

Aerofoils in Low-Speed Incompressible Flow

11.1   Introduction

In this chapter we investigate the flow around two-dimensional aerofoils and three-dimensional wings using irrotational incompressible flow models, i.e. solutions to Laplace's equation.[1] The flow around a solid body depends markedly on the shape of the body. In particular the drag force may be greatly reduced by a careful choice of shape. The general shape of a low-drag body is established by some simple principles. Strong drag at large Reynolds numbers is generated by the presence of a large turbulent wake behind the body, which must therefore be avoided. During the flow around the body, the streamlines must leave the body at some point. This occurs along the line of separation. The flow behind the line of separation is rotational and often turbulent having been modified by viscosity in the boundary layer adjacent to the surface. The pressure in this region is normally much less than that which would be predicted if the flow were able to remain irrotational. As a result separation leads to substantial modifications of the forces on the surface of the body calculated from ideal flow. Consequently d'Alembert's paradox is no longer valid, and drag is generated. To reduce the drag, the wake formed by the separated flow should be made as narrow as possible. The shape of the body must avoid establishing the conditions for separation, in particular strongly increasing pressure gradients and sharp corners. Therefore a streamlined body has a rounded leading edge, where a stagnation point forms at all working incoming

---

[1] An excellent fuller introduction to aerodynamics is given by Anderson (2007).

Introductory Fluid Mechanics for Physicists and Mathematicians, First Edition. Geoffrey J. Pert.

flow directions. Downstream the pressure varies slowly and smoothly up to the point at which separation is required. The profile is consequently elongated downstream, tapering to a point (or line) at which the flows along the two surfaces meet smoothly without having to turn any sharp corner or large angle. Such bodies are called *streamlined*. Separation is avoided except where desired at the trailing edge, provided the body is slender (thickness small compared with length) and the long axis (*chord*) is at a small angle to incoming flow; (*angle of attack*). The body then offers only a small disturbance to the incoming flow; shapes are familiar in many areas such as fast-swimming fish and aircraft.

The flow around the streamlined body cannot be entirely irrotational. Viscosity plays a role close to the surface, even with materials such as air, whose viscosity is very small. When the viscosity is small the region in which vorticity is induced is very narrow, and the flow around the surface may be accurately described by an ideal irrotational flow model. Rotational flow is confined to a thin boundary layer around the surface and a narrow wake leaving the trailing edge. If the flow velocity is small compared with the sound speed, we showed in Chapter 8 that the flow is nearly incompressible. At velocities comparable with the sound speed a scale transformation (the Prandtl–Glauert approximation) (Chapter 12) extends the range of the incompressible flow approximation as an accurate approximation.

As we showed in Chapter 2, the flows in two or three dimensions have a marked difference in their connectivity. In two dimensions the configuration is multiply connected and therefore cyclic, so that a circulation must be specified in order to achieve a unique solution to the flow around an aerofoil. In practice the circulation is determined by the Kutta condition, which ensures that the flow remains finite around the boundary discontinuity at the trailing edge. The flow therefore leaves the wing surface at the trailing edge. In contrast three-dimensional flow is simply connected and therefore acyclic. A unique flow around the three-dimensional wing is therefore possible. However, the presence of the trailing edge discontinuity gives rise to a vortex sheet leaving the wing at the trailing edge. The Kutta condition is required to ensure that the sheet leaves only at the trailing edge.

11.1.1   Aerofoils

A typical aerofoil profile as used for a wing section is shown in Figure 11.1.[2] It has an asymmetric streamlined profile. At the leading edge the profile is blunt and a stagnation point is developed. There is thus no corner at any angle of

---

[2]An aerofoil or wing section is the two-dimensional cross-section chordwise through the three-dimensional wing.

Figure 11.1: A typical aerofoil profile inclined at an angle of attack of $\alpha = 10°$ to the incoming flow U as indicated by the arrow.

attack,[3] which would be the consequence if the leading edge were angled. In contrast the trailing edge is a sharp angle, nearly a cusp, which forces the flow from both the upper (vacuum) and lower (pressure) surfaces to separate and leave the body at this point. If the aerofoil is not too thick, and not operated at too large an angle of attack, separation will not occur over either the upper or lower surface before the trailing edge is reached. In this configuration lift will be developed, and drag will be relatively small.

The flow around an aerofoil calculated directly within the ideal irrotational flow conditions will *not* leave the surface at the trailing edge. Since ideal flow allows the non-physical condition of flow around an outside corner, we might expect such behaviour to occur at the trailing edge. In fact experimental evidence shows this is not the case, and provided the aerofoil is operating in a streamlined flow mode, the streamlines from both the upper and lower surfaces will meet smoothly, and leave at the trailing edge. As the velocities on the upper and lower surfaces are different, there may be in consequence a surface of tangential discontinuity between the two flows in ideal flow.[4] When viscosity is included, this problem is resolved and the discontinuity surface is replaced by a narrow wake. The non-uniqueness inherent in ideal flow is thus resolved by this condition.

The profile itself is characterised by its chord (width) and thickness (Figure 11.2), and in three dimensions by the span b, the distance from wing tip to wing tip. The plan area of the wing S is the projection of the wing onto the plane containing the span and the chord. We define the aspect ratio of the wing as the ratio of b^2/S. It is approximately equal to the ratio of span to the chord b/c. Typical values are about 6.

The characteristic aerodynamic parameters of the wing are:

1. The lift coefficient, which is the dimensionless form for the lift force:

$$C_L = \frac{F_y}{\frac{1}{2} \rho U^2 S} \tag{11.1}$$

---

[3]The angle of attack is the angle between the incoming flow and the axis of the profile.

[4]In two-dimensional flow, the pressure on both upper and lower surfaces at the trailing edge must be equal. Hence pressure continuity across the wake ensures that the velocities on both the top and bottom surfaces are the same. This is not the case for three-dimensional flow, where the directions of flow differ.

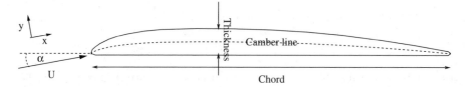

Figure 11.2: Aerofoil profile indicating chord and thickness showing an incoming wind at $10°$ incidence. The x direction is taken parallel to the wind and the y normal to it, showing the direction of the lift force.

where F_y is the total lift force normal to the incoming flow, and S the wing plan area, defined as the projection of the area onto the surface containing the chords of all sections (note that this not the same as the surface area). For thin wings, it can be shown that $C_L \approx 2\pi\alpha$ where α is the angle of attack.

2. Drag, which is minimised by the streamlined profile. It is characterised by the drag coefficient

$$C_D = \frac{F_x}{\frac{1}{2}\rho U^2 S} \tag{11.2}$$

The ratio of lift to drag, C_L/C_D, is typically about 100 for a well-designed wing.

3. The remaining wing parameter, namely the pitching moment M of the lift force about some specified point, e.g. the leading edge. Conventionally the moment is taken as positive when the moment is such as to raise the leading and depress the trailing edge:

$$C_m = \frac{M}{\frac{1}{2}\rho U^2 \ell S} \tag{11.3}$$

where ℓ is the specified reference length, typically the chord.

The *centre of pressure* is defined as the point on the chord through which the lift force is effectively applied. Since the lift is at right angles to the incoming stream and the angle of attack small, the point of zero moment is approximately C_m/C_L.

11.2   Two-Dimensional Aerofoils

As noted earlier, the aspect ratio of a typical wing is large. The airflow over the wing is therefore almost two dimensional in the plane of the chord and the thickness. The transverse flow in the direction of the span is relatively small, although as we shall show it plays an important role in determining the characteristics of the wing. We may therefore approximate the flow over the wing in terms of a series of sections of slightly differing cross-sections and obtain the final aerodynamic characteristics by integrating along the wing span. This procedure requires a detailed study of the two-dimensional wing section and its properties.

The aerofoil causes only a small deviation in a uniform flow originating at $-\infty$ and flowing out at $+\infty$. Far from the body the disturbance will die away and the flow return to its incoming uniform condition. Thus if the incoming flow has velocity $\mathbf{U}$ the asymptotic flow at a large distance r from the body must be expressed by the Laurent series for the complex velocity

$$\frac{\mathrm{d}w}{\mathrm{d}z} = U\mathrm{e}^{-\imath\alpha} + \frac{A}{z} + \frac{B}{z^2} + \dots \tag{11.4}$$

the values of A and B being determined by the geometrical form of the stream-lined body.

If the body is not a source, the flux through any surface enclosing it must be zero. Hence the radial component $\Re(A) = 0$, and the circulation around the body $\Gamma = 2\pi\,\Im(A)$. Furthermore B can be interpreted as the vector strength of a doublet at the body, also known as a *bicirculating flow* (Section 2.10.1).

11.2.1 Kutta Condition

As we have seen, it is only possible to construct a unique solution for the flow around a body in two dimensions if both the flow at infinity and the circulation around the body are known. Thus it is not possible a priori to calculate the lift from the wing without some prior knowledge of the circulation.

This condition is established by a hypothesis first propounded by Kutta (1902) and independently by Zhukovskii (1906),[5] which allows ideal flow solutions to successfully represent the flow around the wing. This is known as the *Kutta condition*[6] and can be expressed in a number of equivalent ways:

- The flow is finite at the trailing edge, i.e. no velocity singularity associated with flow around a corner occurs.

- If the upper and lower surfaces at the trailing edge meet at a finite angle, the trailing edge is a stagnation point, since the two flows must be along their respective surface.

---

[5]The spelling of Zhukovskii is often anglicised to Joukowski or Joukowsky.
[6]This condition is frequently called the Kutta–Zhukovskii condition.

- If the trailing edge is a cusp, i.e. the angle between the upper and lower surfaces is zero, the flow velocities on the upper and lower surfaces are equal in magnitude and direction.

- The pressures on the upper and lower surfaces at the trailing edge are equal.

- The surface vorticities on the upper and lower surfaces at the trailing edge are equal.

- The streamlines from both top and bottom surfaces leave the wing smoothly at the trailing edge.

The circulation is provided by an additional vortex of the type discussed in Section 2.10 around the wing (equation 2.69), which is necessary to generate lift. Physically we may see the origin of this vortex in the increased acceleration needed on the vacuum (upper) surface of the wing to compensate for the displacement of the fluid induced by the wing. The displacement required on the pressure (lower) surface is much less. The generation of lift is due to the pressure differences between the upper and lower surfaces resulting from Bernoulli's equation. In the next section, the lift is calculated by the Kutta–Zhukovskii formula, equation (2.105), from this circulation, and is in good agreement with experiment.

In ideal flow Kelvin's theorem requires constancy of circulation. Thus since the ambient state is one of zero circulation, the circulation must be balanced by a vortex of opposite rotation. This *starting vortex* is established on starting, as shown in Figure 11.3. In ideal flow with no circulation and a finite angle of attack, the body streamline will leave from the upper surface having passed around the trailing edge. As we have seen, this is unphysical at the trailing edge and leads to a separation of the flow at that point. In the initial phase immediately after the airflow over the wing has started, the separated flow forms a vortex over the wing as sketched in Figure 11.3a. As the airflow flows over the wing, the vortex detaches and moves downstream. At this stage, the

(a)   (b)

Figure 11.3: Development of the starting vortex initially at the trailing edge and convecting downstream with its connecting vortex sheet. (a) Initial vortex, and (b) Detached vortex.

circulation around the wing is just that required to bring the body streamline to the trailing edge. The circulation of the vortex is equal to and opposite in sign to that around the wing. The detached vortex moves continuously downstream in the free stream as shown in Figure 11.3(b). If the circulation around a line enclosing both the wing and the starting wing is calculated, the total is zero, consistent with Kelvin's theorem, the circulation around the wing being exactly balanced by that of the starting vortex. On a three-dimensional wing it is connected with the flow around the wing by trailing vortices from the wing tips, tip vortices, the whole forming a horseshoe pattern (see Section 11.8), consistent with Helmholtz's second theorem (theorem ii, p. 28). This structure, although it is slowly damped out by viscosity, plays an important role in determining a component of the drag.

11.3   Generation of Lift on an Aerofoil

In Section 2.11 we showed that the lift force on a wing section was given by the Kutta–Zhukovskii formula, equation (2.105), as

$$L = -\rho U \Gamma \tag{11.5}$$

where L is the lift per unit span and Γ the circulation.

The flow around the wing is calculated in ideal flow by a solution to Laplace's equation, which is linear. We may therefore imagine the flow to be made up of the superposition of two different flows. Thus consider the case of a flow with unit velocity along the chord giving circulation Γ', and separately one with unit velocity normal to it giving circulation Γ'_2.[7] Then if we have a wind of velocity U at an angle α to the chord, the total circulation is

$$\Gamma = \Gamma'_1 U \cos \alpha + \Gamma'_2 U \sin \alpha$$

since ideal flow is linear, and the Kutta condition holds for the combined flow. Thus the lift per unit span is

$$L = -\rho U \left(\Gamma'_1 U \cos \alpha + \Gamma'_2 U \sin \alpha \right)$$

We may introduce the total circulation for unit velocity as

$$\Gamma' = \sqrt{\Gamma'^2_1 + \Gamma'^2_2} \quad \Gamma'_1 = -\Gamma' \sin \alpha_0 \quad \Gamma'_2 = \Gamma' \cos \alpha_0 \quad \alpha_0 = -\arctan\left(\frac{\Gamma'_1}{\Gamma'_2}\right) \tag{11.6}$$

---

[7]In reality the flow normal to the chord cannot exist in this form. However, the solution to Laplace's equation satisfying the Kutta condition does exist. Since this is only used in the superposition with the solution for an incoming flow parallel to the chord satisfying the Kutta condition, the resultant flow will also be a solution to Laplace's equation again satisfying the Kutta condition and will be satisfactory for small angles of attack.

Substituting for the components of the circulation we obtain

$$L = -\rho \Gamma' U^2 \left(\sin \alpha \cos \alpha_0 - \cos \alpha \sin \alpha_0 \right) = -\rho \Gamma' U \sin \left(\alpha - \alpha_0 \right) \quad (11.7)$$

The angle α_0 therefore defines the flow direction of zero lift–*first axis*–of the profile relative to the chord. Since the lift is proportional to U^2, we define the lift coefficient for the wing section as[8]

$$c_L = -\frac{L}{\frac{1}{2} c \rho U^2} = a \sin \left(\alpha - \alpha_0 \right) \approx a \, \alpha' \quad (11.8)$$

where $a = -2\Gamma'/c$ and c is the chord, and $\alpha' = \alpha - \alpha_0$ is the effective angle of attack.[9] The variation of the coefficient of lift as $\sim \sin \alpha'$ is a very general result, provided the effective angle of attack α' is sufficiently small. The *lift slope* $a \approx 2\pi$ for a wide range of profiles as confirmed by experiment.

11.4   Pitching Moment about the Wing

The anti-clockwise pitching moment per unit span on the wing taken about the leading edge of the wing follows directly from pressure on the surface of the wing

$$M = \oint p \left(x \, dx + y \, dy \right) \quad (11.9)$$

As with the lift we may consider the moment in terms of the angular momentum communicated to the wing by the flow. Consider a cylindrical surface of radius R far from the body so that the asymptotic velocity distribution (11.4) is applicable. The velocity components for the asymptotic flow, equation (11.4), are

$$u_r = U \cos \theta - \frac{B \cos \left(\theta - \beta \right)}{2\pi R^2} + \dots \quad u_\theta = -U \sin \theta + \frac{\Gamma}{2\pi R} - \frac{B \sin \left(\theta - \beta \right)}{2\pi R^2} + \dots$$

$$(11.10)$$

where β is the direction of the vector $\mathbf{B}$ and the angles θ and β are measured with respect to the incoming flow.

The product is

$$u_r \, u_\theta = -U^2 \sin \theta \cos \theta + U \frac{\Gamma}{2\pi R} \cos \theta + U \frac{B}{2\pi R^2}$$
$$\times \left[\sin \theta \cos \left(\theta - \beta \right) - \cos \theta \sin \left(\theta - \beta \right) \right] + \dots$$

---

[8]We shall use lower case c to identify coefficients for wing sections and upper case C for the complete wing.

[9]See footnote 13 on page 67 where the change of sign is discussed.

The torque on the wing is equal to the angular momentum per unit time entering through the surface, which is

$$M = - \int_0^{2\pi} \rho \, R \, u_r \, u_\theta \, R \, d\theta$$

Retaining only those terms which are non-zero on integration, the moment is

$$M = -\rho R^2 \int_0^{2\pi} \frac{BU}{R^2} \sin \beta \, d\theta = -2\pi \rho B U \sin \beta = -2\pi \rho B_n U \qquad (11.11)$$

where B_n is the component of $\mathbf{B}$ normal to the incoming flow.[10]

Similar to our analysis of the lift we take two components $\mathbf{B}_1'$ and $\mathbf{B}_2'$ for the components of $\mathbf{B}$ for unit velocity parallel to the axis of zero lift and normal to it.[11] We note that the magnitudes B_1' and B_2', and angles β_1 and β_2, are determined by the profile of the wing section alone. The pitching moment at an effective angle of attack α' is

$$M_0 = -2\pi \rho U^2 \left[B_1' \cos \alpha' \sin \left(\alpha' + \beta_1 \right) + B_2' \sin \alpha' \sin \left(\alpha' + \beta_2 \right) \right]$$

If we now consider the moment about the point (x_0, y_0) the pitching moment becomes

$$M_0 = \; -2\pi \rho U^2 \Big[B_1' \cos \alpha' \sin \left(\alpha' + \beta_1 \right) + B_2' \sin \alpha' \sin \left(\alpha' + \beta_2 \right)$$
$$+ \frac{1}{2\pi} \Gamma' \sin \alpha' \left(x_0 \cos \alpha' - y_0 \sin \alpha' \right) \Big]$$

Expanding the trigonometric expressions we obtain

$$M_0 = -2\pi \rho U^2 \left(a \sin \alpha' \, \cos \alpha' + b \sin^2 \alpha' + b' \cos^2 \alpha' \right) \qquad (11.12)$$

where

$$a \; = \; B_1 \cos \beta_1 + B_2 \sin \beta_2 + \frac{\Gamma'}{2\pi} x_0$$

$$b \; = \; B_2 \cos \beta_2 - \frac{\Gamma'}{2\pi} y_0 \quad b' = B_1 \sin \beta_1$$

The section pitching moment coefficient about the point (x_0, y_0) is therefore

$$c_m = -\frac{M_0}{\frac{1}{2}\rho U^2 c}$$

The change of sign is discussed in footnote 13 on page 67.

---

[10]Unlike the circulation, the bicirculation vector $\mathbf{B}$ is not independent of the origin at which is set, i.e. the point about which the moment is taken.

[11]See footnote (7) on page 301.

We may choose the point

$$x_c = -\frac{2\pi}{\Gamma'}\left(B_1 \cos \beta_1 + B_2 \sin \beta_2\right) \qquad y_c = -\frac{2\pi}{\Gamma'}\left(B_1 \sin \beta_1 - B_2 \cos \beta_2\right)$$

so that $a = b = 0$ and

$$\mathcal{M} = -2\pi\rho U^2 B_1 \sin \beta_1$$

the pitching moment, is independent of the angle of attack α'. This point, which is found for all profiles, is known as the *aerodynamic centre* or *focus*. Its existence has been confirmed by experiment. However, its location can only be found by more detailed calculation for specific profiles. In general it is found to be close to the quarter chord point, $c/4$ from the leading edge. The corresponding pitching moment coefficient is

$$c_\mu = 4\pi B_1 \sin \beta_1 \tag{11.13}$$

The lift may be imagined to act through a point rather than distributed over the surface of the wing, namely the *centre of pressure*. The position of the centre of pressure is defined as the point where the line of the resultant lift meets the chord. If the effective angle of attack is small, the position is given by $x_p = M_0/L = c\,c_m/c_L$. Clearly the centre of pressure changes as the angle of incidence varies.

Referring to equation (11.12), it can be seen either by substituting for a, b and b' or directly that the moment about any general point (x, y) is the sum of two components, one of which is independent of the angle of attack $\mathcal{M}$, and one which varies with the lift:

$$c_m = c_\mu + c_L \left[(x - x_c) \cos \alpha - (y - y_c) \sin \alpha\right] \tag{11.14}$$

If $c_\mu = 0$ the centre of pressure is independent of the angle of attack. If $c_\mu > 0$ the centre of pressure travels towards the leading edge with increasing angle of attack, which promotes instability. Therefore for a wing without a tailplane, the condition for stability is $c_\mu < 0$.

11.5 Lift from a Thin Wing

In the last two sections, 11.3 and 11.4, we have obtained general expressions for the lift and pitching moment of a wing section. These have enabled us to obtain some important general principles concerning the properties of a wing. However, unfortunately the expressions obtained thereby involve terms which require a detailed knowledge of the flow about the profile before quantitative values of the principal quantities are obtained, namely the circulation and the bicirculation vector. There are two classical approaches to this problem, which allow analytic values to be calculated. The conformal transform approach has now been superseded by numerical calculations. However, the thin wing approximation is still useful for rapid calculations.

We define a thin wing as one whose thickness is much less than the chord. The wing may be then considered as a thin plate, whose profile lies along the mean or camber line, i.e. mid-way between the upper and lower surfaces of the wing. The airflow about the wing is then that associated with a flow in which the normal velocity at the mean line is zero. The problem of calculating the flow around such a wing therefore reduces to finding a flow whose velocity normal to the mean line is zero, and thereby satisfying the necessary boundary condition.

The problem of calculating the flow about a thin wing may be tackled in two different ways. In a previous chapter, Case study 2.II, we derived the velocity using the analytic nature of the solution and the Cauchy integral theorem to analytically continue the flow away from the surface of the wing. In this section we use a more direct approach using a straightforward model based on the surface vorticity to calculate the flow. The result yields expressions which are more amenable to numerical evaluation. In this section we will neglect the effect of finite thickness, which as we have previously seen can be treated independently of the camber and angle of attack, which alone give rise to lift.

As we discussed earlier, this aim may be achieved by replacing the wing by a vortex sheet. Since the flow is two dimensional, the sheet comprises a series of rectilinear vortices with axes parallel to the span of the wing. Furthermore, since the wing is thin, the two sets of vortices on the top and bottom surfaces of the wing coalesce into a single set along the mean camber line. The strength of these vortices is adjusted so that the induced velocity normal to the mean line cancels that due to the component of the incoming flow.

Since the wing is thin and the angle of attack small, the flow around the wing can be considered to be a small perturbation to the incoming flow. As before we introduce the perturbation velocity $\mathbf{u} = \mathbf{v} - \mathbf{U}$. From Bernoulli's equation the pressure perturbation $p = p_0 - \rho \mathbf{U} \cdot \mathbf{u}$. Since the wing is thin it is treated as a vortex sheet with vorticity $\gamma(s) = u_2 - u_1$ where u_1 and u_2 are the perturbation velocities at the point s along the top and bottom surfaces of the wing respectively. Therefore the lift force on an element δs of the wing is

$$\delta F_y = (p_2 - p_1)\, \delta s = -\rho U\, \gamma(s)\, \delta s \qquad (11.15)$$

Since the circulation $\Gamma = \int \gamma(s)\, ds$ along the wing, the Kutta–Zhukovskii formula follows directly.

The problem may be further simplified if the camber is small, so that $(dy/dx)^2 \ll 1$

$$\left(\frac{ds}{dx}\right)^2 = 1 + \left(\frac{dy}{dx}\right)^2 \approx 1$$

where s is the distance along the mean line, x is measured along the chord from the leading edge and y perpendicular to it. Similarly subject to the above constraint, the vortex sheet may be treated as lying along the chord rather than the camber line.

The velocity components parallel to and perpendicular (upwards) through the surface due to the incoming flow U at the angle of attack α are

$$u_\parallel = U \cos\left[\alpha - \arctan\left(dy/dx\right)\right] \approx U$$

$$u_\perp = U \sin\left[\alpha - \arctan\left(dy/dx\right)\right] \approx U\left(\alpha - dy/dx\right)$$

and the perturbation velocity components velocity components due to the surface vorticity

$$u_\parallel \approx 0$$

$$u_\perp \approx \int_0^c \frac{\gamma(x')}{2\pi(x - x')}dx'$$

neglecting the small induced velocity component parallel to the surface. since the sheet is treated as lying along the chord. This integral has a singularity at $x' = x$. However, since the velocity induced by a vortex sheet at its surface has no normal component, the integral is in fact the principal value integral discussed earlier. Since the total normal component of velocity must be zero

$$U\left(\alpha - \frac{dy}{dx}\right) + \int_0^c \frac{\gamma(x')}{2\pi(x - x')}dx' = 0 \tag{11.16}$$

Hence we obtain the governing equation of thin wing aerodynamics, namely Glauert's equation (Glauert, 1947), obtained earlier (2.91):

$$\int_0^c \frac{\gamma(x')}{2\pi U(x - x')}dx' = \frac{dy}{dx} - \alpha \tag{11.17}$$

To solve this equation we introduce the eccentric angle θ defined by

$$x = \frac{1}{2}c\left[1 - \cos(\theta)\right] \quad \text{and} \quad y = c\,f(\theta) \tag{11.18}$$

Hence we obtain

$$\frac{1}{2\pi U}\oint_0^\pi \frac{\gamma(\theta')\sin(\theta')}{(\cos(\theta')-\cos(\theta))}\,d\theta' = \frac{dy}{dx} - \alpha \tag{11.19}$$

To solve this equation Glauert introduced a Fourier series expansion for the vorticity

$$\gamma(\theta) = -2U\left\{ A_0\frac{[1+\cos(\theta)]}{\sin(\theta)} + \sum_{n=1}^\infty A_n\sin(n\theta)\right\} \tag{11.20}$$

At the trailing edge the vorticity is easily shown to be zero. In this way satisfaction of the Kutta condition is ensured, and the flow is well behaved. Substituting this form into equation (11.19) and performing the integrals using the results from Appendix 11.A yields

$$\frac{dy}{dx} = (\alpha - A_0) + \sum_{n=1}^\infty A_n\cos(n\theta) \tag{11.21}$$

This equation directly yields the coefficients A_n in a similar manner to that of a Fourier series, namely multiplication by $\cos(n\theta)$ and integration from 0 to π in turn, to give

$$A_0 = \alpha - \frac{1}{\pi}\int_0^\pi \frac{dy}{dx}\,d\theta$$
$$A_n = \frac{2}{\pi}\int_0^\pi \frac{dy}{dx}\cos(n\theta)\,d\theta \tag{11.22}$$

To calculate the lift we must calculate the total circulation along the wing, namely

$$\begin{aligned}
\Gamma &= \int_0^c \gamma(x)\,dx = \frac{1}{2}c\int_0^\pi \gamma(\theta)\sin(\theta)\,d\theta\\
&= -cU\int_0^\pi \left\{ A_0\frac{(1+\cos(\theta))}{\sin(\theta)} + \sum_{n=1}^\infty A_n\sin(n\theta)\right\}\sin\theta\,d\theta\\
&= -\pi cU\left(A_0 + \frac{1}{2}A_1\right) \tag{11.23}
\end{aligned}$$

using the standard integrals. It is easy to show using equations (11.18) and (11.22) that equation (11.23) is identical to equation (2.93) obtained earlier in Case study 2.II. Hence the lift coefficient is

$$\begin{aligned}
c_L &= \pi(2A_0 + A_1)\\
&= 2\pi\left[\alpha + \frac{1}{\pi}\int_0^\pi \frac{dy}{dx}(\cos(\theta)-1)\,d\theta\right] \tag{11.24}
\end{aligned}$$

For thin aerofoils, the slope of the lift profile takes the simple form

$$\frac{dc_L}{d\alpha} = 2\pi \tag{11.25}$$

The axis of zero lift is defined by the angle

$$\alpha_0 = \frac{1}{\pi} \int_0^\pi \frac{dy}{dx} \left(1 - \cos(\theta)\right) d\theta$$

displaced from the geometrical axis defined by the chord.

The moment of the lift about the leading edge is calculated from the local lift force (11.15) integral

$$
\begin{aligned}
M &= -\rho U \int_0^c x\gamma(x)\,dx = \frac{1}{4}\rho U c^2 \int_0^\pi \gamma(\theta) \left[1 - \cos(\theta)\right] \sin(\theta)\,d\theta \\
&= \frac{1}{2}\rho U^2 c^2 \int_0^\pi \left\{ A_0 \frac{(1 + \cos(\theta))}{\sin(\theta)} + \sum_{n=1}^{\infty} A_n \sin(n\theta) \right\} \left[1 - \cos(\theta)\right] \sin(\theta)\,d\theta \\
&= \frac{\pi}{4}\rho U^2 c^2 \left(A_0 + A_1 - \frac{1}{2} A_2 \right)
\end{aligned}
\tag{11.26}
$$

The pitching moment as calculated above is defined with positive sense anti-clockwise with respect to the x and y axes as for the circulation, i.e. a positive couple tending to lift the trailing edge with respect to the leading one. Conventionally the moment is defined with opposite sense, i.e. tending to lower the trailing edge. Thus in conformity with normal practice, we must introduce a negative sign into the pitching moment coefficient (see footnote 13 on page 67) and the pitching moment coefficient about the leading edge

$$C_{M,0} = -\frac{\pi}{2} \left(A_0 + A_1 - \frac{1}{2} A_2 \right) \tag{11.27}$$

The value about the quarter chord point

$$C_{M,1/4} = \frac{\pi}{4} \left(A_2 - A_1 \right) \tag{11.28}$$

is independent of the angle of attack. The quarter chord point is therefore the *aerodynamic centre* of the wing profile. For thicker wings the aerodynamic centre has been shown to exist and is found to lie close to the quarter chord point.

11.6   Application of Conformal Transforms to the Properties of Aerofoils

The thin wing approximation provides a useful solution to many of the problems of wing design, particularly as wings are generally fairly thin. However it is not able to account for the effects of the thickness of the wings. Thus although the general behaviour predicted in the previous section 11.5 is a good approximation, it does not provide an accurate estimate of the lift or pitching moment.

As we have noted in Section 2.12, conformal transforms of a circle can be used to generate an aerofoil section. However, as noted there, the profiles obtained are rather inconvenient for engineering applications. Nonetheless a number of useful and fairly general properties can be derived from considerations of the flow about the generating circle.

11.6.1   Blasius's Equation

When using complex functions for the analytic calculation of the properties of an aerofoil in irrotational incompressible flow, it is particularly convenient to express the aerodynamic force and pitching moment directly in terms of the complex velocity.

The pressure forces and moment on an element of the surface are

$$\mathrm{d}F_x = -p\,\mathrm{d}S_x = -p\,\mathrm{d}y \quad \mathrm{d}F_y = -p\,\mathrm{d}S_y = p\,\mathrm{d}x$$
$$\mathrm{d}M = -p\,(x\,\mathrm{d}S_y - y\,\mathrm{d}S_x) = p\,(x\,\mathrm{d}x + y\,\mathrm{d}y) \tag{11.29}$$

Hence

$$\mathrm{d}F_x - \imath\,\mathrm{d}F_y = \frac{1}{2}\rho v^2\,(\mathrm{d}y + \imath\,\mathrm{d}x) \quad M = -\frac{1}{2}\rho v^2\,(x\,\mathrm{d}x + y\,\mathrm{d}y)$$

Since

$$\mathrm{d}y + \imath\,\mathrm{d}x = \imath\,\mathrm{d}z^* \quad x\,\mathrm{d}x + y\,\mathrm{d}y = \Re\,(z\,\mathrm{d}z^*)$$

and the surface of the body is a streamline $\mathrm{d}\psi = 0$

$$\mathrm{d}w^* = \mathrm{d}\phi - \imath\,\mathrm{d}\psi = \mathrm{d}\phi + \imath\,\mathrm{d}\psi = \mathrm{d}w = \frac{\mathrm{d}w}{\mathrm{d}z}\,\mathrm{d}z$$

Hence integrating over the surface we obtain

$$F_x - \imath F_y = \frac{1}{2}\imath\rho \oint \left(\frac{\mathrm{d}w}{\mathrm{d}z}\right)^2 \mathrm{d}z \quad M = -\Re\left\{\frac{1}{2}\rho \oint z\left(\frac{\mathrm{d}w}{\mathrm{d}z}\right)^2 \mathrm{d}z\right\} \tag{11.30}$$

These contour integrals are taken over the surface of the aerofoil. However Cauchy's theorem allows the contour to be extended arbitrarily provided no new poles are included, i.e. the flow does not include any new sources or vortices. The integrals may be evaluated by summing over the residues of the poles in the usual way.

Extending the contour to one far from the body, the Kutta–Zhukovskii formula for the lift (2.105) follows immediately if the second-order terms in the perturbation velocity are neglected since

$$(v_x - \imath v_y)^2 \, (\mathrm{d}x + \imath \, \mathrm{d}y) \approx -2\,U \, (u_x \, \mathrm{d}x + u_y \, \mathrm{d}y)$$

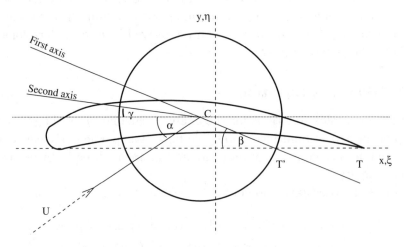

Figure 11.4: Conformal transformation of a circle in the ζ plane into an aerofoil section in the z plane. The two planes are shown plotted on the same set of axes. The centre of the circle C, being transformed at ζ_0, is conventionally taken to be the centre of the aerofoil. The trailing edge point T' in the ζ plane transforms into the trailing edge of the wing T, the line CT' defining the first axis, or axis of zero lift, of the profile at angle β to the x, ζ axes . The second axis of the profile is defined by the condition of zero pitching moment and at angle γ to the x, ζ axis. The incoming airflow of velocity U is at an angle α to the x, ζ axes.

11.6.2   Conformal Mapping of a Circular Cylinder

It is found (see Appendix 11.B) that a number of conformal transformations of a circle have a streamlined profile similar to an aerofoil structure (Figure 11.4). Unfortunately only a limited number of such profiles are available, and no simple transformation is available for the inverse process. However, these special cases are valuable in that they offer an insight through analytic modelling into the flow around realistic wing sections. Although this method of calculating the aerodynamic properties of wings is obsolete nowadays, having been replaced by more general direct calculation using computers, the method is nonetheless valuable in providing an exact analytic solution for a restricted set of profiles. These are useful for validating and benchmarking more sophisticated methods of calculation, which although accurate are nonetheless approximate.

As discussed above, there is a one-to-one correspondence between points on the wing with those on the circle. The properties of the resulting aerofoil section are set by the position ζ_0 of the centre of the circle in ζ space and its radius. The thickness of the profile is determined by the real part of the displacement of centre $\Re(\zeta_0)$ and the 'camber' or slope of the mean line by the imaginary part $\Im(\zeta_0)$. Normally the circle is chosen to pass through only one singular point of the transformation, which corresponds to the trailing edge. Any other singularities are placed inside the circle and do not contribute to the profile. Thus we may identify the point T' in ζ space on the circle with the equivalent point T in z space at the trailing edge on the wing. At the trailing edge the wing generally has a sharp turning point, where the tangent changes its direction discontinuously. As has been shown, if the transform behaves as

$$z - z_T = (\zeta - \zeta_T)^{(2-\tau/\pi)}$$

in the neighbourhood of T', the tail angle passes through an angle τ.

The mapping is normally defined so that the singularity lies on the real ζ axis at T'. The potential for flow around the circle of radius R with circulation Γ and velocity U inclined at an angle α to the x axis in ζ space is

$$F(\zeta) = U \exp(-\iota\alpha)\,(\zeta - \zeta_0) + \frac{UR^2 \exp(-\iota\alpha)}{(\zeta - \zeta_0)} - \frac{\iota\Gamma}{2\pi} \log(\zeta - \zeta_0) \qquad (11.31)$$

Hence the velocity is

$$F'(\zeta) = v_\eta - \iota v_\nu = U \exp(-\iota\alpha) - \frac{UR^2 \exp(-\iota\alpha)}{(\zeta - \zeta_0)^2} - \frac{\iota\Gamma}{2\pi\,(\zeta - \zeta_0)} \qquad (11.32)$$

The conditions to be applied to the mapping require:
1. $dz/d\zeta$ is non-zero everywhere in the ζ plane outside the circle.
2. $dz/d\zeta \to 1$ when $\zeta \to \infty$.
The velocity in the z plane is

$$\frac{dw}{dz} = v_x - \iota v_y = \frac{dF}{d\zeta}\frac{d\zeta}{dz} \qquad (11.33)$$

At the trailing edge if the velocity is finite

$$\frac{dF}{dz} = 0 \qquad (11.34)$$

and the trailing edge point in the circle (ζ) plane must be a stagnation point. If

$$\zeta_T = \zeta_0 + R\exp(-\iota\beta)$$

so that $-\beta$ is the angle between the line from the centre C' of the circle to the trailing edge T' in the ζ plane, then

$$\left(\frac{\mathrm{d}F}{\mathrm{d}\zeta}\right)_T = U\exp(-\iota\alpha) - U\exp[-\iota(\alpha + 2\beta)] - \frac{\iota\Gamma}{2\pi R}\exp(-\iota\beta)$$

and therefore

$$\Gamma = -2\,\iota\,\pi\,U\,R(\exp\left[\iota(\alpha + \beta)\right] - \exp\left[-\iota(\alpha + \beta)\right]) = 4\pi\,U\,R\,\sin\left(\alpha + \beta\right) \quad (11.35)$$

determines the circulation Γ for a well-behaved flow as required by the Kutta condition. The circulation is just that required to 'drag' the streamlines round to move the stagnation point to the trailing edge T'.

11.6.3 The Lift and Pitching Moment of Aerofoils Generated by Transformations of a Circle

It is very convenient that the major properties of aerofoils generated by the conformal transformation of a circle may be generated from those of the governing circle by an application of the Kutta condition, as exemplified by equation (11.35), and described in more detail in Appendix 11.B.

As noted earlier the integral in the Blasius expression for lift and moment may be extended out to large values of $|\zeta|$, where we may use the general asymptotic expression for transformation (11.B.2). The complex velocity in the z plane is therefore

$$\frac{\mathrm{d}w}{\mathrm{d}z} = \frac{\mathrm{d}F}{\mathrm{d}\zeta}\frac{\mathrm{d}\zeta}{\mathrm{d}z} = \left[U\exp(-\iota\alpha) - \frac{U\exp(\iota\alpha)}{(\zeta - \zeta_0)^2} + \frac{2\iota\,U\,R\,\sin(\alpha + \beta)}{(\zeta - \zeta_0)}\right]\frac{\mathrm{d}\zeta}{\mathrm{d}z} \quad (11.36)$$

and

$$\frac{\mathrm{d}z}{\mathrm{d}\zeta} = 1 - \frac{k_1}{\zeta^2} - \frac{2k_2}{\zeta^3} - \cdots \quad (11.37)$$

Substituting these values into the Blasius expression for the lift we obtain

$$L = 4\,\pi\,\rho U^2\,R\,\sin(\alpha + \beta) \quad (11.38)$$

where α and β are the angles that the incoming flow, and the line between the centre C' and the trailing edge T', make with the ξ direction respectively in the circle plane ζ (see Figure 11.4). Since β defines the angle of the axis of zero lift, the angle $\alpha' = \alpha + \beta$ is the effective angle of attack. Furthermore, since angles are unchanged by the transformation, we may conclude that α' is the effective angle of attack in the plane of the wing z.

Taking the centre of the profile to be at the centre of the generating circle, $z_0 = \zeta_0$, the pitching moment about the centre is

$$
M_C = -\Re\left\{\frac{\rho}{2}\oint\left[(\zeta - \zeta_0) + \frac{k_1}{(\zeta - \zeta_0)} + \cdots\right]\right.
$$

$$
\left. \times\left[U\exp(-\iota\alpha) - \frac{U\exp(\iota\alpha)}{(\zeta - \zeta_0)^2} + \frac{2\iota U R \sin(\alpha + \beta)}{(\zeta - \zeta_0)}\right]\frac{\mathrm{d}\zeta}{\left(1 - \dfrac{k_1}{\zeta^2} - \cdots\right)}\right\}
$$

Collecting terms of order ζ^{-1} we obtain

$$
M_C = \Re\left\{\oint\left[2U^2 a\exp(-2\iota\alpha) - 2U^2 R^2 - 4U^2 R^2 \sin^2(\alpha + \beta)\right]\frac{\mathrm{d}\zeta}{\zeta}\right\}
$$

$$
= 2\pi\rho U^2 \Re\left[\iota k_1 \exp(-2\iota\alpha)\right] \tag{11.39}
$$

If $k_1 = k^2\exp(2\iota\gamma)$ then

$$
M_C = -2\pi\rho U^2 k^2 \sin[2(\gamma - \alpha)] \tag{11.40}
$$

The second axis, which is defined as the angle of attack when the pitching moment is zero, is found at $\alpha = \gamma = \frac{1}{2}\arg(k_1)$, and therefore directly by the conformal transform employed. An important case is where the first and second axes coincide, $\beta = \gamma$, and the wing has neutral stability. This is clearly not possible with either the Zhukovskii or Karman–Treffetz transformations (Appendix 11.B) where k_1 is real and $\gamma = 0$ and the second axis is parallel to the chord.

We may rewrite equation (11.40) in terms of the effective angle of attack

$$
M_C = 2\pi\rho U^2 k^2 \sin 2(\beta + \gamma) - 4\pi\rho U^2 k^2 \sin(\alpha + \beta)\cos(2\gamma + \beta - \alpha)
$$

$$
= 2\pi\rho U^2 k^2 \sin 2(\beta + \gamma) - L\frac{k^2}{R}\cos(2\gamma + \beta - \alpha) \tag{11.41}
$$

Clearly the first term

$$
M_F = 2\pi\rho U^2 k^2 \sin 2(\beta + \gamma) \tag{11.42}
$$

is independent of the lift and represents the moment about the focus.

The second term identifies the position of the focus with respect to the centre. Since the lift may be assumed to act through the focus normal to the angle of attack, the focus may be seen to lie on the line CF at an angle $2(\beta + \gamma)$ to the first axis and a distance k^2/R from the centre. Since the angle between the first and second axes is $(\beta + \gamma)$, the second axis bisects the angle between the first axis and the line from the centre to the focus.

11.7   The Two-Dimensional Panel Method

In Section 2.10.2 we showed that the two-dimensional flow around a body may be described by a series of distributed sources and vortices on the surface of the body. We also briefly introduced the general idea of the panel method in the discussion of the uniqueness of the representation of the surface of the body by a vortex sheet. The surface of the body is subdivided into a finite number of small elements, namely *panels*. In general, each panel contains either a finite source or a rectilinear vortex (or both) in the form of either a point or distributed source/vortex. A boundary condition is applied at specified *collocation points* on the surface of the body within the panels, the flow at each point being linearly proportional to the source/vortex strength, and the constant of proportionality being a geometrical factor known the *influence factor*. As we have seen in the discussion of the thin wing (case study 2.II), simple sources are appropriate to take account of the thickness of the wing, and vortices for the camber. In the absence of circulation (symmetric bodies) the flow may be uniquely specified by a set of single sources and control points in each panel. However, if circulation is present (asymmetric bodies) vorticity must be introduced, and an additional boundary element is required. For example, one of the earliest panel methods for calculating the characteristics of aerofoil sections uses a set of sources and a single constant vorticity at the centre of each panel, each taking the form of a straight line segment of the aerofoil. The condition of zero normal velocity at each collocation point in the panel centre is supplemented by one based on the Kutta condition. Thus for N panels, there are $(N + 1)$ unknowns and $(N + 1)$ non-singular simultaneous equations, whose coefficients are the $(N + 1) \times (N + 1)$ influence factors. This ensures the uniqueness of the solution, which may be calculated by standard matrix methods.

The induced flow at the centre of each panel may be readily written in terms of the influence factors and source/vortex strengths, and hence the total velocity due to the sum of the incoming flow and the induced velocity may also be written down. The boundary conditions may be written in two different, but equivalent forms (Section 2.7.1):

1. Neumann condition. The normal component of the total velocity through the panel must be zero, since the surface is solid.

2. Dirichlet condition. The potential just inside the vortex sheet corresponding to the body must be constant.

As noted above, the set of equations is singular. Thus we must provide an additional constraint to allow the singularity to be removed. This is provided by the Kutta condition, which takes the simple form for the surface vorticity; that is, the total surface vorticity at the trailing edge of the wing must be zero. Thus at the trailing edge the vorticity in the panels on the upper $(\ell - 1)$ and lower ℓ surfaces must be equal and opposite in sign $\gamma_{\ell-1} + \gamma_\ell = 0$ provided they are short and of equal length.

The incoming flow velocity and the induced velocity produced by any panel j at any panel $i \neq j$ are both continuous across the vortex sheet. The only velocity discontinuity is that due to vorticity of panel i itself. Hence the circulation around the wing is

$$\Gamma = \sum_{i=1}^{N} \gamma_i \tag{11.43}$$

from which the lift may be calculated by the Kutta–Zhukovskii formula.

Since a system of doublets can be used to introduce circulation, combinations of sources and doublets may also be used. Panel methods based purely on the vorticity alone (vortex panel methods) are difficult to implement and infrequently used due to the zero diagonal elements of the influence matrix. For higher accuracy, extended sources within each panel (e.g. linearly distributed) should be used. The size and distribution of the panels themselves is important in improving the accuracy and speed of calculation. For more details on the implementation of these methods and a full discussion of the panel method reference should be made to Katz and Plotkin (2001).

11.8   Three-Dimensional Wings

Experiments have shown that the basic two-dimensional airflow over a wing section closely follows the picture established by the Zhukovskii hypothesis. At low speed the characteristic properties of lift and pitching moment are well described by the appropriate models based on ideal flow. Clearly drag, which involves viscosity, cannot be calculated by the ideal flow approximation, and falls outside the scope of this section.

However, in three dimensions the mathematical structure of the problem of airflow around the wing is markedly different.[12] This follows immediately from

---

[12]**Historical note**   Prandtl and his co-workers developed the three-dimensional model of the flow around a wing during the First World War, which delayed its general publication. The work was eventually released in two papers entitled 'Airfoil theory' published in *Reports of the Society of Science in Göttingen*. A summary of this work (in English) was subsequently issued as Prandtl (1921) and is described in Prandtl and Tietjens (1957). Lanchester (see Ackroyd *et al.*, 2001, chap. 7) had earlier correctly identified the physical picture of vortex motion resulting from flow around the wing tip, but without the mathematical structure developed by Prandtl.

the change in the connectivity resulting from the finite extent of the body (Section 2.6.1). In two dimensions the aerofoil section is doubly connected and the circulation around it is therefore arbitrary until specified by Zhukovskii's condition. In contrast, in three dimensions the wing is simply connected, and there exist reducible loops which may be contracted to a point in the irrotational flow. Consequently the circulation around an arbitrary wing section must be balanced by the total flux of vorticity through a surface enclosing the wing tip with base at the wing section. The flows in the two cases are therefore markedly different. The reconciliation of these differences caused major problems in the early investigations of aerofoil behaviour. The resolution was early proposed by Lanchester (see Ackroyd *et al.*, 2001, pp.57–69), but in an obscure form. Independently Prandtl (1921) generalised this approach casting it in a more mathematical form, as outlined in this section.

The flow around a simply connected, regular solid[13] in three dimensions is smooth and uniquely determined by the incoming flow (Section 2.6). Thus the flow around a sphere (Section 2.8.3) is smooth and has no circulation. However, in this case if the pressure of the incoming flow is too low, the flow will cavitate. Quite generally, as we have seen, the solution to the flow around a corner in the ideal approximation (Section 2.10.5) always leads to cavitation. Thus we conclude that the flow must separate at a sharp trailing edge in the manner discussed in Section 2.3.

Since the flow is irrotational, Bernoulli's equation ensures that in two dimensions the flow velocities over the top and bottom surfaces are the same at the trailing edge; consequently there can be no discontinuity in the velocity leaving the trailing edge. In contrast, in three dimensions the speeds must be equal, but their directions may be different; consequently a tangential discontinuity or vortex sheet may leave the trailing edge, forming the wake.

The relationship between the two- and three-dimensional flows is now explicable. We consider the case of a wing of large aspect ratio, i.e. span much larger than the chord, $b \gg c$. The flow around a particular section of the wing is therefore approximated by the two-dimensional model. Since there is a pressure difference between the lower (pressure) and upper (vacuum) surfaces, the streamlines are bent outwards on the lower surface and inwards on the upper as flow is permitted around the tip. We therefore expect that, in addition to the two-dimensional flow in the plane of the chord, there is a cross-flow along the span outwards on the under surface and inwards on the top. As the flow leaves the trailing edge, the tangential discontinuity or vortex sheet is formed, with the lines of vorticity approximately normal to the trailing edge (Figure 11.5).

---

[13] A regular surface in this context is smooth and has no discontinuities in gradient. Thus a wing section with an angular or cusp trailing edge is not regular.

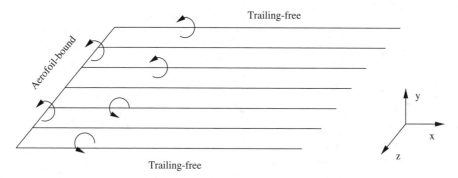

Figure 11.5: Schematic diagram of the vortex sheet formed at the trailing edge. For simplicity the wing is represented by a line, and the sheet by a set of finite vortex lines. The sense of rotation of the vortices is shown in conformity with the convention used earlier.

This sheet is swept out from the wing by the flow increasing in length, and forms the wake.

Wings with no taper across the span have constant circulation along the span. At the wing tip the strict condition of irrotationality is violated as the flux of vorticity along the wing due to the circulation over its surface must be balanced by a flux of vorticity through a surface enclosing the tip (since $\nabla \cdot \zeta = 0$). This gives rise to a vortex leaving the tip (*tip vortex*) which is required to allow the vortex lines to be continuous (Helmholtz's second theorem, *ii* page 28).

The wing profile in general will taper towards the wing tips, and the circulation around the wing consequently decrease as the tip is approached. The vortex lines associated with the circulation, which form a bound vortex around the wing section, are approximately parallel to the trailing edge, and their number decreases outwards. Since Helmholtz's second theorem requires that vortex lines be continuous, we see that the decrease in the number of vortex lines along the span must be balanced by those leaving the trailing edge in the vortex sheet. In the ideal flow picture, the lines are completed by the *starting vortex* established when the flow is initiated, and moving continuously downstream. Since the vortex sheet is unstable it rolls up into two helicoidal *trailing vortices* leaving the wing tips before reaching the starting vortex over a distance of several spans, the vortex lines forming a spiral within the roll. The total circulation, however, remains unchanged, consistent with Helmholtz's fourth theorem (*iv*, page 28). In practice this structure is damped by viscosity far downstream of the wing. The vortex structure thus forms a complete pattern of loops behind the wing–a *horseshoe vortex* (Figure 11.6).

The flow pattern induced by the trailing vortices gives rise to a downward velocity between the vortex pair (*downwash*), and an upward flow outside them

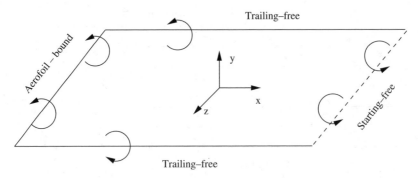

Figure 11.6: Sketch of the general horseshoe pattern of vortices about the trailing edge of a wing following the rolling up of the vortex sheet. The sense of rotation of the vortices is shown in conformity with the convention used earlier.

(*upwash*). This upward flow is made use of by flocks of geese flying in their characteristic V formation during migration. It is also used by aircraft flying in either a V or echelon formation. In each case the lift required is reduced and consequently, as we shall show, the resulting drag.

At the wing tip $z = -b/2$ the circulation around the wing is small and increases towards the median line of the wing $z = 0$, before decreasing towards the other wing tip $z = b/2$. In conformity with our convention, the circulation is taken as positive through the small angle. The strength of the vortex sheet is determined by the rate of rotation $x \to y$, the axis of the circulation being along the z axis. Clearly the strength of the vortex sheet is determined by the rate of change of circulation along the span. The direction of the surface vorticity or strength is parallel to the x axis for $z > 0$ and anti-parallel for $z < 0$. Thus from Helmholtz's second theorem we estimate the surface vorticity as

$$\Sigma(z) \approx -\frac{\mathrm{d}\Gamma(z)}{\mathrm{d}z}\,\hat{\mathbf{i}}$$

where $\hat{\mathbf{i}}$ is the unit vector in the x direction and Γ the circulation of the bound vortex.

11.8.1 Velocity at the Wing Surface

As we saw in Section 2.7.1 we may represent an arbitrary body by a vortex sheet whose surface matches that of the body. Three-dimensional flow around a wing can therefore be considered to be composed of two sets of vortex sheets:

1. The bound vortices around the wing sections, whose strength decreases along the span.
2. The free vortices leaving the trailing edge of the wing and closed by the starting vortex far downstream, effectively at infinity.

The velocity at any point in the flow can therefore be calculated as the sum of three different velocity terms:

1. The incoming velocity $\mathbf{U}$.
2. The velocity induced by the bound vortex sheet around the wing $\mathbf{v}_s$ due to the vorticity $\boldsymbol{\Sigma}_s$.
3. The velocity induced by the wake vortex sheet $\mathbf{v}_w$ with vorticity $\boldsymbol{\Sigma}_w$.

In this picture the wing is imagined to be an internal region of air separated from the external stream by the discontinuity of the sheet. The normal velocity of the fluid at the boundary of the sheet is zero. It follows from the Neumann boundary condition in a simply connected space (Section 2.6) that the velocity inside the sheet must be zero.

Hence we may write the velocities external $+$ and internal $-$ to the sheet as

$$\mathbf{v}^{\pm} = \mathbf{U} + \mathbf{v}_s^{\pm} + \mathbf{v}_w^{\pm}$$

where $\mathbf{v}_s^{\pm}$ are the velocities induced by the sheet.

The flow velocities due to incoming flow and induced by the wake are continuous across the sheet. The velocity induced by the remainder of the bound vortex sheet, excluding the element at which the calculation is made, is given by $\mathbf{v}_s = \frac{1}{2}\lim\left(\mathbf{v}_s^{+} + \mathbf{v}_s^{-}\right)$ (Section 2.2.1). The total velocity inside the sheet $\mathbf{v}^{-}$ is zero. The velocity on the surface is

$$\frac{1}{2}\mathbf{v} = \mathbf{U} + \mathbf{v}_s + \mathbf{v}_w \tag{11.44}$$

11.8.2 The Force on the Wing

The force on the wing is due to the pressure on the surface of the wing, which in turn is given by Bernoulli's equation. The total force is therefore

$$\begin{aligned}
\mathbf{F} &= -\iint_S p\,\mathrm{d}\mathbf{S} \\
&= -\rho \iint_S \left\{\frac{p_0}{\rho} + \frac{1}{2}\left(U^2 - v^2\right)\right\} \mathrm{d}\mathbf{S} \\
&= \frac{1}{2}\rho \iint_S v^2 \mathrm{d}\mathbf{S} \\
&= \frac{1}{2}\rho \iint_S \mathbf{v} \wedge (\mathbf{n} \wedge \mathbf{v})\,\mathrm{d}S
\end{aligned}$$

where $\mathbf{n}$ is the unit normal to the surface element $\mathrm{d}\mathbf{S}$ since the constant terms integrate to zero over the closed surface S of the wing, $\mathbf{n} \cdot \mathbf{v} = 0$ since $\mathbf{n}$ is normal to $\mathbf{v}$ and we have used the expansion of the vector triple product

$$\mathbf{v} \wedge (\mathbf{n} \wedge \mathbf{v}) = (\mathbf{v} \cdot \mathbf{v}) \, \mathbf{n} - (\mathbf{n} \cdot \mathbf{v}) \, \mathbf{v} = v^2 \, \mathbf{n}$$

Since the velocity inside the sheet is zero, $\mathbf{v}$ is the velocity difference across the sheet, and $\mathbf{n} \wedge \mathbf{v}$ is the surface vorticity Σ_s so that

$$\mathbf{F} = \frac{1}{2}\rho \iint_S \mathbf{v} \wedge \Sigma_s \, \mathrm{d}S \tag{11.45}$$

Substituting the total velocity outside the wing (11.44), we obtain three distinct components of the force:

1. **Lift**
$$\mathbf{L} = \rho\, \mathbf{U} \wedge \iint_S \Sigma_s \, \mathrm{d}S \tag{11.46}$$

   We shall show that this is approximately given by the integral of the local lift force, as determined by the Kutta formula, along the span.

2. **Drag**
$$\mathbf{D} = \rho \iint_S \mathbf{v}_w \wedge \Sigma_s \, \mathrm{d}S \tag{11.47}$$

3. **Cross-coupled**
$$\mathbf{C} = \rho \iint_S \mathbf{v_s} \wedge \Sigma_s \, \mathrm{d}S \tag{11.48}$$

   In the usual case, when when the aspect ratio of the wing is large (the span is much larger than the chord), the vortex lines are nearly parallel, the cross-coupled term is small and can be neglected.

11.8.3   Prandtl's Lifting Line Model–Downwash Velocity

In order to proceed, Prandtl made a set of simplifying assumptions about the nature of the form of the wing and the flow known as the *lifting line approximation*:

1. The wing has a median plane of symmetry.
2. The chord parallel to the plane of symmetry is everywhere small compared with the span.
3. The trailing edge may be treated as a straight line.
4. The bound vortex lines are all parallel to the span.
5. The velocity induced at any point P by the trailing vortices is equal to the velocity induced at the point on the trailing edge on the same section as P.
6. All the trailing vortices leave the wing parallel and parallel to the incoming velocity.

It follows from Helmholtz's second theorem that the number of lines of vorticity leaving the bound vortex sheet over the wing in an element δz must determine the strength of the free vortex sheet leaving the trailing edge $\Sigma \, \delta z$ (Figure 11.5). The strength of the free vortex sheet a distance z from the median line is thus

$$\mathbf{\Sigma}(z) = -\frac{\mathrm{d}\Gamma(z)}{\mathrm{d}z} \, \hat{\mathbf{i}} \qquad (11.49)$$

in agreement with our earlier picture.

Within the lifting line model the bound vortex is assumed to lie along the trailing edge and therefore its contribution to the induced velocity at the trailing edge is zero. However, the trailing vortex sheet is an assembly of rectilinear vortices running away from the trailing edge parallel to the median plane, each of width δz and with circulation $\delta \Gamma = \Sigma \, \delta z$. These vortices make a total contribution, given by equation (2.30) with $\theta_1 = 0$ and $\theta_2 = \pi/2$, to the velocity at a point z

$$\mathbf{u} = \frac{1}{4\pi} \int_{-b/2}^{b/2} \frac{1}{(z - z')} \left. \frac{\mathrm{d}\Gamma}{\mathrm{d}z} \right|_{z'} \mathrm{d}z' \; \hat{\mathbf{i}} \wedge \hat{\mathbf{k}}$$

Hence there is only a component in the y direction

$$u_y = -\frac{1}{4\pi} \fint_{-b/2}^{b/2} \frac{1}{(z - z')} \left. \frac{\mathrm{d}\Gamma}{\mathrm{d}z} \right|_{z'} \mathrm{d}z' \qquad (11.50)$$

where $\Gamma(z')$ is the circulation around the section at z'. Since the circulation, $\Gamma(z)$, about the wing is a negative quantity,[14] this velocity is downwards at the trailing edge. The integral is evaluated as a principal value integral. This term is known as the *downwash velocity*.

The forces follow directly:

1. **Lift**   Substituting for the strength of the vortex from (11.49) in equation (11.46) we obtain

$$\mathbf{L} = \rho \, \mathbf{U} \wedge \iint_S \mathbf{\Sigma}_s \, \mathrm{d}S = -\rho \, U \int_{-b/2}^{b/2} \Gamma(z) \, \mathrm{d}z \, \hat{\mathbf{j}} \qquad (11.51)$$

   where the circulation[15]

$$\mathbf{\Gamma}(z) = \left. \oint \Sigma \, \mathrm{d}\ell \right|_z \hat{\mathbf{k}} \qquad (11.52)$$

   Hence the lift is normal to the incoming flow velocity, as expected. This result is an extension of the Kutta–Zhukovskii formula from two dimensions, as might be expected if the aspect ratio is large.

---

[14]See footnote 13 on page 67 where this change of sign is discussed.
[15]$\hat{\mathbf{i}}$, $\hat{\mathbf{j}}$ and $\hat{\mathbf{k}}$ are the unit vectors in the x, y and z directions respectively.

2. **Drag**   Similarly substituting for the strength of the vortex from (11.49) in equation (11.47), we obtain

$$\mathbf{D} = \rho \int_{-b/2}^{b/2} u_y \, \Gamma \, \mathrm{d}z \, \hat{\mathbf{i}} \tag{11.53}$$

The drag is therefore a force in the direction of the incoming flow. Substituting for the downwash velocity (11.50) in equation (11.47), we obtain the *induced drag* due to the transverse flow along the span:

$$\mathbf{D} = \frac{\rho}{4\pi} \int_{-b/2}^{b/2} \oint_{-b/2}^{b/2} \frac{1}{(z - z')} \Gamma(z) \left.\frac{\mathrm{d}\Gamma}{\mathrm{d}z}\right|_{z'} \mathrm{d}z' \, \mathrm{d}z \, \hat{\mathbf{i}} \tag{11.54}$$

Integrating by parts we may write this expression in the alternative form

$$\mathbf{D} = -\frac{\rho}{4\pi} \int_{-b/2}^{b/2} \oint_{-b/2}^{b/2} \left.\frac{\mathrm{d}\Gamma}{\mathrm{d}z}\right|_z \left.\frac{\mathrm{d}\Gamma}{\mathrm{d}z}\right|_{z'} \ln|z - z'| \, \mathrm{d}z' \, \mathrm{d}z \, \hat{\mathbf{i}} \tag{11.55}$$

Although formed in irrotational flow, the extended vortex sheet leaving the trailing edge of the wing is included with the wake, thereby violating d'Alembert's paradox. Together with the *parasitic* (viscosity-induced skin) friction it contributes to the total momentum transfer through the wake, and thus the drag.

3. **Cross-coupled**   Substituting for the downwash velocity (11.50), the cross-coupled term (11.48) can be written as

$$\mathbf{C} = \frac{\rho}{8\pi} \iint_S \iint_S \frac{[\boldsymbol{\Sigma}_s(\mathbf{r}) \wedge \boldsymbol{\Sigma}_s(\mathbf{r}')] \wedge (\mathbf{r} - \mathbf{r}')}{|\mathbf{r} - \mathbf{r}'|^3} \, \mathrm{d}S \, \mathrm{d}S' = 0 \tag{11.56}$$

since in the model the vortex lines are everywhere parallel to the trailing edge. This condition, i.e. that the vortex lines on the surface of the wing are nearly parallel, is obeyed quite generally. The cross-coupled term is consequently normally small.

The incident velocity must be corrected for the downwash, so that the effective incoming flow onto the wing has velocity $\mathbf{U}' = \mathbf{U} + \mathbf{u}$. The change in the magnitude of the velocity is relatively small, but the effective angle of incidence is rotated by a small angle, the *angle of downwash*,

$$\alpha_d = \arctan\left(\frac{u_y}{U}\right) \tag{11.57}$$

which decreases the angle of attack. If the two-dimensional expression for the lift around a section (11.8) is used, the effective angle of attack should be taken as $\overline{\alpha} \approx \alpha' - \alpha_d = \alpha - \alpha_0 - \alpha_d$. As a result the lift slope is decreased. The lift coefficient is

$$C_L = a\overline{\alpha} = \overline{a}\alpha' \tag{11.58}$$

where $\overline{a}$ is the effective lift slope and a that of the wing section treated as a two-dimensional flow. We note that the lift slope a is a characteristic parameter of the wing section determined by its circulation and specifies its lift parameter.

The induced drag can be interpreted as the rotated component of the lift due to the downwash $\sim \alpha_d L$ so that the total force remains normal to the effective flow.

We note that the induced drag does not violate d'Alembert's paradox as the flow is no longer completely steady, the wake (vortex sheet) increasing in length as the trailing vortex moves progressively downstream with the flow.

The induced drag force can be seen from equation (11.54) to be independent of the value of the span. The lift on the other hand increases linearly with the span. The ratio of drag to lift for an infinite wing becomes small limited only by the unavoidable *parasitic drag* due to viscosity. Ultralow-drag wings, e.g. for human-propelled flight, are therefore long and thin. On many airliners small upturned *winglets* are fitted to the wing tips to prevent the cross-flow around the tip and thereby reduce induced drag, since fuel efficiency is an important issue.

Despite its approximations, the lifting line model is remarkably accurate and is still valuable, providing a rapid analytic model for the calculation of lift and induced drag.

11.8.4 Lift and Drag as Properties of the Wake

It is instructive to see how the lift and drag are determined by the flow in the wake. Consider two infinite planes normal to the incoming flow direction x, namely (y, z) planes, one far upstream of the body x_1 and one downstream x_2. The force on the body must then be given by the difference in the total momentum flux through the two planes. Making the same assumptions as previously, the flow is only weakly perturbed by the body, and the lift force per unit width is determined by the loss of momentum of fluid flowing through a closed surface consisting of the two planes normal to the incoming flow direction and closed at infinity:

$$F_y = -\rho U \iint_{\text{back}} u_y \, dy \, dz$$

where u_y is the y component of the perturbation velocity. The flow, which is perturbed by viscosity, passes through the back plane in a narrow region of the flow, which is not irrotational, known as the wake.[16] The fluid in the wake has passed close to the surface of the body and acquired vorticity due to the action of viscosity. The wake is narrow as the body is streamlined and the separation, which must occur, takes place only at that trailing edge. However, the irrotational velocity perturbation, which occurs in ideal flow, will be much more extensive. Thus we may write

$$F_y = -\rho U \left(\int_{y_+}^{\infty} + \int_{-\infty}^{y_-} \right) u_y \, dy$$

where y_+ and y_- are the upper and lower boundaries of the wake respectively. In the region outside the wake there is potential flow, so that $u_y = \partial\phi/\partial y$ and

$$F_y = -\rho U \int_{-b/2}^{b/2} (\phi_- - \phi_+) \, dz$$

But the circulation around the body is

$$\Gamma = \oint \mathbf{u} \cdot d\mathbf{l} = \oint \nabla\phi \cdot d\mathbf{l} = \phi_- - \phi_+ \tag{11.59}$$

Hence, substituting, we recover the Kutta–Zhukovskii lift formula, equation (2.105), for the lift per unit width, F_y.[17]

The drag force per unit width is similarly evaluated by comparing the momentum flow through the front and back planes far from the body where the pressure $p' = p - p_0$ and velocity $(\mathbf{u} = \mathbf{v} - \mathbf{U})$ perturbations are small. To lowest order in the perturbation the drag force is

$$F_x = \left(\iint_{\text{front}} - \iint_{\text{back}} \right) (p' + \rho U \, u_x) \, dy \, dz \tag{11.60}$$

From Bernoulli's equation it follows that, neglecting terms of second order,

$$p' \approx -\rho \, \mathbf{U} \cdot \mathbf{u}$$

The only contribution to the drag in first order is therefore from the wake where the flow is no longer irrotational

$$F_x = -\rho U \iint_{\text{wake}} u_x \, dy \, dz \tag{11.61}$$

---

[16]In ideal flow the wake is the tangential discontinuity across which the tangential component of velocity is discontinuous, i.e. a vortex sheet.

[17]This derivation is strictly only valid for a three-dimensional flow around a wing with large aspect ratio. As we have discussed earlier, the discontinuity at the trailing edge is not formed in two-dimensional flow and the circulation is established as a boundary condition by Zhukovskii's condition.

However, in the absence of viscosity the wake is a thin discontinuous layer across which u_x is approximately continuous. The above equation therefore indicates that the drag is small, which is indeed the case since $F_x \ll F_y$. In ideal flow this term is zero as the wake is a pure discontinuity. In practice, due to viscosity the wake has a finite width and the contribution of this term to the momentum loss represents the *parasitic drag* referred to earlier.

However, there is also transport of energy downstream associated with the net kinetic energy gain resulting from the cross-flow in the wake vortex sheet, which we have neglected thus far and takes place over an area of dimensions typically of the order of the span, much larger than the wake. This energy flow must be supplied by the work done by a force which the wing exerts on the fluid in the direction of motion. This force is equal to and opposite to the drag force applied by the fluid to the wing, namely *induced drag*. The drag force is calculated from the increase in the kinetic energy of the air at the downstream plane

$$\frac{1}{2} \iint \rho \, (\nabla \phi)^2 \, U \, \mathrm{d}y \, \mathrm{d}z$$

per unit time, since a mass $\iint \rho U \, \mathrm{d}y \, \mathrm{d}z$ is swept out. The drag force[18] is given by the work done $F_x U$. Since the flow is incompressible, $\nabla^2 \phi = 0$, F_x may be written as[19]

$$
\begin{aligned}
F_x &= -\frac{1}{2} \iint \rho \, (\nabla \phi)^2 \, \mathrm{d}y \, \mathrm{d}z \\
&= -\frac{1}{2}\rho \left\{ \int_{-b/2}^{b/2} \left[\phi \frac{\partial \phi}{\partial y} \right]_{+} \, \mathrm{d}z - \int_{-b/2}^{b/2} \left[\phi \frac{\partial \phi}{\partial y} \right]_{-} \, \mathrm{d}z \right\}
\end{aligned}
$$

where $+$ and $-$ refer to the top and bottom of the wake respectively, and the span is $z = -b/2$ to $z = +b/2$.

Since $u_y = \partial \phi / \partial y$ is continuous across the wake and $\phi_- - \phi_+ = \Gamma(z)$ as before, we obtain

$$F_x = \frac{1}{2}\rho \int_{-b/2}^{b/2} \Gamma(z) \, u_y \, \mathrm{d}z \tag{11.62}$$

which is identical to the result obtained earlier (11.53).

It follows from either equation (2.106) or equation (11.59) that the circulation is a first-order term in the perturbation (u/U). It therefore follows that the

---

[18]This term is obtained by including the next order terms from Bernoulli's equation in equation (11.60).

[19]Using Green's theorem

$$\int_V (\nabla \phi)^2 \, \mathrm{d}V = \int_S \phi \nabla \phi \cdot . \mathrm{d}\mathbf{S} - \int_V \phi \nabla^2 \phi \, \mathrm{d}V$$

lift coefficient is of first order in (u/U), but the induced drag coefficient second order. Since the circulation is proportional to the incoming flow velocity, it follows that both the lift and induced drag vary as U^2 and that, provided the lift is linear as in Section 11.3, the lift and induced drag coefficients depend only on the geometrical configuration. Hence $C_D \ll C_L$. Furthermore, since the circulation scales linearly with the incoming flow speed, $\Gamma \sim U$ and approximately linearly with the angle of attack α, a useful general relation between the induced drag and the lift for wings of geometrically similar profiles can be seen from equations (11.52) and (11.54): that is, for a given wing profile, the drag and lift coefficients are related as

$$\frac{C_D}{C_L{}^2} \approx \text{const} \tag{11.63}$$

as a consequence of the dependence of each term on the circulation, the constant being determined solely by the geometric form of the wing profile. It follows that the induced drag scales with the angle of attack approximately as $C_D \sim \alpha^2$.

Referring back to equation (11.54) for the induced drag, we see if that all dimensions in the z direction are increased in proportion, the circulation $\Gamma(z)$ remaining constant for the scaled values of z, the drag force is unchanged and the drag coefficient reduced. On the other hand, the lift is increased by the scale factor, the lift coefficient remaining constant.

In ideal flow the wake is vanishingly thin, and drag resulting from the rotational (viscous) components in the flow is therefore vanishingly small. However, when viscosity is included there is a significant region of rotational flow where viscosity has played an important role. The total drag is therefore the sum of the *induced drag* and the *parasitic drag*, which results from the action of viscosity in the boundary layer at the surface of the wing, as we saw in Chapter 6. However, the relationship of the induced drag with equation (11.61) for the loss of momentum in the direction of flow in the wake is not immediately obvious. The flow which contributes to the induced drag is irrotational and may extend far outside the wake, as noted earlier. But it was argued earlier on the basis of Bernoulli's equation that the only contribution to the drag is due to rotational flow in the wake. In fact, as we have already noted, induced drag is a second-order effect, which is neglected in the derivation of equation (11.61). Indeed we may conclude that across the wake there must be a small (second-order) discontinuity in u_x induced by the vortex sheet, which contributes to the general integral of the momentum transfer rate parallel to incoming flow, F_x, across the back plane.

Case study 11.I Calculation of Lift and Induced Drag for Three-Dimensional Wings

11.I.i Wing loading

The lift and induced drag given by equations (11.52) and (11.53) may be considered to be the integrals over the wing of contributions from elements dz of the span, namely $\rho U \Gamma(z) dz$ and $\rho u_y \Gamma(z) dz$ respectively. These terms therefore represent the elementary loading of the wing from the different elements.

In Section 11.3 we showed that the lift coefficient of two-dimensional sections was approximately linearly proportional to the effective angle of attack. Generalising this result for three-dimensional wing profiles by including the upthrust, this result becomes

$$C_L = a \left(\alpha - \alpha_0 - \alpha_d \right) = a' \left(\alpha - \alpha_0 \right) = a' \, \alpha' \qquad (11.64)$$

where α_d is the angle of downwash (11.57), $\alpha' = \alpha - \alpha_0$ is the effective angle of attack (11.8) and we assume that the downwash vanishes at the angle of zero lift. The lift slope a, namely the constant of proportionality (11.8) between the lift coefficient and the effective angle of attack, in the absence of downwash is determined from the two-dimensional theory for sections along the span. The inclusion of the downwash modifies this value to a'. For thin wings, we recall $a = 2\pi$.

Since the lift coefficient

$$C_L = -\frac{\rho U \Gamma}{\frac{1}{2} \rho U^2 c} \qquad (11.65)$$

where c is the chord, equating the values of the lift coefficient and substituting for the angle of downwash α_d, we obtain

$$\frac{\Gamma(z)}{\frac{1}{2} c(z) a(z) U} = -\alpha'(z) - \frac{1}{4\pi U} \int_{z'=-b/2}^{z'=b/2} \frac{d\Gamma(z')}{(z - z')} \qquad (11.66)$$

which provides an integral equation from which $\Gamma(z)$ may be determined, provided $c(z)$, $a(z)$ and α' are known. The two-dimensional flow about the wing section at z is specified by the lift slope $a(z)$.

Prandtl's integral equation (11.66) provides the connection between the two-dimensional theory of wing sections and the realistic flow about a three-dimensional wing. The three-dimensional profile of the wing is specified through the chord and slope angle. We may consider the problem in two separate ways:

1. Given the loading, i.e. $\Gamma(z)$, find the form of wing profile and the induced drag.
2. Given the wing profile, calculate its circulation and aerodynamic properties.

The latter is the more difficult. The first case being relatively straightforward, we consider only the second.

To solve this equation we write the distribution of the circulation along the span as a Fourier series

$$\Gamma(z) = -2bU \sum_{n=0}^{\infty} A_{(2n+1)} \sin[(2n + 1)\theta] \qquad (11.67)$$

where θ is the polar angle defined by $z = -\frac{1}{2} b \cos\theta$, so that the wing tips are respectively $\theta = 0$ and $\theta = \pi$. The sum contains only sine terms, since the circulation is zero at the tips, and only contains even terms since the wing is symmetric about the median line $\theta = \pi/2$. Substituting in the integral equation we obtain

$$\frac{4b}{a(\theta)\,c(\theta)} \sum_{n=0}^{\infty} A_{(2n+1)} \sin[(2n+1)\,\theta] = \alpha'(\theta)$$

$$+ \frac{b}{2\pi} \int_{0}^{\pi} \frac{\sum_{n=0}^{\infty} (2n+1)\, A_{(2n+1)} \, \cos[(2n+1)\theta']}{\frac{1}{2} b(\cos\theta - \cos\theta')} \, d\theta'$$

The integral is obtained from equation (11.A.5) to yield

$$\sum_{n=0}^{\infty} A_{(2n+1)} \left[(2n+1)\,\mu(\theta) + \sin\theta\right] \sin[(2n+1)\theta] = \mu(\theta)\alpha'(\theta)\sin\theta \qquad (11.68)$$

where $\mu(\theta) = a(\theta)\,c(\theta)/4\,b$. The angle of attack relative to the axis of zero lift, namely $\alpha'(\theta) = \alpha - \alpha_0(\theta)$, will vary along the span if the wing has *aerodynamic twist*. This need not arise from a constructional twist in the wing, but may be due to changes in the section along the span, which give rise to a rotation of the zero-lift axis.

In a practical case, the series is truncated to a finite set of terms, say $n \leq m$. The m coefficients A_{2n+1} are then evaluated as solutions of the set of simultaneous equations (11.68) taken at m independent points.

If the wing does not have an aerodynamic twist, it follows from equation (11.68) that the set $A_{(2n+1)}$ is linearly proportional to the effective angle of attack α'.

Lift   The lift is directly calculated from the loading using the integral of the Kutta–Zhukovskii formula (11.52) along the span. Since

$$\Gamma(z)\,dz = b^2\,U \sum_{0}^{\infty} A_{(2n+1)} \, \sin(2n+1)\theta \, \sin\theta \, d\theta$$

and

$$\int_{0}^{\pi} \sin m\theta \, \sin n\theta \, d\theta = \delta_{m,n}$$

it follows that the lift and the lift coefficient for the entire wing are

$$L = \frac{1}{2}\, \pi\,\rho\,U^2\,b^2\,A_1$$

$$C_L = \frac{\pi\,b^2 A_1}{S} = \pi\,A\,A_1 \qquad (11.69)$$

where $A = b^2/S$ is the aspect ratio of the wing.

Downwash velocity The downwash velocity directly from equations (11.50) and (11.A.5) is

$$u_y = \frac{1}{4\pi} \int_0^\pi \frac{2bU\sum(2n+1)A_{2n+1}\cos(2n+1)\theta'}{\frac{1}{2}b(\cos\theta' - \cos\theta)}\,d\theta'$$

$$= U\sum_0^\infty (2n+1)\,A_{2n+1}\frac{\sin(2n+1)\theta}{\sin\theta} \tag{11.70}$$

Induced drag The induced drag is given by equations (11.53) and (11.70)

$$u_y\Gamma(y)\,dy = b^2\,U^2\sum_{m=0}^\infty (2m+1)A_{2m+1}\sin(2m+1)\theta\sum_{n=0}^\infty A_{2n+1}\sin(2n+1)\theta\,d\theta$$

Hence integrating over the span, the total induced drag and the corresponding drag coefficient are

$$D = \frac{1}{2}\pi\rho b^2\,U^2\sum_0^\infty (2n+1)A_{2n+1}^2$$

$$C_D = \pi A\sum_0^\infty (2n+1)A_{2n+1}^2 = \pi A A_1^2\,(1+\delta) \tag{11.71}$$

where $\delta = \{\sum_0^\infty(2n+1)A_{2n+1}^2\}/A_1 \geq 0$ may be considered to be the correction to the case of elliptic loading.

11.I.ii Elliptic loading

Since the total lift depends only on the first Fourier component A_1, it follows from (11.71) that the minimum induced drag subject to constant lift is obtained when $A_1 = C_L/\pi A$ and $A_n = 0$ $(1 \leq n \leq \infty)$. In this case the downwash velocity is constant along the span. Since

$$\Gamma = -2\,bU\,A_1\sin\theta = \Gamma_0\,\sin\theta$$

$$z = -\frac{1}{2}\,b\,\cos\theta$$

it follows that

$$\left(\frac{\Gamma}{\Gamma_0}\right)^2 + \left(\frac{z}{b/2}\right)^2 = 1 \tag{11.72}$$

where $\Gamma_0 = -2\,bU\,A_1$ is the circulation on the median line.

The total induced drag coefficient for elliptic loading may be expressed in terms of the lift coefficient as

$$C_D = \pi A A_1{}^2 = \frac{1}{\pi A}C_L{}^2 \tag{11.73}$$

Thus the induced drag varies as the square of the lift, and in this particular case of elliptic loading, the approximation of equation (11.63) is exact. Induced drag is therefore strong when the lift is high, e.g. at take-off or at high altitude. Its value under these conditions can be larger than that due to parasitic drag resulting from the action of viscosity. Gliders and high-flying aircraft are therefore designed with wings of high aspect ratio, despite the difficulty of manufacture. Aircraft with elliptic wings, such as the Spitfire, minimise the induced drag. However, such wings are difficult to fabricate, and tapered wings with an appropriate taper ratio, approximating to the elliptical, usually provide a satisfactory compromise. It can be shown that the loading is elliptic at all angles of incidence only if the chord function $c(z)$ is such that $a(0)\,c(0) = a(\theta)\,c(\theta)\,\sin\theta$. Hence if the profiles of all sections are geometrically similar, $a(\theta) \approx$ const, the profile of the chord $c(z)$ must also be an ellipse.

11.9   Three-Dimensional Panel Method

In Section 11.7 we outlined briefly the use of the panel method for calculating the flow around aerofoil sections. However, the method achieves its greatest applicability for the calculation of the flow around complete aeroplane surfaces in three dimensions.

The basic method is the same as in two dimensions in that the body surface is broken up into a large number of small panels. The flow over the surface is calculated in terms of the strengths of the singularities assigned to each panel. Applying the boundary conditions at the surface in either the Neumann or Dirichlet forms leads to a set of simultaneous equations for the strengths. However, in this case there is the additional complication of the wake, which takes the form of a vortex sheet leaving the trailing edge of all the streamlined surfaces. This will introduce additional induced velocities, leading to downwash and induced drag as we have seen. In contrast to the two-dimensional case, we may need to invoke a Kutta condition, with the necessary support being given by the vorticity of the wake, namely $\gamma_{\ell-1} + \gamma_\ell = \gamma_w$, so that the deficit in vorticity along the wing equals that in the trailing wake. As a result the set of simultaneous equations is non-singular and no additional condition needs to be invoked. To take account of the wake vorticity, we must introduce additional panels in the wake. This usually takes the form of the wake being assigned by a user-defined surface leaving the trailing edge of the body surfaces. Panels are designated in the same manner on the wake as on the body. It will be appreciated that in the choice of panels in three dimensions, experience and intuition play important roles.

The nature of the singularities which are used can vary with choices from sources, doublets and vortex loops. The singularities may be point or may be distributed over the panel, the latter giving higher accuracy.

The panel method is widely used, not only for the study of wing sections, but in other areas of fluid mechanics such as the design of racing yachts, high-speed racing cars, etc. The method is considerably more sophisticated than indicated in this brief account. For an extensive discussion of the panel method in two or three dimensions the text by Katz and Plotkin (2001) is recommended.

Appendix 11.A Evaluation of the Principal Value Integrals

In many problems of fluid dynamical importance, principal value integrals are encountered, which may be solved by Fourier techniques exemplified by equation (11.23) to calculate the circulation around a thin wing, and subsequently in the calculation of induced drag. This requires the evaluation of the integral (11.19) which contains the terms

$$I_n = \int_0^\pi \frac{\cos(n\theta)}{\cos\theta - \cos\theta'} \, d\theta \qquad (11.A.1)$$

The above integrals have a pole at $\theta = \theta'$, which corresponds to the velocity induced by a vortex at itself. We therefore exclude the singularity from the integral evaluating only the principal value:

$$I_n = \lim_{\epsilon \to 0} \left\{ \int_0^{\theta' - \epsilon} \frac{\cos(n\theta)}{\cos\theta - \cos\theta'} \, d\theta - \int_{\theta' + \epsilon}^\pi \frac{\cos(n\theta)}{\cos\theta' - \cos\theta} \, d\theta \right\}$$

We first evaluate I_0 noting that

$$\frac{1}{\cos\theta - \cos\theta'} = \frac{1}{2\sin\theta'} \left\{ \frac{\cos[\frac{1}{2}(\theta' + \theta)]}{\sin[\frac{1}{2}(\theta' + \theta)]} + \frac{\cos[\frac{1}{2}(\theta' - \theta)]}{\sin[\frac{1}{2}(\theta' - \theta)]} \right\}$$

Hence

$$I_0 = \lim_{\epsilon \to 0} \left\{ \frac{1}{\sin\theta'} \log \left[\frac{\theta' - \frac{1}{2}\epsilon}{\theta' + \frac{1}{2}\epsilon} \right] \right\} = 0 \qquad (11.A.2)$$

The integral I_1 is

$$I_1 = \int_0^\pi \frac{\cos\theta}{\cos\theta - \cos\theta'} \, d\theta = \int_0^\pi \frac{\cos\theta - \cos\theta'}{\cos\theta - \cos\theta'} \, d\theta = \pi \qquad (11.A.3)$$

and finally since

$$\cos(n+1)\theta - \cos(n+1)\theta' + \cos(n-1)\theta - \cos(n-1)\theta'$$
$$= 2\cos n\theta \, (\cos\theta - \cos\theta') + 2\cos\theta' \, (\cos n\theta - \cos n\theta')$$

we have the recurrence for $I_{(n+1)}$

$$I_{(n+1)} + I_{(n-1)} = 2\cos\theta' I_n \qquad (11.A.4)$$

whose solution is easily seen to be

$$I_n = A \cos n\theta' + B \sin n\theta' = \pi \frac{\sin n\theta'}{\sin\theta'} \qquad (11.A.5)$$

since $A = 0$ and $B = \pi/\sin\theta'$ are determined by the known values of I_0 and I_1.

Appendix 11.B The Zhukovskii Family of Transformations

Zhukovskii's transformation

$$z = \zeta + \frac{\ell^2}{\zeta} \tag{11.B.1}$$

and the related modifications formed the basis for much of the theoretical calculation of the flow around aerofoils before rapid numerical computing became available. Although this method is no longer used, it provided the basis for the calculation of the flow around two-dimensional wing sections up to the 1960s. However, as it still gives a nice example of the power and the limitations of calculations using conformal transforms, it remains of interest. The transformation is used to relate the flow around a cylinder to that around an aerofoil section. A circle in the ζ plane passes through one of the singularities in the transformation at $\zeta = \pm\ell$, which transforms to give the sharp trailing edge at $z = \pm 2\ell$. The simple transform is somewhat limited by conditions at the trailing edge. However, modifications to the transform and the introduction of iterative corrections allowed accurate calculations to made of the pressure distribution around the wing in the incompressible limit.

The complex potential for a wind directed onto the aerofoil at an angle α from $x = -\infty$ is $U \exp(-\imath\alpha)z$. Therefore we require a transformation which yields $z = \zeta$ for large values of $|\zeta|$ and $|z|$. Such a form is

$$z = \zeta + \frac{k_1}{\zeta} + \frac{k_2}{\zeta^2} + \frac{k_3}{\zeta^3} + \ldots = F(\zeta) \tag{11.B.2}$$

where the series converges for large $|\zeta|$.

As we have noted, the mapping ceases to be conformal at a zero of $\mathrm{d}z/\mathrm{d}\zeta$. At this singular point the mapping of a circle in the ζ plane will yield two tangents enclosing either a finite angle or a cusp (touching or zero angle). To demonstrate this result consider a small region enclosing the singularity at ζ_0

$$\frac{\mathrm{d}z}{\mathrm{d}\zeta} = F'(\zeta) = (\zeta - \zeta_0)^{\varkappa-1} f(\zeta)$$

where $f(\zeta) \approx \mathrm{const} = A$. Hence

$$z - z_0 \approx \frac{A}{\varkappa}(\zeta - \zeta_0)^{\varkappa}$$

As the path of ζ passes around the singularity, $\arg(\zeta - \zeta_0)$ changes by π and consequently $\arg(z - z_0)$ by $\varkappa\pi$. The angle between the two arms of the transform is therefore $2\pi - \varkappa\pi$ (Figure 11.B.1). The angle between the tangents is therefore $\pi(2 - \varkappa)$, and is a cusp if $\varkappa = 2$.

The centre of the profile is defined by the centre of the generating circle. Two axes for the profile are specified:

1. The *first axis* is defined by the direction of the incoming flow when no lift is developed. It is the line joining the centre of the generating circle to the point representing the trailing edge. We define the angle β as that between the chord and the first axis. For a symmetric profile $\beta = 0$.

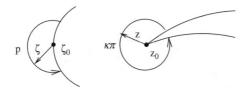

Figure 11.B.1: Transformation around a singularity.

2. The *second axis* is defined as the direction of the incoming flow when the moment of the lift over the wing (*pitching moment*) about the centre is zero. The sense of the pitching moment is taken to be positive when the sense of the moment is such as to lift the leading edge of the wing. We define the angle γ as that between the chord and the second axis. If the second axis lies above the first, i.e. the angle $\gamma > \beta$, the wing is stable, and if $\gamma < \beta$ it is unstable.[20] We have shown in Section11.6.3 that the parameter k_1 in the expansion (11.B.2) determines the angle of the second axis through $\gamma = \frac{1}{2} \arg(k_1)$

11.B.1 Zhukovskii Transformation

The basic properties of the Zhukovskii transformation are seen in problems #36 and #39. In the first a circle centred at the origin generates a symmetric profile, and not passing through either singularity gives a finite thickness. In the second example the circle generates a laminar section; the centre is located on the imaginary axis generating an asymmetric profile and passes through both singularities generating cusps at both ends and a profile of zero thickness. More general profiles pass through one singularity only to give a blunt leading edge and sharp trailing edge. The symmetric thickness function $g(x)$ of Case study 2.II is determined by the real part of the centre in ζ space and the mean line term $h(x)$ by the imaginary part.

11.B.1.1 Transformation of a circle to a streamlined symmetric body

If the centre of the circle is displaced along the real (ξ) axis and passes through one of the singular points, we obtain a symmetric streamlined body (Figure 11.B.2), typical of a fast-swimming fish, or the tailplane and fin of an aircraft. The transformed point corresponding to the singularity is a cusp, where the profile turns through 2π radians as for the laminar section. This is the trailing edge of a wing section. The circle passes closes to, but not through, the second singularity. Consequently the profile is blunt at the leading edge, as for the ellipse.

11.B.1.2 Transformation of a circle to a streamlined asymmetric body

If the centre of the circle is further displaced in the imaginary (ν) direction whilst passing through one discontinuity only, the symmetric profile is bent around the

---

[20]Figure 11.4 shows an unstable configuration where the second axis lies below the first.

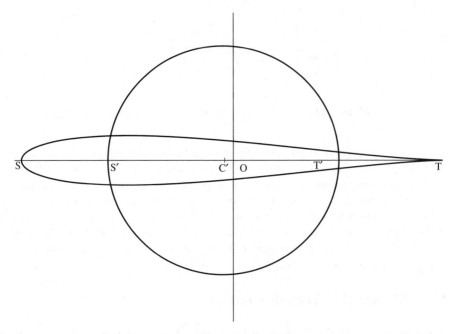

Figure 11.B.2: The geometrical arrangement of the ζ and z planes for the Zhukovskii transformation of a circle to a symmetric streamlined body, the centre C' of the circle of radius (1.1) being at $(-0.1,\ 0)$ passing through the singularity at $(1,\ 0)$. The point S on the leading edge is the transformation of S' on the circle and similarly on the trailing edge T and T'.

skeleton formed by the laminar transformation passing through both discontinuities (Figure 11.B.3). The centre of the circle for the laminar skeleton is taken on the line passing through the centre of the generating circle and the discontinuity. This yields the first axis of the profile, which is the direction of zero lift.

From these results we see that the position of the centre of the circle in the ζ plane determines the shape of the profile in two ways:

- The real part of the displacement determines the thickness. If the wing is not too thick, the ratio of the thickness to the *chord*, the width of the wing between the leading and the trailing edges, known as the *thickness ratio*, is given approximately by $1.3\,\Re(d)/\ell$.

- The imaginary part of the displacement determines the camber of the transformed profile. As we have seen, the ratio of the maximum displacement of the centre line from the chord, known as the *camber*, to the chord is given by $\Im(d)/2\ell$.

11.B.2 Karman–Treffetz Transformation

Zhukovskii profiles are unsatisfactory as practical wing profiles because they suffer from a number of deficiencies. Firstly the leading edge of the profile is too massive

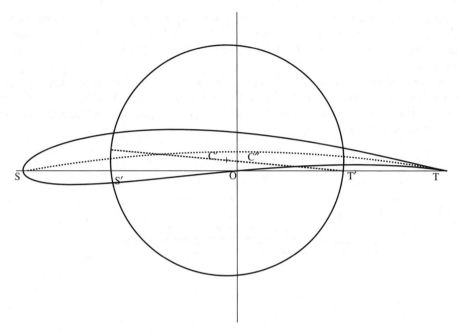

Figure 11.B.3: The geometrical arrangement of the ζ and z planes for the Zhukovskii transformation of a circle to an asymmetric streamlined body, the centre of the circle C' being at (0.1, 0.1) and passing through the singularity at (1.0, 0.0). Also shown (dotted) is the skeleton of the profile given by a laminar transformation by a circle centred at C'' on the axis of zero lift (dashed).

compared with the thin tail. Secondly the maximum camber lies too close to the centre of the profile. Furthermore, the cusp at the trailing edge is very difficult to construct. Realistic profiles have a finite angle between the two surfaces at the trailing edge.

Generalising the Zhukovskii form, we may write the transformation as

$$\frac{z - \varkappa\ell}{z + \varkappa\ell} = \left(\frac{\zeta - \ell}{\zeta + \ell}\right)^{\varkappa} \tag{11.B.3}$$

where $\varkappa = 2$ for the Zhukovskii transformation. The general behaviour, and in particular the aerodynamic properties, of this transformation are very similar to that of the Zhukovskii one. We may therefore follow a similar analysis to identify the various forms, and we shall simply identify the major differences.

The Karman–Treffetz transformation may be expanded as

$$z = \zeta + \frac{(\varkappa^2 - 1)\,\ell^2}{3}\frac{1}{\zeta} + \ldots$$

Hence k_1 is real and $\gamma = 0$. The second axis is therefore parallel to the chord and, like the Zhukovskii profile, unstable.

The discontinuities at $\pm\ell$ transform to $\pm\varkappa\ell$, showing the change in scale. However, the major difference in the final profile is at the discontinuity. From Section 11.B the change in the argument of the term $z - \varkappa\ell$ defines the angle at the trailing edge, which is therefore $\varkappa\pi$. Clearly the range of values of $\varkappa$ is $1.5 \leq \varkappa \leq 2$. The transformation scale of that for the Karman–Treffetz is less than for the Zhukovskii as the points $\pm\ell$ change to $\pm\varkappa\ell$.

The transformation involving the circle passing through both singularities, which led to a laminar section for the Zhukovskii transformation, now has finite angles at the ends, and the resultant has a finite thickness. Thus the skeleton of a Karman–Treffetz profile is thick (Figure 11.B.4).

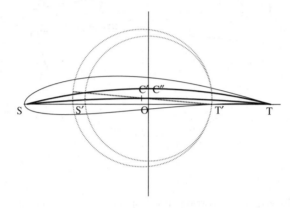

Figure 11.B.4: The geometrical arrangement of the ξ and z planes for the Karman–Treffetz transformation of a circle to a symmetric streamlined body, the centre of the circle C' at $(-0.1, \ 0.1)$ passing through the singularity at $(1.0, \ 0.0)$. Also shown (dotted) is the skeleton of the profile given by a laminar transformation by a circle centred at C'' on the axis of zero lift (dashed). The parameter $\varkappa = 1.95$ corresponds to a trailing edge angle of $9°$.

11.B.3 Von Mises Transformation

The Zhukovskii and Karman–Treffetz transformations are relatively inflexible and cannot reproduce many of the desirable properties of real aerofoils. In particular, as we have seen, the aerofoil sections are always unstable. To this end von Mises proposed a generalisation of the Zhukovskii transformation including an extended number $(n+1)$ of singularities $(-\ell, u_1, u_2, \ldots, u_n)$ whose location could be adjusted to give a range of properties within relatively non-restrictive limits, provided they all lay within or on the generating circle:

$$\frac{dz}{d\zeta} = \left(1 + \frac{\ell}{\zeta}\right)\left(1 - \frac{u_1}{\zeta}\right)\left(1 - \frac{u_2}{\zeta}\right)\cdots\left(1 - \frac{u_n}{\zeta}\right) \qquad (11.\text{B}.4)$$

Clearly the Zhukovskii transformation corresponds to the case $u_1 = -\ell$, $u_2 = \cdots = u_n = 0$. The only condition required is that all the singularities $u_1, u_2, \ldots, k_n$ lie inside the generating circle in the ζ plane.

Expanding the product we may write

$$\frac{dz}{d\zeta} = 1 + \frac{k_0}{\zeta} - \frac{k_1}{\zeta^2} - \frac{2k_2}{\zeta^3} - \cdots - \frac{nk_n}{\zeta^{(n+1)}}$$

and hence, integrating,

$$z = \zeta + k_0 \ln \zeta + \frac{k_1}{\zeta} + \frac{k_2}{\zeta^2} + \cdots + \frac{k_n}{z^n}$$

Since we require that $z \to \zeta$ as $|\zeta| \to \infty$, it follows that $k_0 = \ell - \sum u_m = 0$, and the singularities must satisfy the condition

$$\sum_{m=1}^{n} u_m = \ell \qquad (11.\text{B}.5)$$

Thus the 'centre of mass' of the singularities u_i must equal the value at the trailing edge ℓ, or alternatively the origin must be at the centre of mass of all the singularities.

The second axis is no longer along the chord, and the angle $\gamma = \frac{1}{2}\arg(k_1)$. Its direction can be flexibly set by an appropriate choice of the singularities.

There is a cusp at the trailing edge. If $n \leq 3$, the three terms in the expansion k_1 are easily obtained:

$$k_1 = \ell^2 - \frac{1}{2}\sum_{i=1}^{n}\sum_{\substack{j=1 \\ i \neq j}}^{n} u_i u_j \qquad k_2 = \frac{1}{2}\left[\prod_{i=1}^{n} u_i - \ell \sum_{i=1}^{n}\sum_{\substack{j=1 \\ i \neq j}}^{n} u_i u_j\right] \qquad k_3 = \frac{1}{3}\ell\prod_{i=1}^{n} u_i$$

In general only a limited number of singularities are used. We consider the two cases $n = 2$ (three singularities) and $n = 3$ (four singularities):

- In the case $n = 2$ the sum of two singularities, u_1 and u_2, must be real and equal to $u_1 + u_2 = \ell$. Thus writing

$$u_{1,2} = \frac{\ell}{2} \pm \mu\ell \exp(\iota\theta)$$

  hence

$$k_1 = \ell^2 - u_1 u_2 = \ell^2\left(\frac{3}{4} + \mu^2 \cos 2\theta + \iota\mu^2 \sin 2\theta\right) \qquad (11.\text{B}.6)$$

  The angle of the second axis is therefore given by

$$\gamma = \frac{1}{2}\arctan\left\{\frac{\mu^2 \sin 2\theta}{3/4 + \mu^2 \cos 2\theta}\right\}$$

- In the case $n = 3$, it is normal to choose one of the singularities to be at $u_1 = +\ell$. The remaining two are then arranged symmetrically about O so that $u_2 = -u_3$. Hence

$$u_{2,3} = \pm\mu\ell \exp(\iota\theta)$$

  and

$$k_1 = \ell^2\left(1 + \mu^2 \exp(2\iota\theta)\right) \qquad k_2 = 0 \qquad k_3 = -\frac{1}{3}\mu^2 \ell^4 \exp(2\iota\theta) \quad (11.\text{B}.7)$$

The angle of the second axis is therefore given by

$$\gamma = \frac{1}{2} \arctan \left\{ \frac{\mu^2 \sin 2\theta}{1 + \mu^2 \cos 2\theta} \right\}$$

In the present form the profile will have a cusp at the trailing edge. We may remove this by modifying equation (11.B.3) to give a finite angle $\varkappa\pi$ between the tangents as before

$$\frac{\mathrm{d}z}{\mathrm{d}\zeta} = \left(1 + \frac{\ell}{\zeta}\right)^{(\varkappa-1)} \left(1 - \frac{u_1}{\zeta}\right)\left(1 - \frac{u_2}{\zeta}\right) \cdots \left(1 - \frac{u_n}{\zeta}\right) \tag{11.B.8}$$

and hence the governing condition becomes

$$\sum_{m=1}^{n} u_m = (\varkappa - 1)\ell \tag{11.B.9}$$

Figure 11.B.5 shows a typical profile generated by an $n = 4$ transformation. The profile shown has an S-shaped (*relevé*) profile. The skeleton is formed by transforming the circle passing through the two singularities $(\pm\ell, 0)$ with centre on the first axis lying almost along the x axis.

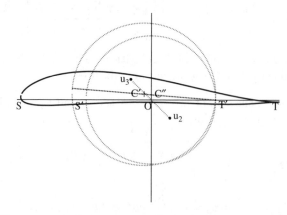

Figure 11.B.5: The geometrical arrangement of the ξ and z planes for the $n = 4$ von Mises transformation of a circle to a streamlined body, the centre of the circle C' of radius 1.1135 being at $(-0.11, 0.088\,324)$ passing through the singularity at $(1.0, 0.0)$. The singularities are $(1.0, 0.0)$, $\pm(-0.282\,843, 0.282\,843)$. Also shown (dotted) is the generating circle of the profile centre C' and the circle passing through the singularities at $\pm(1,0)$ centred at C'' on the axis of zero lift (dashed).

11.B.4 Theodorsen's Solution for an Arbitrary Profile

The conformal transforms identified above are very useful for identifying the general properties of wings, but are difficult to apply to actual designs, which may have been identified by experiment and experience. To overcome this problem, iterative methods were developed to treat the inverse problem: that is, given a wing profile, calculate the velocity and pressure around it. These were based on the use of Zhukovskii or Karman–Treffetz transformations, the latter having the advantage of allowing a finite trailing edge angle.

A map is made of the wing section from the wing plane (ζ) into an approximating plane (z'), using a standard transformation. The resulting form is nearly circular. Using a constant area the approximating circle is converted into a true circle (z). Three steps are involved:

1. Derivation of the relations between the flow in the plane of the wing (ζ) and the plane of the near (approximating) circle (z').
2. Derivation of the relations between the flow in the near circle plane of the wing (z') and the plane of the true circle (z).
3. Combination of the above relations to obtain the velocity distribution in the plane of the wing section (ζ).

At the conclusion of the first iteration, the velocity profile does not exactly match that required at the surface of the wing, but the differences are small. A perturbation solution is therefore used to correct this. In most situations only one iteration was found to be needed. Further details can be found in the texts by Abbott and Doenhoff (1959) and Ashley and Landhal (1986), and of an earlier method in the text by von Karman and Burgers (1935).

We shall not discuss these transformations further as they are well covered in many standard texts. Methods based on transformations have been largely superseded for aerofoil design by direct computer modelling, principally using the panel method.

Chapter 12

Aerofoils in High-Speed Compressible Fluid Flow

12.1 Introduction

Thus far we have treated supersonic flow by exact solutions of the equations of fluid dynamics. As we have seen, in expansion the flow is the consequence of a series of non-dissipative pressure waves each acting independently in an ideal fluid. Each wave propagates more slowly through the medium modified by its predecessors. In contrast, in compression the successive pressure waves travel more rapidly and consequently accumulate directly to generate a shock in which the entropy increases. The behaviour of the fluid in expansion or compression is therefore markedly different.

However, we have also found that if the disturbance of the flow is small, it is accomplished by a simple pressure wave throughout the fluid travelling at the local sound speed along a characteristic Section 8.2. Furthermore we have seen that weak shocks have a very small entropy increase of only third order and therefore propagate as sound waves.

These considerations suggest that we examine whether small disturbances in compressible gas flow can be treated as a set of simple pressure waves, which account for flows in both compression and expansion within a single consistent framework. With such a goal in mind, we examine when the steady flow of the fluid maintains irrotationality in the presence of shocks, although, of course, the entropy differs upstream and downstream of the shock.

Introductory Fluid Mechanics for Physicists and Mathematicians, First Edition. Geoffrey J. Pert.
© 2013 John Wiley & Sons, Ltd. Published 2013 by John Wiley & Sons, Ltd.

Since Bernoulli's equation holds across the shock, then

$$(\mathbf{v} \cdot \nabla)\,\mathbf{v} = \frac{1}{2}\nabla v^2 - \mathbf{v} \wedge (\nabla \wedge \mathbf{v}) = -\frac{\nabla p}{\rho}$$

Hence making use of the second $T\,\mathrm{d}S$ equation

$$\nabla\left(h + \frac{1}{2}v^2\right) - \mathbf{v} \wedge (\nabla \wedge \mathbf{v}) = T\,\nabla s \qquad (12.1)$$

However, as we have seen in Section 2.3, Crocco's theorem shows that in steady flow $h + \frac{1}{2}v^2$ is constant everywhere in an irrotational flow provided the flow is homo-energetic and homo-entropic. Since Bernoulli's equation holds everywhere, including across the shock, the entropy change across the shock is related to the vorticity change

$$\mathbf{v} \wedge (\nabla \wedge \mathbf{v}) = \mathbf{v} \wedge \boldsymbol{\zeta} = -T\,\nabla s \qquad (12.2)$$

where $\boldsymbol{\zeta} = \nabla \wedge \mathbf{v}$ is the vorticity.

In general the entropy change across the shock varies with the intensity of the shock. However, we may identify two important cases in which the vorticity remains unchanged across the shock:

1. The shock is weak. From equations (10.16) and (10.14), $\Delta s \sim \Delta p^3$ and $\Delta v \sim \Delta p$. Hence the vorticity jump across the shock scales as $\Delta\zeta \sim T\Delta s/v \sim \Delta p^3 \to 0$. If $\Delta p \to 0$ we may neglect the vorticity behind the shock and treat the downstream flow as irrotational. This is the case treated in this chapter.

2. The shock has constant intensity along its surface $\nabla s = 0$. Hence either $\boldsymbol{\zeta} = 0$ or $\mathbf{v} \parallel \boldsymbol{\zeta}$. The latter cannot be the case as can be shown as follows. The vorticity component normal to the shock is formed by the derivatives in the shock surface of the tangential velocity component. However, since the flow is irrotational upstream of the shock and the tangential velocity is continuous across the shock, the downstream tangential derivatives must also be zero, and therefore the normal component of the vorticity is zero. But the normal component of the velocity cannot be zero, therefore $\mathbf{v} \nparallel \boldsymbol{\zeta}$. Consequently the vorticity is zero and the flow remains irrotational.

In these two cases, if the upstream flow was irrotational, the downstream flow behind the shock is also. Since the overall flow is irrotational we may simplify the equations of fluid dynamics by introducing the velocity potential ϕ, where $\mathbf{v} = \nabla\phi$. In steady flow, Euler's equation may be expressed as

$$(\mathbf{v} \cdot \nabla)\,\mathbf{v} = -\frac{1}{\rho}\nabla p = -\frac{c^2}{\rho}\nabla\rho$$

and the equation of continuity as

$$\nabla \cdot (\rho \mathbf{v}) = \rho \nabla \cdot \mathbf{v} + \mathbf{v} \nabla \rho = 0$$

hence

$$c^2 \nabla \mathbf{v} - \mathbf{v} \cdot (\mathbf{v} \cdot \nabla) \mathbf{v} = \left(c^2 - \phi_x{}^2\right) \phi_{xx} + \left(c^2 - \phi_y{}^2\right) \phi_{yy} + \left(c^2 - \phi_z{}^2\right) \phi_{zz}$$
$$- 2 \left(\phi_x \phi_y \phi_{xy} + \phi_y \phi_y \phi_{yz} + \phi_z \phi_x \phi_{zx}\right) = 0 \qquad (12.3)$$

where the suffices denote partial derivatives, e.g. $\phi_x \equiv \partial\phi/\partial x$.

If the effect of an obstacle introduced into the flow is a minor disturbance to the uniform incoming flow, we may seek to reduce the complexity of this equation (12.3) by neglecting terms of higher order in the perturbation. Thus if the incoming flow velocity is $\mathbf{U}$ in the x direction and the perturbation $\mathbf{v}$, the total velocity is $\mathbf{U} + \mathbf{v}$. The corresponding potential is $U x + \phi$, where ϕ is the potential of the perturbation. Neglecting the terms containing cubic powers of ϕ, we obtain from equation (12.3)

$$\left(1 - M_1{}^2\right) \frac{\partial^2 \phi}{\partial x^2} + \frac{\partial^2 \phi}{\partial y^2} + \frac{\partial^2 \phi}{\partial z^2} \approx 0 \qquad (12.4)$$

where the Mach number $M_1 = U/c_1$, c_1 being the sound speed of the incoming flow.

From Bernoulli's equation, the enthalpy

$$h = h_1 - \frac{1}{2} \left[(\mathbf{U} + \mathbf{v})^2 - U^2\right] \approx h_1 - \frac{1}{2} \left(v_y{}^2 + v_z{}^2\right) - U v_x \qquad (12.5)$$

yields the pressure

$$p = p_1 - \rho_1 U v_x - \frac{1}{2} \rho_1 \left(v_y{}^2 + v_z{}^2\right) \qquad (12.6)$$

where ρ_1 is the density in the incoming flow, and from the thermodynamics

$$dh = \cancel{T ds} + \frac{1}{\rho} dp$$

since $ds = 0$.

It is clear that equation (12.4) takes different forms if the flow is:

Subsonic $M_1 < 1$. The equation has an elliptic form. In this case the flow pattern is a solution of a modified form of Laplace's equation in which one dimension is increased by the scale factor $\sqrt{1 - M_1{}^2}$.

Supersonic $M_1 > 1$. The equation has a hyperbolic form. In this case the flow pattern is described by a series of waves leaving the source of the perturbation at the Mach angle $\arcsin(c_1/U)$ to the incoming flow. These are the characteristics or pressure waves discussed earlier.

Clearly the perturbation limit is not valid if the Mach number is too large $M_1 \gg 1$ or if the flow transonic $M_1 \approx 1$.

12.2 Linearised Theory for Two-Dimensional Flows: Subsonic Compressible Flow around a Long Thin Aerofoil – Prandtl–Glauert Correction

The flow around an aerofoil in compressible flow in the linearised approximation may be directly compared with that around the same aerofoil in incompressible flow by a simple scale transformation to reduce the linearised perturbation equation (12.4) for the potential to Laplace's equation (2.20). In order to do this the potential function for the potential must be scaled to satisfy the boundary conditions at the wing surface.

Since the wing is a streamlined body in subsonic flow, the perturbation to the flow is small and we may use equation (12.4) for the potential

$$\left(1 - M_1{}^2\right) \frac{\partial^2 \phi}{\partial x^2} + \frac{\partial^2 \phi}{\partial y^2} + \frac{\partial^2 \phi}{\partial z^2} = 0 \tag{12.4}$$

At the surface of the wing the flow velocity component normal to the surface is zero

$$\left(U + \frac{\partial \phi}{\partial x}\right) n_x + \frac{\partial \phi}{\partial y} n_y + \frac{\partial \phi}{\partial z} n_z = 0 \tag{12.7}$$

where the unit vector $\hat{\mathbf{n}}$ is the outward normal to the wing. Let the wing section C be given by $y = f_\pm(x)$, where $+$ refers to the upper and $-$ to the lower surface. The wing is thin, the angle of attack small $|n_y| \sim 1$, n_x and n_z are nearly zero and, as the span is large, the z component of velocity may be neglected. The boundary condition reduces to

$$U n_x \pm \left. \frac{\partial \phi}{\partial y} \right|_C = 0 \tag{12.8}$$

where the $+$ sign is taken for the upper surface and the $-$ for the lower one.

Now consider the incompressible flow about the same wing section in the scaled co-ordinate system

$$
\begin{array}{llll}
x' = x & y' = \sqrt{1 - M_1{}^2}\, y & z' = \sqrt{1 - M_1{}^2}\, z & \\
U' = U & \phi' = \sqrt{1 - M_1{}^2}\, \phi & \rho'_1 = \rho_1 &
\end{array}
\tag{12.9}
$$

where the x co-ordinates are unchanged and the y and z ones reduced by a factor $\beta = \sqrt{1 - M_1{}^2}$. The incoming flow at speed U and density ρ_1 is unchanged. The wing section C' remains unchanged, $y' = f_\pm(x')$. Equation (12.4) transforms to equation (2.20) in the scaled frame. The boundary conditions in the scaled frame become

$$U'n'_x \pm \left.\frac{\partial\phi'}{\partial y'}\right|_{C'} = U'\left.\frac{\mathrm{d}y'}{\mathrm{d}x'}\right|_{C'} \pm \left.\frac{\partial\phi'}{\partial y'}\right|_{C'} = U\left.\frac{\mathrm{d}y}{\mathrm{d}x}\right|_C \pm \left.\frac{\partial\phi}{\partial y}\right|_C = Un_x \pm \left.\frac{\partial\phi}{\partial y}\right|_C = 0$$

$$(12.10)$$

The unit vector component $n_x \approx \pm\mathrm{d}y/\mathrm{d}x$ along the contour C, namely $y(x)$, the sign depending on whether the upper $(+)$ or lower $(-)$ surface is considered. The choice of scaling for the potential ϕ is therefore made to match that required to satisfy the boundary conditions in the incompressible system.

The perturbation velocity components are

$$v_x = \frac{\partial\phi}{\partial x} = \frac{1}{\sqrt{1 - M_1{}^2}}\frac{\partial\phi'}{\partial x'} = \frac{1}{\sqrt{1 - M_1{}^2}}v'_x$$

$$v_y = \frac{\partial\phi}{\partial y} = \frac{\partial\phi'}{\partial y'} = v'_y \qquad v_z = \frac{\partial\phi}{\partial z} = \frac{\partial\phi'}{\partial z'} = v'_z$$

$$(12.11)$$

The pressure increment around the profile (12.6) neglecting second-order terms is given by

$$p = -\rho_1 U v_x = -\frac{1}{\sqrt{1 - M_1{}^2}}\rho_1 U v'_x = \frac{1}{\sqrt{1 - M_1{}^2}}p' \qquad (12.12)$$

Defining the pressure coefficient as

$$C_p = \frac{p}{\frac{1}{2}\rho_1 U^2} = \frac{1}{\sqrt{1 - M_1{}^2}}\frac{p'}{\frac{1}{2}\rho'_1 U'^2} = \frac{1}{\sqrt{1 - M_1{}^2}}C'_p \qquad (12.13)$$

we obtain the *Prandtl–Glauert* correction for compressible flow: that is, the pressure coefficient is increased by the factor $1/\sqrt{1 - M_1{}^2}$ over the incompressible value by compression. The section lift coefficient c_L is therefore also increased by the same factor $c_L = c'_L/\sqrt{1 - M_1{}^2}$.

It is easy to show that the Kutta–Zhukovskii lift formula (2.105) is unchanged in compressible flow within the perturbation approximation (see Section 11.8.4). Thus the lift is

$$F_y = -\rho_1 U \int\limits_{-l_z/2}^{l_z/2} \Gamma(z)\,\mathrm{d}z$$

where l_z is the wing span. Transforming between the compressible and incompressible flows we have that the circulation is

$$
\Gamma = \oint \mathbf{v} \cdot d\ell = \oint \frac{\partial \phi}{\partial x} dx + \frac{\partial \phi}{\partial y} dy
$$

$$
= \frac{1}{\sqrt{1 - M_1{}^2}} \oint \frac{\partial \phi'}{\partial x'} dx' + \frac{\partial \phi'}{\partial y'} dy' = \frac{1}{\sqrt{1 - M_1{}^2}} \oint \mathbf{v}' \cdot d\ell' = \frac{1}{\sqrt{1 - M_1{}^2}} \Gamma'
$$

$$(12.14)$$

and the lift force transforms as

$$
F_y = -\rho_1 U \int_{-l_z/2}^{l_z/2} \Gamma \, dz = -\frac{1}{(1 - M_1{}^2)} \rho_1' U' \int_{-l_z'/2}^{l_z'/2} \Gamma' \, dz' = \frac{1}{(1 - M_1{}^2)} F_y'
$$

$$(12.15)$$

Although the dimensions of the wing section $l_x l_x'$ and $l_y = l_y'$ are unchanged by the transformation, the span changes in conformity with the z scaling $l_z = l_z'/\sqrt{1 - M_1{}^2}$. The total lift coefficient for the complete wing C_L in compressible flow is directly related to that of the transformed wing in incompressible flow by

$$
C_L = \frac{F_y}{\frac{1}{2}\rho_1 U^2 l_x l_z} = \frac{F_y'}{\frac{1}{2}\rho_1' U'^2 l_x' l_z' \sqrt{(1 - M_1{}^2)}} = \frac{C_L'}{\sqrt{(1 - M_1{}^2)}}
$$

$$(12.16)$$

consistent with equation (12.13).

It is simple to see that the expression for the induced drag derived in Section 11.8.4, equation (11.62) (and therefore (11.53)), carries over into linearised compressible flow. Since the circulation is unchanged for the transformation from wing C to C', the relationship between the induced drag force on the wing in compressible flow and the transformed one in incompressible flow is

$$
F_x = \frac{\rho_1}{4\pi} \int_{-l_z/2}^{l_z/2} \oint_{-l_z/2}^{l_z/2} \frac{1}{(z - z')} \Gamma(z) \frac{d\Gamma}{dz}\bigg|_{z'} dz' \, dz
$$

$$
= \frac{\rho_1'}{4\pi (1 - M_1{}^2)} \int_{-l_z'/2}^{l_z'/2} \oint_{-l_z'/2}^{l_z'/2} \frac{1}{(z - z')} \Gamma'(z) \frac{d\Gamma'}{dz}\bigg|_{z'} dz' \, dz = \frac{1}{(1 - M_1{}^2)} F_x'
$$

$$(12.17)$$

Remembering that the drag force is independent of the span, we may compare the drag coefficient from the compressible flow with that from an incompressible one over an identical wing, i.e. one of span l_z, namely C_{D0}. Hence the drag coefficient scales as

$$
C_D = \frac{F_x}{\frac{1}{2}\rho_1 U^2 l_x l_z} = \frac{F_x'}{\frac{1}{2}\rho_1' (1 - M_1{}^2) U'^2 l_x' l_z} = \frac{C_{D0}}{(1 - M_1{}^2)}
$$

$$(12.18)$$

Similarly the lift coefficient from the identical wing $C_{L0} = C'_L$ so that the relationship $C_D/C_L{}^2 = C_{D0}/C_{L0}{}^2 \approx \text{const}$ is retained in compressible flow.

12.2.1 Improved Compressibility Corrections

The Prandtl–Glauert correction (12.13), which was derived above, is a linear correction to a nonlinear problem. This is a reasonably accurate approximation provided the Mach number of the incoming flow is low, $M_1 < 0.7$. Subsequent work has included some of the nonlinearities to generate corrections which are applicable at greater (but still subsonic) Mach numbers.

The earlier, but widely used, correction is due to von Karman and Tsien

$$C_p = \frac{C'_p}{\sqrt{1 - M_1{}^2} + \left[M_1{}^2 / \left(1 + \sqrt{1 - M_1{}^2} \right) \right] C'_p / 2} \tag{12.19}$$

A subsequent form is due to Laitone

$$C_p = \frac{C'_p}{\sqrt{1 - M_1{}^2} + \left(M_1{}^2 \left\{ 1 + [(\gamma - 1)/2] M_1{}^2 \right\} / 2\sqrt{1 - M_1{}^2} \right) C'_p} \tag{12.20}$$

12.3 Linearised Theory for Two-Dimensional Flows: Supersonic Flow about an Aerofoil – Ackeret's Formula

The flow past a thin wing in a supersonic stream in two dimensions is only slightly disturbed from the ambient uniform state and the perturbation approximation, equation (12.4), may be used to calculate the flow. However, when the incoming flow is supersonic, $M_1 > 1$, the form of the equation is changed

$$\frac{\partial^2 \phi}{\partial y^2} - \left(M_1{}^2 - 1 \right) \frac{\partial^2 \phi}{\partial x^2} = 0 \tag{12.21}$$

from an elliptic to a hyperbolic partial differential equation. We therefore expect the solution to be expressed in terms of a set of waves originating at the surface of the wing. Since the body only presents a perturbation to the flow, the waves, either rarefactions or shocks, are weak (Figure 12.1) and far from the body degenerate into sound waves. The outgoing sound waves carry significant energy away from the body and therefore generate a form of drag, known as *wave drag*. The mechanism is therefore very similar to that of wave drag of ships due to surface waves, which were discussed earlier on page 99.

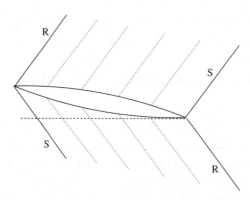

Figure 12.1: Sketch of the rarefactions and shocks around an aerofoil in a supersonic flow. All the waves are weak and at the Mach angle to the incoming flow.

The wave nature of this system is easily seen if we write 'time' as $t = x/U$ and the 'wave velocity' as $c = U/\sqrt{M_1{}^2 - 1}$ to give

$$\frac{\partial^2 \phi}{\partial y^2} + \frac{\partial^2 \phi}{\partial z^2} - \frac{1}{c^2}\frac{\partial^2 \phi}{\partial t^2} = 0$$

the two-dimensional equation for the propagation of waves away from the wing. It is easily seen that the 'time' $\delta t = \delta x/U$ is the interval over which the wing moves a distance δx and that the wavefronts are propagating out at the Mach angle $\alpha = \arcsin(1/M_1)$. The picture is therefore identical to that discussed in Section 8.2.

Since the wing is assumed to be two dimensional the governing equation for the flow reduces to

$$\frac{\partial^2 \phi}{\partial y^2} - \beta^2 \frac{\partial^2 \phi}{\partial x^2} = 0 \tag{12.22}$$

where $\beta = \sqrt{M_1{}^2 - 1}$, whose general solution is expressed in terms of incoming and outgoing waves

$$\phi = f_1(x - \beta y) + f_2(x + \beta y) \tag{12.23}$$

subject to a boundary condition on the wing surface:

$$\left.\frac{\partial \phi}{\partial y}\right|_{\xi(x)} = \mp U n_x \approx U\dot{\xi}(x)$$

where $\mathbf{n}$ is the outward unit normal and $\xi(x)$ the profile of the wing section. The $-$ sign refers to the upper surface and $+$ to the lower.

Since only outgoing waves can be established the correct solution must have $f_1(x - \beta y)$ for the upper surface $y > 0$ and $f_2(x + \beta y)$ for the lower $y < 0$.

Furthermore, since the wing is thin and the velocity must be parallel to the wing surface, the boundary condition takes the form

$$
\begin{aligned}
\left.\frac{\partial\phi}{\partial y}\right|_{\xi_1(x)} &= -\beta\dot{f}_1(x) \approx U\dot{\xi}_1(x) && \text{upper} \\
\left.\frac{\partial\phi}{\partial y}\right|_{\xi_2(x)} &= \beta\dot{f}_2(x) \approx U\dot{\xi}_2(x) && \text{lower}
\end{aligned}
\tag{12.24}
$$

Matching the boundary conditions to the solution we obtain

$$
\phi = \begin{cases}
-\dfrac{U}{\beta}\,\xi_1(x-\beta y) & \text{upper} \\[2mm]
\dfrac{U}{\beta}\,\xi_2(x+\beta y) & \text{lower}
\end{cases}
\tag{12.25}
$$

At the leading and trailing edges the functions $\xi(x)$ are discontinuous. However, as the treatment allows only weak discontinuities, weak rarefactions and weak shocks leave the edges. The flow is constant along the lines $x \mp \beta y$, which are the Mach lines or characteristics of the flow. Between the discontinuities the flow is a simple wave.

The pressure distribution is given by Bernoulli's equation

$$
p - p_1 = -\rho_1 U \frac{\partial\phi}{\partial x} = \pm \frac{\rho_1 U^2}{\beta}\,\dot{\xi}(x \mp \beta y)
\tag{12.26}
$$

for the upper and lower surfaces respectively.

The section lift coefficient follows directly from the pressure differential between the upper and lower surfaces

$$
c_L = -\frac{2}{\beta\,\ell_x}\int_0^{\ell_x}\left[\dot{\xi}_1(x) + \dot{\xi}_2(x)\right]\,\mathrm{d}x = \frac{4\,\ell_y}{\beta\,\ell_x} = \frac{4\,\alpha}{\sqrt{M_1{}^2 - 1}}
\tag{12.27}
$$

where $\alpha = \ell_y/\ell_x$ is the angle of attack (Figure 12.2).

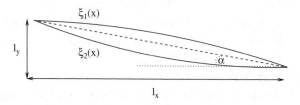

Figure 12.2: Sketch of an aerofoil in a supersonic flow.

The drag coefficient is similarly calculated from the pressures over the upper and lower surfaces

$$c_D = \frac{1}{\frac{1}{2}\rho_1 U^2 \ell_x} \int_0^{\ell_x} \left[p_1 \, n_{1x} + p_2 \, n_{2x} \right] \mathrm{d}x = \frac{2}{\beta \ell_x} \int_0^{\ell_x} \left[\dot{\xi}_1^2(x) + \dot{\xi}_2^2(x) \right] \mathrm{d}x$$

Defining the angles subtended by the wing surfaces at the trailing edge $\theta_{1,2}$ $\approx \dot{\xi}_{1,2} + \alpha,$[1] we obtain

$$c_D = \frac{\left[4\alpha^2 + 2 \left(\overline{\theta_1^2} + \overline{\theta_2^2} \right) \right]}{\sqrt{M_1^2 - 1}} \tag{12.28}$$

The minimum drag therefore is found from a flat plate, $\theta_1 = \theta_2 = 0$. The drag depends on the mean square deviation of the wing surface from a flat plate, and is due to the wave drag set up by waves leaving rugosities on the surface. Efforts must be made to ensure that any roughness of the surfaces of an aeroplane designed for supersonic flight is reduced to a minimum.

For a flat plate when the Mach number $M_1 \gg 1$ the lift to drag ratio is identical to equation (12.53) found later for weakly supersonic flow.

12.4 Drag in High-Speed Compressible Flow

12.4.1 Swept Wings

At speeds at or above the sound speed, shock waves form at the wing surface. This leads to a marked increase in the forces experienced by the wing, due to the formation of a blunt body shock at the leading edge of the wing. In addition to the danger of structural failure in wings designed for subsonic flight, it means a substantial increase in the drag, and therefore the power required to drive the aeroplane forward. At speeds just below but close to Mach 1, shock waves will form on the thickest section of the upper surface due to the accelerated velocity. All these effects are damaging unless special wing profiles designed specifically for high-speed flight are used.

The simplest solution to avoiding this problem is to reduce the normal component of the velocity on-to the wing leading edge, for, as we have seen, it is this velocity which determines whether a shock is established. Thus most modern aircraft designed to fly at speeds of about $M_1 \approx 0.8$ are built with wings swept to an angle of about 30°.

A serious problem with swept wings is the *one-wing stall* caused by the cross-flow, due to the spanwise pressure differentials along the wing, especially on

---

[1]Note that $\dot{\xi}(x) < 0$.

the upper surface. This leads to a thickening of the boundary layer towards the tip, and possibly stalling–particularly at high angles of attack. It is prevented by decreasing the thickness of the wings towards the tip, and by fitting wing fences to prevent the cross-flow.

12.4.2   Drag in Supersonic Flow

In supersonic flight there are three forms of drag:

1. Skin friction. This is drag arising from the boundary layer in streamlined flow. The magnitude is essentially given by the methods discussed in Chapter 6 taking into account the compressibility of the fluid. It may contribute a significant fraction ($\sim 30\%$) of the total drag.

2. Vortex drag. This is drag which results from the shedding of the trailing vortices behind the wing. These form within the trailing wake behind the aerofoil and are effectively stationary with respect to the fluid. Their contribution can therefore be fairly accurately estimated from the incompressible flow theory of induced drag, Section 11.8.4.

3. Wave drag. This is due to the energy carried by pressure (sound and shock) waves generated at the surface of the wing by the disturbance. As we have seen, it is specific to supersonic flow, and is a consequence of the hyperbolic nature of fluid mechanics when the flow velocity exceeds the sound speed. Since it can make an appreciable contribution to the overall drag, efforts are made to reduce or, if possible, eliminate its contribution.

The most familiar method to offset the introduction of supersonic flow conditions and thus reduce wave drag is to sweep the aircraft wing. As we have seen, the onset of a shock is determined by velocity normal to the wing surface, not by the total velocity (Section 10.9). Most aircraft operating at Mach numbers $M_1 \gtrsim 0.7$ are built with swept wings despite the inherent cross-flow problems associated with them (Section 12.4.1).

12.4.3   Transonic Flow

Transonic flow is the case where, although the incident flow is subsonic, regions of supersonic flow occur. Typical of this case are flows at incident Mach number $M_1 \gtrsim 0.7$ where the Mach number may increase to values about 1 due to the acceleration introduced by the divergence induced by the body.

Even if the wing is sufficiently swept to guarantee that the incident flow is subsonic, the fluid is accelerated by the increased thickness to provide the pressure reduction on the vacuum surface of the wing. At the *critical Mach number* the consequent acceleration leads to the flow becoming sonic. Clearly it is important to be able to estimate the critical Mach number of the incoming flow M_{crit} at which the sonic condition is reached on the wing surface. Assuming that the position of maximum flow velocity or equivalently the minimum pressure coefficient is known in incompressible flow, we may use one of the subsonic corrections, equation (12.13), (12.19) or (12.20), to estimate the Mach number of the incoming flow at which the flow over the wing becomes sonic. To do this we need to relate the pressure coefficient at the pressure minimum to the flow speed. From Bernoulli's equation we may obtain the pressure coefficient at an arbitrary point on the wing surface from the Mach number of the flow at that point M and the incoming flow Mach number M_1

$$ C_p = \frac{2}{\gamma M_1{}^2} \left\{ \left[\frac{1 + \frac{(\gamma-1)}{2} M_1{}^2}{1 + \frac{(\gamma-1)}{2} M^2} \right]^{\gamma/(\gamma-1)} - 1 \right\} \tag{12.29} $$

The pressure coefficient at the point at which the flow just becomes sonic is therefore

$$ C_{p\text{crit}} = \frac{2}{\gamma M_{\text{crit}}{}^2} \left\{ \left[\frac{1 + \frac{(\gamma-1)}{2} M_{\text{crit}}{}^2}{1 + \frac{(\gamma-1)}{2}} \right]^{\gamma/(\gamma-1)} - 1 \right\} \tag{12.30} $$

Assuming the pressure coefficient over the wing surface in incompressible flow is known, either from wind tunnel measurements or calculations, one of the corrections may be used to estimate the critical Mach number of the incoming flow M_c. From the ratio of the pressure coefficient of the compressible and incompressible flows, the value is obtained most simply by graphical methods (Figure 12.3).

Once the flow becomes supersonic, wave drag sets in behind the sonic surface, enhancing the normal subsonic induced and parasitic drag. This increase becomes marked at a Mach number M_d slightly above the critical value M_c at which the flow first becomes sonic, known as the *drag divergent Mach number*, sketched in Figure 12.4. In particular, as the aerofoil thins towards the trailing edge, the flow again becomes subsonic, requiring a passage through a shock (Figure 12.5(a)). As a result there is a substantial increase in the drag and a change in the pressure distribution over the wing. The shock frequently gives rise to separation in the boundary layer, known as the *shock stall*. Although the flow is nearly sonic and the linearised approximation fails, the lift continues to increase until the effects of separation associated with the shock stall cause a decrease in the lift. Once the flow is fully supersonic the drag decreases again.

The substantial increase in the drag found in the transonic region is known as the *sound barrier*. It therefore requires significant engine power to overcome this drag and achieve supersonic flight, which led to major difficulties in the

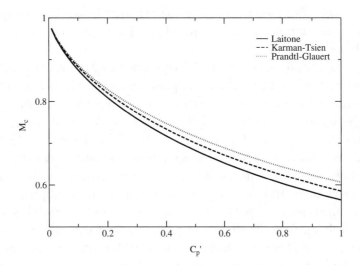

Figure 12.3: Critical Mach number for generating supersonic flow as a function of the incompressible pressure coefficient C'_p.

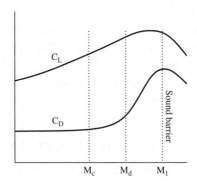

Figure 12.4: Sketch of the increase in lift and drag as the Mach number approaches the value associated with the sound barrier.

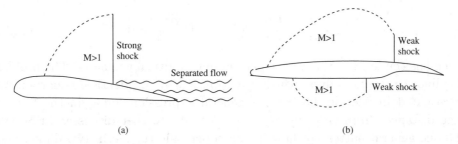

Figure 12.5: Sketches of the sonic transition line and shock wave for the flow about a conventional and a supercritical wing section. Note the separated flow generated by the shock on the conventional wing. (a) Conventional wing at Mach no. $M_1 \sim 0.7$, and (b) Supercritical wing at Mach no. $M_1 \sim 0.8$.

early days to achieve Mach numbers greater than 1. In addition to the increase in drag, there are significant control problems due to the separation associated with the shock.

Although the conditions of applicability are not well obeyed, the lift increases as predicted by the Prandtl–Glauert expression as the Mach number increases up to a limit, after which it decreases. Comparing the variation of the lift coefficient with the drag coefficient (Figure 12.4) we find that the lift maximum often occurs at a larger Mach number than the drag divergent Mach number. The lift to drag ratio is therefore a maximum just above the drag divergent Mach number, which in turn is larger than the critical Mach number. The detailed behaviour in this regime has important financial implications for the design of commercial transport aircraft where speed, determining utilisation, and drag, involving fuel efficiency, must be balanced against each other.

From Figure 12.3 it is clear that the critical Mach number may be increased by making the incompressible pressure coefficient small, i.e. using a thin wing. However, there are clear limits to the advantage to be gained by reducing the drag in this manner as the lift is also reduced, and a specified lift is required. An alternative method of reducing drag is to allow the flow over the wing to become weakly supersonic and the shock consequently weak. This is achieved by designing a wing profile which has a nearly flat upper surface with negative camber over the forward part of the wing (Figure 12.5(b)) (Whitcomb and Clark, 1965). As a result the rear section of the wing must have a strong positive camber, which gives rise to a 'cusplike' shape of the lower surface.

12.5   Linearised Theory of Three-Dimensional Supersonic Flow–von Karman Ogives and Sears–Haack Bodies

The flow around a three-dimensional body is described by

$$\frac{\partial^2 \phi}{\partial y^2} + \frac{\partial^2 \phi}{\partial z^2} - \left(M_1{}^2 - 1\right)\frac{\partial^2 \phi}{\partial x^2} = 0 \tag{12.31}$$

which has the form of the wave equation in cylindrical geometry. The problem of cylindrical sound waves, which are governed by an identical equation, was investigated in Section 8.5, with a general expression found in the limit of large distances from the axis of the source. We further discussed in Section 12.3 the general pattern of flow in the supersonic regime in two dimensions, when the disturbance is sufficiently small that perturbation theory may be used. We saw that it was represented by a series of pressure waves leaving the surface of the body at the Mach angle and carrying with them significant energy, which gives rise to wave drag. The same behaviour is found in three dimensions, where we may consider the disturbance in the asymptotic limit to be represented by a set of cylindrical waves formed by the body, as described by equation (12.31). This is identical to the cylindrical sound wave equation with the distributed source determined by the gradient of the area of cross-section $\dot{S}(x)$, time represented by $t \to x/U$ and the sound speed by $c \to U/\sqrt{M_1^2 - 1}$. Using the results from Section 8.5 we obtain the velocity potential at radius R as given by equation (8.34) (von Karman and Moore, 1932)

$$\phi = \frac{U}{2\pi} \int\limits_0^{x-\beta R} \frac{\dot{S}(\xi)\,\mathrm{d}\xi}{\sqrt{(x-\xi)^2 - \beta^2 R^2}} \tag{12.32}$$

where one end is at $x = 0$ and the other at $x = \ell$. For convenience we have set $\beta = \sqrt{M_1^2 - 1}$. The waves are propagated from the surface of the body within the cones $x - \beta R = 0$ and $x - \beta R = \ell$, whose surfaces are weak discontinuities.

We consider a thin cylindrical body, whose area $S(x)$ is zero at one end, and aligned with the flow. To calculate the drag we need to determine the momentum carried away by the waves. Thus we calculate the integral of the x component of the momentum flux tensor over a surface enclosing the body, e.g. a very large cylinder of radius $R \to \infty$. The appropriate component of the momentum flux tensor is

$$\Pi_{xr} = \rho\,v_r\,(v_x + U) \approx \rho_1 \frac{\partial \phi}{\partial r}\left(U + \frac{\partial \phi}{\partial x}\right) \tag{12.33}$$

Since there is no mass source, the mass flux through the surface at R, and consequently the term containing U, must both be zero. The force is therefore

$$F_x = -2\pi R \int_{-\infty}^{\infty} \Pi_{xr}\,\mathrm{d}x = -2\pi R\,\rho_1 \int_{-\infty}^{\infty} \frac{\partial \phi}{\partial r}\frac{\partial \phi}{\partial x}\,\mathrm{d}x \tag{12.34}$$

As we are only interested in the asymptotic region as $R \to \infty$, we need consider only the lowest order terms (in $1/R$). The values of $(x - \xi) \sim \beta R$ give the largest contribution to the integral, and we may approximate

$$(x - \xi)^2 - \beta^2 R^2 \approx 2\,\beta\,R\,(x - \xi - \beta\,R)$$

and retain only terms of lowest order to obtain

$$\phi \approx -\frac{c}{2\pi\sqrt{2R}} \int\limits_0^{x-\beta R} \frac{\dot{S}(\xi)\,\mathrm{d}\xi}{\sqrt{x - \xi - \beta R}}$$

$$\frac{\partial \phi}{\partial r} = -\frac{1}{\beta}\frac{\partial \phi}{\partial x} \approx \frac{U}{2\pi}\sqrt{\frac{\beta}{2R}} \int\limits_0^{x-\beta R} \frac{\ddot{S}(\xi)\,\mathrm{d}\xi}{\sqrt{x - \xi - \beta R}} \tag{12.35}$$

Substituting for $\partial \phi / \partial r$ yields the drag as the triple integral

$$F_x = \frac{\rho_1 U^2}{4 \pi} \int_{-\infty}^{\infty} \int_0^X \int_0^X \frac{\ddot{S}(\xi_1) \ddot{S}(\xi_2)}{\sqrt{(X - \xi_1)(X - \xi_2)}} \, d\xi_2 \, d\xi_1 \, dX \qquad (12.36)$$

where $X = x - \beta R$.

To proceed we interchange the order of integration taking care with the limits. The upper limit for the second integral (ξ_2) is along the length being ℓ and that for the first ($\xi_1 < \xi_2$) being ξ_2, which is the lower limit for X. Thus we obtain

$$F_x = \frac{\rho_1 U^2}{4 \pi} \int_0^\ell \int_0^{\xi_2} \int_{\xi_2}^L \frac{\ddot{S}(\xi_1) \ddot{S}(\xi_2)}{\sqrt{(X - \xi_1)(X - \xi_2)}} \, dX \, d\xi_1 \, d\xi_2$$

where L is a large distance tending to infinity. The integral over X is a standard form, evaluated using either the third Euler substitution $t^2 = (X - \xi_1)/(X - \xi_2)$ or by completing the square with a cosh substitution to give

$$F_x = -\frac{\rho_1 U^2}{2 \pi} \int_0^\ell \int_0^{\xi_2} \ddot{S}(\xi_1) \ddot{S}(\xi_2) \left[\ln(\xi_2 - \xi_1) - \ln(4L) \right] d\xi_1 \, d\xi_2 \qquad (12.37)$$

If the source gradients $\dot{S}(0) = \dot{S}(\ell) = 0$ are zero at the ends of the body, the term in L vanishes. Hence

$$\begin{aligned} F_x &= -\frac{\rho_1 U^2}{2 \pi} \int_0^\ell \int_0^{\xi_2} \ddot{S}(\xi_1) \ddot{S}(\xi_2) \ln(\xi_2 - \xi_1) \, d\xi_1 \, d\xi_2 \\ &= -\frac{\rho_1 U^2}{4 \pi} \int_0^\ell \int_0^\ell \ddot{S}(\xi_1) \ddot{S}(\xi_2) \ln |\xi_2 - \xi_1| \, d\xi_1 \, d\xi_2 \qquad (12.38) \end{aligned}$$

The scaling of wave drag is clearly $F_x \sim \rho_1 U^2 S^2 / \ell^2$ and the drag coefficient correspondingly scales as $C_D \sim S^2 / \ell^4$, proportional to the square of the cross-sectional area.

Comparing this result (12.38) with equation (11.55) for induced drag, we note the formal equivalence of the two forms. Of course the physical behaviour which leads to the drag is different in the two cases. Induced drag is associated with subsonic vortex flow around the tips of the wings and involves an integral along the span. Wave drag is due to the momentum removed from the flow by waves generated at the wing surface in supersonic flow, and involves an integral over the chord. As a consequence of this mathematical equivalence, we may apply the Fourier series method (case study 11.I.i) used for induced drag in incompressible flow in a similar manner to this problem also. Thus following equation (11.67) and defining

$$x = \frac{1}{2}\ell\left(1 + \cos\theta\right) \qquad\qquad 0 \le \theta \le \pi$$

we write

$$\dot{S}(x) = -\ell \sum_{n=1}^{\infty} A_n \sin n\theta \qquad\qquad (12.39)$$

the terms $A_0 = 0$ being a consequence of the boundary condition on S at the ends of the body. The relevant integrals are evaluated using the methods as in case study 11.I.i. Hence the drag force

$$F_x = \frac{\pi \rho_1 U^2 \ell^2}{8} \sum_{n=1}^{\infty} n\, A_n{}^2 \qquad\qquad (12.40)$$

The cross-sectional area is obtained by integration

$$S(\theta) = \frac{\ell^2}{4}\left\{ A_1\left(\pi - \theta + \frac{1}{2}\sin 2\theta\right) \right.$$
$$\left. + \sum_{n=2}^{\infty}\left[\frac{\sin\left(n+1\right)\theta}{n+1} - \frac{\left(n-1\right)\sin\theta}{n-1}\right] \right\}$$

and the volume of the body is easily calculated

$$V = \frac{\pi}{8}\ell^3\left(A_1 - \frac{1}{2}A_2\right)$$

The required condition on the source/area function may be satisfied in two ways:

Von Karman ogive has one pointed end and a cylindrical base with sides parallel to the axis. It is clear that the term A_1 contributes to the base area $S(\ell)$

$$A_1 = \frac{4\,S(\ell)}{\pi\,\ell^2}$$

The drag is clearly a minimum if $A_n = 0$ for $n \ne 1$ and has value

$$F_x = \frac{2\,\rho_1 U^2}{\pi}\frac{S(\ell)^2}{\ell^2}$$
$$C_D = \frac{4}{\pi}\frac{S(\ell)}{\ell^2} \qquad\qquad (12.41)$$

with area distribution

$$S(\theta) = \frac{S(\ell)}{\pi} \left(\pi - \theta + \frac{1}{2} \sin 2\theta \right) \qquad (12.42)$$

This body is known as a von Karman ogive. It is a member of a general family of such bodies pointed at one end and flat at the other, and that find wide application as nose cones for aircraft and rockets.

Sears–Haack body   is symmetric with two pointed ends. In this case $A_1 = 0$ and

$$A_2 = -\frac{16\, V}{\pi\, \ell^3}$$

The drag force and drag coefficient are

$$F_x = \frac{64\, \rho_1\, U^2}{\pi} \frac{V^2}{\ell^4}$$

$$C_D = 24\, \frac{V}{\ell^3} \qquad (12.43)$$

and the area distribution is

$$S(\theta) = \frac{4V}{\pi\, \ell} \left(\sin\theta + \frac{1}{3} \sin 3\theta \right) \qquad (12.44)$$

The radius varies along the axis as

$$R(x) = R_{\max} \left(4x - 4x^2 \right)^{3/4} \qquad R_{\max} = \sqrt{\frac{16V}{3\, \pi^2\, \ell}} \qquad (12.45)$$

Sears–Haack bodies are typically used as a model for projectiles.

It is found that quite generally drag is not very sensitive to small departures from the optimum shape provided the area distribution is smooth.

12.5.1   Whitcomb Area Rule

In many cases it is found that the drag experienced by an asymmetric body is equal to that of an equivalent body of revolution, having the same cross-sectional area distribution. In the asymptotic limit, far away from the body, the flow becomes axisymmetric and equal to that around the equivalent body. This may be understood in terms of the multi-pole expansion noted in Section 8.4, where the lowest order term is the simple axisymmetric monopole source. At the very large distances $R \to \infty$ used to calculate drag, the contributions from higher order terms become vanishingly small, in accord with more detailed analysis.

Independently, as a result of experimental wind tunnel testing, Whitcomb (1956) unexpectedly found shock formation could occur at Mach numbers as low as 0.7, leading to strong drag. Following a suggestion by Busemann that, at near sonic speeds, stream tubes become inflexible and no longer able to expand or contact to accept the changes in aircraft cross-section (see page 22), Whitcomb realised that the fuselage must be able to accommodate the stream-lines displaced by the wings and tailplanes to reduce shock formation and wave drag. To minimise drag, the aircraft cross-section as a whole must more closely approximate a Sears–Haack body, consistent with the above theoret-ical result. He therefore proposed the important area rule that the drag of a slender wing transonic aeroplane could be reduced by indenting the fuselage ('Coke-bottle' shape) so that the overall cross-section was smoothly varying and more closely matched the ideal Sears–Haack form. More recent anti-shock design employs careful arrangement of components, and/or adding additional features (Küchemann carrots). This has proved to be extremely successful and is extensively used, particularly for commercial transport aircraft where eco-nomic considerations require drag minimisation.

Case study 12.I Hypersonic Wing

At very high Mach numbers a very thin wing is used to reduce the drag. A thin plate serves as a model of such a wing, The flow around the wing is shown in Figure 12.6 inclined at an angle α. On the pressure side of the wing, the flow is compressed by a shock parallel to the wing surface and is attached to the leading edge of the wing provided the angle of attack α is not large. This shock generates an increased pressure on the underside. At the trailing edge, the flow expands through a centred rarefaction back to a direction nearly parallel to the incoming flow. On the top side this configuration is reversed. A centred rarefaction at the trailing edge expands the

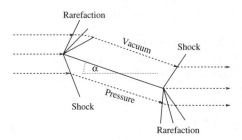

Figure 12.6: Sketch of the shock waves, rarefactions and streamlines (dashed) around a thin wing in hypersonic flow at an angle of attack α.

flow parallel to the surface of the wing, reducing the pressure. At the trailing edge a shock turns the flow back nearly parallel to the incoming flow.

Lift is generated by the difference between the pressures on the upper and lower surfaces. Drag, however, is caused both by the pressure difference and by waves leaving the upper and lower surfaces, and is enhanced by any surface deviations from the plane, which lead to additional characteristics, and waves, leaving the surface.

We can estimate the lift from a flat plate aerofoil, as depicted in Figure 12.6 by calculating the pressures over the upper and lower surfaces. Over the upper surface, the flow expands in a Prandtl–Meyer expansion, whose properties were evaluated earlier. The flow is rotated around a corner of angle α. The angle of attack α is generally small and the Mach number of the incoming flow M_1 large. Using the notation of Section 9.4.2 the pressure ratio between the upper surface and the incoming flow is

$$\frac{p_3}{p_1} = \left[\frac{\cos^2(k\psi + \varepsilon)}{\cos^2 \varepsilon}\right]^{2\gamma/(\gamma-1)}$$

where ψ is the angle of rotation of the characteristics C_- measured from the one on entry, p_3 is the pressure behind the rarefaction, and ε is given in Section 9.4.3. On exit the angle $\psi = \alpha + \mu_1 - \mu$, where μ is the Mach angle. Since α is small and the Mach number $M_1 \gg 1$, $\mu \approx \mu_1$ is small. Hence using the results from Section 9.4.3,

$$
\begin{aligned}
\mu - \mu_1 &\approx \left[\frac{1}{v_r}\frac{\partial v_\psi}{\partial \psi} - \frac{v_\psi}{v_r^2}\frac{\partial v_r}{\partial \psi}\right]\psi \\
&= v_{\max}\left(-\frac{k^2}{v_r}\sin^2\varepsilon - \frac{k\,v_\psi}{v_r^2}\cos\varepsilon\right)\psi \\
&= \left(-k^2 - k^2\frac{\cos^2\varepsilon}{\sin^2\varepsilon}\right)\psi \\
&\approx -k^2\psi
\end{aligned}
\tag{12.46}
$$

since $M_1 \gg 1$, $\varepsilon \approx \pi/2 - 1/k\sqrt{M_1^2 - 1} \approx \pi/2$.

Hence we obtain the pressure ratio across the rarefaction

$$
\begin{aligned}
\frac{p_3}{p_1} &\approx \left[1 - \frac{k^2}{(1 - k^2)}M_1\,\alpha\right]^{2\gamma/(\gamma-1)} \\
&= \left[1 - \frac{2}{(\gamma - 1)}M_1\,\alpha\right]^{2\gamma/(\gamma-1)}
\end{aligned}
\tag{12.47}
$$

The pressure on the lower surface is obtained from the pressure ratio across a shock inclined so as to rotate the flow through the small angle α. Since the Mach number is large, the shock polar becomes a circle. For small angles of deflection, the weak shock, which occurs in practice, has

$$v_{2x} \approx U \qquad \text{and} \qquad v_{2y} \approx U\alpha \tag{12.48}$$

and therefore it follows from the shock polar that

$$|\mathbf{U} - \mathbf{v}_2| = |(U - v_{2x})\hat{\mathbf{i}} + v_{2y}\hat{\mathbf{j}}| \approx v_{2y} \approx U\alpha \tag{12.49}$$

where $\hat{\mathbf{i}}$ and $\hat{\mathbf{j}}$ are unit vectors in the x and y directions respectively. Hence

$$|\mathbf{U} - \mathbf{v}_2| = |U_n\hat{\mathbf{n}} + U_t\hat{\mathbf{t}} - v_{2n}\hat{\mathbf{n}} - v_{2t}\hat{\mathbf{t}}| = V_{1n} - v_{2n} = U\alpha$$

where $\hat{\mathbf{n}}$ and $\hat{\mathbf{t}}$ are the normal and tangential unit vectors respectively. Using equation (10.8) we obtain

$$(U\alpha)^2 = (p_2 - p_1)(U - V_2)$$

where $V = 1/\rho$ is the specific volume. Using equation (10.8b) we eliminate the specific volume in terms of the pressure ratio $\Pi = p_2/p_1$ to yield the quadratic equation

$$(M_1\alpha)^2 = \frac{2(\Pi - 1)^2}{\gamma\left[(\gamma - 1) + (\gamma + 1)\Pi\right]} \tag{12.50}$$

whose solution is

$$\frac{p_2}{p_1} = \Pi = 1 + \frac{\gamma(\gamma + 1)}{4}(M_1\,\alpha)^2 + \gamma(M_1\alpha)\sqrt{1 + \frac{(\gamma + 1)(M_1\alpha)^2}{16}} \tag{12.51}$$

The section lift coefficient is therefore

$$
\begin{aligned}
c_L &= \frac{(p_2 - p_3)}{\frac{1}{2}\rho U^2} \\
&= \frac{2}{\gamma M_1{}^2}\left\{ 1 + \frac{\gamma(\gamma + 1)}{4}(M_1\,\alpha)^2 \right. \\
&\quad \left. + \gamma(M_1\alpha)\sqrt{1 + \frac{(\gamma + 1)(M_1\alpha)^2}{16}} - \left[1 - \frac{(\gamma - 1)}{2}(M_1\alpha)^2\right]^{2\gamma/(\gamma-1)} \right\}
\end{aligned} \tag{12.52}
$$

If $\sqrt{(\gamma - 1)/2}\,(M_1\,\alpha) > 1$, the flow forms a void on the upper surface and the corresponding pressure term is zero.

If $\alpha \ll M_1\alpha \ll 1$, the flow is perturbed weakly by the wing, and we may alternatively apply the perturbation method discussed in the next chapter. The lift coefficient takes the simpler form $c_L \approx 4\alpha/M_1$ in agreement with the earlier approach (Section 12.3).

We may also estimate the drag due to the pressure difference across the wing as

$$c_D \approx c_L\,\alpha \tag{12.53}$$

This term is due to supersonic flow alone and is the result of the energy flow generated by the waves, i.e. wave drag. The source of the drag can be seen in the energy transport away from the wing by the waves. The finite expansion of the rarefaction wave from the vacuum surface inherent in the fan plays an important role in this regard. In accordance with Section 10.8, as the two waves move away from the shock, the rarefaction interacts with the shock and progressively weakens it. Thus at large distances from the object the two waves destroy each other, and the net energy loss is limited. Similarly for the rarefaction and shock from the pressure surface. If the shocks were allowed to progress unchecked to infinity, the drag would become infinite in order to sustain the shock out to very large distances. This drag is of course in addition to that due to skin friction and vortex generation.

   In the course of this calculation we assumed that the angle of attack, and there-fore the angle of the deflection of the flow, is small. Without this approximation, the calculation can be carried in the same manner using the properties of the full Prandtl–Meyer expansion for the rarefaction and the (θ, β, M) plot for the shock, but not in an analytic form. For a general profile, the calculation may be continued in a similar fashion by considering the upper and lower surfaces made up of a series of short plane segments each inclined at an appropriate angle. The pressure on each segment is then calculated directly, and hence the lift, drag and moment.

Chapter 13

Deflagrations and Detonations

13.1 Introduction

Thus far we have not considered the situation when the energy of the fluid is changed either by a chemical reaction or by an external source, such as laser irradiation. Two distinct modes of flow are possible:

1. *Deflagration* The heat is released whilst the fluid expands, the density decreasing through the deflagration. The flow is continuous, and similar to a rarefaction with additional heat input. Deflagrations are a familiar feature as (usually controlled) flames. They also occur during the heating and ablation of a solid target by imposed laser irradiation.

2. *Detonation* The heat is released rapidly and the fluid compressed. The flow is essentially discontinuous, resembling a shock with heat input. We shall show later that a detonation may be imagined as a deflagration initiated by a shock. Detonations are also familiar as the source of explosions. The breakdown of gas by laser radiation is a detonation process.

13.1.1 Deflagrations

The speed of a chemical reaction varies with temperature, increasing rapidly as the temperature is raised. If the reaction is exothermic (energy is released) heat will spread to undisturbed regions of the flow, where reactions are initiated. There are many different mechanisms by which heat is transferred from the front to the undisturbed fluid, specific to the individual process.

Introductory Fluid Mechanics for Physicists and Mathematicians, First Edition. Geoffrey J. Pert.
© 2013 John Wiley & Sons, Ltd. Published 2013 by John Wiley & Sons, Ltd.

13.1.1.1   Propagating burn

As an example to illustrate the deflagration mechanism, we consider the propagation of a simple burn through a combustible mixture, which occurs by
the following process. Burning fluid, heated by the reaction, heats the unburnt
fluid ahead of it by thermal conduction to start the reaction. The reaction front
therefore propagates steadily through the undisturbed fluid, leaving behind a
region of hot burnt combustion products (Figure 13.1). Due to the increased
temperature and therefore pressure, the fluid behind the front is accelerated
and flows from the combustion zone. In this situation the thermal diffusion
and the burn time are matched and the structure propagates as a steady flow
pattern. In most cases fluid flow within the burn zone is small and the region
has nearly constant pressure.

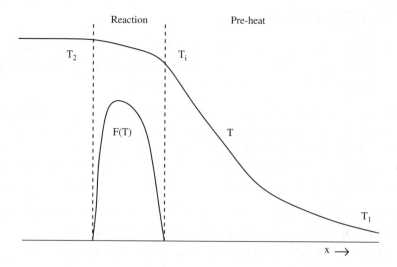

Figure 13.1: Sketch of the structure of the flame zone.

Within the flame zone, chemical reactions of the type

$$\sum_\nu x_\nu M_\nu \rightarrow \sum_\nu x'_\nu M'_\nu \tag{13.1}$$

between the molecules M_ν with the stoichiometric coefficients x_ν chemically react to produce product molecules M'_ν with the coefficients x'_ν (e.g.
$2\,H_2 + O_2 \rightarrow 2\,H_2O$). For an elementary reaction (i.e. a single step reaction),
the number of reactions per unit volume (the reaction rate) at temperature T
is given by the Arrhenius expression

$$R(T) = A\,\rho^{\sum_\nu x_\nu} \prod_\nu C_\nu{}^{x_\nu} \exp\left(-U/kT\right) \tag{13.2}$$

where C_ν is the mass concentration of species ν and U the reaction activation energy. It is evident that the reaction has a strong temperature dependence, switching on at the ignition temperature T_i and switching off as the fuel $\nu = f$ is exhausted, $C_f \to 0$, through the zone. The heat liberated per second per unit volume is given by $F(T) = \Delta q\, R(T)$ where Δq is the heat released in a single reaction. Assuming that the fuel f of molecular mass m_f is completely consumed, $q = (C_f/m_f\, x_f)\,\Delta q$ is the heat generated by burning of unit mass of the input gas. The exponential form of the rate ensures a rapid growth of the burning phase, which is limited by depletion of the fuel. The burning phase is therefore short with rapid onset once the ignition temperature is achieved and an equally rapid fall as the reaction consumes the fuel (Figure 13.1).

In the deflagration process, since the temperature gradients are large, the main heat transfer from the combustion zone to the undisturbed gas is by thermal diffusion. In addition since the gradients of the reaction constituents are equally large, particle diffusion also plays an important role, which is neglected in this simple picture. If τ is the duration of the reaction given approximately by an energy balance, then

$$\rho_1\,(h_2 - h_1) \approx \tau \int_{T_1}^{T_2} F(T)\,dT \,/\, (T_2 - T_1)$$

where $F(T)$, the rate of energy release by the chemical reaction per unit volume per second at temperature T, is averaged over the temperature range, and T_2 and h_2 are the temperature and enthalpy at the end of the reaction respectively. T_1 and h_1 are the temperature and enthalpy in the unignited fuel. The initial density ρ_1 is used as the mass heated remains constant, determined by the initial condition. The reaction rate is a strong function of temperature, being nearly zero below the ignition temperature and rapid above. The reaction zone tends to be quite short once ignition is established and is preceded by a relatively long pre-heat zone heated by thermal conduction (Figure 13.1). The final temperature behind the flame is given later by equation (13.4). The thickness of the reaction zone is determined by the heat transport from the reaction given by the expression for heat diffusion from a point source

$$\delta \approx \sqrt{4\chi\tau}$$

where $\chi = \kappa/\rho\,c_p$ is the thermal diffusivity. Substituting for δ and τ we obtain an approximation for the *laminar flame speed*, the velocity at which the front moves forward. A more accurate value is given by the Zel'dovich formula (problem #47)

$$S \approx \frac{\delta}{\tau} \approx \frac{1}{\rho_1\,(h_2 - h_1)} \left\{ 2 \int_{T_1}^{T_2} \kappa\, F(T)\,dT \right\}^{1/2} \tag{13.3a}$$

In estimating this speed it is assumed that the fluid motion is laminar.[1] In most cases the structure of the flame is more complex. Molecular diffusion plays an important role as the molecular number densities are of an equal magnitude to the thermal ones, the relative strength being expressed by the dimensionless Lewis number, the ratio of the thermal to the mass diffusion coefficients $Le = \chi/D$. If the Lewis number is unity, the mass and thermal transport equations become similar and the solution may be analysed as by Semenov, and by Zel'dovich and Frank–Kamenetskii, by a simple generalisation of the method of problem #47 (see Glassman and Yetter, 2008, Chap. 4). In intense flames turbulent mixing becomes an important mixing agent. This important area from an engineering standpoint lies outside the scope of this book and one of the standard references on combustion (e.g. Glassman and Yetter, 2008) should be consulted.

If the length scale in the experiment ℓ is much larger than the width δ we may regard the flow as changing discontinuously across the deflagration, although, as we have argued, the flow has a finite structure within this zone. From kinetic theory the thermal diffusivity of a gas $\chi \sim \lambda \bar{c} \sim \bar{\tau} \overline{c^2}$, where λ is the mean free path, $\overline{c^2}$ is the mean squared thermal velocity and $\bar{\tau}$ the mean free time between collisions. Hence we may alternatively express the velocity of the deflagration front as

$$S/\bar{c} \sim \delta/\bar{c}\tau \sim \sqrt{\chi/\overline{c^2}\tau} \sim \sqrt{\bar{\tau}/\tau} \tag{13.4}$$

In practice there is only a small probability of a reaction in each collision, so that $\bar{\tau} \ll \tau$, and therefore $S \ll \bar{c} \sim c$. Deflagrations (or burn waves) therefore propagate subsonically into undisturbed gas at rest.

In the rest frame of the front variations with time are usually much slower than the characteristic time, τ, so that the front may be considered to propagate as a steady state structure. Outside the burn zone, dissipative processes are weak, and therefore the integrated conservation laws generating the mass, momentum and energy relations, namely the Rankine–Hugoniot equations, must hold across the deflagration with the condition that the energy input from the reaction is accounted for. This is easily done by including the heat of formation g in the total enthalpy of the molecules

$$H = h + g = \epsilon + pV + g = E + g \tag{13.5}$$

---

[1] An alternative form of the speed using an average of the product of rate coefficient and thermal diffusivity $\bar{R}$ and $\bar{\chi}$ is

$$S = \sqrt{2\bar{\chi}\,\bar{R}\,\frac{T_2 - T_i}{T_i - T_1}} \tag{13.3b}$$

This result, due to Mallard and Le Chatelier, requires a value of the poorly defined ignition temperature.

where E is the total internal energy including the heat of formation. If g_1 and g_2 are the heats of formation before and after burn, the heat of reaction $q = g_1 - g_2$. Clearly the change in the enthalpy per unit mass across the flame front is $h_2 - h_1 = q$.

Since the flows are strongly subsonic $v_1 \ll c_1$, $p_1 \approx p_2$ and $H_1 \approx H_2$, we obtain by treating the reactants and products as perfect gases with polytropic indices γ_1 and γ_2

$$p_1 \approx p_2 \qquad T_2 \approx \frac{c_{p1}}{c_{p2}} T_1 + \frac{1}{c_{p2}} q \qquad p_2 \approx \left\{ \frac{\gamma_2 \, (\gamma_1 - 1)}{\gamma_1 \, (\gamma_1 - 1)} \right\} \rho_1 \left/ \left\{ 1 + \frac{q}{c_{p1} \, T_1} \right\} \right.$$

c_{p1} and c_{p2} are the specific heats at constant pressure in the undisturbed and burnt fluid respectively.

Furthermore, since the downstream flow behind the burn is hotter than the upstream in front, its density is less. The flow is therefore expanding, and in the rest frame the downstream velocity is larger than the upstream one.

Propagating burn in the form of a flame is a familiar feature, although in many cases the input flow velocity is limited by the supply of the reactants, e.g by a controlled flow as in a torch, or by the gaseous release from a solid as in a fire.

13.1.1.2   Deflagration propagating in a closed tube

We note that since the deflagration is subsonic relative to the upstream gas, characteristics leave it moving upstream into the unburnt gas, and modify the flow upstream of the front. Since the front velocity is determined by the heat of reaction, this allows the flow to satisfy the prescribed boundary conditions imposed by the experimental conditions.

For example, consider a burn initiated at the closed end of a cylinder. Since the deflagration is propagated from a wall, the downstream flow is brought to rest; that is, since the front speed v_1 is given, then so is v_2, the exit velocity. The flow speed downstream in the rest frame relative to that upstream is $v_1 - v_2 < 0$, i.e. towards the end if the deflagration is at rest. However, in the laboratory frame this velocity must be zero relative to the end. Therefore the gas must flow with velocity $v_2 - v_1$ along the pipe upstream of the front. Far upstream from the front the gas is at rest. Hence a compression wave must travel down the tube, which will steepen to a shock separating the fluid entering the deflagration from the ambient upstream gas.

13.1.2   Detonations

If the chemical reaction is initiated by the density and temperature increase in a shock resulting from the heat released in the reaction, for example, the shock in the preceding problem is sufficiently strong. The front will be supersonic. When the shock has passed and initiated the reaction it continues for a time τ as before. The front velocity is the shock velocity and the thickness of the reaction zone therefore

$$\delta' = \tau \times \text{Shock velocity}$$

Typically the thickness is small compared with experimental scale length ℓ and we may also treat the detonation as a discontinuity.

Hence the modified Rankine–Hugoniot relations hold for this case also.

13.2   Detonations, Deflagrations and the Hugoniot Plot

In the rest frame of the burn zone, the Rankine–Hugoniot relations are modified to take account of the energy release

$$\rho_1 v_1 = \rho_2 v_2$$
$$p_1 + \rho_1 v_1{}^2 = p_2 + \rho_2 v_2^2 \tag{13.6}$$
$$H_1 + \frac{1}{2}v_1^2 = H_2 + \frac{1}{2}v_2^2$$

Hence we obtain the Hugoniot adiabat relations

$$H_2 - H_1 = \frac{1}{2}(V_1 + V_2)(p_2 - p_1)$$
$$E_2 - E_1 = \frac{1}{2}(V_1 - V_2)(p_2 + p_1) \tag{13.7}$$
$$\epsilon_2 - \epsilon_1 = \frac{1}{2}(V_1 - V_2)(p_2 + p_1) + q$$

In a similar manner to Section 10.9 we may construct the Hugoniot curve for the end points (p_2, V_2) from a given initial point (p_1, V_1), illustrated in Figure 13.2. In contrast to the shock case, which corresponds to $q = 0$, this curve no longer passes through the initial point. The curve is displaced to larger values of p and V by the reaction energy q. Since

$$\frac{p_2 - p_1}{V_2 - V_1} = -\rho_2^2 v_2^2 = -\rho_1^2 v_1^2 = -j^2 < 0$$

where j is the mass flux, it follows that either $p_2 > p_1$ and $V_2 < V_1$, or $p_2 < p_1$ and $V_2 > V_1$. There are therefore two branches of the Hugoniot curve:

Compression flow $V_2 < V_1$ corresponding to detonation
Expansion flow $V_2 > V_1$ corresponding to deflagration

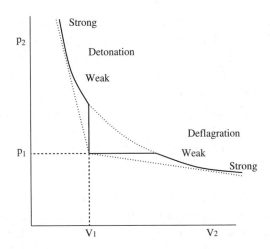

Figure 13.2: Sketch of a typical Hugoniot plot for flows with chemical reaction.

The enthalpy of a polytropic gas $h = \gamma/(\gamma - 1)\, pV$ substituted into the Hugoniot relation yields the equation of the corresponding Hugoniot plot

$$\frac{(\gamma_2 + 1)}{(\gamma_2 - 1)}\, p_2\, V_2 - \frac{(\gamma_1 + 1)}{(\gamma_1 - 1)}\, p_1\, V_1 + p_1\, V_2 - p_2\, V_1 = 2\,q \qquad (13.8)$$

where the initial and final adiabatic indices may be different due to the change in chemical composition of the gas.

Clearly the plot of p_2 versus V_2 is a rectangular hyperbola. Furthermore, if the pressure ratio across the deflagration becomes very large, $p_2/p_1 \to \infty$, the density ratio tends to a finite limit $V_2/V_1 \to (\gamma_2 - 1)/(\gamma_2 + 1)$, as in a strong shock. The maximum compression in a detonation is therefore $(\gamma_2 + 1)/(\gamma_2 - 1)$.

Under most conditions of burn the initial pressure p_1 and internal energy ϵ_1 are small compared with those generated by the heat release, and may be neglected. Equation (13.8) simplifies to

$$\left[\frac{(\gamma_2 + 1)}{(\gamma_2 - 1)} V_2 - V_1 \right] p_2 = 2\,q \qquad (13.9)$$

More generally, there exist a family of Hugoniot curves lying regularly between the shock curve and that with energy release q. The curves in the set are approximately parallel, each corresponding to an energy $\varepsilon\, q$. It is clear that no two curves of the set may intersect, for to do so would correspond to a non-unique value of the energy release. Clearly $\varepsilon = 0$ corresponds to the shock transition and $\varepsilon = 1$ to the energy release q.

Equation (10.7) also holds for deflagrations and detonations so that the gradient of the line from final state ② to initial state ① defines the flux through the front. Inspection of Figure 13.2 shows that for each branch there are two possible intersections of this line with the same flux. These are separated by the singular cases where the flux line is tangent to the adiabat, known as Chapman–Jouget detonations or deflagrations respectively. In both cases the intersection with the smaller change in pressure or density is known as the weak transition and the larger one as the strong transition (Figure 13.2).

The following rules, known as *Jouget's rules*, are easily established from the geometry of the Hugoniot plot in Figure 13.2 recalling that the case $\varepsilon = 0$ corresponds to the shock flow:

Strong detonation: flow supersonic in front, subsonic behind
Weak detonation: flow supersonic in front, supersonic behind
Strong deflagration: flow subsonic in front, supersonic behind
Weak deflagration: flow subsonic in front, subsonic behind

The Chapman–Jouget process plays an important role in the study of deflagrations and detonations. At the Chapman–Jouget point the flux line is tangent to the Hugoniot plot, therefore

$$\frac{dp_2}{dV_2}\bigg|_H = \frac{p_2 - p_1}{V_2 - V_1}$$

Differentiating the Hugoniot relation,

$$d\epsilon_2 - \frac{1}{2}\left[(V_1 - V_2)\,dp_2 - (p_1 + p_2)\,dV_2\right] = 0$$

but

$$d\epsilon = T\,ds - p\,dV$$

hence

$$T_2\,ds_2 = \frac{1}{2}\left[(V_1 - V_2)\,dp_2 - (p_1 - p_2)\,dV_2\right] = 0 \tag{13.10}$$

At the Chapman–Jouget point the entropy is an extremum and is constant along the Hugoniot curve, which is therefore parallel to the adiabatic plot. Therefore

$$\rho_2{}^2 v_2{}^2 = \frac{dp}{dV}\bigg|_H = \frac{\partial p}{\partial V} = \rho_2{}^2 c_2{}^2 \tag{13.11}$$

The flow speed of the burnt gas is therefore sonic at the Chapman–Jouget point.

Similarly the flow speed of the incoming flow is given by

$$v_1{}^2 = -V_1{}^2 \frac{(p_2 - p_1)}{(V_2 - V_1)}$$

and hence

$$dv_1{}^2 = -\frac{V_1{}^2}{(V_2-V_1)^2}\left[(V_2-V_1)\,dp_2 - (p_2-p_1)\,dV_2\right] = 0 \qquad (13.12)$$

The incoming flow velocity is therefore an extremum. This result confirms the result already apparent, that the flux term j^2 has a minimum at the Chapman–Jouget point. It also follows from the behaviour of the Hugoniot adiabat (Section 10.3) that at the Chapman–Jouget point the entropy is a minimum/maximum on the detonation/deflagration branch.

It is clear that the intermediate Hugoniot plots form a continuous sequence between the shock ($\varepsilon = 0$) and the developed deflagration ($\varepsilon = 1$). At the shock limit, $\varepsilon \to 0$, the two Chapman–Jouget points merge, the weak fronts being lost, to give an extremum and point of inflexion. Noting the behaviour along the shock Hugoniot and the continuation referred to above, the following rules are established by analogy from Figures 13.2 and 13.3 (below):

Chapman–Jouget deflagration: velocity and entropy are maxima
Chapman–Jouget detonation: velocity and entropy are minima

Chapman–Jouget flow in a polytropic gas

At the Jouget point the downstream velocity is equal to the sound speed $v_2 = \sqrt{\gamma_2\,p_2\,V_2}$ and the flux $j^2 = \gamma_2\,p_2/V_2$ which, when substituted in (10.7), gives

$$
\begin{aligned}
p_2 &= \frac{(p_1 + j^2 V_1)}{(\gamma_2 + 1)} & V_2 &= \frac{\gamma_2(p_1 + j^2 V_1)}{(\gamma_2 + 1)j^2} \\
&= \frac{v_1{}^2 + (\gamma_1 - 1)\,\epsilon_1}{(\gamma_2 + 1)(\gamma_1 - 1)\,\epsilon_1} & &= \frac{\gamma_2\left[v_1{}^2 + (\gamma_1 - 1)\,\epsilon_1\right]}{(\gamma_2 + 1)\,v_1{}^2}
\end{aligned}
\qquad (13.13)
$$

Substituting these relations in equation (13.8) yields a fourth-order equation for the incoming velocity v_1

$$v_1{}^4 - 2v_1{}^2\left[(\gamma_2{}^2 - 1)\,q + (\gamma_2{}^2 - \gamma_1)\,\epsilon_1\right] + \gamma_2{}^2\,(\gamma_1 - 1)\,\epsilon_1{}^2 = 0 \qquad (13.14)$$

since the internal energy $\epsilon = pV/(\gamma - 1)$. This is a quadratic equation in $v_1{}^2$ whose two roots correspond to the detonation and deflagration branches of the Hugoniot adiabat. The former is the larger of the two. This equation is the standard form $x^4 - 2bx^2 + c = 0$ whose solution is

$$x = \sqrt{b \pm \sqrt{b^2 - c}} = \sqrt{\frac{1}{2}\left(b + \sqrt{c}\right)} \pm \sqrt{\frac{1}{2}\left(b - \sqrt{c}\right)}$$

and whose reciprocal is

$$x^{-1} = \sqrt{\left[b \mp \sqrt{b^2 - c}\right]/c} = \sqrt{\frac{1}{2}\left(b + \sqrt{c}\right)/\sqrt{c}} \mp \sqrt{\frac{1}{2}\left(b - \sqrt{c}\right)/\sqrt{c}}$$

The corresponding solution for the inflow velocity is

$$v_1 = \left\{ \frac{1}{2} (\gamma_2 - 1) \left[(\gamma_2 + 1) q + (\gamma_1 + \gamma_2) \epsilon_1 \right] \right\}^{1/2}$$

$$\pm \left\{ \frac{1}{2} (\gamma_2 + 1) \left[(\gamma_2 - 1) q + (\gamma_2 - \gamma_1) \epsilon_1 \right] \right\}^{1/2} \tag{13.15}$$

The outflow velocity which follows from the relation $v_2 = V_2 \, v_1 \, / \, V_1$ is also equal to the sound speed

$$c_2 = v_2 = \left\{ \frac{1}{2} (\gamma_2 - 1) \left[(\gamma_2 + 1) q + (\gamma_1 + \gamma_2) \epsilon_1 \right] \right\}^{1/2}$$

$$\pm \frac{(\gamma_2 - 1)}{(\gamma_2 + 1)} \left\{ \frac{1}{2} (\gamma_2 + 1) \left[(\gamma_2 - 1) q + (\gamma_2 - \gamma_1) \epsilon_1 \right] \right\}^{1/2} \tag{13.16}$$

The velocity difference, which is the relative velocity of the outflow with respect to the incoming fluid, is

$$v_1 - v_2 = \pm \left\{ 2 \left[(\gamma_2 - 1) q + (\gamma_2 - \gamma_1) \epsilon_1 \right] \right\}^{1/2} \tag{13.17}$$

As noted earlier the energy released by the chemical reaction is frequently much larger than that of the incoming fluid, $q \gg \epsilon_1$, which greatly simplifies the above results. However, we must separate the values for a detonation from those of a deflagration:

Detonation   The velocities reduce to

$$v_1 = \sqrt{[2 (\gamma_2{}^2 - 1) q]} \quad v_2 = c_2 = \frac{\gamma_2}{(\gamma_2 + 1)} v_1 \quad v_1 - v_2 = \frac{1}{(\gamma_2 + 1)} v_1 \tag{13.18}$$

and the thermodynamic state of the gas to

$$\frac{V_2}{V_1} = \frac{\gamma_2}{(\gamma_2 + 1)} \quad \frac{p_2}{p_1} = \frac{2(\gamma_2 - 1)}{(\gamma_1 - 1)} \frac{q}{\epsilon_1} = \frac{\gamma_1 v_1{}^2}{(\gamma_2 + 1)c_1{}^2} \quad \epsilon_2 = \frac{2\gamma_2 q}{(\gamma_2 + 1)} \tag{13.19}$$

Deflagration   The velocities reduce to

$$v_1 = \frac{(\gamma_1 - 1) \gamma_2 \epsilon_1}{\sqrt{[2 (\gamma_2{}^2 - 1) q]}} \quad v_2 = c_2 = \sqrt{\left[\frac{2 (\gamma_2 - 1) q}{(\gamma_2 + 1)} \right]} \tag{13.20}$$

and the thermodynamic state of the gas to

$$\frac{p_2}{p_1} = \frac{1}{(\gamma_2 + 1)} \quad \epsilon_2 = \frac{2q}{(\gamma_2 + 1)} \tag{13.21}$$

As can be seen, the velocity ratio across a Chapman–Jouget deflagration is large. Such a front propagating in a closed tube therefore moves rapidly away from the end and generates a strong shock. We therefore conclude that Chapman–Jouget deflagrations are unlikely to be established in practice unless the burnt gas can be rapidly removed, e.g. by flow into a vacuum.

13.2.1   The Structure of a Deflagration

If we assume that during a deflagration process thermal conduction, mass diffusion and viscosity are all weak and can be neglected, i.e. the heat deposited at each point in the flow is almost entirely due to the chemical reaction, then the conservation laws must hold at each point in the deflagration. Thus if a fraction ε of the reaction is complete, we have at each point on the trajectory

$$\rho_1 v_1 = \rho_\varepsilon v_\varepsilon = \rho_2 v_2$$

$$p_1 + \rho_1 v_1{}^2 = p_\varepsilon + \rho_\varepsilon v_\varepsilon{}^2 = p_2 + \rho_2 v_2{}^2$$

$$E_1 + p_1 V_1 + \frac{1}{2} v_1{}^2 + q = E_\varepsilon + p_\varepsilon V_\varepsilon + \frac{1}{2} v_\varepsilon{}^2 + (1 - \varepsilon) q = E_2 + p_2 V_2 + \frac{1}{2} v_2{}^2$$

Since the relation

$$\frac{p_\varepsilon - p_1}{V_\varepsilon - V_1} = -j^2$$

holds at each point ε, each point lies on the chord from (p_1, V_1) to (p_2, V_2) and its position is given by the Hugoniot corresponding to a fraction ε of the overall reaction. The flow through the deflagration takes the form shown in Figure 13.4.

There are two possible end points corresponding to the weak (2) and the strong (2′) end points in Figure 13.3. A passage to the lower (strong) end point can only take place if the flow passes through the region where $\varepsilon > 1$, which is clearly impossible. Hence only weak deflagrations are allowed.

When thermal conduction and diffusion are included, but viscosity is negligible, the intermediate curves include the additional energy deposited by conduction, but still lie between the initial ($\varepsilon = 0$) and final ($\varepsilon = 1$) curves for all physically realistic situations.

Deflagrations in which viscosity dominates the heating process are improbable. We therefore conclude that in practice *only weak deflagrations occur.*

If we consider a typical deflagration, the experimental parameters determining the flow are the upstream physical state of the fluid and the heat released by the chemical reaction. As we have seen in Section 13.1.1.2, the head of the front is sometimes preceded by a shock wave, which must be taken into

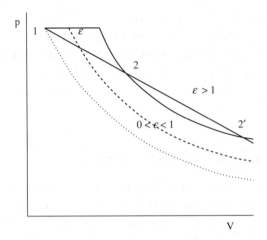

Figure 13.3: Sketch of the path of a deflagration through the family of Hugoniot curves.

account in modifying the initial conditions. Although we have concluded that only weak deflagrations are possible, this leaves a large range of possible flows from the very slow to the Chapman–Jouget flow. There is therefore a degree of indeterminacy in the flow at this stage. This uncertainty is removed when we include thermal conduction and diffusion necessary to ensure the initiation of the chemical reaction at the head of the front. This condition determines the flow in a particular situation, limiting it to one path only through the set of adiabats. In Section 13.1.1 we argued that the distance penetrated by the head of the heat front is determined by the thermal and mass diffusivity and the duration of the reaction. The velocity of the heat front into stationary gas is therefore determined by the speed of the head of the heat front which initiates the reaction, namely $\sim \sqrt{\chi/\tau}$. It is easy to see from the discussion of Section 13.1.1 how this situation may be formulated more formally. The solution will depend on the detailed behaviour of the reaction rate, the thermal conductivity, the mass diffusion coefficient and the viscosity with temperature and density through the deflagration in direct analogy with the analysis of shock structure, Section 10.5.

13.2.1.1 The Shvab–Zel'dovich model of a deflagration

A relatively simple model of the structure of a deflagration is given the Shvab–Zel'dovich approximation. Consider a one-dimensional laminar flame front moving with flame velocity S into unburnt gas with initial concentration (by mass) C_ν of molecules M_ν at density ρ_1, pressure p_1 and enthalpy h_1 in its rest frame. After passage through the burn zone the density is ρ_2, pressure $p_2 \approx p_1$ and $h_2 = h_1 + q$ (Section 13.1.1). The intermediate flow is governed by the one-dimensional flow equations modified to take into account molecular diffusion and the chemical reaction. Since the diffusion coefficient for each molecular species may be different and molecules are either destroyed or created by the reaction, each species has a separate conservation equation

$$\frac{\mathrm{d}(\rho\, C_\nu\, S)}{\mathrm{d}x} + \frac{\mathrm{d}}{\mathrm{d}x}\left\{ D_\nu \frac{\mathrm{d}(\rho\, C_\nu)}{\mathrm{d}x}\right\} = \pm x_\nu\, m_\nu\, R(T) \tag{13.22a}$$

$$\frac{\mathrm{d}(\rho\, h\, S)}{\mathrm{d}x} + \frac{\mathrm{d}}{\mathrm{d}x}\left\{ \chi \frac{\mathrm{d}(\rho\, h)}{\mathrm{d}x}\right\} = \Delta q\, R(T) \tag{13.22b}$$

where the $+$ sign is taken for the reaction products and the $-$ sign for the reactants, m_ν is the mass of molecule M_ν and Δq is the energy liberated per reaction. Since the flow is nearly isochoric (constant pressure), the momentum equation can be omitted, provided the viscosity is small.

Dividing each equation of the set (13.22a) by $\pm m_\nu\, x_\nu$ and equation (13.22b) by Δq, the complete set of equations (13.22) reduces to the form

$$\frac{\mathrm{d}}{\mathrm{d}x}\left\{ \eta\, S - D\frac{\mathrm{d}\eta}{\mathrm{d}x}\right\} = R(T) \tag{13.23}$$

where $\eta_\nu = \{(\rho\, C_\nu / m_\nu\, x_\nu)$ and $(\rho\, h/\Delta q)\}$, provided the diffusivity of every molecule has the same value $D_\nu = D$ and the Lewis number χ/D is unity, $\chi = D$. The approximate solution of this equation is obtained in problem #47. The boundary conditions on entry to the flame are the initial density, concentrations and enthalpy (i.e. initial values of the set η_ν), with zero gradients at entry and exit at $\pm\infty$. Knowing the concentrations and temperature through the zone, the rate coefficient may be evaluated and hence the complete structure obtained numerically.

13.2.1.2 Detonations as deflagrations initiated by a shock

As we have seen, a deflagration may generate a shock wave ahead of it as it progresses. If the shock is sufficiently strong it will heat the fluid and initiate the combustion process. In this case the burn forms a detonation. Behind the shock, the flow is subsonic and the burn itself identical to a deflagration with its head in the shocked gas.

Since the flux is given by

$$j^2 = -\frac{p - p_1}{V - V_1}$$

for both the shock and the deflagration, both the initial state (1), the shocked state ($*$) and the final state (2) following the burn lie on the chord from (p_1, V_1). This line makes two intersections with the final detonation Hugoniot, Figure 13.4, corresponding to a strong (2) and a weak (2') detonation. Treating the shocked state ($*$) as the starting point for the deflagration we see that:

A strong detonation corresponds to a weak deflagration
A weak detonation   corresponds to a strong deflagration

Since we have concluded that only weak deflagrations exist, it follows that *only strong detonations occur.*

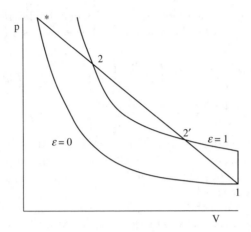

Figure 13.4: Sketch of the path of a detonation initiated by a shock.

13.2.2   Chapman–Jouget Hypothesis

In many experimental situations the flow in a detonation is specified only by the initial undisturbed rest state of the fluid and the energy release. In contrast to a shock wave (Section 10.2), where the downstream pressure or the Mach number is also given, this leaves a deficit of one parameter to uniquely identify the flow. Thus the boundary conditions on the flow are insufficient to uniquely specify the flow. For example, if we consider a detonation propagating into undisturbed gas in a known state with a specified energy release per unit mass, there remain a range of strong detonations possible depending on the downstream condition. To obtain a unique solution it is often convenient to use the Chapman–Jouget hypothesis, which states that the flow takes a Chapman–Jouget form, where the downstream flow velocity is sonic in the stationary frame.

Thus consider a detonation propagating in a tube closed at one end by a piston which is slowly withdrawn (Figure 13.5). At the piston the flow must be outwards. But since the detonation starts at the piston, it is travelling inwards into the undisturbed fluid. The detonation is compressive, and therefore the

Figure 13.5: Detonation propagating in a tube with an expanding piston.

flow velocity of the combustion products relative to the unburnt gas is

$$v_1 - v_2 = v_1 \left(1 - \frac{\rho_1}{\rho_2}\right) > 0$$

i.e. travelling in the direction of the detonation away from the piston.

Since the detonation is strong the flow is either subsonic or sonic in the rest frame of the front, but the flow velocity must be changed to match the piston. This can take place in either a shock or a rarefaction. A shock is supersonic relative to the flow and will therefore catch up with the detonation, where the downstream flow is subsonic, and become a part of it. Similarly the head of a rarefaction is locally sonic and will also be absorbed unless the downstream flow from the detonation is itself sonic.

Consequently the detonation must have a Chapman–Jouget form.

In contrast, as we have seen, the flow in a deflagration is weak or Chapman–Jouget established by the detailed structure determined by the distributed heat release, thermal conduction and viscosity. In fact a Chapman–Jouget flow is extremely unlikely to occur. Since the downstream flow velocity is greater than the upstream one, the deflagration must move upstream in order to accommodate the fluid behind it if the cylinder is closed by either a stationary piston or one moving upstream. As we have seen, this leads to a preceding shock wave with the result that the downstream one is not sonic. Only in special cases can a Chapman–Jouget deflagration be established, as discussed in the next section.

Case study 13.I Deflagrations and Detonations in Laser–Matter Breakdown

When an intense laser pulse is focused into gas or onto the surface of a solid, the material is ionised and rapidly forms a plasma in which strong absorption of the laser radiation occurs. The resulting high-temperature gaseous medium then undergoes rapid hydrodynamic motion with heat input from the absorbed laser pulse. A good approximation is obtained by treating the plasma as a polytropic gas with $\gamma = 5/3$ and constant ionisation Z. There are a range of possible flows depending on the wavelength and intensity of the laser and the nature of the material with which it interacts. These different flows take the form of steady and time-dependent deflagrations and detonations depending on the conditions.

In order to have a broad understanding of this behaviour we need to appreciate some basic properties of laser–plasma interaction, which are governed almost completely by the electron density and temperature:

- Energy is absorbed within the body of the front from the laser beam, whose irradiance (energy flux) is Φ. The energy release per unit mass is therefore Φ/j where j is the mass flux.

- Radiation can only propagate in plasma when the electron density is less than the critical density at which the laser frequency equals the characteristic oscillation frequency of the electrons in the plasma. A simple expression for the critical mass density in kg/m^3 is

$$\rho_{\text{crit}} \, (\text{kg/m}^3) \approx 1.66 \, \frac{A}{Z} \, \lambda \, (\mu\text{m})^{-2}$$

where Z is the fractional ionisation, A the mass number of the ions and $\lambda \, (\mu\text{m})$ the wavelength of the radiation in microns.

- Provided the electron density is markedly less than the critical density, the absorption coefficient is due to electron collisions with ions, *inverse bremsstrahlung*, with value

$$\mu = b \, \rho^2 \, \bar{c}^{-3} \tag{13.24}$$

where the temperature is represented by the sound speed, b is a constant depending on the wavelength and plasma composition, and $\bar{c}$ the mean thermal velocity.

- In the neighbourhood of the critical density, absorption is due to a variety of non-linear mechanisms. Experimentally this is found to result in strong absorption of typically 30% locally at the critical density of the incident radiation.

High-intensity laser radiation breaks down cold unionised material initially by a multi-photon release of electrons leading to a low density of electrons. These are rapidly heated by inverse bremsstrahlung and in turn cause further ionisation. A rapid cascade thus occurs leading to a plasma with a high degree of ionisation. This breakdown takes different forms in gases and solids due to the markedly dissimilar background density with respect to the critical density.

13.I.i Solid targets

In this case the ambient density is typically about 10^2–10^3 critical, i.e. $\rho_0 \gg \rho_{\text{crit}}$. The plume expands into vacuum, so that no downstream limitation on the flow occurs. The ambient temperature is very much less than that generated by the heating $T_1 \ll T_2$ and may be neglected. Absorption takes place in an expanding plasma plume as the material heats and expands. The flow therefore has the form of a deflagration. As we have seen, a shock precedes the heat front penetrating into the solid and probably initiates ionisation. A zone of heating and expansion from the high density follows the shock at the head of the deflagration structure, resulting from thermal conduction, before absorption takes place. In contrast to deflagrations generated by chemical heat release where thermal conduction initiates the reaction, energy deposition is independent of the heat front and depends solely on the local thermodynamic state and the incident (external) energy flux. Since there are two different absorption processes, there are correspondingly two limiting forms of the deflagration depending on which is dominant. We compare and contrast these two systems within a one-dimensional flow model.

13.I.i.a High-intensity irradiation – deflagration model

If the radiation is very intense, the electron temperature is high and inverse bremsstrahlung plays little part in absorption. The energy deposition is therefore localised at the critical density. No energy can be deposited from the laser beam upstream of the critical density, where a zone of thermal conduction heats the fluid behind the shock (Fauquignon and Floux, 1970). Within this region the fluid is progressively heated and expanded to flow smoothly into the zone of local absorption. In this case the simple nature of the deflagration zone makes it time independent.

In contrast to the case of a reaction-supported deflagration, there is no additional condition specified by the initiation of the heat deposition or similar limitation. The flow therefore requires an additional condition to be determinate. This is supplied by the requirement that, similar to a detonation, the deflagration must be a Chapman–Jouget flow with the exit velocity equal to the local sound speed. This model assumes radiation is absorbed only in the neighbourhood of the critical density, no heat is deposited further downstream, and the subsequent flow must be a rarefaction. Hence, following our earlier argument, the downstream flow must be sonic and from equations (13.20) and (13.21) the flow conditions are

$$\rho_2 = \rho_{\mathrm{crit}} \qquad v_2 = c_2 = \sqrt[3]{\frac{2\,(\gamma - 1)\,\Phi}{(\gamma + 1)\,\rho_{\mathrm{crit}}}} \qquad p_2 = \left[\frac{2\,(\gamma - 1)}{(\gamma + 1)}\right]^{2/3} \rho_{\mathrm{crit}}^{1/3}\,\Phi^{2/3}$$

In fact the plasma is likely to be sufficiently hot that the rarefaction is maintained at uniform temperature by thermal conduction; that is, it is an isothermal rarefaction, which is driven by a downstream heat flow $p_2 v_2$ as described in Section 9.3.3 and the downstream exit velocity is the isothermal sound speed $v_2 = \bar{c}_2$. The deflagration is described by the modified Rankine–Hugoniot equations (13.6) with the Chapman–Jouget condition applied to the isothermal sound speed at the downstream end. Neglecting the heat loss to the rarefaction, the solutions for the exit velocity and pressure are then

$$\rho_2 = \rho_{\mathrm{crit}} \qquad v_2 = \bar{c}_2 = \sqrt[3]{\frac{2\,(\gamma - 1)\,\Phi}{(3\,\gamma - 1)\,\rho_{\mathrm{crit}}}} \qquad p_2 = \frac{1}{\gamma}\left[\frac{2\,(\gamma - 1)}{(3\gamma - 1)}\right]^{2/3} \rho_{\mathrm{crit}}^{1/3}\,\Phi^{2/3}$$

Φ, the heat deposited per unit time per unit area in the deflagration, is a fraction of that incident due to reflection at the critical density. We note two characteristic results:

- The downstream velocity is larger for the adiabatic flow, reflecting the general result that for a deflagration the Chapman–Jouget flow velocity is a maximum.

- The downstream pressure of the isothermal Chapman–Jouget deflagration is greater than that of the adiabatic, so that the former is a weak deflagration as required.

Taking into account the heat conducted downstream, namely $\rho_{\mathrm{crit}} v_2 \epsilon_2$,

$$p_2 = \left[\frac{2\,(\gamma - 1)}{(5\gamma - 3)}\right]^{2/3} \rho_{\mathrm{crit}}^{1/3}\,\Phi^{2/3} \tag{13.25}$$

This model represents one limiting case, namely one where the absorption is entirely at the critical density and no distributed absorption occurs. The plasma is very hot and thermal conduction ensures that energy is spatially distributed.

13.I.i.b Low-intensity irradiation – self-regulating model

If the irradiation intensity is relatively weak, the temperature of the plasma in the plume is no longer sufficiently high that the downstream plasma is absorbent. The entire incident energy flux Φ is therefore deposited within the plasma body. The plume is heated by inverse bremsstrahlung, which provides sufficient distributed heat to maintain the expansion without the need for thermal conduction to be active. Absorption at the critical density is negligible and plays no role in the flow. The absorption is determined by the optical depth, namely the product of the absorption coefficient and the length (μx). Since heating at the head of the deflagration must be supported by absorption of the radiation to maintain the flow, the overall optical depth is of order unity. The spatial heat distribution within the plasma varies in time as the scale length of the plasma plume increases. Consequently the density must decrease and the temperature increase with time along the plume. The flow self-regulates to maintain this relation. If the plasma is too hot or too tenuous the optical depth is reduced, more radiation reaches the ablation surface, and increased ablation restores the status quo. Similarly vice versa if the plasma is too cold or dense. The flow is therefore stabilised by the functional form of the absorption coefficient (Afanas'ev et al., 1966; Caruso et al., 1966). Scaling relations for the optical depth, flow speed and energy follow from dimensional analysis

$$b \rho^2 c^{-3} v t \sim 1 \qquad v \sim c \qquad \Phi \sim \rho v c^2$$

The constants of proportionality are easily seen to be of order unity from consideration of momentum and energy conservation. Hence we obtain the general scalings

$$c \sim b^{1/8} \Phi^{1/4} t^{1/8} \qquad\qquad \rho \sim b^{-3/8} \Phi^{1/4} t^{-3/8} \qquad\qquad (13.26)$$

with constants of proportionality of order unity.

Since the problem involves only two characteristic parameters, namely b and Φ, it is expressible in a self-similar form with parameter

$$\xi = \frac{x}{b^{1/8} \Phi^{1/4} t^{9/8}} \sim \frac{x}{ct} \qquad\qquad (13.27)$$

in terms of which the overall flow may be expressed as a set of ordinary differential equations, which can be numerically integrated by standard methods to obtain accurate scalings.

In practical situations the purely one-dimensional solution given above fails when the length of the plasma becomes comparable with the focal spot radius. At this point the flow expands radially and the absorption rapidly decreases to form a steady flow. Since the initial planar flow from the target is subsonic, it cannot pass through the sonic point (Section 1.8.1.2). However, the transition through a sonic point is allowed in a divergent channel. The two- or three-dimensional expansion of the plasma plume therefore allows the sonic transition to occur at a distance of approximately the focal spot radius from the target, Under these conditions, the absorption coefficient falls rapidly with distance and the optical depth is approximately limited to the value at the sonic point. A steady flow is therefore established and an additional parameter, namely the focal spot radius r_0, is introduced into the scaling:

$$\rho \sim b^{-1/3}\Phi^{1/3}r_0^{-(\nu+1)/3} \qquad c \sim b^{1/3}\Phi^{2/9}r_0^{-(2\nu-1)/9} \tag{13.28}$$

where $\nu = 1$ for a line focus and $\nu = 2$ for a spot. As before, the scaling constants are of order unity, and can only be calculated from simplified models or by numerical simulation.

We note that in this case of the self-regulating model, the flow is uniquely specified by the experimental parameters and no additional condition is necessary.

The two models presented here clearly represent opposite limits of the full solution. The limiting condition is readily seen to be that the self-regulating model holds if $\rho < \rho_{crit}$. However, accurate calculations of the flow in the general case involving thermal conduction, and absorption by both inverse bremsstrahlung and nonlinear processes at the critical surface, can be only be generated by numerical modelling. Nonetheless the simple models given above do generate useful estimates of the plasma conditions in their appropriate regime. The values generated are broadly in agreement with those found in experiments within the regimes for which the models are applicable. However, it should be noted that there are a number of plasma effects which are not included in the models and in many cases markedly change the nature of the interaction.

13.I.ii Gaseous targets

Gaseous targets, in contrast to solid targets, have the ambient density less than the critical density. Consequently, stronger absorption results from higher density, and heat deposition is promoted by compression, not expansion. Thus the heat front is found in the shock transition of a detonation. As the detonation propagates towards the laser along the beam, initial ionisation is due to multi-photon breakdown, followed by rapid collisional absorption once the first electrons are released. An ionisation cascade within the shock structure of the detonation completes the breakdown (Raizer, 1977).

A laser-driven detonation, however, differs from one driven by a chemical reaction in one characteristic. In a chemically driven burn, each molecule contributes an equal energy to the total, namely q per unit mass. On the other hand, in a laser-driven explosion, the energy is delivered at a constant rate determined by the laser intensity, Φ. The energy absorbed per unit mass is therefore Φ/j, which depends on the characteristic values of the detonation flow parameters, in particular the detonation velocity or the rest frame velocity into the wave, v_1. The Hugoniot adiabat relation (13.7) therefore becomes

$$\epsilon_2 - \epsilon_1 = \frac{1}{2}(V_1 - V_2)(p_1 + p_2) + \Phi\sqrt{\frac{(V_1 - V_2)}{(p_2 - p_1)}} \tag{13.29}$$

The Hugoniot adiabat for such a detonation is significantly modified from that of a chemical reaction, Figure 13.2. The adiabat for a laser-driven detonation is sketched

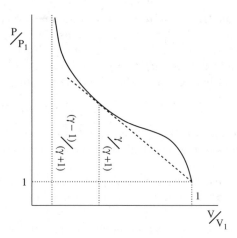

Figure 13.6: Sketch of the Hugoniot for a laser-driven detonation. The Chapman–Jouget point and the asymptotic limit are noted.

in Figure 13.6. We note that if the strength is small, the detonation passes through the initial point, the energy delivered per unit mass being very small. This form of Hugoniot should be contrasted with that in Figure 13.2.

In the limit where $\epsilon_1 \ll q = \Phi/\rho_1 v_1$, the Hugoniot takes the simpler form from equation (13.9)

$$p_2 = \left\{ \frac{2(\gamma-1)\sqrt{V_1 - V_2}}{(\gamma+1)V_2 - (\gamma-1)V_1} \right\}^{2/3} \Phi^{2/3} \tag{13.30}$$

Two useful relations which can be derived from the momentum and energy equations respectively are

$$\epsilon_2 = \frac{V_2\,(V_1 - V_2)}{(\gamma-1)\,V_1{}^2}\,v_1{}^2 = \frac{2V_1 V_2}{[(\gamma+1)V_2 - (\gamma-1)V_1]\,v_1}\,\Phi \tag{13.31}$$

Since the breakdown must be a Chapman–Jouget flow, we obtain the conditions from equations (13.18) and (13.19), in particular the velocity of the front

$$v_1 = \left\{ 2(\gamma^2 - 1)\frac{\Phi}{\rho_1} \right\}^{1/3} \qquad \epsilon_2 = \frac{2^{1/3}\gamma}{(\gamma^2 - 1)^{2/3}(\gamma+1)} \left(\frac{\Phi}{\rho_1} \right)^{2/3} \tag{13.32}$$

being the minimum possible velocity and maximum heating. These results are in good agreement with values measured in experiments.

Chapter 14

Self-similar Methods in Compressible Gas Flow and Intermediate Asymptotics

14.1   Introduction

As we have noted, many flows may be reduced from a multi-dimensional problem to a simpler one using the concept of self-similarity. Although we briefly discussed the underlying idea of self-similarity in an earlier section (3.8.2) dealing with dimensional methods in viscous fluids, the method achieves its greatest power and application in treating problems of compressible flow. In many cases these would otherwise be intractable except by computational simulation. We have already met several problems which could be reduced to self-similar form in previous chapters, e.g. Blasius's solution for the boundary layer over a flat plate, Section 6.3, and Kolmogorov's model of isotropic turbulence, Section 5.2. In each of these a characteristic dimensional parameter was missing from the list required to specify the flow. In fact there are many cases of this behaviour.

Self-similar motion in time-varying flow is one in which the parameters that specify the state and velocity of the flow vary in such a manner that the spatial distribution of these variables is similar to itself as the flow proceeds. The scales of the magnitude of the variable and its spatial variation change with time. Consequently if the distribution in space is known at specified times, then the values at other points in space will be known at times lying on well-defined lines or surfaces.

Introductory Fluid Mechanics for Physicists and Mathematicians, First Edition. Geoffrey J. Pert.
© 2013 John Wiley & Sons, Ltd. Published 2013 by John Wiley & Sons, Ltd.

We have already met a number of simple flows of this type, namely centred rarefactions, Section 9.3.1 and Section 9.4.2, and adiabatic compression, Section 10.10, although we have not drawn attention to their self-similar nature. In this chapter we will show how to use these methods to develop solutions to complex problems.

The method is usually based on dimensional analysis to identify specific functional relationships amongst the variables characterising the problem. To solve a particular problem we identify three classes of parameters:

1. **Independent variables.** The determining variables of the problem, e.g. spatial distance **r** and time t. Most problems we investigate are symmetric and involve one spatial dimensional (Cartesian, cylindrical polar or spherical polar) co-ordinate r and time t.

2. **Dependent variables.** Typically the experimentally measurable quantities of the problem, e.g. pressure p, density ρ and velocity **v**, at points in space and time specified by values of the independent variables.

3. **Constant parameters.** The values which specify the flow, e.g. background pressure p_0, ambient density ρ_0 and incoming velocity v_0. These are the initial values, boundary conditions and fluid parameters of the specific problem, whose number is appropriate to the solution of the differential equations.

The solution must take the form of a relationship in which any one of the dependent variables is expressed in terms of the independent set and the parameters. It follows from Buckingham's Π theorem that this solution must have the form of a function of a complete set of independent dimensionless products, at least one of which contains the appropriate dependent variable. In general the terms of the three classes are all dimensional, expressed typically with respect to the three basic dimensions of mechanics: mass $[M]$, length $[L]$ and time $[T]$.

The expression of a problem in self-similar form requires that the functions representing the dependent variables are separable in the independent variables. Thus if we consider a time-dependent, one-dimensional flow, then the density, for example, must be able to be expressed in a relationship of the form

$$\rho = D(t)\, f\left[\frac{r}{R(t)}\right]$$

where $D(t)$ is the scale factor for density with the dimensions of density, and $R(t)$ the scale factor for length with dimensions of length also dependent on time, the self-similar distribution $f(\xi)$ being expressed in terms of the self-similar variable $\xi = r/R(t)$. The distribution of density in space is constant over lengths on a scale which varies in time; the magnitude of the density varies in time. Typically the characteristic scale length is expressed as a power law in time: $R(t) = A t^{\delta}$, where A is a dimensional constant of dimensions $[L][T]^{-\delta}$ formed from the initial and boundary values of the problem. The self-similar variable is therefore $\xi = r / A t^{\delta}$.

Self-similar solutions are exact solutions of the equations of fluid mechanics usually subject to specific initial conditions. However, they have a much wider application as *intermediate asymptotics* of the general problems (Barenblatt, 1996). Suppose there are two governing parameters R_1 and R_2 with the same dimensions as the variable r; then an intermediate asymptotic is an asymptotic representation of the solution as $r/R_1 \to 0$ whilst $r/R_2 \to \infty$. Thus suppose, for example, that the initial condition of the problem is limited to a region of space of radius R_0. As time progresses, the dimensions of the resulting flow become large compared with the initial radius, the only dimensionless parameter containing R_0, namely $r/R_0 \to \infty$. Since dimensional analysis assumes that if dimensionless parameters are either very large or very small, they may be neglected, the flow becomes self-similar once it is much larger than the initial dimension. Consequently all memory of the initial condition is lost. Similarly it may happen that as the scale becomes very large, additional constraints are introduced. The self-similar solution is only applicable within limits much larger than the initial condition and much smaller than the final one. As we saw in Appendix 6.A we may match the outer (self-similar) solution to the inner (initial phase) by a perturbation expansion using the *method of matched asymptotics*.

Self-similar problems of this type fall into two kinds:

1. **Type 1 problems:** the constants (or boundary values) allow the identification of both the terms A and δ. The value of the similarity constant δ is therefore uniquely specified by dimensional constraints.

   Most problems are of the first kind. We illustrate these with a range of problems dealing with the expansion and compression of gas bodies. In many cases analytic solutions can be found. Self-similar solutions of many problems of this type are given by Sedov (1959) and by Stanyukovich (1960).

2. **Type 2 problems:** the constant parameters do not allow the term δ to be generated by dimensional analysis. It is still possible to identify a similarity parameter $\xi = ar/t^{\delta}$, where a is an unknown dimensional constant. The value of δ is, however, determined by the nature of the singularities in equations (14.33) since the resulting functions representing physical quantities must be well behaved. The value of δ is thus

equivalent to an eigenvalue of equations (14.33). Several examples of this type of self-similar flows are to be found in Zel'dovich and Raizer (1967).

Problems of the second kind do not admit analytic solution and require a numerical solution to find the 'eigenvalue'. To illustrate this case we will examine the problem of a collapsing shock wave.

14.2 Homogeneous Self-similar Flow of a Compressible Fluid

Homogeneous flows are ones in which the volume of every fluid element is changed by the same fraction in every time interval, i.e. the dilation rate is constant in space. Alternatively the velocity of a fluid element is linearly proportional to its distance from the centre, the scale factor varying as a function of time only (Nemchinov, 1964). Such flows have a particularly simple separable self-similar form, but are nonetheless very useful for estimating the expansion or compression of limited volumes of gas with the intermediate asymptotic approximation. The simple case of the generation of a homogeneous flow in an infinitely small body of gas with power law heating in time is treated in problem #48.

14.2.1 General Homogeneous Expansion of a Compressible Gas

The simple model of problem #48 may be generalised to embrace a large class of flows where the flow is multi-dimensional and the heat pulse has an arbitrary form. Considering a body of gas whose characteristic dimensions $X_i(t)$ are not necessarily equal, the self-similar condition is the generalisation of the equation of homogeneous flow

$$\xi_i = x_i/X_i(t) \tag{14.1}$$

to all ν dimensions of the motion. The dimensionless variables ξ_i are a set of Lagrangian co-ordinates constant in time for a particular fluid particle. The heating of the gas can take an arbitrary (rather than power law) functional form with time, subject to the initial rate of energy deposition being finite. The initial spatial dimensions need not be zero, but it is required that the density and pressure distribution functions, $f(\xi_i)$ and $\phi(\xi_i)$, match the spatial heat deposition function $W(\xi_i)$ at the start of the motion ($t = 0$).

The velocity of a fluid particle ξ_i is

$$v_i = \left.\frac{\mathrm{d}x_i}{\mathrm{d}t}\right|_{\xi_i} = \xi_i \frac{\mathrm{d}X_i}{\mathrm{d}t} = \frac{x_i}{X_i} V_i \tag{14.2}$$

where $V_i = \mathrm{d}X_i/\mathrm{d}t$. It follows from this equation that the flow must have symmetry in the coordinates x_i, namely $v_i(-x_i) = -v_1(x_i)$ etc. The mass of a cell fixed in the fluid of volume $\mathrm{d}\tau_x$ and the elementary volume $\mathrm{d}\tau_\xi$ of the Lagrangian co-ordinate element ξ_i are constant in time. The Jacobian of the transformation between the configuration spaces is

$$\frac{\mathrm{d}\tau_x}{\mathrm{d}\tau_\xi} = J = \frac{\partial(x_1, x_2, \dots)}{\partial(\xi_1, \xi_2, \dots)} = \prod_{i=1}^{\nu} X_i$$

The flow is therefore homogeneous as the ratio of the volume elements in configuration space for fluid elements of identical size depends only on a factor varying in time. Since ρJ is invariant of motion (Section 1.3.1) and J is a function of time alone, the density ρ is a separable function of time and the Lagrangian space co-ordinates (similarity variables)

$$\rho = \rho_0(t)\, f(\xi_i^2)$$

where $f(\xi_1^2, \xi_2^2, \dots)$ is represented as $f(\xi_i^2)$ taking account of the symmetry of the function $f(\xi_i^2)$ noted above.

Euler's equation enables us to find the density (and equivalently pressure) distribution functions. The equation takes the form

$$\frac{\partial p}{\partial \xi_i} = \xi_i\, f(\xi_i^2)\, \rho_0(t)\, X_i(t)\, \frac{\mathrm{d}V_i}{\mathrm{d}t} \tag{14.3}$$

Hence the pressure is also a separable function, and therefore all other thermodynamic state variables are as well:

$$p = p_0(t)\, \phi(\xi_i^2)$$

where $\phi(\xi_1^2, \xi_2^2, \dots)$ is represented by $\phi(\xi_i^2)$. Therefore

$$\frac{\partial[\phi(\xi_i^2)]}{\partial(\xi_i^2)} = -\frac{1}{2}\lambda_i\, f(\xi_i^2) \qquad \text{and} \qquad p_0(t) = \lambda_i^{-1} \rho_0(t)\, X_i\, \frac{\mathrm{d}V_i}{\mathrm{d}t} \tag{14.4}$$

for all i. λ_i are separation constants, which have an arbitrary value and usually determine the relationship of the characteristic scale width X_i to the fluid boundary.

The equation of energy conservation in Lagrangian co-ordinates can be written as

$$\frac{\mathrm{d}\epsilon}{\mathrm{d}t} - \frac{p}{\rho^2}\frac{\mathrm{d}\rho}{\mathrm{d}t} = Q \tag{14.5}$$

where Q is the heat release rate per unit mass and ϵ the specific internal energy. Clearly Q must be separable

$$Q = Q_0(t)\, q(\xi_i^2)$$

For a polytropic gas

$$q(\xi_i^2) = \mu\,\phi(\xi_i^2)/f(\xi_i^2)$$

$$Q_0(t) = \mu^{-1}\left\{\frac{1}{(\gamma-1)}\,p_0(t)\,\dot{p}_0(t) - \frac{\gamma}{(\gamma-1)}\,p_0(t)\,\dot{\rho}(t)/[\rho(t)]^2\right\} \quad (14.6)$$

where μ is a separation constant. This equation together with equation (14.4) defines the pressure and density distributions from $q(\xi_i^2)$.

The mass M and the total energy $E(t)$ in the gas at any time, which is the sum of the kinetic E_k and thermal E_t energies, are easily found:

$$M = \int \rho\,\mathrm{d}\tau = [J\rho_0(t)]\int f(\xi_i^2)\,\mathrm{d}\tau_\xi$$

$$E_k = \int \rho\,\frac{1}{2}\sum_{i=1}^{\nu} v_i^2\,\mathrm{d}\tau = \frac{1}{2}[J\rho_0(t)]\int f(\xi_i^2)\sum_{i=1}^{\nu}\xi_i^2\,V_i^2\,\mathrm{d}\tau_\xi \quad (14.7)$$

$$E_t = \int \rho\,c_v\,T\,\mathrm{d}\tau = \frac{1}{(\gamma-1)}\int p\,\mathrm{d}t = \frac{1}{\nu}[J\,\rho_0(t)]\int f(\xi_i^2)\sum_{i=1}^{\nu}\xi_i^2 X_i\,\frac{\mathrm{d}V_i}{\mathrm{d}\tau}\,\mathrm{d}\tau_\xi$$

since the pressure is zero at the gas boundary.

Ellipsoidal flows are a useful sub-class of flows, which are particularly easy to evaluate. Although the initial shape of the body may be ellipsoidal, rather than spherical, so that X_i and V_i take different values, the heat distribution function and the state variables have no preferred direction in Lagrangian space. Thus these variables are functions of the Lagrangian 'radius'

$$\zeta^2 = \sum_{i=1}^{\nu}\xi_i^2 = \sum_{i=1}^{\nu}\left(\frac{x_i}{X_i}\right)^2$$

The density distribution function f depends on ζ^2 only

$$\int \xi_i^2\,f(\xi_i^2)\,\mathrm{d}\tau_\xi = \frac{1}{\nu}\int \zeta^2\,f(\zeta^2)\,\mathrm{d}\tau_\xi$$

Defining

$$\Psi = \frac{1}{2}\frac{\int \zeta^2\,f(\zeta^2)\,\mathrm{d}\tau_\xi}{\int f(\zeta^2)\,\mathrm{d}\tau_\xi}$$

we obtain

$$\sum_{i=1}^{3}\left\{V_i^2 + \frac{2}{\nu(\gamma-1)}X_i\,\frac{\mathrm{d}V_i}{\mathrm{d}t}\right\} = \nu\,\frac{E(t)}{\Psi M} \quad (14.8)$$

where $E(t)$ is the total energy of the gas.

This equation taken together with equation (14.4) in the form

$$X_1 \frac{dV_1}{dt} = X_2 \frac{dV_2}{dt} = \cdots = \lambda \frac{p_0(t)}{\rho_0(t)} \tag{14.9}$$

uniquely defines the flow, in a convenient form for numerical integration. The fluid may have a finite initial energy $E(0)$, and the temporal form of the heating pulse $Q_0(t)$ is subject to the condition that $\lim_{(t\to 0)} Q_0(t) \to A\, t_n, n > -1$ (see problem #48).

14.2.1.1   Adiabatic flow

If the fluid is initially hot and not subsequently further heated, $E(t) = \text{const}$, the flow is adiabatic, and the entropy of each fluid particle is constant in time although not necessarily the same throughout the flow, the temporal distributions of pressure and density must satisfy the adiabatic equation of state

$$\frac{p_0(t)}{\rho_0(t)^\gamma} = \frac{p_0(0)}{\rho_0(0)^\gamma} \tag{14.10}$$

If in addition the fluid is isentropic, i.e. the entropy is everywhere constant, then it follows from equation (14.4) that

$$\gamma f(\xi_i^2)^{(\gamma-2)} \frac{\partial f}{\partial(\xi_i^2)} = -\lambda_i \tag{14.11}$$

where the set λ_i are the separation constants which determine the relationship of the variables X_i to the boundary. Considering ellipsoidal flow and taking a suitable value of λ we obtain

$$f^{(\gamma-1)} = f_0^{(\gamma-1)} \left(1 - \zeta^2/\zeta_0^2\right)$$
$$\phi^{(\gamma-1)/\gamma} = \phi_0^{(\gamma-1)/\gamma} \left(1 - \zeta^2/\zeta_0^2\right) \tag{14.12}$$

subject to $\phi(\zeta_0) = 0$, i.e. the pressure at the edge is zero. We may set $\zeta_0 = 1$ so that the set $X_i(t)$ are the edge of the distribution, and $f_0 = 1$ so that $\rho_0(t)$ is the density at the centre.

14.2.1.2   Isothermal flow

In many cases where a small mass of gas is heated (for example) by a laser, the fluid internal energy per particle (temperature) is approximately constant throughout the body of gas. A condition which may be due either to strong thermal conduction or to the nature of the heat deposition is $q(\zeta^2) = \text{const}$.

The spatial distribution of the density and pressure follows immediately from equations (14.4) and (14.6) to be

$$f(\zeta^2) = f_0 \exp(-\zeta^2/\zeta_0)^2 \tag{14.13}$$

where f_0 and ζ_0 are appropriately chosen. We may define $\zeta_0 = 1$ so that the sets $X_i(t)$ are the $1/e$ half widths of the distribution and $f_0 = 1$ so that $\rho_0(t)$ is the density at the centre.

14.2.2   Homogeneous Adiabatic Compression

If the separation constants $\lambda_i < 0$, the flow represents an inward motion of the fluid with the sign of the terms $(\zeta^2/\zeta_0{}^2)$ in equations (14.12) positive. The flow therefore represents the adiabatic collapse of a mass of gas towards the centre driven by an inward pressure gradient.[1] The temporal dependence of the external pressure applied at the edge determines whether a shock is formed, and therefore the extent of the adiabatic implosion. The velocity $V_i = dX_i/dt < 0$ is seen to be inwards. This flow should be compared with that in Section 10.10, where compressive flow is also adiabatic and not accompanied by a shock, provided the temporal dependence of the externally applied pressure is such as to avoid the intersection of the inward characteristics. In this case very high compression may in principle be achieved provided the pressure pulse is carefully structured.

14.2.2.1   Homogeneous collapse of spheres

We consider the self-similar adiabatic compression of an isentropic spherical body of gas of radius $R(t)$ by an external pressure applied at the outer edge. This case is treated by a simple extension of the above model using equations (14.4). Defining $h(t) = R(t)/R(0)$ as a measure of the compression, it follows that since the motion is adiabatic and the mass constant

$$\frac{\rho_0(t)}{\rho_0(0)} = \left(\frac{R(t)}{R(0)}\right)^3 = h(t)^3 \quad \text{and} \quad \frac{p(t)}{p(0)} = \left(\frac{R(t)}{R(0)}\right)^{-3\gamma} = h(t)^{-3\gamma} \tag{14.14}$$

Writing the separation constant as

$$\lambda = -\frac{\rho(0)\,R(0)^2}{p(0)}\frac{1}{t_0{}^2} = \gamma\left(\frac{R_0}{c_0\,t_0}\right)^2 \tag{14.15}$$

---

[1]The application of homogeneous flow to compression is due to Kidder (1975).

where c_0 is the speed of sound at time $t = 0$ at the origin $r = 0$, the equation of motion is

$$t_0{}^2 h^{(3\gamma-2)} \frac{d^2 h}{dt^2} = -1 = -2\gamma \left(\frac{R_0}{c_0 t_0}\right)^2 \frac{1}{f(\xi^2)} \frac{d\phi(\xi^2)}{d(\xi^2)} \tag{14.16}$$

If $\gamma = 5/3$ the solution is obtained subject to the boundary condition $h(0) = 1$

$$h^2 = (1 + b\tau)(1 - \tau) \tag{14.17}$$

where

$$b = \frac{1-q}{1+q} \qquad \tau = \frac{t}{t_c} \qquad t_c = \frac{t_0}{(1+q)} \qquad q = -t_0\,\dot{h}(0) = -t(0)\frac{V(0)}{R(0)} \geq 0 \tag{14.18}$$

We note that t_c is the collapse time, i.e. the time taken for the radius $R(t)$ to become zero.

When $b = 1$ the flow is initially stationary, the dimension factor $h = \sqrt{1 - \tau^2}$ and the surface pressure scales as $p(R, t) = p(R, 0)/(1 - \tau^2)^{5/2}$. When $b = 0$ the Mach number is constant, the dimension factor $h = \sqrt{1 - \tau}$ and the surface pressure scales as $p(R, t) = p(R, 0)/(1 - \tau)^{5/2}$.

Solving for the spatial part of the equation of motion, subject to the boundary conditions $\phi(0) = f(0) = 1$, we obtain

$$\phi(\xi) = \left(1 + \beta\,\xi^2\right)^{5/2} \qquad \text{and} \qquad f(\xi) = \left(1 + \beta\,\xi^2\right)^{3/2} \tag{14.19}$$

where

$$\beta = \frac{1}{3}\left(\frac{R(0)}{c_0 t_c}\right)^2 \tag{14.20}$$

where c_0 is the sound speed at the centre $c_0 = \sqrt{\gamma\, p_0\, \rho_0}$.

When the gas is initially at rest $q = 0$ and the collapse time $t_c = t_0$, the flow velocity and the sound speed are easily shown to be

$$v = \frac{dr}{dt}\bigg|_\xi = -\frac{c_0\tau}{\sqrt{1 - \tau^2}}\sqrt{3\beta\xi^2} \qquad \text{and} \qquad c = \sqrt{\frac{\gamma\, p}{\rho}} = \frac{c_0}{\sqrt{1 - \tau^2}}\sqrt{1 + \beta\xi^2} \tag{14.21}$$

for a gas with $\gamma = 5/3$. The Mach number is

$$M = \frac{|v|}{c} = \sqrt{\left\{\frac{3\beta\,\xi^2}{1 + \beta\,\xi^2}\right\}}\quad \tau < \sqrt{3} \tag{14.22}$$

The path of the characteristics is given by the usual expression

$$\frac{dr}{dt}\bigg|_\pm = v \pm c$$

The Mach lines (trajectory of the characteristics) can be expressed in analytic form[2] which shows that the arrival time at the centre of the ingoing characteristic leaving the surface at τ_0 is given by

$$\tau_1 = \frac{(1-g)}{(1+g)} \quad \text{where} \quad g = \left(\sqrt{\beta+1} - \sqrt{\beta}\right)^{2\sqrt{3}} \frac{(1-\tau_0)}{(1+\tau_0)} \tag{14.24}$$

For small values of $\beta \ll 1$, the arrival time of the characteristic τ_0 at the centre, $\tau_1 \approx \tau_0$, is nearly the same as the time it leaves the surface. For weakly driven compression, the characteristics only converge slowly, and sequentially reach the centre (Figure 14.1(a)), the fluid slowly compressing adiabatically.

For large values of $\beta \gg 1$, the arrival time of the characteristic τ_0 is given by $\tau_1 \approx 1 - 2\left[(1-\tau_0)/(1+\tau_0)\right][4\beta]^{\sqrt{3}}$, which approaches 1 for very large β. The characteristics therefore converge towards the centre for very strongly driven collapses (Figure 14.1(b)). However, the arrival time of a later characteristic is always after that of an earlier one. The characteristics do not intersect.

It can be shown quite generally that in no case do the characteristics intersect before convergence at the centre. Consequently no shock is formed during

---

[2]The solution is obtained by the substitution

$$r_\pm = R_0 \sqrt{\frac{(1-\tau^2)}{\beta}} f_\pm(\tau)$$

Making the further substitution

$$f_\pm(\tau) = \frac{1}{2}\left(F_\pm - F_\pm^{-1}\right)$$

we obtain the velocity of the characteristics

$$v \pm c = -\frac{\sqrt{3}}{2} c_0 \left[F_\pm(\tau) - F_\pm^{-1}(\tau)\right] \frac{\tau}{\sqrt{1-\tau^2}} \pm \frac{1}{2} c_0 \left[F_\pm(\tau) + F_\pm^{-1}(\tau)\right]$$

$$= \frac{dr_\pm}{dt}\bigg|_\pm = -\frac{\sqrt{3}}{2} c_0 \left[F_\pm(\tau) - F_\pm^{-1}(\tau)\right] \frac{\tau}{\sqrt{1-\tau^2}} \pm \frac{\sqrt{3}}{2} c_0 \sqrt{1-\tau^2} \left[\frac{dF_\pm(\tau)}{d\tau} - \frac{dF_\pm^{-1}(\tau)}{d\tau}\right]$$

the first term being the flow velocity v, and the second the sound speed c. Equating terms we obtain the differential equation

$$-\sqrt{3}\left(1-\tau^2\right)\left[\frac{dF_\pm}{d\tau} - \frac{dF_\pm^{-1}}{d\tau}\right] = \pm\left[F_\pm(\tau) + F_\pm^{-1}(\tau)\right]$$

Noting that $dF_\pm^{-1}/d\tau = -F_\pm^{-2} \, dF_\pm/d\tau$, the solution is readily obtained for a Mach line passing through the surface of the sphere at time τ_0 at $r = R_0\sqrt{1-\tau_0^2}$:

$$F_\pm = \left[\sqrt{\beta+1} + \sqrt{\beta}\right]\left[\frac{(1+\tau)}{(1+\tau_0)}\frac{(1-\tau_0)}{(1-\tau)}\right]^{\pm 1/2\sqrt{3}} \tag{14.23}$$

since $f_\pm(\tau_0) = \sqrt{\beta}$.

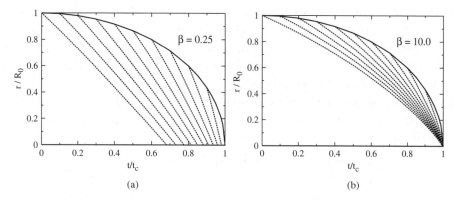

Figure 14.1: Plots of the inward Mach lines for small and large values of the parameter β. Note the contrasting convergence of the lines to the centre. The Mach lines (shown dashed) plotted start at intervals of $t_c/10$ at the outer surface. The path of the outer surface is shown by the full line. (a) $\beta = 0.25$, and (b) $\beta = 10$.

the compression phase. This is a consequence of the slowly increasing time history of the pressure pulse applied to the outer surface of the sphere, namely $p \sim (1 - \tau^2)^{-5/2}$. If the pressure increases more rapidly a shock will form converging onto the centre, destroying the adiabatic nature of the collapse.

14.2.2.2 Homogeneous collapse of shells

Consider an isentropic hollow shell with inner radius R_i and outer radius R_o collapsing into a void driven by a pressure pulse on the outer surface. It is easily shown that, if the collapse is homogeneous, equations (14.14) and (14.17) apply in this case as well. The time dependence of the scale factor $h(t)$ is therefore given in this case also by

$$h^2 = 1 - \tau^2 \qquad \text{where} \qquad \tau = t/t_c \tag{14.25}$$

where t_c is the time to collapse. If the shell is initially at rest the pressure on the outer surface scales as $p(R_o, t) = p(R_o, 0)/(1 - \tau^2)^{5/2}$, whereas that on the inner surface is always zero, $p(R_i, t) = 0$. If the adiabatic index of the gas $\gamma = 5/3$, the spatial distributions of pressure and density through the shell are

$$p(r,t) = p(R_o, 0) \left(\frac{r^2 - R_i^2}{R_o^2 - R_i^2} \right)^{5/2} \quad \text{and} \quad \rho(r,t) = \rho(R_o, 0) \left(\frac{r^2 - R_i^2}{R_o^2 - R_i^2} \right)^{3/2} \tag{14.26}$$

If c_0 is the initial sound speed in the gas, the collapse time is

$$t_c^2 = \left(R_o^2 - R_i^2 \right) / 3\, c_0^2 \tag{14.27}$$

Unlike the homogeneous compression of a sphere, there is a single solution for the shell. The introduction of a second boundary condition, namely at the inner surface in addition to that at the outer, removes the flexibility introduced by the parameter β to the solution for a sphere. As a result the ratio of the time to collapse t_c to the sound transit time R_0/c_0 is fixed.

Figure 14.2 shows a typical example of such an idealised collapse. Notice how rapidly the final stages of the collapse occur as the external pressure increases rapidly.

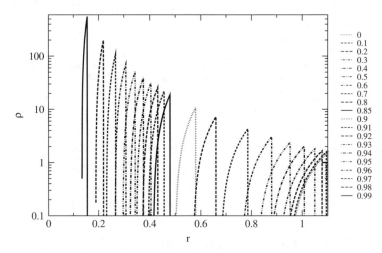

Figure 14.2: Variation of density with distance at various times relative to the collapse time. The initial condition is prescribed by a uniform shell of inner radius 1 unit and thickness 0.1 units. The initial density and sound speed are both 1 unit.

It may appear that the value of this solution is limited by the specific density distribution, since most practical problems involve uniform shells. However, the concept of intermediate asymptotics proves useful here, as we may represent a uniform shell of thickness Δ and outer radius R_o by the equivalent homogeneous form with the same mass, total internal energy and total entropy. If the shell is thin, $\Delta \ll R_o$, this requires the parameters of the homogeneous flow to take the values

$$R_o - R_i = \frac{5}{2}\left(\frac{5}{7}\right)^{3/2}\Delta \qquad t_c = \left(\frac{5}{3}\right)^{1/2}\left(\frac{5}{7}\right)^{5/4}\frac{\sqrt{R_o \Delta}}{c_0}$$

$$p(R_o,0) = \left(\frac{7}{5}\right)^{5/2}p_0 \qquad \rho(R_o,0) = \left(\frac{7}{5}\right)^{3/2}\rho_0 \tag{14.28}$$

where p_0, ρ_0 and c_0 are the initial pressure, density and sound speed in the uniform shell. The approximation is valid provided the inner surface of the shell does not reach the centre before convergence. Making use of the results from

Section 9.3.1, the time taken for the inner surface to reach the centre is approximately $(\gamma - 1)\, R_i/2\, c_0 = R_i/3\, c_0$. From equation (14.28) this requires that the shell thickness $\Delta/R_o < (7/5)^{5/2}/15 = 0.1546$. This condition also ensures that the first characteristic propagates through the shell before convergence.

14.3 Centred Self-similar Flows

The self-similar picture of flow can be set in a more formal context using dimensional analysis. The resultant flow equations are more complex than those derived for the homogeneous flows but are less restrictive in their application. It will be found that they fall into two distinct classes specified by the nature of the similarity variable. We will consider as examples a problem from each class.

Since the density and pressure both contain the dimension of mass, it is clear that at least one of the parameters must also contain mass. Let this term be a and its dimensions

$$[a] = [M]\,[L]^k\,[T]^s$$

We may introduce dimensionless forms for the velocity, density and pressure based on the dimensions of a and the independent variables

$$v = \frac{r}{t}V \qquad \rho = \frac{a}{r^{(k+3)}t^s}D \qquad p = \frac{a}{r^{(k+1)}t^{(s+2)}}P \qquad (14.29)$$

where V, D and P depend on dimensionless combinations of r, t and the other parameters.

In general we can find from the parameters individual combinations with the dimensions of length and time separately. However, in some circumstances this may not be possible, and it is only possible to find a combination b with dimensions containing both length and time, e.g. $c_0 = \sqrt{\gamma p_0/\rho_0}$

$$[b] = [L]^m\,[T]^n$$

In this case the spatial variable and time can only enter the problem through the dimensionless combination

$$\xi = \frac{r}{b^{1/m}\,t^{-n/m}} \qquad (14.30)$$

Such a problem is called self-similar and ξ is known as the self-similar variable. Since V, D and P can only be functions of ξ and any constant dimensionless parameter, it follows that the system exhibits similarity with itself for constant values of ξ for differing values of r and t.

For one-dimensional time-dependent flow, the partial differential equations in two variables are replaced by a single one in ξ with considerable improvement in the ease of solution. The equations of one-dimensional time-dependent flow of a polytropic gas may be written as

$$\frac{\partial \rho}{\partial t} + \frac{1}{r^{(\nu-1)}} \frac{\partial}{\partial r} \left(r^{(\nu-1)} \rho \right) = 0$$

$$\frac{\partial v}{\partial t} + v \frac{\partial v}{\partial r} + \frac{1}{\rho} \frac{\partial p}{\partial r} = 0 \tag{14.31}$$

$$\frac{\partial}{\partial t} \left(\frac{p}{\rho^\gamma} \right) + v \frac{\partial}{\partial r} \left(\frac{p}{\rho^\gamma} \right) = 0$$

where ν is the dimensionality parameter

$$\nu = \begin{cases} 1 & \text{in planar Cartesian geometry} \\ 2 & \text{in cylindrically symmetric geometry} \\ 3 & \text{in spherically symmetric geometry} \end{cases}$$

Substituting for the velocity, density and pressure in terms of the dimensionless forms (14.29), and introducing the sound speed as a variable through $c^2 = Zr^2/t^2$ so that $Z = \gamma P/D$, we obtain

$$\xi \left[(\delta - V)\dot{V} - \frac{1}{\gamma} \left(\dot{Z} + Z\frac{\dot{D}}{D} \right) \right] = V^2 - V - \frac{(k+1)}{\gamma} Z \tag{14.32a}$$

$$\xi \left[-\dot{V} + (\delta - V)\frac{\dot{D}}{D} \right] = -s - (k - \nu + 3)V \tag{14.32b}$$

$$\xi (\delta - V) \left[\frac{\dot{Z}}{Z} - (\gamma - 1)\frac{\dot{D}}{D} \right] = -s(1 - \gamma) - 2 - [k(1 - \gamma) + 1 - 3\gamma]V \tag{14.32c}$$

where $\delta = -m/n$. The equations in this form are somewhat inconvenient in that they form a simultaneous set of three ordinary differential equations, whose properties are not easy to identify. However, they may be cast into a single ordinary differential equation plus two further equations, which may be integrated by quadrature:

$$\frac{\mathrm{d}Z}{\mathrm{d}V} = \frac{Z \left\{ \begin{array}{l} [2(V-1) + \nu(\gamma-1)V](V-\delta)^2 - \\ (\gamma-1)V(V-1)(V-\delta) - [2(V-1) + \kappa(\gamma-1)]Z \end{array} \right\}}{(V-\delta)[V(V-1)(V-\delta) + (\kappa - \nu V)Z]} \tag{14.33a}$$

$$\frac{\mathrm{d}\ln\xi}{\mathrm{d}V} = \frac{Z - (V-\delta)^2}{[V(V-1)(V-\delta) + (\kappa - \nu V)Z]} \tag{14.33b}$$

$$(V - \delta)\frac{\mathrm{d}\ln D}{\mathrm{d}\ln\xi} = [s - (k - \nu + 3)V] - \frac{[V(V-1)(V-\delta) + (\kappa - \nu V)Z]}{[Z - (V-\delta)^2]} \tag{14.33c}$$

where $\kappa = [s + 2 + \delta(k+1)]/\gamma$.

These equations identify a number of singularities in their solutions, which must be taken into account in finding the solution, as discussed in detail by Sedov (1959). The general solution is found by a numerical solution of the first equation (14.33a) followed by two quadratures for the remaining equations. Fortunately in a number of cases integrals based on the conservation laws may be used to identify one or two integrals of the equations.

Equation (14.33a) contains a number of singularities, where either or both of the numerator and denominator go to zero. Some of these points play a significant role in the problems we investigate. In particular the singularity at the origin $(V = 0, Z = 0)$ where $\xi \to \infty$ plays a critical role in many flows. It is easily seen from equation (14.33a) that in this neighbourhood

$$Z = C V^2 \tag{14.34}$$

This singularity is a node. Since $\xi \to \infty$ it is clear that this singularity represents the flow in the limit of convergence $t \to 0$. The density tends to a constant value.

Singularities play an important role in the examples we investigate. Z is the dimensionless square of the sound speed, and $V - \delta$ the flow speed relative to the self-similar wave $\xi = \text{const}$. The line $Z - (V - \delta)^2 = 0$ represents the sonic condition relative to the wave, and therefore separates the region of subsonic flow from supersonic flow. Transition across this line can only take place at a singularity or in a shock wave, where the adiabatic equations break down.

14.4   Flow Resulting from a Point Explosion in Gas – Blast Waves

In Section 14.2 we considered the expansion of a mass of heated gas into vacuum. We now turn our attention to the situation where a large quantity of energy is suddenly released at a point (or a line or a surface depending on the geometry) into gas at a uniform density.[3] The gas at the point is instantaneously heated to a high temperature and pressure. As a result a shock wave will be driven into the ambient gas followed by an expanding flow. The only dimensional characteristic parameters specifying the flow are the ambient density ρ_1 and the energy E (or energy per unit length or energy per unit area). The ambient gas is assumed to be cold, so that its pressure and temperature may be neglected. The speed of sound in the background gas is therefore zero.

---

[3]This problem was solved by several workers independently: Taylor (1941), von Neumann (1941) and, slightly later, Landau (1959, §99), Stanyukovich (1960, §64) and Sedov (1959, §4.11), Taylor and Landau and Stanyukovich giving numerical solutions and von Neumann and Sedov the analytic form followed here.

The parameters cannot be combined to form a dimensional quantity with the dimensions of length or time. The problem is therefore self-similar. The parameter a may be conveniently taken as the ambient density ρ_1 so that $k = -3$ and $s = 0$. The parameter b not containing mass is E/ρ_1 with dimensions

$$[b] = \left[\frac{E}{\rho_1}\right] = [L]^{(\nu+2)}[T]^{-2}$$

Thus $m = \nu + 2$ and $n = -2$. The value of the self-similar power coefficient is $\delta = -n/m = 2/(\nu + 2)$ and we form the self-similar variable

$$\xi = r\left(\frac{\rho_1}{E t^2}\right)^{1/(\nu+2)} \tag{14.35}$$

At this stage the energy E is a quantity with the dimensions of energy a fraction α of the energy released E_0, α being determined subsequently. The dimensionless forms of the variables are obtained in terms of the parameters ρ_1 and E/ρ_1

$$\xi = \left(\frac{E}{\rho_1}\right)^{-1/2\delta}\frac{r}{t^\delta} \quad v = \frac{r}{t}V(\xi) \quad \rho = \rho_1 D(\xi) \quad p = \rho_1 \frac{r^2}{t^2}P(\xi) \quad \mathfrak{M} = \rho_1 r^\nu M(\xi) \tag{14.36}$$

where $\delta = 2/(\nu + 2)$ and $\mathfrak{M}$ is the mass contained between the surfaces at the origin and at a distance r.

The value of E is taken such that the position of the shock front is determined by a constant value $\xi = 1$ of the self-similar variable

$$R(t) = \left(\frac{E}{\rho_1}\right)^{1/(\nu+2)} t^{2/(\nu+2)} \tag{14.37}$$

The shock velocity $U(t)$ is therefore

$$U(t) = \frac{dR}{dt} = \frac{2}{(\nu + 2)}\left(\frac{E}{\rho_1}\right)^{1/(\nu+2)} t^{-\nu/(\nu+2)}$$

$$= \frac{2}{(\nu + 2)}\left(\frac{E}{\rho_1}\right)^{1/2} R^{-\nu/2} \tag{14.38}$$

Since the energy release is large, the shock is strong. The downstream state behind the shock is therefore described by the approximate Rankine–Hugoniot forms (10.6)

$$\rho_2 \approx \rho_1 \frac{\gamma + 1}{\gamma - 1} \qquad p_2 \approx \frac{2}{\gamma + 1}\rho_1 U^2 \qquad u_2 \approx \frac{2}{\gamma + 1}U$$

u_2 being the velocity behind the shock in a laboratory frame. The downstream density immediately behind the shock is therefore constant and the pressure decreases as

$$p_2 \sim \rho_1 \, U^2 \sim \rho_1 \left(\frac{E}{\rho_1} \right)^{2/(\nu+2)} \sim \frac{E}{R^\nu}$$

This scaling is readily understood as the total energy is a constant of the motion, so that the energy density (per unit volume) scales as E/R^ν. Since it follows from thermodynamics that the pressure scales as the energy density, this result follows.

The gas flow behind the shock is adiabatic, described by equations (14.31). The dimensionless variables take the values behind the shock

$$V_2 = \frac{4}{(\nu+2)\,(\gamma+1)} \qquad D_2 = \frac{(\gamma+1)}{(\gamma-1)} \qquad P_2 = \frac{8}{(\nu+2)^2\,(\gamma+1)} \qquad \xi = 1$$

$$(14.39)$$

The calculation is greatly simplified by the identification of three integrals of the motion, which are essentially expressions of the differential equations expressed in conservation law form:

1. **Mass Integral** Since the flow contains no mass source at the origin, and is symmetric and one dimensional, the mass $\mathfrak{M}$ between any two surfaces moving with the fluid is invariant of the motion. Taking the surfaces at the origin $r = 0$ and at a distance $r(t)$, the mass may be treated as a Lagrangian variable, i.e. one which remains constant for a particular fluid element. The mass contained by the surface $r(t)$ is

$$\mathfrak{M} = \sigma_\nu \int_0^{r(t)} \rho \, r^{(\nu-1)} \, \mathrm{d}r$$

with the area parameter

$$\sigma_\nu = \begin{cases} 2 & \text{in planar Cartesian geometry} \\ 2\pi & \text{in cylindrically symmetric geometry} \\ 4\pi & \text{in spherically symmetric geometry} \end{cases}$$

The mass $\mathfrak{M}$ may be written in dimensionless form as

$$\mathfrak{M} = \rho_1 \, r^\nu \, M = \rho_1 \, \xi^\nu \, t^{\nu\delta} \, M \tag{14.40}$$

Since the total mass of fluid is conserved, the rate of change of the mass inside the surface ξ is balanced by the flow through the surface. The velocity of the surface outwards is simply $\mathrm{d}r/\mathrm{d}t|_\xi$ and the net outwards flow velocity through it therefore $[v - dr/dt|_\xi]$. Consequently

$$\frac{d\mathfrak{M}}{dt} = \sigma_\nu \, \rho \, r^{(\nu-1)} \left(\frac{dr}{dt}\Big|_\xi - v \right) \tag{14.41}$$

Expressing this result in dimensionless form we obtain

$$-\nu \, \delta \, M - \sigma_\nu \, D \, (V - \delta) = 0 \tag{14.42}$$

which is an integral of the equation of continuity, namely the second equation of the set (14.32) due to the conservation of mass. Note that since in general $V < \delta$, M is positive.

We may use this result to identify the state of a specific fluid element as it moves in time. Since $\mathfrak{M}$ is constant it follows from equations (14.40) and (14.42) that

$$\frac{\mathfrak{M}}{\mathfrak{M}_0} = \frac{t^{\nu \delta} \, \xi^\nu \, D \, (\delta - V)}{t_0^{\nu \delta} \, \xi_0^\nu \, D_0 \, (\delta - V_0)} = 1 \tag{14.43}$$

where the subscript 0 refers to values at the initial position of the element at time t_0. The current position follows immediately from $R = R_0 \, (t/t_0)^\delta$.

2. **Adiabatic Integral**   Behind the shock the flow is adiabatic, so that the entropy of a fluid particle is unchanged through the flow. Since the motion is one dimensional we may identify a fluid particle by the mass from the centre to the surface r, namely $\mathfrak{M}$, which therefore acts as a Lagrangian variable. The only variables which can determine the entropy of a particle in the flow are the imposed parameters E and ρ_1 and that identifying the fluid particle $\mathfrak{M}$. Considering the function $p/\rho^\gamma = F(\mathfrak{M}, E, \rho_1)$ representing the value of the entropy, we may use dimensional analysis to determine the form of the function $F(\mathfrak{M}, E, \rho_1)$ as

$$\frac{p}{\rho^\gamma} = C \, \frac{\rho_1 E}{\mathfrak{M}} \, \rho_1^{-\gamma} \tag{14.44}$$

where C is an unknown dimensionless constant. Substituting the dimensionless forms for p, ρ and $\mathfrak{M}$ from equation (14.36), we obtain

$$\frac{P}{D^\gamma} = C M(\xi)^{-1} \xi^{-(2+\nu)} \tag{14.45}$$

This equation is similarly an integral of the third equation of the set (14.32) determined by the conservation of entropy in an ideal flow.

The trajectory of an adiabat is therefore given by $M(\xi) \, \xi^{(\nu+2)}$, where $M(\xi)$ satisfies the mass integral (14.42).

3. **Energy Integral**   If the energy is constant we may also seek an integral of the motion based on energy conservation. It is assumed that the explosion is so strong that the energy released is much larger than that in the background gas, i.e. we can assume the background gas is very cold and neglect its pressure and internal energy. The total energy contained with the surface embraced by the shock wave is therefore constant and equal to E_0, the energy of the explosion. Furthermore, if we consider the energy of the gas within the surface ξ defined by the similarity variable, its value must also remain constant provided the surface lies within the shock-heated gas. The velocity of a point on this surface is simply $v' = \delta r/t$.

In a short time interval δt, energy $\sigma_\nu \, r^{(\nu-1)} \, \rho v (h + \frac{1}{2} v^2) \, \delta t$ flows out through the surface ξ. On the other hand, the volume increases in that time by $\sigma_\nu \, r^{(\nu-1)} \, v' \, \delta t$ and the associated energy increase is $\sigma_\nu \, r^{(\nu-1)} \, v' \, (\epsilon + \frac{1}{2} v^2) \, \delta t$. Equating these two terms we obtain

$$Z = \gamma \frac{P}{D} = \frac{[(\gamma - 1) V^2 (V - \delta)]}{2 [\delta/\gamma - V]} \tag{14.46}$$

This result provides an integral of the set of equations (14.31). Substituting for Z in the second equation of the set (14.33b), we obtain

$$\frac{d \ln \xi}{dV} = \frac{[(\gamma - 1) V^2 - 2(\delta/\gamma - V)(V - \delta)]}{2 V (\delta/\gamma - V) [V \{1 + (\gamma - 1) \nu/2\} - 1]} \tag{14.47}$$

which can be directly integrated by expressing the terms in V in partial fractions, subject to the boundary condition at the shock $\xi = 1$, namely $V = 2\delta/(\gamma + 1)$

$$\frac{r}{R} = \xi = \left[\frac{(\gamma + 1)}{2\delta} V \right]^{-\delta} \left[\frac{(\gamma + 1)}{(\gamma - 1)} \left(\frac{\gamma}{\delta} V - 1 \right) \right]^{-\alpha_1}$$

$$\times \left[\frac{(\gamma + 1)}{(\gamma + 1) - \delta \{2 + \nu(\gamma - 1)\}} \left(1 - \frac{2 + \nu(\gamma - 1)}{2} V \right) \right]^{-\alpha_2} \tag{14.48}$$

where

$$\delta = \frac{2}{\nu + 2} \qquad \alpha_1 = \frac{1 - \gamma}{2(\gamma - 1) + \nu} \qquad \alpha_2 = \frac{(\nu + 2) \gamma}{2 + \nu(\gamma - 1)} \left[\frac{2\nu (2 - \gamma)}{\gamma (\nu + 2)^2} - \alpha_1 \right] \tag{14.49}$$

provide a parametric solution for ξ in terms of V. The ratio Z, namely the square of the sound speed, is found from the energy integral equation (14.46). Eliminating the mass from the adiabatic integral (14.45) using equation (14.42) yields

$$Z = C D^{(\gamma-2)} (V - \delta)^{-1} \xi^{-(2+\nu)} \tag{14.50}$$

the constant C being obtained from the values behind the shock. Hence we obtain the density D, and thus the pressure P to complete the solution. It is clear from equation (14.49) that $V = V_0 = \delta/\gamma$ at $\xi = 0$, i.e. the flow velocity is zero at the centre, and $V = V_2 = 4/\delta\,(\gamma + 1)$ at $\xi = 1$, i.e. immediately behind the shock. Thus V is limited in the range $V_0 \leq V \leq V_2$.

Detailed tabulation of the results can be found in the book by Sedov (1959). The dimensionless energy parameter $\alpha = E/E_0$ is found by evaluating the total energy

$$E_0 = \int_0^R \left(\frac{\rho v^2}{2} + \frac{p}{(\gamma - 1)} \right) \sigma_\nu\, r^{(\nu-1)}\, dr \tag{14.51}$$

which reduced to its dimensionless form becomes

$$\frac{8\,\sigma_\nu}{(\nu + 2)^2(\gamma^2 - 1)} \int_0^1 \left(D\,V^2 + P \right) \xi^{(\nu+1)}\, d\xi = 1 \tag{14.52}$$

Figure 14.3 shows plots of the velocity, density and pressure in a spherical blast wave normalised to their values at the shock wave, as functions of the distance relative to the shock wave. The value of the polytropic constant $\gamma = 1.4$. Some interesting features are immediately apparent. The density at the centre tends to zero, but the pressure is finite. The temperature at the centre is therefore very large, the energy remaining after the explosion. The pressure is nearly constant and the velocity almost linearly dependent on the radius until the shock wave is approached. In this case the parameter $\alpha \approx 1.175$.

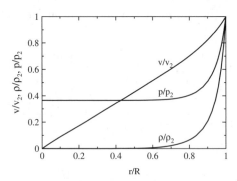

Figure 14.3: Plots of the velocity, density and pressure normalised to their values behind the shock at different distances from the shock for a spherical blast wave with $\gamma = 1.4$.

14.5 Adiabatic Collapse of a Sphere

Consider the case of a sphere of uniform density ρ_0 and pressure p_0 under the action of a piston with an appropriate pressure history. As we have already seen in Sections 10.10 and 14.2.2, strong compression of the ambient gas may be generated if the coalescence of the characteristics can be avoided. An adiabatic self-similar motion of class 1 is easily found. In this case no analytic solution exists. The dimensional constants are $a = \rho_0$ and $b = c_0 = \sqrt{\gamma p_0/\rho_0}$, the sound speed. Clearly $\delta = 1$ and $\kappa = 0$.

The simple value of $\delta = 1$ simplifies the problem. The sonic line becomes $Z = (1 - V)^2$, which the flow must avoid during the collapse. The solution involves three singular points.[4] The solution starts at the point $V = 0$, $Z = 1$ lying on the sonic line, which is a singularity A (a focus) where the solutions touch the Z axis.. The point of convergence O at the origin where $V = 0$, $Z = 0$ and $\xi \to \infty$ is a focus where all the solutions touch the V axis. The point B, $V = 2/(3\gamma - 1)$, $Z = 3[(\gamma - 1)/(3\gamma - 1)]^2$, is a saddle point. This limits the solutions which can start from A at $(0, 1)$ and reach O at $(0, 0)$ to Mach numbers $M < M_{0\text{crit}} \approx 1.30$ near O.[5] Figure 14.4 shows the typical trajectories of the collapse for three different values of the Mach number at convergence. The approach and reversal near the saddle point B can be clearly seen. The larger the Mach number M_0 at convergence, the closer the integral curve approaches

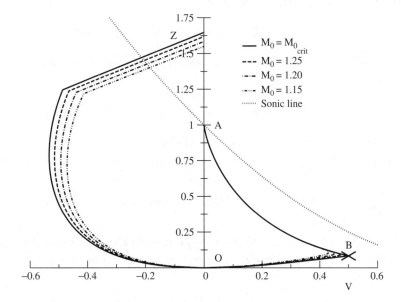

Figure 14.4: The trajectory of the collapse of a uniform sphere and the subsequent shock bringing it to rest. The singularities are shown together with the sonic line. The calculation is for a gas with $\gamma = 5/3$, and convergence Mach numbers M_0 of 1.15, 1.2, 1.25 and $M_{0\text{crit}}$.

---

[4] The singular points are identified using Sedov's notation.
[5] This value may be compared with the value of $\sqrt{3}$ in homogeneous collapses, equation (14.22).

the saddle point B, and the more rapid the subsequent motion. In the early stages of compression the flow is only weakly dependent on M_0 before the neighbourhood of B is reached. In the later stages after convergence, the weak reflected shock bringing the gas to rest can be seen crossing the sonic line.

Measuring the time from the instant of convergence, the flow divides into two separate phases:

- The collapse phase when the time $t < 0$, and we must take the self-similar variable $\xi = r/\{c_0(-t)\}$. The dimensionless velocity $V > 0$.

- The post-collapse phase when the time $t > 0$, and the self-similar variable $\xi = r/\{c_0 t\}$. The flow continues to move inwards, $V < 0$, until meeting the shock reflected at the collapse point, which brings the flow to rest.

The flow is continuous through the convergence where $\xi \to \infty$. Consequently we have in the neighbourhood of the point $Z = V = 0$, $Z_- = Z_+$, $V_- = -V_+$ and $D_- = D_+$, where the subscripts indicate values at equal values of ξ, before and after the convergence.

At convergence it is easily seen that

$$Z \sim V^2 \qquad V\xi \sim \text{const} \qquad \text{and} \qquad D \sim \text{const} \qquad (14.53)$$

Therefore, transforming back into laboratory variables we have near convergence

$$M = v/c \quad v = \text{const} \quad c = \text{const} \quad \rho = \text{const} \quad \text{and} \quad p = \text{const} \qquad (14.54)$$

and the Mach number $M = V/\sqrt{Z}$ is well defined.

The dimensionless forms appropriate to this problem are easily identified

$$\xi = \frac{r}{c_0 t} \quad v = \frac{r}{t} V(\xi) \quad \rho = \rho_1 D(\xi) \quad p = \rho_1 \frac{r^2}{t^2} P(\xi) \quad \mathfrak{M} = \rho_1 r^\nu M(\xi)$$
$$(14.55)$$

The derivation of the mass integral, equation (14.42), given earlier, may be applied with $\delta = 1$. The adiabatic integral is simply identified as

$$\frac{p}{\rho^\gamma} = \rho_1{}^{-(\gamma-1)} c_0{}^{-2} \frac{P}{D^\gamma} \xi^2 = C' \qquad (14.56)$$

Substituting in terms of Z we obtain

$$Z = C\,D^{(\gamma-1)}\,\xi^{-2} \qquad (14.57)$$

where the constant $C = 1$ is determined by the initial state of the gas at A. This result allows a direct calculation of the density from the sound speed parameter

Z. The calculation therefore proceeds as an integration of equation (14.33a) to obtain V as a function of Z starting at O with prescribed Mach number. A quadrature using equation (14.33b) gives $\ln \xi$ followed by the application of the adiabatic integral to give D.

The integration is continued into the post-convergence phase until the upstream conditions for a shock bringing the flow to rest are reached, i.e. the values of V_u and Z_u are such that the downstream value $V_d = 0$ is given by the Rankine–Hugoniot equations

$$V_d = 1 + \left\{ (V_u - 1) + \frac{2}{(\gamma + 1)} \frac{[Z_u - (V_u - 1)^2]}{(V_u - 1)} \right\} = 0 \qquad (14.58)$$

The final state of the gas is given by the value of Z_d:

$$Z_d = \left(\frac{\gamma - 1}{\gamma + 1} \right)^2 \left\{ (V_u - 1)^2 + \frac{2 Z_u}{(\gamma - 1)} \left[\frac{2\gamma}{(\gamma - 1)} (V_u - 1)^2 - Z_u \right] \right\} \qquad (14.59)$$

The compression generated by these flows depends on how closely the solution approaches the singularity B (Figure 14.5) The Mach number at convergence is also limited by the nature of the saddle point to a value $M_{0_{crit}} \approx 1.292\,12$. As expected the compression is weaker for smaller values of M_0.

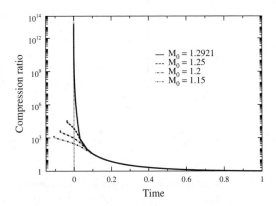

Figure 14.5: Compression ratio of the gas during the collapse of a sphere as shown in Figure 14.4 for Mach numbers M_0 at convergence of 1.15, 1.2, 1.25 and $M_0 = 1.2911$. The time is measured as a ratio of the collapse time.

The compression may be considered to be driven by a piston whose radius is $R(t)$. Since the flow velocity at the piston must equal the piston speed, the position of the piston or alternatively the time into the compression may be simply calculated from equation (14.43) or by simple quadrature from

$$\frac{d\xi}{\xi} = \frac{dr}{r} - \frac{d(-t)}{(-t)} = \left(\frac{-v}{r/(-t)} - 1 \right) \frac{d(-t)}{(-t)} = (V - 1) \frac{dt}{t} \qquad (14.60)$$

since $V = v/[r/(-t)]$ is taken to be positive, as the motion is inward. Figure 14.6 shows the time history of the dimensionless flow parameters at the piston for various values of the Mach number at convergence for a gas with polytropic index $\gamma = 5/3$. As the Mach number at convergence is increased towards the limiting value $M_{0\text{crit}}$, it can be seen that a correspondingly larger part of the motion is spent in the neighbourhood of the singular point B. In this region both Z and V have nearly constant values approximately equal to those at the singularity B, namely $Z \approx 1/12$ and $V \approx 1/2$. The Mach number is therefore $\sqrt{3}$ in this region. Making use of this result we may derive the solution for the behaviour of the flow at the limit of convergence for gas with $\gamma = 5/3$ as follows.

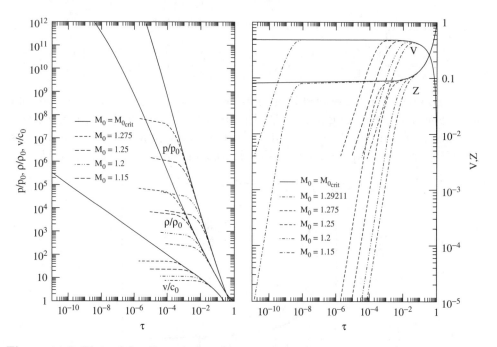

Figure 14.6: Plots of the dimensionless flow parameters: pressure ratio p/p_0, compression ratio $D = \rho/\rho_0$ and inwards velocity v/c_0, and Z and V at the piston as functions of the fractional collapse time $\tau = c_0(-t)/R$ into the collapse ($\gamma = 5/3$).

The values of the laboratory variables at the piston $R = 1$ are therefore

$$v = \tfrac{1}{2}\alpha\, c_0\, \tau^{-1/2} \qquad \rho = 2\,\alpha^{-3/2}\, \rho_0\, \tau^{-3/2} \qquad p = 2^{5/3}\,\alpha^{-5/2}\, p_0\, \tau^{-5/2} \quad (14.61)$$

where $\tau = c_0(-t)/R$ is the time to collapse as a fraction of the total collapse time. These scalings are in excellent agreement with the values shown in Figure 14.6.

An integral solution, given in problem #49, can be shown to be an exact solution of the equations of motion, and passes through the singular point $V = 1/2$, $Z = 1/12$. It is therefore the limit solution leading to infinite compression. However, it is unable to satisfy the initial conditions $V = 0$, $Z = 1$, which must therefore lie on the other branch of the solution through the saddle point B (Figure 14.4).

From Figure 14.6 it can be seen that once the flow is far from the singularity B, the velocity of the piston becomes nearly constant and the variable $V \sim \xi^{-1}$. In this case $Z \approx (M_0 V)^2 \sim \xi^{-2}$. The density $\rho \sim$ const and pressure $p \sim$ const scalings follow immediately.

14.6   Convergent Shock Waves – Guderley's Solution

A spherical shock wave is generated, converging to the centre in a system with spherical symmetry ($\nu = 3$). The shock is initially generated at much larger distances from the centre than those at which we are observing, e.g. by a spherical piston. Converging into the centre, the energy becomes concentrated at the shock, and the 'memory' of its initiation becomes lost. The flow achieves some form of limiting process in which the structure of the wave repeats itself. Since the initial conditions have been 'lost', the only external parameters are those of the ambient gas into which the shock propagates, namely the density ρ_1 and the pressure p_1 of the ambient gas. As the shock radius becomes very small, the strength of the shock increases so that it may be treated as strong and both the initial pressure p_1 and enthalpy h_1 may be neglected in the calculation of the flow behind it. The only dimensional terms remaining in the problem are the background density ρ_1 together with the independent variables r and t. Referring to Section 14.3, the dimensional powers of the parameter a are $k = -3$ and $s = 0$; b does not exist and m and n therefore undefined. Since the problem is self-similar, it must be of type 2 where we cannot a priori determine the value of the scaling power δ, but must calculate it from the condition that the solution is well defined.

The shock collapses towards the centre becoming progressively stronger for times $t < 0$, with the flow behind the shock moving inwards. At the instant of collapse at time $t = 0$, the shock converges on the centre. Following convergence, a reflected shock moves outwards through already shocked gas, which is still moving inwards, bringing it nearly to rest for $t > 0$. The final state of the gas is calculated from the conditions following the reflected shock (Guderley, 1942).

During the phase of convergence $(t < 0)$ we seek to find the state of the flow at and behind the shock front as functions of the self-similar variable:

$$\xi = \frac{Ar}{(-t)^\delta} \tag{14.62}$$

where δ is an unknown constant and A a dimensional constant whose value (if required) may be found from knowledge of an initial shock front location (radius R at time $-t$). The flow is found as before from the strong shock jump conditions and the adiabatic gas flow equations (14.31). The shock trajectory is described by the self-similar form with a constant value of ξ, say $\xi = 1$. The shock velocity is

$$\dot{R} = -\delta \frac{R}{(-t)}$$

the $-$ sign reflecting the inward motion of the shock. Since the velocity of the flow into the shock in the shock frame is equal to the velocity of the shock into the stationary gas in front of it, the parameters behind the shock are

$$V_2 = \frac{2}{(\gamma+1)}\delta \qquad D_2 = \frac{(\gamma+1)}{(\gamma-1)} \qquad Z_2 = \frac{2\gamma(\gamma-1)}{(\gamma+1)^2}\delta^2 \qquad \xi = 1 \tag{14.63}$$

as in the previous case, equation (14.39). After passing through the shock at $\xi = 1$, the fluid flow is continuous outwards to infinity, i.e. for increasing ξ as $\xi \to \infty$ where $V \to 0$ and $Z \to 0$, no additional discontinuities (such as shocks) occurring. It is easily seen that the flow behind the shock wave is subsonic, since $Z - (V - \delta)^2 > 0$. However, at the point of convergence $(0,0)$ the flow is supersonic. The flow must therefore pass through the sonic line at a singularity [P3][6] as described below. This motion is described by equations (14.33), which are subject to the requirement of continuity throughout the solution, which in turn determines the value of the power δ.

Following the same arguments as those leading to equation (14.45), it may be shown that the adiabatic integral may be generalised for a general power δ as follows:

$$\frac{p}{\rho^\gamma} = \frac{1}{\rho_1^{(\gamma-1)}}\frac{P}{D^\gamma}\frac{r^2}{t^2} = \frac{1}{\rho_1^{(\gamma-1)}}\left[\frac{\rho_1 r^\nu}{\mathfrak{M}}\right]^{2(1-\delta)/\nu\delta}\left[\frac{r}{t^\delta}\right]^{2/\delta}$$

$$= \frac{B}{\rho_1^{[(\gamma-1)-2(1-\delta)/\nu\delta]}}M^{-2(1-\delta)/\nu\delta}\xi^{-2} = \text{const}$$

where B is an unknown constant with appropriate dimensions. Substituting for the dimensionless mass from equation (14.42) we obtain

$$Z = C\,D^{[(\gamma-1)-2(1-\delta)/(\delta\nu)]}\,(\delta - V)^{-2(1-\delta)/(\delta\nu)}\,\xi^{-2/\delta} \tag{14.64}$$

---

[6]We use Guderley's numbering of the singularities for this problem.

The constant C is evaluated at the downstream state of the appropriate shock. It is easily shown that if $\delta = 2/(\nu + 2)$, appropriate to the blast wave, equation (14.45) is obtained.

As we have seen, the general equations of one-dimensional fluid motion may be cast into the dimensionless forms (14.32). Solving the three simultaneous equations, these equations may be reduced to the set (14.33). Using Cramer's rule the determinantal forms of the solutions are written as

$$\frac{dZ}{d\ln\xi} = \frac{\Delta_1}{\Delta} \qquad \frac{dV}{d\ln\xi} = \frac{\Delta_2}{\Delta} \qquad \frac{d\ln D}{d\ln\xi} = \frac{\Delta_3}{\Delta} \qquad (14.65)$$

where the determinants Δ_1, Δ_2 and Δ_3 are formed from the coefficients of the terms in equations (14.32). The determinant is

$$\Delta = \begin{vmatrix} -\dfrac{1}{\gamma} & (\delta - V) & -\dfrac{Z}{\gamma} \\ 0 & -1 & (\delta - V) \\ \dfrac{1}{Z} & 0 & -(\gamma - 1) \end{vmatrix} = 1 - \frac{(V-\delta)^2}{Z} \qquad (14.66)$$

Suppose that $\Delta = 0$; that is, on the parabola $(V - \delta)^2 = Z$ corresponding to sonic flow relative to the surface $\xi = \text{const}$, the process of moving across this line is the sonic transition from subsonic flow (behind the shock) to supersonic (towards infinity). If Z, V and D are continuous crossing this line, as we require, it follows that $\Delta_1 = \Delta_2 = \Delta_3 = 0$ where the solution crosses this line. Referring back to equations (14.63), we see that behind the shock $V_2 = [2/(\gamma + 1)]\,\delta$ and $Z_2 = [2\gamma(\gamma - 1)/(\gamma + 1)^2]\,\delta^2$, and at infinity $V(\infty) = Z(\infty) = 0$. The solution must therefore intersect the parabola at some point, the flow initially being subsonic and becoming supersonic. The condition of singularity, namely that the derivatives are finite, determines the allowed value of δ, which is found by trial and error integrating equation (14.33a) up to the parabola from the shock. Integration of equations (14.33) is straightforward using standard numerical packages. An initial starting value can found from approximations using expressions from a modified set of equations derived from (14.33).[7] At the singular point, the numerator and denominator of equation (14.33b) must both be zero

$$(V - \delta)^2 - Z = 0 \qquad \text{and} \qquad Z\left[2(\delta - 1)/\gamma + \nu V\right] - V(V - 1)(\delta - V) = 0$$

which has three roots

$$\overline{V}_{\pm} = \frac{\gamma\nu\delta - \gamma + 2(1 - \delta) \pm \sqrt{[\gamma\nu\delta - \gamma + 2(1 - \delta)]^2 + 8\gamma(\nu - 1)\delta(\delta - 1)}}{2\gamma(\nu - 1)}$$

$$\overline{V}_0 = \delta \qquad (14.67)$$

---

[7] To illustrate this process it is convenient to use (14.33b).

It can be shown that $\overline{V}_-$ and $\overline{V}_0$ are not possible singularities, so that $(\overline{V}_+, \overline{Z}_+)$ is the required point where $\overline{Z}_+ = (\overline{V}_+ - \delta)^2$. Unfortunately, without knowing δ we cannot determine $\overline{V}_+$ and vice versa. If the surd in equation (14.67) is zero, the resulting values of V and Z lie on the parabola but not on the integral curve from the shock. However, if this is used to give an initial estimate of δ it is remarkably accurate (Stanyukovich, 1960, p.522). Finally direct integration up to the point where either Δ_1 or Δ_2 changes sign is used iteratively to find the value of δ.

Convergence $(t = 0)$ occurs at the singular point [P4] where $V = Z = 0$, $\xi = \infty$. Although $V = 0$, the inward flow is not brought to rest, but continues inwards until a reflected shock wave generates a slow outward motion which decreases with time and eventually comes to rest. The similarity variable now changes $\xi \to Ar/t^\delta$. The variables Z and D are unchanged by passage through the convergence point at $t = 0$. However, as a consequence of the redefinition of ξ, the continuing inward flow corresponds to values $V < 0$. Equations (14.32) are still valid.

The reflected shock cannot be considered strong, so that the full Rankine–Hugoniot equations must be used. Expressed in terms of the variables V, Z and D these are

$$D_u (V_u - \delta) = D_d (V_d - \delta)$$

$$(V_u - \delta) + \frac{Z_u}{\gamma (V_u - \delta)} = (V_d - \delta) + \frac{Z_d}{\gamma (V_d - \delta)} \qquad (14.68)$$

$$(V_u - \delta)^2 + \frac{2Z_u}{(\gamma - 1)} = (V_d - \delta)^2 + \frac{2Z_d}{(\gamma - 1)}$$

where subscript u refers to the upstream flow before passage through the shock (inwards) flow and d to the downstream after.

The final flow condition is determined by the requirement that $\xi \to 0$ regularly, which requires a transition through a further singularity [P6] where $V = \kappa/\nu$, $Z \to \infty$. Thus the flow through the reflected shock must be matched to the solution from [P6]. As described by Guderley this is accomplished by progressively integrating the upstream solution to calculate the upstream values of V and Z and hence the downstream values, at each step progressing the downstream solution to the new value of Z until the values of V match. The solution for $\gamma = 1.4$ is plotted in Figure 14.7.

An important quantity is the compression ratio at convergence (ρ_4/ρ_1) and behind the reflected shock (ρ_R/ρ_1), whose values are given in Table 14.1. The most characteristic observable feature is the rapid increase of the total compression behind the reflected shock as the value of γ for the gas is reduced.

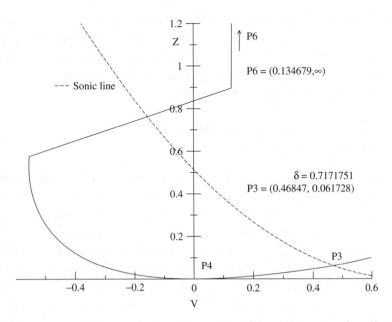

Figure 14.7: Plot of the (V, Z) history of the flow behind a spherical shock wave in a gas with $\gamma = 1.4$.

Table 14.1: Compression ratio for a collapsing shock.

γ	δ	ρ_4/ρ_1	ρ_R/ρ_1	γ	δ	ρ_4/ρ_1	ρ_R/ρ_1
5/3	0.688 38	9.55	30.92	1.6	0.694 19	11.06	41.11
1.5	0.704 43	14.39	69.87	7/5	0.717 18	20.07	143.06
1.3	0.733 78	31.27	402.37	9/7	0.736 65	33.76	439.65

The differential equation for an incoming characteristic C_-, namely $dr/dt = v - c$, is transformed in dimensionless units to

$$\frac{dr}{dt} = V\frac{r}{t} + \sqrt{Z}\frac{r}{t} = \frac{\xi \dot{R}}{\delta}\left(V + \sqrt{Z}\right)$$

where R is the radius of the shock wave. Since the sound speed c is always positive and the time negative during convergence, it is necessary to set $c = -r/t \sqrt{Z}$. If we consider the characteristics which pass through the line ξ_3 in the (r, t) plane corresponding to the singularity [P3], then

$$\sqrt{Z(\xi_3)} - \delta + V(\xi_3) = 0$$

Hence, substituting, we find the slope of the C_- characteristics to be

$$\frac{dr}{dt} = \frac{\xi_3 \dot{R}}{\delta}\left[V(\xi_3) + \sqrt{Z(\xi_3)}\right] = \xi_3 \dot{R}$$

which is the slope of the line ξ_3 itself. Therefore either the line ξ_3 is a characteristic C_- or it is an envelope of them. In fact it must be a characteristic, and it is therefore the C_- characteristic converging self-similarly to the centre corresponding to the singular point ξ_3. Thus the characteristic $C_-(\xi_3)$ bounds the region of influence of the incoming flow.

14.6.1 Compression of a Shell and Collapse of Fluid into a Void

The collapse of fluid into a void (bubble) or equivalently the driven collapse of a shell may also be tackled by the method outlined above for the solution of the convergent shock wave. In contrast to the collapse of a sphere, Section 14.5, the solution is started at the point $(V = 1, Z = 0)$, corresponding to a cold shell with an incoming velocity. The solution involves a problem of type 2 in which the exponent δ is determined by the nature of the solution and the singularity in the governing differential equation at the crossing of the sonic line. However, in contrast to the converging shock wave, the value of δ is not unique – at least for values of $\gamma \lesssim 8.47$. This leads to a set of solutions for different Mach numbers at convergence as for the sphere treated in Section 14.5. Only one of these solutions is analytic in that the gradient through the singularity (a node) is continuous. In the case of $\gamma = 5/3$ this occurs at a value of $\delta = 0.9396$. In the case of a collapsing bubble in water, treated by Hunter (1960), where $\gamma \approx 7$, the range of δ is very small and the solution approximately unique. However, for a gas shell with $\gamma = 5/3$, the range is quite large. The solution is consequently more complex than the examples treated earlier, and will not be pursued further. For more details, the reader is referred to the paper by Brushlinskii and Kazhdan (1963), where the full solution is discussed.

The collapse of shells is more easily treated by direct simulation, which allows greater flexibility in the conditions imposed externally on the collapse. This may lead to non-adiabatic collapse and the formation of shock waves if the external conditions are badly applied. To this end we may consider two distinct modes of collapse:

1. **Initially imposed velocity** Consider a shell of uniform density with inner and outer radii 1 and 1.1 units respectively, imploding with a uniform velocity 1 unit. The shell is assumed to be very cold and the pressure consequently small. The fluid particles therefore retain their initial velocity and the width of the shell remains constant. The density of a fluid particle, whose initial radius is R, therefore increases as R^2/r^2 (Figure 14.8). This compression continues until the inner surface reaches the centre, when an outgoing shock is formed. Near the centre the r^{-2} density

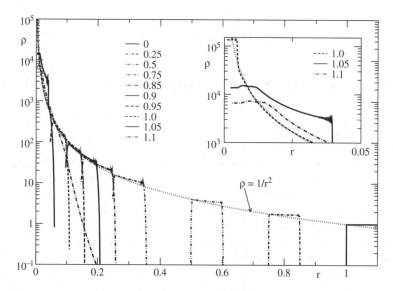

Figure 14.8: Convergence of a simple cold shell of uniform density 1 unit moving inwards with velocity 1 unit. The spatial distribution of density is shown at different times before and after convergence which occurs at approximately 1 unit of time. The density follows a $1/r^2$ dependence during the collapse.

scaling fails as the initially small temperature is adiabatically increased to give rise to a variation in velocity. After collapse the outgoing shock precedes a rapid expansion. It can be seen that the compression is nearly adiabatic except very close to the centre, where a shock is formed. In this region close to the centre, the self-similar solution described above becomes applicable. However, the complexity of the latter solution outweighs any value it may have for all practical purposes.

2. **Pressure driven**   As we have seen (Section 14.2.2.2) it is possible to adiabatically compress a shell from rest by the application of a piston at the outside surface. However, the initial density profile and the temporal development of the pressure must be carefully matched if shocks are to be avoided. In practice, shells are less sensitive to pressure pulse mismatch than solid spheres and therefore easier to drive to high compression. Once the transit of the initial characteristics is complete, the density profile settles to a form similar to that of the self-similar motion, Figure 14.2.

In any realistic collapse, both these effects must occur together. The shell must be initially driven to attain the inward velocity. Similarly if the shell is driven, the effects of radial convergence and inward momentum will increase the compression near the centre. A judicious choice of the temporal profile of the applied pressure can clearly improve the compression achievable.

Case study 14.I The Fluid Dynamics of Inertial Confinement Fusion

14.I.i Basic principles

In this section we briefly discuss the underlying physics of inertial confinement fusion (ICF). This is essentially a problem of applied hydrodynamics and involves many topics covered earlier. Plasma physics also plays a role, but is generally subordinate to the hydrodynamics at the introductory level outlined here.

Inertial confinement fusion relies on the ability of a pellet of fusion material (usually a 50:50 mixture of deuterium and tritium, DT) to undergo fusion before disassembling by expansion. This requires the fluid to be at high temperature $\gtrsim 10\,\mathrm{keV} \approx 10^7\,\mathrm{K}$. Making use of equation (14.8) we may estimate the disassembly time of a sphere of uniformly heated plasma of sound speed c and radius R from the time taken for the density to decrease by half as

$$\tau \approx R/3\,c \qquad (14.69)$$

since the fluid is a plasma with equal numbers of ions and electrons under these conditions $(E/M = 3c^2)$.

The rate at which fusion reactions occur in the plasma is determined by the average of the rate product $\langle \sigma v \rangle$, where σ is the cross-section for fusion and v the random thermal velocity, over the velocity distributions of the deuterium and tritium ions. It is found that in the temperature range $20 - 50\,\mathrm{keV}$, the ratio of $\langle \sigma v \rangle$ and c is approximately constant and consequently the fractional burn-up of the fusion constituents is

$$f \approx \rho R/(6 + \rho R) \qquad (14.70)$$

taking into account the depletion during the burn, where the density ρ is measured in g/cm^3 and the radius R in cm. In practice, depletion limits the fractional burn to about 35%. To achieve these values requires a density/radius product $\rho R > 0.3\,\mathrm{g/cm}^2$. The burn time can be increased by restricting the expansion of the fuel by enclosing it in a heavy shell, known as a *tamp*. However, this requires considerable input energy and is consequently inefficient, although other design constraints may introduce a measure of tamping.

The thermonuclear gain is defined as

$$g = \frac{\text{Fusion energy yield}}{\text{Initial thermal energy}} \qquad (14.71)$$

A simple uniformly heated sphere has a gain of only about 50, which is insufficient to overcome the losses inherent in the pumping power source and generator. Typically values $\sim 10^4$ are required for an effective power plant.

The parameter ρR is a measure of the collision probability for particles within the plasma body. If a region with $\rho R > 0.3$ is generated at the centre of the pellet, escaping

α-particles and photons suffer collisions and transfer energy to heat a colder surround of DT mixture to fusion temperatures and thereby induce a burn, which propagates through the pellet, namely *hot spot ignition*. In this case only the central region need be heated to fusion temperatures, the outer parts remaining relatively cold. The gain may thus be greatly increased.

The energy released by burning 1 g of DT fuel is 3×10^5 J equivalent to about 75 Mt of TNT. The largest amount of energy which can be expected to be handled routinely and safely is about 100 MJ (25 kg of TNT). This corresponds to a pellet mass of 0.3 mg of fuel. Since we have argued that we require a ρR product of not less than 0.3 g/cm^2 and $M \approx \rho R^3 = (\rho R)^3/\rho^2$, we see that the density of the pellet must be $\approx 10^2$ g/cm^3 or a compression over liquid DT of about 5×10^2. Under these conditions the pressure in the hot spot is about 10^{12} atm and even in the cooler outer regions about 10^9 atm. These values are clearly far in excess of those achievable mechanically.

14.I.i.a Hydrodynamic compression

We have seen that it is possible to achieve high compression during the collapse of spheres and shells. The compression achievable in a collapsing shock is only about 30, Table 14.1. On the other hand, if the collapse is achieved adiabatically very large compression may be obtained. This is a consequence of the fact that the entropy generated in a shock leads to a temperature rather than a density increase, by which the pressure is raised. The generation of shocks in the collapsing pellet therefore leads to a reduction in compression. A purely adiabatic collapse leads to a cold core, which is ideal for the outer cold region into which the burn may propagate. However, it is essential to design a central hot spot, e.g. by generating a shock propagating into this region by a suitable design of the drive pulse. Although the original proposal for inertial fusion was based on spheres, it was quickly realised that multi-layered shells were a more effective approach, and have been used subsequently. It can be shown that collapsing spheres are much more sensitive to deviations from the ideal pressure pulse than shells (a consequence of the fact that significant compression can occur during the coasting compression phase) and that, provided it is smooth, the profile of the applied pulse is not too critical. In addition the energy required to compress a sphere is significantly greater than that needed for a shell. Current designs for inertial fusion targets are all based on multiply layered shells.

A typical fusion pellet consists of a thin inner layer of frozen DT on the inner shell. The outer layers of the shells are in two parts. An inner layer of relatively heavy material acts as a shield preventing X-rays or fast electrons penetrating the fuel and raising its temperature to prevent high compression being achieved. An outer layer of lighter material is used to generate the pressure. The outer layers also act as a tamp, helping to increase the disassembly time of DT fuel.

The pellet and drive pulse must be carefully engineered to overcome a number of problems, essentially related to uniformity and stability. The shells and the drive beams must be uniform to a few per cent to allow a uniform collapse and high compression. Hydrodynamic instabilities, principally the Rayleigh–Taylor instability (Section 4.3.2) at interfaces undergoing acceleration, destroy the uniformity of the collapse, and therefore present a serious problem and must be limited or avoided.

In case study 13.I.i we showed that large pressures can be generated by the ablation of material away from a surface. This is essentially the same as the rocket effect where the momentum transfer from the escaping burnt fuel generates the pressure. In principle any source of heat rapidly deposited at the surface generates such a pressure, which may be used to compress the fuel, provided the rate is sufficiently large. However, the pressure generated by these methods is not sufficiently large than needed to balance the pressure generated in the burning fuel. The required pressure multiplication is the result of two factors, namely the convergence of the collapsing shell and the rapid release of momentum accumulated over a long drive period. Several methods have been proposed to provide the drive source:

1. **Direct drive laser heating** In this case the laser directly irradiates the surface of the outer layer of the shell. From equation (13.25) we see that higher pressure is obtained with higher critical density, i.e. shorter wavelength lasers. As a result the recent design of direct drive fusion is based on laser pulses of 0.35 μm wavelength. Due to the fact that the drive is applied at the critical density, which is significantly less than solid, the interface is susceptible to the Rayleigh–Taylor instability. Direct drive also suffers from a number of problems associated with plasma instabilities and care must be taken with the magnitude of the pulse intensity. These effects are also mitigated by the use of short-wavelength lasers.

2. **Indirect drive laser heating** As we noted above, higher pressures are achieved with shorter wavelength radiation. This raises the possibility of using soft X-rays for the drive, where the absorption is due to photo-ionisation and the critical density much greater than solid. The Rayleigh–Taylor instability is thereby avoided. By placing the pellet in a hot enclosure, namely a *hohlraum*, filled with thermal black body radiation, the uniformity of illumination may (in principle) be improved. A major disadvantage of this approach is the inefficiency resulting from the energy required to generate the X-rays.

3. **Light and heavy ion heating** An alternative approach which is less well developed is to use either light or heavy ions, or electrons generated externally, but by different means; however, the necessary generators of suitable ion beams are not currently available.

Problems

Problem 1: Show that in two dimensions the Lagrangian equations for the Jacobian and the derivative may be simplified when $\mathbf{r} = (x, y)$ by introducing the normal vector $\mathbf{s} = (y, -x)$ to give

$$J = \frac{\partial \mathbf{r}}{\partial \lambda} \cdot \frac{\partial \mathbf{s}}{\partial \mu}$$

$$\nabla f = \frac{1}{J} \left\{ \frac{\partial \mathbf{s}}{\partial \mu} \cdot \frac{\partial f}{\partial \lambda} - \frac{\partial \mathbf{s}}{\partial \lambda} \cdot \frac{\partial f}{\partial \mu} \right\}$$

where (λ, μ) are the invariant Lagrangian parameters. Calculate the gradient and divergence in this geometry.

Extend this result to consider an axisymmetric system $\mathbf{r} = (\varrho, z)$ and construct the divergence using the results for two dimensions.

Problem 2: A simple device is used to measure the velocity of flow, in particular the air speed in aircraft. The tube has a blunt end such that the flow is brought to rest at the tip (stagnation point) (Figure P.1). In the free stream, which is re-established down the outside of the tube, the flow velocity returns to the incoming (free stream) flow speed U. Measurement of the pressure between these points yields the flow speed.

Figure P.1: Sketch of the arrangement of a simple Pitot tube to determine the velocity of the flow U, by measuring the pressure difference $p_1 - p_2$.

Assuming the flow may be treated as incompressible, find the relationship between the flow speed and the pressure difference

Introductory Fluid Mechanics for Physicists and Mathematicians, First Edition. Geoffrey J. Pert.
© 2013 John Wiley & Sons, Ltd. Published 2013 by John Wiley & Sons, Ltd.

Problem 3: A venturi is a constriction in a pipe (Figure P.2). Assuming steady incompressible flow, uniform across the duct, derive an expression for the pressure difference between the flow in the pipe and through the constriction. Hence show how the device may be used to measure flow speed.

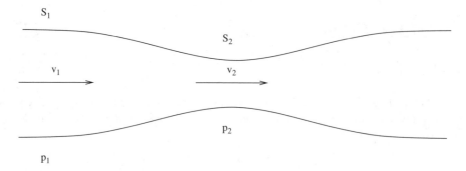

Figure P.2: Sketch of a simple venturi to show the constriction in which the flow velocity is measured.

Problem 4: Consider a large reservoir at a height h connected through a pipe to an outlet with a tap, which can be either on or off. The pressure at the reservoir p_1 is atmospheric, p_a. From the equation of continuity, the flow velocity at the reservoir is approximately zero $v_1 \approx 0$ as the cross-section of the surface is much larger than that of the tap. Hence Bernoulli's equation takes the form

$$\frac{1}{2}\overset{0}{\cancel{v_1^2}} + \overset{\frac{p_a}{\rho}}{\cancel{\frac{p_1}{\rho}}} + \rho g h = \frac{1}{2}v_2^2 + \frac{p_2}{\rho}$$

Show that if the tap is open and the exit pressure is atmospheric, the exit flow velocity is equal to the free fall value. Alternatively, if the tap is closed, and the exit velocity is zero, show that the pressure across the tap is equal to the hydrostatic head.

Problem 5: Two identical jets of incompressible fluid meet at an angle 2θ. Assuming the flow is ideal and laminar during the interaction, show that the resulting flow consists of two jets, one forward going and one backward going. Calculate the mass flow in each.

Problem 6: A Pelton water turbine consists of a series of hemispherical buckets arranged evenly around the rim of a wheel. A jet of water is injected from a nozzle into the centre of the buckets so as to rotate the wheel. The water is 'reflected' in the bucket so as to form a jet lying outside the incoming flow. The buckets are spaced so that one is always within the jet. The water, after striking the bucket and being reflected, leaves in a narrow range of angles, at an average value θ, in the opposite direction to the incoming flow. The incoming flow is tangential to the rim of the wheel. The change of momentum due to the reversal of the flow direction contributes a net force on the wheel rim.

Noting that, in the frame of reference of the stationary bucket, the water flow is quasi-steady, then Bernoulli's theorem may be applied to relate the incoming and outgoing flows. Assuming no viscous loss in the bucket, calculate the force on the bucket in terms of the density and cross-section of the jet, the incoming and outgoing speeds and the angle of deflection θ' in the frame of the buckets. Transform back into the laboratory frame and calculate the torque on the wheel and the power developed.

Correct the power for the time each bucket remains in the flow. Hence calculate the maximum efficiency of the device.

Problem 7: The combination of a two-dimensional uniform flow with velocity U and the flow from a combination of a line source and a line sink of strength m separated by a distance d (Figure 2.1) yields the cylindrical form of the Rankine oval. Show that the streamfunction at a point P may be written as

$$\psi = \frac{m}{2\pi}(\theta_- - \theta_+) + U\,r\,\sin\theta \qquad\qquad \text{(P.1)}$$

where r and θ are the polar co-ordinates of P relative to the mid-point between the sources, and $(r_-,\ \theta_-)$ and $(r_+,\ \theta_+)$ the position of P relative to the sink and the source.

Hence find the form of the oval.

Show that as the separation of the source and sink becomes small the flow around a circular cylinder is obtained.

Problem 8: Define the elliptic co-ordinate system

$$x = c\cosh\xi\cos\eta \quad y = c\sinh\xi\sin\eta \qquad\qquad \text{(P.2)}$$

so that in complex form

$$z = x + \imath y = c\,\cosh(\xi + \imath\eta) = c\,\cosh\zeta \quad \{\approx \tfrac{1}{2}\exp(\zeta) \quad \text{as} \quad \zeta \to \infty\}$$

where $\zeta = \xi + \imath\eta$ is the complex form of the elliptic co-ordinates $(\xi,\ \eta)$. Thus there is a conformal mapping $z \rightleftarrows \zeta$.

Show that the lines $\xi = $ const are a family of ellipses

$$\frac{x^2}{c^2\cosh^2\xi} + \frac{y^2}{c^2\sinh^2\xi} = 1 \qquad\qquad \text{(P.3)}$$

with semi-major axis $a = c\cosh\xi$ and semi-minor axis $b = c\sinh\xi$.

The lines $\eta = $ const are a family of hyperbolas

$$\frac{x^2}{c^2\cosh^2\eta} - \frac{y^2}{c^2\sinh^2\eta} = 1 \qquad\qquad \text{(P.4)}$$

Show that:

- As $\xi \to 0$ the ellipse collapses to a plate along the x axis of length $2c$.

- As $\xi \to \infty$, the plots of constant values of the co-ordinates $(\xi,\ \eta)$ become circles of radius $\frac{1}{2}ce^\xi$ and straight lines of gradient $\tan\eta$ through the origin.

Problem 9: Using the results of the above problem #8, show that the velocity potential
$$w = C \cosh(\zeta - \zeta_0) \qquad (P.5)$$
represents the flow about the ellipse described by ξ_0 with incoming velocity $(C/c)\exp(-\xi_0)$ at an angle $-\eta_0$ with respect to the major axis, where $\zeta_0 = \xi_0 + i\eta_0$.

Verify that the stagnation points lie on the surface at the points (ξ_0, η_0) and $(\xi_0, \eta_0 + \pi)$.

Show that the maximum velocity on the surface, which must lie at the minor axis, $\eta = \pi/2$ or $3\pi/2$, is given by
$$v_{\max} = \frac{4\,a}{a+b} U \qquad (P.6)$$

Problem 10: Using elliptic co-ordinates show that the flow described by the complex potential
$$w = i\,k\,\zeta/2\,\pi \qquad (P.7)$$
is a vortex centred at the origin about one of the family of ellipses $\xi = $ const width circulation $-k$.

Problem 11: An ellipse with semi-major axis a and semi-minor axis b is subjected to a stream of velocity U at an angle of attack α with reference to the major axis. Using results obtained from problems #8, #9 and #10 show that the additional circulation required to achieve a stagnation point on the major axis is $-2\,\pi\,U\,\sin\alpha$.

Problem 12: Using the velocity potential, calculate the flow in the neighbourhood of a stagnation point at the surface of a blunt body in irrotational incompressible flow. Note that the incoming flow is normal to the surface, and the resultant flow is symmetric about the streamline to the stagnation point.

Problem 13: Calculate the second derivative of the velocity potential at the stagnation point of the flow about a circular cylinder with no rotation. Hence, using the results from problem #12 deduce the coefficient in the flow near a stagnation point in two dimensions in terms of the radius of curvature at the stagnation point.

Obtain the behaviour near the stagnation point of an axisymmetric flow in a similar way.

Problem 14: Calculate the shear strain rate tensor for a cylindrical vortex in fluid with viscosity μ azimuthally rotating with angular velocity $w(r)$. Hence calculate the torque exerted by neighbouring surfaces on each other.

Problem 15: Two coaxial cylinders of radius R_1 and R_2 rotate independently with angular velocities Ω_1 and Ω_2 respectively. The gap between the cylinders is filled with a fluid of viscosity μ. Using the results from problem #14 calculate the torque per unit length exerted by one cylinder on the other, assuming laminar flow in the fluid.

In the limit as $\Omega_2 \to 0$ and $R_2 \to \infty$ show that the flow is identical to that of the external flow in a Rankine vortex. (This result is the basis for the rotating cylinder viscometer.)

Problem 16: Calculate the total kinetic energy of a Rankine vortex with core radius R.

Noting that the central core rotates as a solid cylinder, replace it with a 'spindle' which supplies a torque sufficient to maintain the motion. Hence calculate the power required to support the motion against the energy loss by viscosity.

Noting that this energy must be that damped by the vortex, estimate the rate of decay of the vortex before it starts to dissipate and hence the growth of the core.

Problem 17: Consider a simple two-dimensional vortex of circulation Γ, centred at the origin, with irrotational flow elsewhere. Noting that the vorticity only has a single component, simplify equation (3.14) to obtain the familiar equation of diffusion. Using the standard solution for a line source, calculate the radial distribution of vorticity at a later time, t.

Problem 18: A viscous incompressible fluid flows slowly along a capillary tube of radius a due to a pressure drop Δp along the length ℓ of the tube. The flow rapidly achieves a steady uniform state over a distance of a few tube diameters. Noting that the flow is axisymmetric and that, since it is uniform, the axial velocity, v_z, is constant along the tube, calculate the total mass flow through the tube.

Problem 19: Consider problem #17 of vorticity diffusion as one of dimensional analysis. Noting that the problem is linear in the circulation Γ, show that it leads to a self-similar form with self-similar variable $\eta = R^2/\nu t$, and hence to the solution found in problem #17.

Problem 20: The drag on a ship's hull W is due to two factors: wave drag due to the displacement and frictional drag on the hull. The physical quantities involved are the length of the ship L, the volume of water displaced D, the speed of the ship U, the viscosity μ and density ρ of water, and the acceleration due to gravity g. Show that the complete set of dimensionless products is $W/\rho U^2 L^2$, $L/\sqrt[3]{D}$, $\rho \ell U/\mu$ and $U/\sqrt{\ell g}$. Comment on the application of these results to ship tank testing to measure the drag of a ship.

Problem 21: Water is confined in a tank of rectangular cross-section with sides X and Y to a depth h. Write down the boundary conditions at the walls of the tank. Identify the normal modes of oscillation of the water in the tank.

The surface of the water is disturbed by a small-amplitude perturbation. Show that the disturbance can be expressed in terms of the normal modes of surface waves in the tank.

Problem 22: Derive the modified condition for stability against the Rayleigh–Taylor instability for a heavy fluid of density ρ_+ above a lighter fluid ρ_- resulting from the surface tension σ at the interface.

A square mesh is fitted across the open end of a container filled with water. Calculate the maximum size of mesh that will prevent the inverted jar of water from emptying due to the Rayleigh–Taylor instability at the air/water surface.

Problem 23: An impulsive acceleration is applied to a light/heavy interface, which is therefore unstable for a very brief time. Show that a constant velocity perturbation is imposed on the surface, which grows linearly with time (Richtmyer–Meshkov instability)

Problem 24: The general solution of the Kelvin–Helmholtz problem, equation (4.37), for the complex frequency contains two terms $\omega = \omega_0 \pm \imath \omega_1$. Consider the development of the perturbation in the limit as both the velocity shear $(U_+ - U_-) \to 0$ and the density difference $(\rho_+ - \rho_-) \to 0$, and therefore $\omega_1 \to 0$ and $\omega \to \omega_0$ is approached. Show that this leads to a steady propagation of the original perturbation and a purely growing wave.

Problem 25: The velocity on the tube axis given by the Blasius correlation for pipe flow may be written in terms of the scaling parameter $\mathcal{C}$ for power law n, equation (5.36). The Fanning friction factor f, wall shear stress and friction velocity may each be written in terms of dimensionless variables $\mathcal{F}$, $\mathcal{T}$ and $\mathcal{V}$, which themselves depend on $\mathcal{C}$ for different values of the parameter n:

$$\bar{u} = \mathcal{C}(n) \{v^* a/\nu\}^{1/n} v^* \qquad f = \mathcal{F}(n) (\bar{u} a/\nu)^{-2/(n+1)}$$

$$\tau_0 = \mathcal{T}(n) \{\bar{u} a/\nu\}^{-2/(n+1)} \rho \bar{u}^2 \qquad \qquad \text{(P.8)}$$

$$v^* = \mathcal{V}(n) \{\bar{u} a/\nu\}^{-1/(n+1)} \bar{u}$$

Show that these parameters are given by

$$\frac{1}{2} \mathcal{F}(n) = \mathcal{V}(n)^2 = \mathcal{T}(n) = \mathcal{C}(n)^{-2n/(n+1)} \qquad \text{(P.9)}$$

Problem 26: An estimate of the friction coefficient over an extended range of Reynolds numbers may be made by using the power law formula appropriate to its value with highest accuracy. An estimate of the best value of the power n for a given range is obtained by equating the wall stress at the upper and lower limits of neighbouring values to give a piecewise continuous function for the wall shear stress. Show that the bounding value of the Reynolds number is given by

$$\mathcal{R}(n, (n+1)) = \mathcal{C}(n)^{-n(n+2)} \mathcal{C}(n+1)^{(n+1)^2} \qquad \text{(P.10)}$$

where $\mathcal{R}(n, (n+1)) = a\bar{u}/\nu$ is the bounding Reynolds number between powers n and $(n+1)$.

Problem 27: Consider the boundary layer in the neighbourhood of a stagnation point in a two-dimensional flow, where the free stream velocity along the surface varies linearly with the distance along the surface x. Comparing the magnitudes of the various terms in the two-dimensional boundary layer equations (6.7) and (6.8), show that the boundary layer thickness is constant and scales approximately as $\sqrt{\nu/c}$ where c is the constant of proportionality of the free stream velocity along the surface $U = cx$.

Problem 28: Using the result of problem #27 deduce that the streamfunction in a two-dimensional boundary layer can be written as

$$\psi = \sqrt{\nu c}\, x\, f(\eta) \quad \text{where} \quad \eta = \sqrt{c/\nu}\, y \qquad \text{(P.11)}$$

Substituting into the x component of the boundary layer equations, derive a differential equation for $f(\eta)$, and deduce the necessary boundary conditions.

Problem 29: Use the approximate Karman–Pohlhausen boundary layer method (Section 6.4.1) to derive values for the displacement and momentum thicknesses of the boundary layer near a stagnation point.

Problem 30: Derive the expressions for the flow in a centred rarefaction obtained in Section 9.3.1 using self-similar methods.

Problem 31: Derive the Prandtl–Meyer solution, obtained in Section 9.4.2 for the supersonic flow around a corner using a self-similar method.

Problem 32: Using the Rankine–Hugoniot equations and the equation of state for a polytropic gas, show that the jump relations are given by the set of equations (10.3).

Problem 33: The entropy per unit mass of a perfect gas is given by

$$s = c_V \ln\left(p/\rho^\gamma\right) + \text{const} \qquad \text{(P.12)}$$

Using equation (10.4a) for the density ratio across the shock, show that the entropy jump across a discontinuity is positive in compression and negative in expansion. Conclude that, quite generally, only compressive shocks occur in polytropic gases.

Problem 34: Using the geometry of the shock polar curve (Figure 10.14), show that the total velocity increment for a small deflection angle is given by equation (10.75).

Problem 35: When the Mach number of the incoming flow is large show that the shock polar equation (10.82) reduces to that of a circle, centred on $(\gamma/(\gamma_1 + 1)\,v_1, 0)$ with radius $1/(\gamma + 1)\,v_1$. Hence show that the limit angle is given by $\arcsin(1/\gamma)$. Show also that the downstream flow is sonic.

Problem 36: Show that the Zhukovskii transform of the circle $\zeta = R\exp(\imath\theta)$ centred at the origin, but not passing through the singularities, generates the ellipse

$$\frac{x^2}{(R^2 + \ell^2)^2} + \frac{y^2}{(R^2 - \ell^2)^2} = \frac{1}{R^2} \qquad \text{(P.13)}$$

Problem 37: Using the results of problems #9 and #10, derive the circulation required to cause the body streamline to leave the surface at the major axis. Hence derive the lift coefficient for an elliptic wing section.

Problem 38: Using the results of problem #36 and Section 11.6.2, calculate the circulation necessary to give a stagnation point at the major axis. Confirm that this value is identical to that obtained in problem #37. Hence show that lift coefficient is

$$c_L \approx 2\pi\,\alpha\,(1 + b/a) \tag{P.14}$$

Show that if $b = 0$ the lift reduces to that found from a flat plate, equation (11.25), and if $b = a$ that from a circular cylinder, equation (11.38).

Problem 39: Show that Zhukovskii transformation of the circle radius R centred on the imaginary axis at $\imath\,d$ and passing through both singularities generates the arc of a circular lamina centred on the imaginary axis at $(\ell^2 - d^2)/d$ and radius $(\ell^2 + d^2)/d$, subtending an angle $2\chi = \mathrm{arccot}(d/\ell)$. Deduce the camber of the lamina treated as an aerofoil.

Problem 40: Using the results from Section 11.6.2 show that the axis of zero lift of a Zhukovskii lamina is at an angle $\beta = \pi/2 - \chi$ to the chord line between the leading and trailing edges.

A uniform wind of velocity U is introduced at angle of attack α to the chord line of the wing. Deduce the necessary circulation about the wing to ensure that the flow at trailing edge is finite (Zhukovskii condition). Hence calculate the lift coefficient.

Problem 41: The arrangement of the sails on a 'fore-and-aft' rigged sailing craft, a sloop, is shown in Figure P.3. The wind blowing from the side billows the sails out into an approximately aerofoil section with chord varying up the sail. The keel, also acting as an aerofoil, prevents sideways motion of the boat. Considering the forces on the boat due to the wind and the keel, show how the craft is able to sail into the wind. Explain why in principle the maximum drive is when the wind is normal to the axis of the boat.

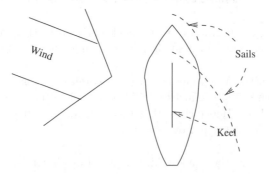

Figure P.3: Sketch of the arrangement of sails on a 'fore-and-aft' rigged sailing craft.

Problem 42: Using expressions (9.50) and (10.76) for the pressure drop across a rarefaction and a shock in which the deflection is small, show that the lift from a flat plate in hypersonic flow is given by

$$c_L \approx 4\alpha/M_1$$

Problem 43: Generalising the previous problem #42, consider an aerofoil section in supersonic flow made up of a series of straight line segments, each of length δs_i, inclined at an angle α_i to the direction of the incoming flow. The deflection at each junction gives rise to a rarefaction or shock depending on the sign of the angle $[\alpha_i - \alpha_{(i-1)}]$. Assuming that the angles α_i are small and noting that the pressure changes across both shocks and rarefactions in the perturbation limit are given by the same simple expression common to both equations (9.50) and (10.76), derive the pressure coefficient on each surface. By summing the pressure differences over the upper and lower surfaces obtain Ackeret's formulae for the lift and drag coefficients, equations (12.27) and (12.28), when the segments become infinitesimally small.

Problem 44: The hypersonic wing may be treated by an alternative approach, which leads to a simpler calculation. The velocity behind the shock parallel to the incoming flow is almost unchanged, and the transverse velocity just matches that needed to provide the flow along the surface. Transform to the frame moving with the incoming flow, where the planar flow normal to the wing surface has velocity $v_1 \alpha$ away from and towards the surface. Treating the surface therefore as a piston driving a normal shock downwards (10.2) and a centred rarefaction (9.3) upwards into the gas, with the flow velocity at the piston surface $v_2 = v_1 \alpha$, derive equation (12.51).

Problem 45: Using Bernoulli's equation for a compressible gas, derive the pressure coefficient/Mach number relation, equation (12.29).

Problem 46: The critical Mach numbers for the flow around a sphere and a cylinder are respectively
For a cylinder $M_{\text{crit}} = 0.418\,135$.
For a sphere $M_{\text{crit}} = 0.5675$.
Using equations (11.4) and (2.59), verify these values using the Prandtl–Glauert correction.

Although the flows around the cylinder and sphere are qualitatively similar, there is a clear difference in compressible flow. Account for this distinction.

Problem 47: Derive the Zel'dovich formula for the laminar flame speed S by making the following approximations in the rest frame of the burn:
1. Neglect particle diffusion.
2. Due to the strongly increasing burn rate with temperature after ignition and the subsequent rapid decrease due to fuel depletion, the burn has a short-lived intense narrow distribution.
3. The burn zone may be divided into two parts.
4. In the first zone the reactants are pre-heated from the ambient temperature T_1 to an intermediate ignition temperature T_i, no reaction taking place. In this region heat conduction is balanced by convection of the fluid at the flame speed.
5. In the second zone the reaction takes place with heat extracted by thermal conduction. The zone extends in temperature from T_i to T_2 when the reaction is complete.
6. The detailed structure of the burn zone may be neglected.

Problem 48: A mass of cold gas M initially of zero spatial width in a space of ν dimensions expands into vacuum. Each fluid element (ζ) of the body of gas is heated at a rate which depends on a power of time

$$e = a\,t^n = A\,W(\zeta)\,t^n$$

where A is a dimensional constant, $W(\zeta)$ the (dimensionless) heat distribution function and $n \geq 0$.

Using dimensional analysis show that the motion is self-similar, and further conclude that the motion is homogeneous.

Finally, using dimensional analysis find the forms of the density and pressure.

Problem 49: The self-similar collapse of a uniform sphere, Section 14.5, under the action of an applied pressure in the limit of infinite convergence can be calculated in a simple manner by considering the self-similar form derived from the Lagrangian element R, the initial radius of a fluid element. The self-similar variable is defined by the dimensionless quantity $s = c_0(-t)/R$ with the subsidiary quantities $f = r/R$ and $\epsilon = \rho_0/\rho$. Using the equation of continuity in the form $\rho r^2\,\mathrm{d}r = \rho_0 R^2\,\mathrm{d}R$ and Euler's equation show that

$$\epsilon = \frac{\rho_0}{\rho} = f^2\left(f - s\frac{\mathrm{d}f}{\mathrm{d}s}\right) = \frac{1}{2}\alpha^3\,s^{3/2} \tag{P.15}$$

and

$$\frac{\mathrm{d}^2 f}{\mathrm{d}s^2} = \epsilon^{-(\gamma-1)}\,s\,f^2\,\frac{\mathrm{d}\epsilon^{-1}}{\mathrm{d}s} \tag{P.16}$$

Hence show that for $\gamma = 5/3$

$$f = 2^{2/3}\,3^{1/4}\,s^{1/2} \approx 2.089\,s^{1/2} \quad \text{and} \quad \epsilon = 2\,3^{3/4}\,s^{3/2} \approx 4.559\,s^{3/2} \tag{P.17}$$

Compare this result with the limit passing through the singularity in Section 14.5. Show that the Mach number is $1/\sqrt{3}$.

Problem 50: The self-regulating model (case study 13.I.i.b) for the one-dimensional time-dependent laser heating of a solid target can be expressed in self-similar form. Consider the case when the input intensity is constant. Using dimensional analysis, identify the self-similar variable (13.27) for the problem. Hence, using the equations of mass, momentum and energy conservation, derive a set of ordinary differential equations to describe the flow in dimensionless form.

Solutions

Problem 1: In two dimensions the Jacobian is

$$J = \begin{vmatrix} \dfrac{\partial x}{\partial \lambda} & \dfrac{\partial x}{\partial \mu} \\[2ex] \dfrac{\partial y}{\partial \lambda} & \dfrac{\partial y}{\partial \mu} \end{vmatrix} = \frac{\partial x}{\partial \lambda}\frac{\partial y}{\partial \mu} - \frac{\partial x}{\partial \mu}\frac{\partial y}{\partial \lambda} = \frac{\partial \underline{r}}{\partial \lambda} \cdot \frac{\partial \underline{s}}{\partial \mu} \tag{S.1}$$

and the gradient and divergence

$$\frac{\partial f}{\partial x} = \frac{1}{J}\begin{vmatrix} \dfrac{\partial f}{\partial \lambda} & \dfrac{\partial y}{\partial \lambda} \\[2ex] \dfrac{\partial f}{\partial \mu} & \dfrac{\partial y}{\partial \mu} \end{vmatrix} = \frac{1}{J}\left\{ \frac{\partial f}{\partial \lambda}\frac{\partial y}{\partial \mu} - \frac{\partial f}{\partial \mu}\frac{\partial y}{\partial \lambda} \right\} \tag{S.2}$$

and

$$\frac{\partial f}{\partial y} = \frac{1}{J}\begin{vmatrix} \dfrac{\partial x}{\partial \lambda} & \dfrac{\partial f}{\partial \lambda} \\[2ex] \dfrac{\partial x}{\partial \mu} & \dfrac{\partial f}{\partial \mu} \end{vmatrix} = \frac{1}{J}\left\{ \frac{\partial f}{\partial \mu}\frac{\partial x}{\partial \lambda} - \frac{\partial f}{\partial \lambda}\frac{\partial x}{\partial \mu} \right\} \tag{S.3}$$

Collecting these terms together we obtain

$$\nabla f = \frac{1}{J}\left\{ \frac{\partial f}{\partial \lambda}\frac{\partial \underline{s}}{\partial \mu} - \frac{\partial f}{\partial \mu}\frac{\partial \underline{s}}{\partial \lambda} \right\} \tag{S.4}$$

and

$$\nabla \cdot \underline{A} = \frac{1}{J}\left\{ \frac{\partial A_x}{\partial \lambda}\frac{\partial y}{\partial \mu} + \frac{\partial A_x}{\partial \mu}\frac{\partial x}{\partial \lambda} - \frac{\partial A_y}{\partial \mu}\frac{\partial y}{\partial \lambda} - \frac{\partial A_y}{\partial \lambda}\frac{\partial x}{\partial \mu} \right\} = \frac{1}{J}\left\{ \frac{\partial \underline{A}}{\partial \lambda}\cdot\frac{\partial \underline{s}}{\partial \mu} - \frac{\partial \underline{A}}{\partial \mu}\cdot\frac{\partial \underline{s}}{\partial \lambda} \right\} \tag{S.5}$$

In an axisymmetric system $\underline{r} = (\varrho, z)$ we evaluate the divergence by replacing $\underline{A}$ by $\varrho \underline{A}$ and the normal vector $\underline{s} = (z, -\varrho)$

$$\nabla \cdot \underline{A} = \frac{1}{\varrho J}\left\{ \frac{\partial(\varrho \underline{A})}{\partial \lambda}\cdot\frac{\partial \underline{s}}{\partial \mu} - \frac{\partial(\varrho \underline{A})}{\partial \mu}\cdot\frac{\partial \underline{s}}{\partial \lambda} \right\} \tag{S.6}$$

Introductory Fluid Mechanics for Physicists and Mathematicians, First Edition. Geoffrey J. Pert.
© 2013 John Wiley & Sons, Ltd. Published 2013 by John Wiley & Sons, Ltd.

Problem 2: Let the point ① be taken at the tip $v_1 = 0$, and point ② in the free stream $v_2 = U$. Applying Bernoulli's theorem to the streamline adjacent to the body, and assuming incompressible flow,

$$\frac{1}{2}v_1^{2\,0} + \frac{p_1}{\rho} = \frac{1}{2}v_2^{2\,U^2} + \frac{p_2}{\rho}$$

Hence

$$U = \sqrt{\frac{2(p_1 - p_2)}{\rho}} \tag{S.7}$$

Although useful, this equation is only an approximation as there are several factors not taken into account, such as compressibility, finite size of the pressure sensor at the tip and effect of viscosity. In consequence the device is normally calibrated experimentally before use.

Problem 3: Let the cross-section of the pipe be S_1 and that of the constriction S_2. Assuming incompressible flow, and that the flow velocity is constant across the pipe, the equation of continuity in steady flow yields

$$\rho\, v_1\, S_1 = \rho\, v_2\, S_2$$

where v_1 and v_2 are the flow speeds in the pipe and the constriction respectively. Measuring the pressures in the pipe p_1 and in the constriction p_2 gives

$$\frac{1}{2}(v_2^2 - v_1^2) = \frac{1}{2}\left[\left(\frac{S_1}{S_2}\right)^2 - 1\right]v_1^2 = \frac{(p_1 - p_2)}{\rho}$$

Hence

$$v_1 = \sqrt{\frac{2\,(p_1 - p_2)}{\rho\left[(S_1/S_2)^2 - 1\right]}} \tag{S.8}$$

which yields the flow speed if the pressure differential $p_2 - p_1$ is measured.

Problem 4: Bernoulli's equation including gravity takes the form

$$\frac{1}{2}v_1^{2\,0} + \frac{p_1}{\rho}^{(p_a/\rho)} + \rho g h = \frac{1}{2}v_2^2 + \frac{p_2}{\rho}$$

If the tap is open the exit pressure is atmospheric, $p_2 = p_a$, and the exit flow velocity $v_2 = \sqrt{2gh}$, the free fall value.
If the tap is closed, the exit velocity is zero, $v_2 = 0$, and $p_2 = p_a + \rho g h$, the hydrostatic head.

Problem 5: Let the density of the fluid be ρ, and the velocity and cross-section of each jet be v and A respectively. The total mass flow rate in each jet is $m = \rho\, A\, v$.

Since the pressure on the jets before and after the interaction is the constant atmospheric pressure, the flow speed of the fluid is unchanged.

The conservation of momentum and symmetry normal to the mutual axis of the flows ensures that after the interaction the motion is along the axis. The momentum flow along the axis of flow is

$$2m\,v\,\cos\theta = m_f\,v_f - m_b\,v_b \tag{S.9}$$

where m_f and m_b are the total mass flow rate in the forward and backward jets and v_f and v_b are the forward and backward flow speeds respectively, both of which equal v. The total mass flow into and out of the interaction is

$$2m = m_f + m_b \tag{S.10}$$

Hence

$$m_f = 2\,m\,(1 + \cos\theta) \quad\text{and}\quad m_b = 2\,m\,(1 - \cos\theta) \tag{S.11}$$

Since the velocity of the incompressible fluid is constant before and after the collision of the jets, the change in the mass flow rate must be associated with a change in area of the jets. In a real flow the interaction is clearly complex and viscosity and particularly turbulence must be expected to play a major role.

Problem 6: In the frame of reference of the stationary bucket, the water flow is quasi-steady. We apply Bernoulli's theorem to relate the incoming and outgoing flows. Since both experience the same atmospheric pressure we conclude that the flow speed in the frame of the bucket is unchanged, namely

$$v'_{out} = v'_{in} \tag{S.12}$$

Remaining in the bucket frame we calculate the force on the bucket by the rate of change of the momentum flux in the direction normal to the plane of the bucket:

$$F = \rho\,v'_{in}\,A\,(v'_{in} - v'_{out}\cos\theta') \tag{S.13}$$

where v', θ' refer to values in the bucket frame, ρ is the density and A is the cross-section of the incoming jet. To transform back into the laboratory frame:

$$
\begin{aligned}
v_{in} &= v'_{in} + v_b \\
v_{out}\cos\theta &= v'_{out}\cos\theta' - v_b \\
v_{out}\sin\theta &= v'_{out}\sin\theta'
\end{aligned}
$$

where $v_b = \omega R$ is the bucket speed, R the radius and ω the angular velocity of the wheel. The exit speed in the laboratory frame is therefore

$$v_{out} = \sqrt{[v_{in}^2 - 2v_{in}\omega R + (\omega R\cos\theta)^2]} - \omega R\cos\theta \tag{S.14}$$

and the exit angle in the bucket frame takes the rather complex form

$$\theta' = \arcsin\left\{\frac{v_{out}\sin\theta}{v_{in} - \omega R}\right\} \tag{S.15}$$

Hence we obtain the torque on the wheel:

$$\tau = \rho \left(v_{in} - \omega R\right)^2 A R \left[1 + \cos \theta'\right] \tag{S.16}$$

The work done in the time Δt that the bucket is in the jet is thus

$$\rho \left(v_{in} - \omega R\right)^2 A R \left[1 + \cos \theta'\right] \omega \Delta t \tag{S.17}$$

The actual time a bucket spends in the water jet is slightly increased because there is a short time when two buckets are simultaneously struck by the jet, due to the rotation of the buckets. During the interval Δt the mass striking the bucket is $\rho v_{in} A \Delta t$ whereas the mass flow used above in equation (S.13) for the force was $\rho \left(v_{in} - \omega R\right) A$. Thus power is developed for a time $\left[v_{in}/(v_{in} - \omega R)\right] \Delta t$ by each bucket, allowing for the fact that for a short period of time, two buckets simultaneously are struck by the jet due to their spacing. The average power delivered by the wheel is therefore

$$P = \rho \left(v_{in} - \omega R\right) v_{in} A \omega R \left[1 + \cos \theta'\right] \tag{S.18}$$

Maximum power is developed when the bucket speed $\omega R = \frac{1}{2} v_{in}$, and has a fraction $\frac{1}{2}[1 + \cos \theta']$ of that delivered by the incoming water jet $\frac{1}{2} \rho A v_{in}^3$.

Problem 7: The total streamfunction is the sum of that due to the incoming flow $U y$ and that due to each of the sources $(m/2\pi)\theta$ which may be written as

$$\psi = \frac{m}{2\pi} \left(\theta_+ - \theta_-\right) + U r \sin\theta \tag{S.19}$$

where the angles

$$\theta_- = \arctan[y/(x + d/2)] \quad \text{and} \quad \theta_+ = \arctan[y/(x - d/2)] \tag{S.20}$$

Hence the oval is given by the curve for which $\psi = 0$

$$U y + \frac{m}{2\pi} \left\{ \arctan\left[\frac{y}{x - d/2}\right] - \arctan\left[\frac{y}{x + d/2}\right] \right\} = 0 \tag{S.21}$$

Since

$$\tan(\arctan A - \arctan B) = \frac{A - B}{1 + A B}$$

it follows that the surface may be expressed as

$$U y = -\frac{m}{2\pi} \arctan\left[\frac{d y}{(x^2 + y^2 - d^2/4)}\right] \tag{S.22}$$

If the sink/source separation is very small, $d \to 0$, the surface becomes

$$U y \to -\frac{m}{2\pi} \frac{d y}{(x^2 + y^2)} \tag{S.23}$$

Noting that m is negative and writing the doublet strength $\mathcal{M} = md$, we obtain a circle with radius $R = \sqrt{|\mathcal{M}|/(2\pi U)}$ and streamfunction

$$\psi \approx Uy \left\{ 1 - \frac{\mathcal{M}}{2\pi U \ (x^2 + y^2)} \right\} \tag{S.24}$$

to recover the solution for a circular cylinder (2.94), which can be seen to be the summation of the uniform flow with that from a doublet of appropriate strength.

The length of the oval is found by setting $y \to 0$ as

$$x_0{}^2 = (d/2)^2 - (m\,d/2\pi\,U) \tag{S.25}$$

and the width, when $x = 0$, is the solution of

$$y_0{}^2 = (d/2)^2 - d\,y_0 \cot(2\,\pi U\,y_0/m) \tag{S.26}$$

Problem 8: The complex number z is represented as

$$z = x + \imath y = c \cosh \zeta = c \cosh \xi \ \cos \eta + \imath c \sinh \xi \sin \eta$$

Equating real and imaginary parts, and eliminating η, we obtain

$$\frac{x^2}{c^2 \cosh^2 \xi} + \frac{y^2}{c^2 \sinh^2 \xi} = 1 \tag{S.27}$$

i.e. an ellipse if $\xi = \text{const}$ with semi-major axis $c \cosh \xi$ and semi-minor axis $c \sinh \xi$. Similarly, eliminating ξ we obtain

$$\frac{x^2}{c^2 \cos^2 \eta} + \frac{y^2}{c^2 \sin^2 \eta} = 1 \tag{S.28}$$

i.e. a hyperbola if $\eta = \text{const}$ with semi-major axis $a = c \cos \eta$ and semi-conjugate axis $b = c \sin \eta$.

If $\xi = 0$,

$$z = x + \imath y = c\,(\cos \eta + \imath 0 \sin \eta) \tag{S.29}$$

which corresponds to a flat plate of length $2c$.

If $\xi \to \infty$,

$$z = x + \imath y = \cosh \zeta \to \frac{1}{2} c \exp(\zeta) = \frac{1}{2} c \exp(\xi) \ \{\cos \eta + \imath \sin \eta\} \tag{S.30}$$

Therefore at large distances lines of constant ξ become circles of radius $\frac{1}{2} c \exp(\xi)$ and lines of constant η straight lines of gradient $\tan \eta$.

Problem 9: The real and imaginary parts of the complex potential are

$$w = \phi + \imath \psi = C \left\{ \cosh(\xi - \xi_0) \ \cos(\eta - \eta_0) + \imath \ \sinh(\xi - \xi_0) \ \sin(\eta - \eta_0) \right\} \tag{S.31}$$

Hence the streamline ψ follows the lines $\xi = \xi_0$ and $\eta = \eta_0$. The first of these is the surface of the ellipse, which is therefore a streamline and may be solid. Far from the body, the streamline tends to the straight line $\theta = \arctan \eta_0$ which corresponds to the angle between the incoming flow and the major axis of the ellipse.

The complex velocity is

$$v = v_x - i v_y = \frac{dw}{d\zeta} \Big/ \frac{dz}{d\zeta} = \frac{C \sinh(\zeta - \zeta_0)}{c \sinh \zeta} \tag{S.32}$$

The incoming flow velocity is easily obtained by allowing $|\zeta| \to \infty$

$$U = \frac{C}{c} \exp(-\zeta_0)$$

If $\zeta = \zeta_0$ or $\zeta = \zeta_0 + i\pi$ $\sinh(\eta - \eta_0) = 0$ and the complex velocity is zero – a stagnation point.

When $\eta_0 = 0$ the incoming flow is parallel to the major axis. Since the maximum displacement occurs at the minor axis $x = 0$, $y = \pm b$ the velocity must take its maximum value there and be in the x direction, i.e. $\eta = \pi/2$ or $3\pi/2$

$$v_{\max} = 2 \frac{C}{c} \frac{\exp(-\xi_0)}{1 + \exp(-2\xi_0)}$$

$$= \frac{2}{1 + (a - b)/(a + b)} U = \frac{a + b}{a} U = U \left(1 + \sqrt{1 - e^2}\right) \tag{S.33}$$

since $(a + b) = c \exp(\xi_0)$ and $(a - b) = c \exp(-\xi_0)$; e is the eccentricity.

Problem 10: The complex potential

$$w = \phi + i\psi = i k \zeta 2\pi = i k (\xi + i\eta)/2\pi$$

yields the streamfunction $\psi = k\xi/2\pi$ and is therefore constant on the elliptic surface $\xi = \text{const. } \exp(\xi)$ At large distances $\eta \to \theta$, the polar angle, and the flow therefore rotates about the surface centred on the origin. The potential $\phi = -k\theta/2\pi$. Therefore the circulation is $-k$.

Problem 11: Adding a circulation to the incoming flow, the complex potential is

$$w = C \cosh(\zeta - \zeta_0) + i k / 2\pi \tag{S.34}$$

We require

$$v = \frac{dw}{d\zeta} \Big/ \frac{dz}{d\zeta} = \frac{\{C \sinh(\zeta - \zeta_0) + i k / 2\pi\}}{c \sinh \zeta} = 0$$

at $\zeta = \xi_0$. Therefore, since $C = U(a + b)$,

$$k = -2\pi C \sin(\eta_0) = -2\pi U (a + b) \sin(\eta_0) \tag{S.35}$$

Problem 12: Define a set of co-ordinates x and y in the tangent plane to the body at the stagnation point and z outwards along the incoming streamline. In the neighbourhood of the stagnation point $(x = y = z = 0)$ the flow is symmetric with respect to x and y. Therefore, expanding the velocity potential as a Taylor's series about the origin,

$$\phi = \phi_0 + \frac{\partial \phi}{\partial x_i} x_i + \frac{1}{2} \frac{\partial^2 \phi}{\partial x_i\, \partial x_j} x_i\, x_j + \dots$$

but $\nabla \phi = 0$ at a stagnation point and all terms containing $x_i\, x_j$, $i \neq j$, must also be zero due symmetry.

Since the flow is irrotational and incompressible, $\nabla^2 \phi = 0$ and therefore

$$\phi = \phi_0 + \frac{1}{2} \left[\frac{\partial^2 \phi}{\partial x^2} \left(x^2 - z^2 \right) + \frac{\partial^2 \phi}{\partial y^2} \left(y^2 - z^2 \right) \right] \tag{S.36}$$

In a two-dimensional and axisymmetric flow this result is simplified to

$$\phi = \begin{cases} \phi_0 + \dfrac{1}{2} \dfrac{\partial^2 \phi}{\partial x^2} \left(x^2 - z^2 \right) & \text{two dimensions} \\[2ex] \phi_0 + \dfrac{1}{2} \dfrac{\partial^2 \phi}{\partial \varrho^2} \left(\varrho^2 - 2 z^2 \right) & \text{axisymmetric} \end{cases} \tag{S.37}$$

respectively.

Problem 13: Differentiating the complex potential of the flow around the cylinder,

$$\frac{\partial^2 \phi}{\partial z^2} = \Re \left\{ \frac{d^2 w}{dz^2} \right\} = \Re \left\{ 2 U \frac{R^2}{z^3} \right\} = 2 U \frac{R^2 \exp(-\imath 3\theta)}{r^3}$$

At the stagnation point $\theta = \pi$, the flow is towards the cylinder, and the coefficient is correspondingly $-2/R$. Since the blunt body is tangential to a circle of radius of curvature R, the flow incident at the stagnation point is described by

$$\phi \approx \phi_0 + U \left(x^2 - z^2 \right) / R \tag{S.38}$$

where x is measured in the tangent plane and z outwards along the streamline (as in problem#12).

For a sphere of radius R the derivative along the incoming flow is easily obtained

$$\frac{\partial^2 \phi}{\partial z^2} = \frac{\partial v_r}{\partial r} \bigg|_{\theta = \pi} = -\frac{4\, U}{R^2} = -\frac{\partial^2 \phi}{\partial \varrho^2}$$

and therefore the flow near the stagnation point

$$\phi \approx \phi_0 + 2\, U \left(\varrho^2 - 2 z^2 \right) / R^2 \tag{S.39}$$

Problem 14: Take a local Cartesian co-ordinate set (x, y), with x in the radial direction and y in the azimuthal one, to calculate the local rate of shear strain tensor component $\dot{e}_{xy}$. Consider the changes in velocity induced by a small azimuthal angle shift $\delta\theta$

$$\delta v_x = -v_\theta\, \delta\theta \quad \text{and} \quad \delta v_y = v_\theta\, (1 - \cos \delta\theta) \approx 0 \qquad \therefore\ \dot{e}_{xy} = \frac{1}{2}\left(\frac{dv_\theta}{dr} - \frac{v_\theta}{r}\right) = \frac{1}{2}\, r\, \frac{d\omega}{dr}$$

so the total torque per unit length at the interface is

$$\tau = 2\,\pi\, R^3\, \mu\, \frac{d\omega}{dr} \tag{S.40}$$

where μ is the coefficient of viscosity.

Problem 15: In the steady state, there is no gain in angular momentum between the cylinders. The torque τ is therefore constant across the gap. Integrating equation (S.40),

$$\omega - \Omega_1 = \frac{\tau}{4\,\pi\,\mu}\left(\frac{1}{R_1{}^2} - \frac{1}{r^2}\right)$$

The torque is therefore

$$\tau = 4\,\pi\mu R_1{}^2 R_2{}^2\, \frac{(\Omega_2 - \Omega_1)}{(R_2{}^2 - R_1{}^2)} \tag{S.41}$$

Substituting for τ,

$$\omega = \frac{(\Omega_2 R_2{}^2 - \Omega_1 R_1{}^2)}{(R_2{}^2 - R_1{}^2)} + \frac{(\Omega_1 - \Omega_2)\, R_1{}^2 R_2{}^2}{(R_2{}^2 - R_1{}^2)}\, \frac{1}{r^2} \tag{S.42}$$

If $R_2 \to \infty$ and $\Omega_2 \to \infty$

$$\omega = \Omega_1\, \frac{R_1{}^2}{r^2}$$

in agreement with the Rankine vortex.

Problem 16: The kinetic energy in a Rankine vortex is easily calculated from the velocity profile, equation (2.75),

$$\frac{1}{8}\, 2\,\pi\, \rho \zeta^2 \left\{\int_0^R r^3\, dr + \int_R^{r_{\max}} R^4\, \frac{dr}{r}\right\} = \frac{1}{4\,\pi}\, \rho \Gamma^2 \left\{\frac{1}{4} + \ln\left(\frac{r_{\max}}{R}\right)\right\} \tag{S.43}$$

where an outer limit to the vortex $r_{\max}$ is imposed to avoid the logarithmic divergence at infinity.

The vortex is damped by viscosity due to the shear initially starting at the interface between rotational and irrotational flow, but spreading through the flow. We may estimate the initial rate of dissipation very simply from the viscous torque at the interface. The flow internal to the interface is a uniform rotation, characteristic of a solid, with angular velocity $\omega = \frac{1}{2}\zeta$ acting as a 'spindle' supplying energy via a viscous torque on the interface. At the interface there is viscous force. The work done per unit time by the 'spindle' is therefore $\tau\omega = \mu\Gamma^2/\pi R^2$. In the absence of any such external agent, the energy of the vortex must decrease at this rate due to viscosity. Assuming viscous damping is weak and the flow remains nearly irrotational, the radius R of the interface must increase to accommodate the energy loss. Since the total circulation of the core remains constant

$$R\frac{dR}{dt} = 4\nu \tag{S.44}$$

where $\nu = \mu/\rho$. This represents only an approximation for small times whilst the velocity profile of equation (2.75) is valid. For large times such that initial size is proportionately very small, the radius may be expected to scale as

$$R \sim \sqrt{8\nu t} \tag{S.45}$$

Problem 17: The vorticity associated with a two-dimensional vortex is a vector with a single component perpendicular to the plane of variation. The vorticity dissipation equation (3.14) therefore takes the simpler form characteristic of diffusion. Writing this in polar co-ordinates and neglecting the angular term due to symmetry,

$$\frac{\partial\zeta}{\partial t} = \nu\frac{1}{r}\frac{r\partial}{\partial r}\left(\frac{r\partial\zeta}{\partial r}\right) \tag{S.46}$$

which has the well-known solution, which may be checked by substitution,

$$\zeta = \frac{\Gamma}{4\pi\nu t}\exp\left(-\frac{r^2}{4\nu t}\right) \tag{S.47}$$

since the circulation $\Gamma = 2\pi\int_0^\infty \zeta r\,dr$.

Problem 18: It follows from the equation of continuity

$$\frac{1}{r}\frac{\partial(r\,v_r)}{\partial r} + \overset{0}{\cancel{\frac{\partial v_z}{\partial z}}} = 0$$

that since the radial velocity v_r is zero at the wall, it must be zero everywhere.

Under conditions of slow flow, the dynamics are dominated by viscosity and the inertial term may be neglected. The Navier–Stokes equation takes the form

$$-\nabla p + \mu\nabla^2\underline{v} = 0 \tag{S.48}$$

Taking components we obtain

$$\frac{\partial p}{\partial r} = 0 \quad\text{and}\quad \frac{\partial p}{\partial z} = \mu\left(\frac{d^2 v_z}{dr^2} + \frac{dv_z}{dr}\right) \tag{S.49}$$

The pressure is therefore constant across the tube and depends on the axial distance z alone.

Integrating and noting that the axial velocity v_z must be zero at the wall $r = a$ yields

$$v_z = -\frac{1}{4\mu}\frac{dp}{dz}\left(a^2 - r^2\right) \tag{S.50}$$

A further integration yields the total mass flux through the tube

$$J = \int_0^a 2\pi\, r\, \rho\, v_z\, dr = \frac{\pi\,\rho\,a^4}{8\,\mu}\frac{\Delta p}{\ell} \tag{S.51}$$

This is Poiseuille's equation and forms the basis of a simple method of measuring viscosity.

Problem 19: The variables in the problem are: distance $r = [L]$, time $t = [T]$, vorticity $\zeta = [T]^{-1}$, circulation $\Gamma = [L]^2[T]^{-1}$ and kinematic viscosity $\nu = [L]^2[T]^{-1}$. Since the diffusion is linear, the variable ζ is directly proportional to Γ and may only appear in the combination $\zeta/\Gamma = [L]^{-2}$. A complete set of dimensionless variables is therefore

$$\frac{r^2}{\nu t} \quad \text{and} \quad \frac{\zeta\nu t}{\Gamma}$$

Since r and t can only appear in the dimensionless combination, then

$$\eta = \frac{r^2}{\nu t} \tag{S.52}$$

Hence the solution must take the form

$$\zeta = \frac{\Gamma}{\nu t}f(\eta) \tag{S.53}$$

and the derivatives take the form

$$\frac{\partial}{\partial t} = -\frac{\eta}{t}\frac{d}{d\eta} \quad \text{and} \quad \frac{1}{r}\frac{\partial}{\partial r}\left(r\frac{\partial}{\partial r}\right) = \frac{4}{\nu t}\frac{d}{d\eta}\left[\nu\frac{d}{d\eta}\right] \tag{S.54}$$

Substituting in equation (S.46) for ζ/Γ, and the derivatives, we obtain

$$4\frac{d}{d\eta}\left(\eta\frac{df}{d\eta}\right) + \frac{d}{d\eta}\left(\eta f\right) = 0 \tag{S.55}$$

whose solution with the boundary condition $df/d\zeta = 0$ when $f = 0$ is

$$f = \text{const}\, \exp(-\eta/4) \tag{S.56}$$

As in problem#17 the constant $(4\pi)^{-1}$ is found by integration giving equation (S.47).

Problem 20: The dimensional matrix is

	ρ	μ	g	L	D	U	W
$[M]$	1	1	0	0	0	0	1
$[L]$	-3	-1	1	1	3	1	1
$[T]$	0	-1	-2	0	0	-1	-2

It is easily checked that there exist non-zero determinants of order 3 within the array and the rank of the matrix is therefore 3. A solution array, which may be checked by substitution, is

	k_1	k_2	k_3	k_4	k_5	k_6	k_7
$W/\rho U^2 L^2$	-1	0	0	-2	0	-2	1
$L/D^{1/3}$	0	0	0	1	$-1/3$	0	0
$\rho U L/\mu$	1	-1	0	1	0	1	0
$U/(Lg)^{1/2}$	0	0	$-1/2$	$-1/2$	0	1	0

The area factor L^2 in the dimensionless drag is usually replaced by the wetted area $S \sim D/L$ to define the drag coefficient, which may be written as

$$C_D = \frac{W}{\frac{1}{2}\rho S U^2} = F(\psi, \mathcal{R}, \mathcal{F}) \tag{S.57}$$

where $\psi = L/\sqrt[3]{D}$ is the fineness coefficient, $\mathcal{R} = \rho U L/\mu$ the Reynolds number and $\mathcal{F} = U\sqrt{Lg}$ the Froude number.

In a ship tank experiment it not possible to achieve complete similarity. Although the model and prototype may achieve geometrical similarity, it is not possible to have similarity with respect to both the Reynolds and Froude numbers at the same time. Froude hypothesised that it is possible to separate the drag due to friction in the boundary layer (Section 6.6), which depends on $\mathcal{R}$, and wave drag (case study 4.I), which depends on $\mathcal{F}$ alone, allowing the total drag to be written as

$$W = W_F + W_w = \frac{1}{2}\rho S U^2 C_f(\mathcal{R}) + \rho, g\, D\, c_w(\psi, \mathcal{F}) \tag{S.58}$$

where S is the wetted surface area and using a different dimensionless scaling for the wave drag. The frictional drag coefficient is measured in a separate set of experiments, e.g. with flat plates, or by using scaling relations for turbulent boundary layer flow, thereby separating the wave and frictional drag. Hence measurements of the total hull drag in the model experiments scale to the prototype at constant ψ and $\mathcal{F}$. In engineering practice, extensive scaling relations for the coefficient c_w are available.

Problem 21: At the boundaries $x = 0$ and $x = X$, and $y = 0$ and $y = Y$, the normal component of velocity must be zero, i.e.

$$\frac{\partial \phi}{\partial x} = 0 \quad \text{if } x = 0 \text{ or } X \qquad \text{and} \qquad \frac{\partial \phi}{\partial y} = 0 \quad \text{if } y = 0 \text{ or } Y$$

Assuming the surface is initially $(t = 0)$ at rest, the general surface wave takes a standing wave pattern

$$\phi = \phi_0 \cosh\{k(z + h)\} \, \cos(k_x \, x) \, \cos(k_y \, y) \, \sin(\omega \, t)$$

where $k_x = \pi \, n_x / X$, $k_y = \pi \, n_y / Y$ and $k = \sqrt{k_x^2 + k_y^2}$. These waves form the normal modes of the oscillation with $n_x \geq 0$ and $n_y \geq 0$ integers, the individual frequencies $\omega(k)$ being given by the dispersion relation (4.12).

The general wave is found from the Fourier series expansion of the initial disturbance

$$\eta(t) = \sum_{n_x=0} \sum_{n_y=0} A_{n_x n_y} \cos(\pi \, n_x \, x / X) \, \cos(\pi \, n_y \, y / Y) \cos[\omega(n_x, n_y) \, t] \qquad \text{(S.59)}$$

where the displacement amplitudes $A_{n_x n_y}$ are easily related to those of the potentials.

Problem 22: From equation (4.29) the frequency at a contact surface when the heavy fluid is above the light is real when

$$\sigma \, k^2 > (\rho_+ - \rho_-) \, g$$

The sinusoidal perturbations at the surface must form standing waves where the normal component of velocity at the mesh walls is zero. The most unstable mode, $n_x = 1$, $n_y = 0$, is determined by the mesh spacing d, and the corresponding wavenumber $k = \pi / \sqrt{d}$. Waves with wavenumber less than k cannot exist at the surface. Consequently if

$$d \lesssim \sqrt{\sigma / (\rho_+ - \rho_-) \, g} \approx 9 \, \text{mm} \qquad \text{(S.60)}$$

no unstable waves can grow on the surface.

Problem 23: The basic characteristics of the Richtmyer–Meshkov instability are obtained from equation (4.32). An impulsive acceleration to velocity ΔU is applied in a very short time Δt, so that the acceleration $a = \Delta U / \Delta t$, and

$$\gamma = \sqrt{\frac{(\rho_+ - \rho_-)}{(\rho_+ + \rho_-)} \cdot k \, \frac{\Delta U}{\Delta t}}$$

In the limit $\Delta t \to 0$, $\gamma \Delta t$ is small and we obtain from equation (4.32)

$$\eta \approx \eta_0 \qquad \text{and} \qquad v \approx \eta_0 \, \gamma^2 \, \Delta t = \frac{(\rho_+ - \rho_-)}{(\rho_+ + \rho_-)} \, k \, \Delta U \qquad \text{(S.61)}$$

In practice this instability is normally associated with shock waves moving through a contact discontinuity between two media of different densities. In this case the fluids are compressible, but the above results still give a good approximation to the behaviour.

Problem 24: Writing the complex frequency in the form $\omega = \omega_0 \pm \imath \omega_1$ since they are a conjugate pair,

$$
\begin{aligned}
\eta &= \eta_+ \exp\left\{i\left[k_x\, x - (\omega_0 + \imath \omega_1)\, t\right]\right\} + \eta_-\, \exp\left\{i\left[k_x\, x - (\omega_0 - \imath \omega_1)\, t\right]\right\} \\
&\to \left[\eta_0 \cos\, \mathrm{h}\,(\omega_1\, t) + \eta_1 \sin\, \mathrm{h}\,(\omega_1\, t)\right] \exp\left[i\,(k_x\, x - \omega_0\, t)\right] \\
&\to \left[\eta_0 + \eta_1'\, t\right] \exp\left[i\,(k_x\, x - \omega_0\, t)\right] \qquad\qquad \text{as } \omega_1 \to 0
\end{aligned}
$$

The first term with amplitude η_0 represents the steady propagation of the original perturbation with no change of amplitude downstream with the flow since $\omega_0 = k_x U$. The second term is a purely growing wave also propagating downstream.[1] It is this term which leads to flapping flags and sails as a uniform flow is initially perturbed by the supporting post. In non-ideal flows the vortex sheet induced by the disturbance is diffused by viscosity and the flow returned to its initial condition.

Problem 25: The friction velocity v^* may be written as

$$
v^* = C(n)^{-n/(n+1)}\, (a/\nu)^{-1/(n+1)}\, \overline{u}^{n/(n+1)}
$$

to give the value of $\mathcal{V}(n)$. Using this value the Fanning friction factor may be written as

$$
f = 2\,(v^*/\overline{u})^2 = 2\,C(n)^{-2n/(n+1)}\, (a/\nu)^{-2/(n+1)}\, \overline{u}^{-2/(n+1)}
$$

and we obtain the required result for $\mathcal{F}(n)$. The value of $\mathcal{T}(n)$ follows directly from $\tau_0 = \frac{1}{2}\rho\overline{u}^2 f$. Values of $C(n)$ given by experiment are listed in Table S.1.

Table S.1: Experimental values of the parameter $C(n)$.

n	7	8	9	10
$C(n)$	7.14	8.12	9.04	9.96

Problem 26: From equation (P.8) the boundary is at

$$
\mathcal{T}(n)\,\mathcal{R}(n,(n+1))^{-2/(n+1)} = \mathcal{T}(n+1)\,\mathcal{R}(n,(n+1))^{-2/(n+2)}
$$

Therefore

$$
\mathcal{R}(n,(n+1))^{2/(n+1)\,(n+2)} = \frac{\mathcal{T}(n+1)}{\mathcal{T}(n)^{(n+2)/2}} = \frac{C(n+1)^{2(n+1)/(n+2)}}{C(n)^{2n/(n+1)}}
$$

and equation (P.10) follows.

---

[1] This term is familiar from solutions of a second-order linear differential equation when the roots of the characteristic equation are equal.

Problem 27: From the solution to problem #12, $U = cx$, where c is the velocity gradient. We may therefore assume that $v_x \sim cx$ also. The various terms in equation (6.7) are then

$$u\frac{\partial u}{\partial x} \sim c^2 x \qquad v\frac{\partial u}{\partial y} \sim \frac{cxv}{\delta} \qquad \nu\frac{\partial^2 u}{\partial y^2} \sim \nu\frac{cx}{\delta^2} \qquad U\frac{dU}{dx} \sim c^2 U$$

and from equation (6.8)

$$\frac{\partial u}{\partial x} \sim cu \qquad \frac{\partial v}{\partial y} \sim \frac{v}{\delta}$$

from which we obtain $v \sim c\delta$ and $\nu cx/\delta^2 \sim c^2 x$ or $\delta \sim \sqrt{\nu/c}$. The boundary layer thickness is therefore independent of position along the surface.

Problem 28: The only dimensional quantities in the problem are c and ν. Since the boundary layer thickness varies as $\sim \sqrt{\nu/c}$ and is independent of x, it is clear that the streamfunction must vary as

$$\psi = \sqrt{\nu c}\, x\, f(\eta) \qquad \text{where} \quad \eta = \sqrt{c/\nu}\, y$$

Differentiating,

$$v_x = \frac{\partial \psi}{\partial y} = cx\, \dot{f}(\eta) \qquad \text{and} \qquad v_y = -\frac{\partial \psi}{\partial x} = -\sqrt{c\nu}\, f(\eta) \qquad (S.62)$$

Differentiating again, it is clear that the equation of continuity (6.8) is satisfied. Substituting for v_x and v_y in the x component of either the Navier–Stokes equation (3.13) or the boundary layer equation (6.7), we obtain

$$\dddot{f} + f\ddot{f} - \dot{f}^2 + 1 = 0 \qquad (S.63)$$

subject to the boundary conditions $(f(0) = 0;\ \dot{f}(0) = 0)$ and $\dot{f}(\infty) = 1$, which follow since at the surface the velocity is zero, and at infinity the velocity normal to the surface is $-cy$ at $y = \infty$. This equation cannot be solved in analytic form, but numerical solutions may be obtained by standard methods (Schlichting, 1968, p.90).

Problem 29: The approximate boundary layer flow near a stagnation point is easily obtained from the Pohlhausen modification of the von Karman boundary integral method in Section 6.4.1. The displacement and momentum thicknesses δ_1 and δ_2 respectively are given in terms of the parameter $\Lambda = (\delta^2/\nu)\, dU/dx$, whose value at a stagnation point is $\lambda_0 = 7.052$. Making use of the results in Section 6.4.1, we obtain $\delta_1 = 0.641\sqrt{\nu/c}$ and $\delta_2 = 0.278\sqrt{\nu/c}$. These values may be compared with those obtained from the integration of the differential equation in the previous problem #28, namely $\delta_1 = 0.648\sqrt{\nu/c}$ and $\delta_2 = 0.292\sqrt{\nu/c}$.

Problem 30: The variables in the problem are:

Independent variables	x	t	
Parameters	c_0	ρ_0	
Dependent variables	u	c	ρ

We note that there are no parameters expressing length or time alone, the ambient sound speed being the parameter containing both length and time. Thus the complete set of dimensionless products formed from this group of variables is are

$$\eta = \frac{x}{c_0\,t} \qquad U = \frac{u}{c_0} \qquad C = \frac{c}{c_0} \qquad \mathcal{P} = \frac{\rho}{\rho_0} \tag{S.64}$$

η is the similarity variable. The derivatives are

$$\frac{\partial}{\partial x} = \frac{\partial \eta}{\partial x}\frac{\mathrm{d}}{\mathrm{d}\eta} = \frac{1}{c_0\,t}\frac{\mathrm{d}}{\mathrm{d}\eta}$$

$$\frac{\partial}{\partial t} = \frac{\partial \eta}{\partial t}\frac{\mathrm{d}}{\mathrm{d}\eta} = \frac{\eta}{t}\frac{\mathrm{d}}{\mathrm{d}\eta}$$

The one-dimensional equation of continuity and Euler's equation become

$$\frac{\partial \rho}{\partial t} + v\frac{\partial \rho}{\partial x} + \rho\frac{\partial v}{\partial x} \quad\rightarrow\quad (V - \eta)\frac{\mathrm{d}\mathcal{P}}{\mathrm{d}\eta} + \mathcal{P}\frac{\mathrm{d}V}{\mathrm{d}\eta} = 0 \tag{S.65}$$

$$\frac{\partial v}{\partial t} + v\frac{\partial v}{\partial x} + \frac{c^2}{\rho}\frac{\partial \rho}{\partial x} \quad\rightarrow\quad (V - \eta)\frac{\mathrm{d}V}{\mathrm{d}\eta} + \frac{C^2}{\mathcal{P}}\frac{\mathrm{d}\mathcal{P}}{\mathrm{d}\eta} = 0$$

If these two equations are consistent, the determinant of their coefficients must be zero. Therefore

$$(V - \eta)^2 - C^2 = 0 \tag{S.66}$$

Thus we obtain

$$V = \eta \pm C$$
$$V = \text{const} \mp \int \frac{C\,\mathrm{d}\mathcal{P}}{\mathcal{P}} \tag{S.67}$$

which we recognise as the equations of the characteristics. Thus choosing an appropriate pair of solutions, we obtain

$$\frac{x}{t} = v - c$$
$$\Gamma_+ = v + \int \frac{c\,\mathrm{d}\rho}{\rho} = \text{const} \tag{S.68}$$

as before.

Problem 31: Expansion around a corner involves a problem in two spatial dimensions, but not temporally varying. The problem contains no parameter with dimensions of length. Using polar co-ordinates the characteristic variables in the problem are:

Independent variables	r	θ	
Dependent variables	v_r	v_θ	c
Parameters	U	c_0	ρ_0

From these we may form the dimensionless products

$$\theta, \quad V_r = \frac{v_r}{v_{\max}}, \quad V_\theta = \frac{v_\theta}{v_{\max}}, \quad C = \frac{c}{v_{\max}} \tag{S.69}$$

There is no dimensionless product which can be formed from the variable r using a combination of the dependent variables and the parameters. Consequently all of the above dimensionless variables must be functions of θ alone. As the normalising velocity it is convenient to use the maximum speed $v_{\max} = \sqrt{U^2 + [2/(\gamma - 1)]\, c_0^2}$.

The condition for irrotationality is included by introducing the velocity potential ϕ, whose dimensionless form is

$$\Phi = \frac{\phi}{r\, v_{\max}} \tag{S.70}$$

Hence the velocities become

$$V_r = \Phi, \quad V_\theta = \dot{\Phi} \tag{S.71}$$

Since

$$\frac{c}{c_0} = \left(\frac{\rho}{\rho_0}\right)^{(\gamma-1)/2}$$

the equation of continuity and Euler's equation take the respective forms

$$\Phi C + \dot{\Phi}\dot{C} + \frac{2}{(\gamma - 1)}\dot{\Phi}\dot{C} = 0$$
$$\Phi\dot{\Phi} + \dot{\Phi}\ddot{\Phi} + \frac{2}{(\gamma - 1)}C\dot{C} = 0 \tag{S.72}$$

It is obvious by inspection that these equations have the solution $\dot{\Phi} = C$ or $v_\theta = c$, obtained earlier from a consideration of the characteristics. The solution for Φ is easily obtained by substituting for C

$$\frac{(\gamma + 1)}{(\gamma - 1)}\dot{\Phi}\ddot{\Phi} + \Phi\dot{\Phi} = 0 \tag{S.73}$$

We neglect the trivial solution $\dot{\Phi} = 0$ to obtain

$$\Phi = A\cos(k\theta + \varepsilon) \tag{S.74}$$

where $k^2 = (\gamma - 1)/(\gamma + 1)$ as before. The constant A is easily shown to be unity by the application of Bernoulli's equation, using the normalisation given above.

Thus we obtain the solution found earlier in Section 9.4.2, but without the overt application of characteristics. The latter may be seen through the appearance of θ as the sole independent variable.

Problem 32: Since the enthalpy is given by $h = [\gamma/(\gamma - 1)]\, p/\rho$ for a polytropic gas, we may solve the Rankine–Hugoniot equations directly to obtain explicit values for the downstream flow in terms of the upstream values.

From the first and second Rankine–Hugoniot equations (10.1) we obtain

$$p_1 - p_2 + \rho_1 v_1\,(v_1 - v_2) = 0$$

and hence, using Bernoulli's equation (10.2),

$$\frac{\gamma}{(\gamma - 1)}\left(\frac{p_1}{\rho_1} - \frac{p_2}{\rho_2}\right) + \frac{1}{2}\,(v_1 + v_2)\,(v_1 - v_2) = 0$$

Eliminating p_2 and substituting for ρ_2 we obtain

$$\frac{\gamma}{(\gamma-1)}\frac{p_1}{\rho_1}\left(1-\frac{v_2}{v_1}\right) - \frac{\gamma}{(\gamma-1)}v_2\,(v_1-v_2) + \frac{1}{2}\,(v_1+v_2)\,(v_1-v_2) = 0$$

Hence there is either the trivial solution $v_2 = v_1$ or

$$\frac{p_1}{\rho_1 v_1} - v_2 + \frac{(\gamma-1)}{2\gamma}\,(v_1+v_2)$$

and introducing the Mach number of the incoming flow $M_1 = v_1/c_1 = v_1/\sqrt{\gamma p_1/\rho_1}$, we obtain for the compression (density) and velocity ratios

$$y = \frac{\rho_2}{\rho_1} = \frac{v_1}{v_2} = \frac{(\gamma+1)M_1{}^2}{(\gamma-1)M_1{}^2 + 2} \tag{S.75}$$

The remaining equations are simple to obtain by substitution in equations (10.1):

$$\Pi = \frac{p_2}{p_1} = \frac{2\gamma M_1{}^2 - (\gamma-1)}{(\gamma+1)}$$

$$\frac{T_2}{T_1} = \frac{c_2^2}{c_1^2} = \frac{\left[2\gamma M_1{}^2 - (\gamma-1)\right]\left[(\gamma-1)M_1{}^2 + 2\right]}{(\gamma+1)^2 M_1{}^2}$$

$$\frac{v_2}{c_1} = \frac{(\gamma-1)M_1{}^2 + 2}{(\gamma+1)M_1}$$

$$\frac{v_2}{c_2} = \sqrt{\left\{\frac{(\gamma-1)M_1{}^2 + 2}{2\gamma M_1{}^2 - (\gamma-1)}\right\}}$$

where the ideal gas law $p = R_g\,\rho T$ is used to obtain the temperature T.

Problem 33: The entropy of a perfect gas is given by

$$s = c_V \ln\left(\frac{p}{\rho^\gamma}\right) + \text{const}$$

Using equation (10.4a), the entropy change across the shock is given by the function

$$z = \ln\left\{\frac{(\gamma+1)\,y - (\gamma-1)}{(\gamma+1) - (\gamma-1)\,y}\right\} - \gamma\ln(y)$$

The following properties of z are easily shown

$$z = 1 \quad \dot{z} = \ddot{z} = 0 \quad \dddot{z} = \frac{1}{2}\gamma\,(\gamma^2-1) \qquad \text{for } y = 1$$

and that z has no turning points except at $y = 1$. Hence $\delta s > 0$ if $1 \le y \le (\gamma+1)/(\gamma-1)$ and $\delta s < 0$ if $(\gamma-1)/(\gamma+1) \le y \le 1$. Shocks are therefore allowed in compression, but not in expansion, extending our earlier conclusion for weak shocks more generally, but limited to polytropic gases. The result is in fact applicable to nearly all materials, but with the proviso noted earlier (Section 10.16).

Problem 34: Referring to Figure 10.14 the incoming flow velocity v_1 is represented by the line OB and the outgoing v_2 by the line OP. The normal components of the velocity are represented by BN and velocity increment PN before and after the shock respectively. The velocity increment is therefore BP, which is normal to the shock, i.e. the shock lies along the line ON. From the sine theorem applied to $\triangle OBP$ it follows that

$$BP = \frac{\sin\theta}{\sin\angle OBP}OP \approx \frac{\theta}{\cos\beta}OB \qquad \text{(S.76)}$$

since θ is small and the shock is weak $OP \approx OB$, which is equation (10.75).

Problem 35: When the Mach number M_1 is large we may neglect c_1 in comparison with v_1 so that the critical velocity is given by

$$\frac{1}{2}v_1{}^2 + \frac{1}{(\gamma-1)}c_1{}^2 = \frac{(\gamma+1)}{2(\gamma-1)}c_*{}^2$$

Hence equation (10.82) becomes

$$v_{2y}{}^2 = (v_1 - v_{2x})^2\,\frac{[v_{2x} - (\gamma-1)/(\gamma+1)v_1]}{(v_1 - v_{2x})}$$

which after cancellation gives

$$v_{2y}{}^2 + (v_{2x} - v_1)\,[v_{2x} - (\gamma-1)/(\gamma+1)v_1] \qquad \text{(S.77)}$$

and is the equation of a circle passing through $([\gamma-1]/[\gamma+1]v_1,0)$ and $(1,0)$, i.e. centred on $(\gamma/(\gamma_1+1)\,v_1,0)$ with radius $1/(\gamma+1)\,v_1$. From Figure 10.14 the limiting angle is given by the tangent to the circle through the origin, namely

$$\theta = \arcsin\left\{\frac{[1/(\gamma+1)]\,v_1}{[\gamma/(\gamma_1+1)]\,v_1}\right\} = \arcsin\left(\frac{1}{\gamma}\right) \qquad \text{(S.78)}$$

Problem 36: From problem #9 the complex potential $w_1 = C\cosh(\zeta - \zeta_0)$ has stagnation points at $\zeta = \zeta_0$. From problem #10 the complex potential $w_2 = \imath k\zeta/2\pi$ is a circulating flow around an ellipse. Thus we require a combination of the flows $w = w_1 + w_2$ with an appropriate value of k such that the stagnation point lies at the end of the major axis $\zeta = (\xi_0, 0)$. The complex velocity

$$\frac{dw}{dz} = \frac{dw}{d\zeta}\bigg/\frac{dz}{d\zeta} \qquad \text{(S.79)}$$

is zero if $dw/d\zeta = 0$, provided $dz/d\zeta \neq 0$.
    The complex potential is

$$w = C\cosh(\zeta - \zeta_0) + \imath k\zeta/2\pi \qquad \text{(S.80)}$$

and

$$\frac{dw}{d\zeta} = C \sinh(\zeta - \zeta_0) + i\,\frac{k}{2\,\pi} \tag{S.81}$$

On the surface of the ellipse $\xi = \xi_0$ and therefore the stagnation point occurs when

$$\frac{dw}{d\zeta} = C\left[\underbrace{\sinh(\xi-\xi_0)}_{0}\cos(\eta-\eta_0) - i\underbrace{\cosh(\xi-\xi_0)}_{1}\sin(\eta-\eta_0)\right] + i\,\frac{k}{2\,\pi} = 0 \tag{S.82}$$

The required circulation to give a stagnation point at $\eta = 0$ is therefore

$$k = -2\,\pi\,C\,\sin\eta_0 = -2\,\pi\,U\,c\,\exp(\xi,0)\,\sin\eta_0 = 2\,\pi\,(a+b)\,U\,\sin\eta_0 \tag{S.83}$$

since $a + b = c\exp(\xi,0)$ (problem #8).
   The lift coefficient is therefore

$$c_L = \frac{-\rho\,U\,\Gamma}{\frac{1}{2}\,(2\,a)\,\rho\,U^2} = 2\,\pi\,\frac{(a+b)}{a}\,\sin\eta_0 \approx 2\,\pi\alpha\left(1+\frac{b}{a}\right) \tag{S.84}$$

since the angle of attack $\alpha = \eta_0$ and the chord is equal to the major axis $2\,a$.
   This result is in agreement with the previously obtained values for a thin wing $b = 0$, equation (11.25), and a circular cylinder $b = a$.

Problem 37: Writing the equation of the circle

$$\zeta = R\exp(i\,\theta) \tag{S.85}$$

so that the transformation takes the form

$$z = x + iy = R\exp(i\,\theta) + \frac{\ell^2}{R}\exp(-i\,\theta) = \left[R + \frac{\ell^2}{R}\right]\cos\theta + i\left[R - \frac{\ell^2}{R}\right]\sin\theta \tag{S.86}$$

and hence eliminating θ between the real and imaginary parts

$$\frac{x^2}{(R+\ell^2/R)^2} + \frac{y^2}{(R-\ell^2/R)^2} = 1 \tag{S.87}$$

we obtain an ellipse with semi-major axis $a = (R^2 + \ell^2)/R$ and semi-minor axis $b = (R^2 - \ell^2)/R$.

Problem 38: From problem #36 we obtain the relationships between the transformation parameter ℓ and circle radius R with the semi-major and semi-minor axes of the ellipse

$$R = \frac{a+b}{2} \quad \text{and} \quad \ell = \frac{a-b}{2} \tag{S.88}$$

In the ellipse frame the complex co-ordinate z is represented by its value in the circle $\zeta = R\exp(i\theta)$ transformed by the conformal transformation

$$z = \zeta + \frac{\ell^2}{\zeta}$$

Noting that the angle θ_0 between the incoming flow and the major axis is unchanged by the conformal transformation, the complex velocity potential in the frame of the ellipse is obtained from that in the circle frame

$$w(z) = \zeta \exp(\imath\theta_0) + \frac{\ell^2}{\zeta} \exp(-\imath\theta_0) + \frac{\imath k \ln \zeta}{2\pi} \tag{S.89}$$

where k is the circulation associated with the flow around the ellipse required to place the stagnation point at the end of the major axis. Hence

$$\frac{dw}{dz} = \frac{dw}{d\zeta} \bigg/ \frac{dz}{d\zeta} = \frac{U\left[\exp(-\imath\theta_0) - (R/\zeta)^2 \exp(-\imath\theta_0)\right] - \imath k/2\pi\,\zeta}{1 - (\ell/\zeta)^2} = 0$$

yields the required circulation

$$k = -2\pi UR \cdot 2 \sin\theta_0 = -2\pi U\,(a+b)\sin\theta_0 \tag{S.90}$$

in agreement with the previous problem.

Problem 39: Writing the Zhukovskii transformation as

$$\frac{(z - 2\ell)}{(z + 2\ell)} = \left\{\frac{(\zeta - \ell)}{(\zeta + \ell)}\right\}^2$$

gives

$$\arg(z + 2\ell) - \arg(z - 2\ell) = 2\left\{\arg(\zeta + \ell) - \arg(\zeta - \ell)\right\}$$

It is clear that the chord $-\ell$ to ℓ subtends the angle 2χ at the centre of the circle and therefore χ at the point P on the circumference in the ζ plane (Figure S.1). In the z plane, the corresponding chord -2ℓ to 2ℓ subtends an angle 2χ at the

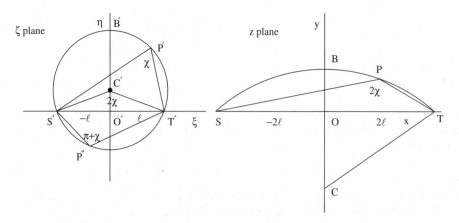

Figure S.1: The geometrical arrangement of the ξ and z plane for the Zhukovskii transformation of a circle to a circular lamina. Both the points P' and P'' transform to P.

corresponding point P'. Thus as the point P moves around the circle in the ζ plane, the angle subtended by the chord remains constant. The path of P' is therefore a circle in the z plane.

The angle χ moves from 0 to π on the upper arc of the ζ circle generating the arc in z from $+2\ell$ to -2ℓ traversed anti-clockwise. On the lower chord in ζ, the angle χ moves from π to 2π and again generates the same arc in z from -2ℓ to $+2\ell$ but traversed clockwise.

Since the circle in the z plane passes through the singularities $\pm\ell$, the transformed circle passes through the equivalent points $\pm 2\ell$. The transformation of the centre in ζ gives the centre at $[-\imath\,(\ell^2 - d^2)/d]$ and the radius $[(\ell^2 + d^2)/d]$ in z. The camber ratio is therefore $2\,d/2\,\ell = \mathrm{arccot}\chi$, consistent with the conformal nature of the transformation.

Problem 40: Referring to Figure 11.4 it can be seen that the axis of zero lift is the line from the centre of the transformation circle to the point corresponding to the trailing edge, i.e. where the circle passes through the point $(\ell, 0)$ on the chord line. The angle between the axis of zero lift and the chord line is $\beta = \pi/2 - \chi$.

From equation (11.35) it follows that since $\beta = \arctan(d/\ell)$ the necessary circulation is

$$\Gamma = 4\pi R \sin(\alpha + \beta) = 4\pi\,(\ell\,\sin\alpha + d\,\cos\alpha)$$

Hence, since the total chord length is 4ℓ, the lift coefficient is

$$c_L = 2\,\pi\,(\sin\alpha + c\cos\alpha) \tag{S.91}$$

where $c = \arctan\beta$ is the camber ratio. Note that if $c = 0$ we recover the result for a flat plate. At zero angle of attack $(\alpha = 0)$ the flow is parallel to the chord line and is finite at the leading edge as well as the trailing edge. At non-zero angles of attack, the flow at the leading edge is no longer finite and a laminar wing of this profile would suffer severe separation. In practical wings this is avoided by the blunt nose to the wing section.

Problem 41:
The sails generate a lift force normal to the direction of the wind, which may be resolved into components parallel $S_L \sin\beta$ and perpendicular $S_L \cos\beta$ to the direction of the boat, where β is the angle between the wind direction and the direction of the boat (Figure S.2). In addition there are components due to drag of the sail due both to parasitic and induced drag S_D. The keel is aligned with the axis of the boat and takes the form of a symmetric aerofoil section. When the flow is inclined at a finite angle θ to the axis, the keel exerts a lift force normal to the direction of motion K_L and the hull, including the keel, a drag K_D parallel to it. As both the keel and the hull are designed as symmetric streamlined bodies to minimise both skin and wave drag, the total hydrodynamic drag on a well-designed racing yacht may be relatively small. Induced drag from the flow of air upwards over the top of the sail adds to the total drag D. Resolving the forces along and perpendicular to the motion gives

$$\begin{aligned} S_L \sin\beta - S_D \cos\beta &= K_D \\ S_L \cos\beta + S_D \sin\beta &= K_L \end{aligned} \tag{S.92}$$

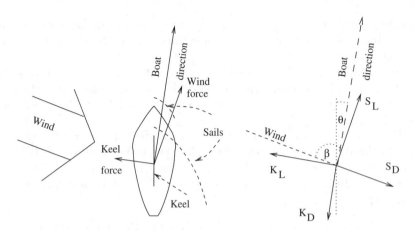

Figure S.2: Sketch of the aerodynamic forces generated by a boat sailing into the wind. The sail generates both lift S_L and drag S_D from the wind. The keel generates lift K_L from the flow of water past the hull, which also generates a drag force K_D.

As with an aerofoil the induced drag on the sail may be minimised by a large span, i.e. a tall mast. The angle of attack of the wind on the sail should not lead to stalling. However, when starting a small boat, such as sail board, there may be an advantage in allowing the airflow to stall to obtain high lift briefly with low hull drag. On large racing yachts, the induced drag on the keel may be reduced by winglets, which may compensate for the additional drag, although they also have other advantages.

The lift force on the sail is determined by the angle of attack α on the billowed sail and is adjusted by the sailor to avoid separation, $S_L \approx \frac{1}{2} C_{L\text{sail}} \rho_{\text{air}} A_{\text{sail}} U^2 \sin\alpha$, where A_{sail} is the area of the sail, ρ_{air} the density of the air, U the wind speed. The lift coefficient $C_{L\text{sail}} \propto \sin\alpha$ is determined by the angle of attack, α, between the wind and the axis of zero lift, taking into account the downwash. Assuming the sail drag S_D is small, and the wind normal to the axis of the boat $\beta = \pi/2$, the motion is along the axis of the boat and the drag is solely due to parasitic drag on the hull (including the keel) alone. This configuration gives maximum drive and speed. The boat speed is determined by $K_D/S_L = \rho_{\text{water}} C_{D\text{hull}} V^2 / \rho_{\text{air}} C_{L\text{sail}} U^2 \approx 1$, where ρ_{water} is the density of water and V the speeds respectively. In principle the boat speed may exceed the wind speed. Clearly there is a limit to how close to the wind the boat may sail upwind, determined by the minimum value of $\sin\beta = K_D/S_L$, which is a function of forces on the sail and hull.

Sailing downwind the wind force on the sail pushes the wind, essentially as a drag force. In this configuration the wing is stalled, and the boat speed is limited to the wind speed.

Problem 42: The flow over the top surface of the plate is conditioned by a rarefaction which 'refracts' the flow through an angle α. As the rarefaction is assumed to be weak the pressure is reduced by

$$\frac{\delta p}{p_1} = \gamma \frac{M_1{}^2}{\sqrt{M_1{}^2 - 1}} \alpha$$

Similarly the flow over the lower surface is 'refracted' by a weak shock giving a pressure increase of the same value. The total pressure difference is therefore $2\delta p$ and the lift force per unit span is approximately $\delta p \, \ell$ where ℓ is the length of the plate. The lift coefficient is correspondingly

$$c_L \approx \frac{2\,\delta p\,\ell}{\frac{1}{2}\rho v_1{}^2 \ell} = \frac{4\alpha}{M_1} \tag{S.93}$$

as given in the text.

Problem 43: When the flow is deflected through an angle $([\alpha_i - \alpha_{(i-1)}]$ at the junction of two elements $i-1$ and i, where α_i is the angle the element i makes with the incoming flow, the pressure increment in the weak rarefaction or shock limit is

$$\frac{[\delta p_i - \delta p_{(i-1)}]}{p_1} = \gamma \frac{M_1{}^2}{\sqrt{M_1{}^2 - 1}}\left[\alpha_i - \alpha_{(i-1)}\right]$$

the sign of the pressure increment being the same as the deflection angle $[\alpha_i - \alpha_{(i-1)}]$, reflecting either a compression (> 0) or expansion (< 0). In the perturbation limit $M_1 \approx 1$ and p_1, the incoming pressure, the excess pressure at the surface i, is therefore

$$\frac{\delta p_i}{p_1} = \gamma \frac{M_1{}^2}{\sqrt{M_1{}^2 - 1}}\,\alpha_i$$

The pressure coefficient c_p on the surface i is therefore

$$c_{p_i} = p_i/1/2\,\rho\,u_1{}^2 \approx 2\,\alpha_i/\sqrt{M_1{}^2 - 1} \tag{S.94}$$

Defining x parallel and y perpendicular to the incoming flow, the force on element i is

$$\delta F_{y_i} \approx \delta p_i\,\delta x \qquad \text{and} \qquad \delta F_{x_i} \approx \delta p_i\,\alpha_i\,\delta x$$

Substituting and noting that, since $\alpha_i = \mp \delta y_i/\delta x_i$, where the plus sign applies to the lower surface (compression) and the minus to the upper (expansion), we must distinguish the surfaces in the sums identifying the forces

$$F_y \approx \rho_1 u_1{}^2 \left\{ \sum_i^{\text{lower}} \left.\frac{\delta y}{\delta x}\right|_i \delta x_i - \sum_i^{\text{upper}} \left.\frac{\delta y}{\delta x}\right|_i \delta x_i \right\}$$

$$F_x \approx \rho_1 u_1{}^2 \left\{ \sum_i^{\text{lower}} \left(\left.\frac{\delta y}{\delta x}\right|_i\right)^2 \delta x_i + \sum_i^{\text{upper}} \left(\left.\frac{\delta y}{\delta x}\right|_i\right)^2 \delta x_i \right\} \tag{S.95}$$

which lead directly to Ackeret's equations (12.27) and (12.28).

Problem 44: In the rarefaction, the flow is determined by the Riemann invariant for the C_+ characteristic from the ambient gas, equation (9.11), i.e.

$$c = c_1 - \frac{(\gamma - 1)}{2}v_1\alpha \tag{S.96}$$

which immediately reduces to equation (12.47) above after substitution for the Mach number and the equation of state.

In the shock, we note that equation (10.8) applies to a normal shock and thus we obtain equation (12.50) as before, The calculation follows the earlier one to yield equation (12.51).

Problem 45: Bernoulli's equation for a compressible fluid obeying the polytropic gas equation of state may be written in terms of the sound speed c as

$$\frac{1}{(\gamma - 1)}c^2 + \frac{1}{2}v^2 = \frac{1}{(\gamma - 1)}c_1{}^2 + \frac{1}{2}v_1{}^2 \tag{S.97}$$

or, introducing the Mach number,

$$\frac{1}{(\gamma - 1)} + \frac{1}{2}M^2 = \frac{c_1{}^2}{c^2}\left[\frac{1}{(\gamma - 1)} + \frac{1}{2}M_1{}^2\right] \tag{S.98}$$

Hence, since $c^2/c_1{}^2 = (p/p_1)^{\gamma/(\gamma-1)}$ we obtain

$$\frac{p}{p_1} = \left\{\frac{1 + \frac{(\gamma-1)}{2}M_1{}^2}{1 + \frac{(\gamma-1)}{2}M^2}\right\}^{\gamma/(\gamma-1)} \tag{S.99}$$

from which equation (12.29) follows directly.

	M_{crit}	$C_{p\text{crit}}$	$C_{p\text{crit}}\sqrt{1 - M_{\text{crit}}}$	$C'_{p\text{max}}$
Cylinder	0.418 135	3.3026	3.00	3.00
Sphere	0.5675	1.584	1.25	1.25

Problem 46: In the table above we compare the pressure coefficient calculated from equation (12.30) multiplied by the Prandtl–Glauert correction factor with the maximum pressure coefficient for the incompressible flow around a cylinder and a sphere.

The large values of the incompressible pressure coefficient C'_p are due to the blunt body shape of both the sphere and the cylinder, whose thickness:chord aspect ratios are both 1. Consequently the streamlines around the body suffer a greater relative displacement compared with a body of low aspect ratio, e.g. a wing section. This results in the generation of higher pressures. This effect is reduced for the sphere where the three-dimensional relieving effect of the flow is due to the streamlines passing around the sides of the object, item (iii) on page 63. This corresponds to the marked change in the critical Mach number, and indeed to its low values for both cylinder and sphere compared with a streamlined aerofoil.

Problem 47: Let the flame speed be S. In the rest frame of the flame the overall energy balance including convection, thermal conduction and heat release is

$$\frac{\mathrm{d}}{\mathrm{d}x}\left\{\rho_1 h S - \kappa\frac{\mathrm{d}T}{\mathrm{d}x}\right\} = F(T)$$

The initial density is used as $\rho_1 S$ corresponds to the constant mass flux which is heated. The boundary conditions are $T \to T_1$, $dT/dx \to 0$ as $x \to -\infty$ and $dT/dx \to 0$ as $x \to \infty$. The final temperature T_2 is obtained as an integral of the equation. The value of the flame speed S is an eigenvalue of this equation. In zone ① $F(T) \approx 0$ and

$$\kappa \left. \frac{dT}{dx} \right|_i = \rho_1 \left(h_i - h_1 \right) S$$

In zone ② the convection term $\rho S \, dh/dx \ll \kappa \, d^2T/dx^2$ and

$$\kappa \frac{d}{dx} \left\{ \kappa \frac{dT}{dx} \right\} = \left\{ \kappa \frac{dT}{dx} \right\} \frac{d}{dT} \left\{ \kappa \frac{dT}{dx} \right\} = \kappa F(T)$$

Integrating subject to $dT/dx \to 0$ as $T \to T_2$ at $x = \infty$,

$$\kappa \frac{dT}{dx} = \sqrt{2 \int_T^{T_2} \kappa F(T) \, dT}$$

Matching the solution at the zone boundary T_i,

$$S = \frac{1}{\rho_1 \left(h_i - h_1 \right)} \sqrt{2 \int_{T_i}^{T_2} \kappa F(T) \, dT}$$

Since $F(T < T_i) \approx 0$, we set T_1 as the lower limit of the integral. As the pre-heat ignition temperature is $T_i \approx T_2$, the upper temperature limit of the pre-heat is nearly T_2. The flame speed is approximately

$$S = \frac{1}{\rho_1 \left(h_2 - h_1 \right)} \sqrt{2 \int_{T_1}^{T_2} \kappa \, F(T) \, dT}$$

where the enthalpy of the pre-heat mixture treated as a perfect gas is $h = c_p T$.

Problem 48: Defining the dimensionless Lagrangian variable ζ as, for example, the fraction of the mass enclosed by the symmetry surface containing the fluid particle, the body of gas is heated at a rate which depends on a power of time:

$$e = a \, t^n = A W(\zeta) \, t^n$$

where A is a dimensional constant, $W(\zeta)$ the (dimensionless) heat distribution function and $n \geq 0$.

The dimensions of the only dimensional parameters in the problem are $[M] = [M]$ and $[A] = [M][L]^2[T]^{-(n+2)}$. It is not possible to construct a dimensionless product containing length without also including time. Hence the motion is self-similar with parameter

$$\xi = \frac{r}{A^{1/2} t^{[(n+2)/2]}} \tag{S.100}$$

The position of the fluid particle ζ must be given by

$$r = \text{const } A^{1/2} t^{(1+n/2)} g(\zeta) \tag{S.101}$$

The body of gas therefore expands homogeneously in that each length is increased by the same factor, since $\mathrm{d}(\delta r)/\mathrm{d}t\big|_\zeta$ is independent of r. Furthermore $\xi = g(\zeta)$, where $g(\zeta)$ is an appropriate unknown function, and since the position vector of an individual fluid element is unique, there is a one-to-one correspondence between ξ and ζ, or equivalently ξ is itself a Lagrangian variable, and consequently the self-similar variable can be expressed as

$$\xi = r/R(t) \qquad \text{where} \qquad R(t) = c\, A^{1/2}\, t^{(1+n/2)} \tag{S.102}$$

where the constant c is determined by an appropriate (arbitrary) condition.

Finally, since the dimensions of density and pressure are

$$[\rho] = [M]\,[L]^{-3} \qquad \text{and} \qquad [p] = [M]\,[L]^{-1}\,[T]^2$$

the density and pressure are separable and can be expressed as

$$
\begin{aligned}
\rho &= M r^{-\nu} f(\xi) = M A^{-\nu/2}\, t^{-\nu(1+n/2)}\, \xi^{-\nu}\, f(\xi)\\
p &= M A^{2/(2+n)}\, r^{[-4/(2+n)-2+\nu]}\, \phi(\xi)\\
&= M A^{(1-\nu/2)}\, t^{[(2-\nu)(1+\nu/2)-2]}\, \xi^{[-4/(2+n)-2+\nu]}\, \phi(\xi) \tag{S.103}
\end{aligned}
$$

The spatial distributions of density $f(\xi)$ and pressure $\phi(\xi)$ are determined by the form of the heat deposition $W(\zeta)$.

The particle velocity is

$$v(\xi) = \frac{\mathrm{d}r}{\mathrm{d}t}\bigg|_\xi = \frac{r}{R(t)}\frac{\mathrm{d}R(t)}{\mathrm{d}t} = \frac{r}{R(t)}V(t) \tag{S.104}$$

where $V(t) = \mathrm{d}R/\mathrm{d}t$ is the velocity of characteristic scale length $R(t)$, and may be used to identify a characteristic speed of expansion.

The solution is applicable for the case of constant energy $M A$, when $n = 0$.

In reality a body of gas may have a small, but finite, initial radius. This solution is applicable once the flow radius becomes much larger than the dimensions of the original body. At such time the memory of any imposed structure has been lost. The analysis given in the text shows that the model can be continued once the homogeneous flow structure is established.

Problem 49: Let R be the initial radius of a fluid element, which is a Lagrangian variable, and define dimensionless variables $s = c_0(-t)/R$, $f = r/R$ and $\epsilon = \rho_0/\rho$; then the velocity

$$\frac{\mathrm{d}r}{\mathrm{d}(-t)} = -v = V\frac{r}{(-t)} = V c_0 \frac{f}{s} = c_0\frac{\mathrm{d}f}{\mathrm{d}s} \tag{S.105}$$

From the equation of continuity it follows that

$$\epsilon = \frac{r^2}{R^2}\frac{\mathrm{d}r}{\mathrm{d}R} = f^2\left(f - \frac{\mathrm{d}f}{\mathrm{d}s}\frac{\mathrm{d}s}{\mathrm{d}R}\bigg|_t\right) = f^2\left(f - s\frac{\mathrm{d}f}{\mathrm{d}s}\right) \tag{S.106}$$

and from Euler's equation

$$\frac{dv}{dt}\bigg|_R = -c_0 \frac{d}{ds}\left(\frac{df}{ds}\right)\frac{ds}{dt}\bigg|_R = c_0 \frac{d^2 f}{ds^2}$$

$$= -\frac{1}{\rho}\frac{\partial p}{\partial r} = \gamma \frac{p_0}{\rho_0}\epsilon^{-(\gamma-1)}\frac{d(\epsilon^{-1})}{ds}\frac{ds}{dr}\bigg|_t = c_0\,\epsilon^{-(\gamma-1)}\,s\,f^2\,\frac{d(\epsilon^{-1})}{ds}$$

since

$$\frac{dr}{ds}\bigg|_t = c_0\,(-t)\left[\frac{1}{s}\frac{df}{ds} - \frac{1}{s^2}\right] = -\frac{R}{s}\frac{\epsilon}{f^2} \tag{S.107}$$

We therefore have the pair of simultaneous differential equations

$$\frac{d^2 f}{ds^2} = \epsilon^{-(\gamma-1)}\,s\,f^2\,\frac{d(\epsilon^{-1})}{ds} \qquad \text{and} \qquad \epsilon = f^2\left(f - s\frac{df}{ds}\right) \tag{S.108}$$

whose solution for $\gamma = 5/3$ is easily shown by the substitution $f = as^{1/2}$ to be

$$f = 2^{2/3}\,3^{1/4}\,s^{1/2} \approx 2.089\,s^{1/2} \qquad \text{and} \qquad \epsilon = 2.3^{3/4}\,s^{3/2} \approx 4.559\,s^{3/2} \tag{S.109}$$

passing through $s = 0$ with $f = 0$ and $\epsilon = 0$. The Mach number is

$$M = \frac{v}{c} = \epsilon^{1/3}\frac{df}{ds} = \frac{1}{\sqrt{3}} \tag{S.110}$$

Problem 50: The characteristic parameters are the absorption constant b and the laser power Φ, whose dimensions are

$$[b] = [M]^{-2}\,[L]^8\,[T]^{-3} \qquad \text{and} \qquad [\Phi] = [M]\,[T]^{-3}$$

It is not possible to form a dimensionless product involving the length x without including the time t. The only dimensionless product is

$$\xi = b^{-1/8}\,\Phi^{-1/4}\,t^{-9/8}\,x \tag{S.111}$$

and the problem is self-similar with variable ξ. The dimensionless forms of the velocity, density, pressure and internal energy are similarly found to be

$$\begin{aligned} v &= b^{1/8}\,\Phi^{1/4}\,t^{1/8}\,V(\xi) & c &= B^{1/8}\,\Phi^{1/4}\,t^{1/8}\,C(\xi) \\ \rho &= b^{-3/8}\,\Phi^{1/4}\,t^{-3/8}\,D(\xi) & p &= b^{-1/8}\,\Phi^{3/4}\,t^{-1/8}\,P(\xi) \\ \epsilon &= b^{1/4}\,\Phi^{1/2}\,t^{1/4}\,E(\xi) & \mu x &= D^2\,C^{-3}\xi \end{aligned} \tag{S.112}$$

The direction x is taken in the direction of the outgoing flow, i.e. in the opposite direction to the laser beam. The laser flux within the plasma can be cast into dimensionless form, $I(\xi) = \Phi(\xi)/\Phi_0$, normalised with respect to the incoming power Φ_0. Clearly several of these quantities are not independent: for a polytropic gas of constant γ

$$E = \frac{1}{(\gamma-1)}\frac{P}{D} = \frac{1}{\gamma(\gamma-1)}\,C^2 \tag{S.113}$$

Noting that

$$x \left. \frac{\partial}{\partial x} \right|_t = \xi \frac{d}{d\xi} \qquad \text{and} \qquad t \left. \frac{\partial}{\partial t} \right|_x = -\frac{9}{8} \xi \frac{d}{d\xi}$$

and that terms such as v, ϵ, ρ and p contain an explicit dependence on t, we may transform the conservation equation equations for mass, momentum and energy to

$$\frac{\partial \rho}{\partial t} + \frac{\partial (\rho v)}{\partial x} \rightarrow \left(V - \frac{9}{8} \right) \xi \frac{dD}{d\xi} + D \left(\xi \frac{dV}{d\xi} - \frac{3}{8} \right) = 0$$

$$\frac{\partial v}{\partial t} + v \frac{\partial v}{\partial x} + \frac{1}{\rho} \frac{\partial p}{\partial x} \rightarrow \left(V - \frac{9}{8} \right) \xi \frac{dV}{d\xi} + \frac{1}{8} V^2 + \frac{1}{D} \xi \frac{dP}{d\xi} = 0$$

$$\frac{\partial \epsilon}{\partial t} + v \frac{\partial \epsilon}{\partial x} - \frac{p}{\rho^2} \frac{\partial \rho}{\partial t} - \mu \Phi(x) \rightarrow \left(V - \frac{9}{8} \right) \xi \frac{dE}{d\xi} + \frac{1}{4} E \qquad\qquad (\text{S.114})$$

$$+ \frac{P}{D^2} \left(\frac{9}{8} \xi \frac{dD}{d\xi} + \frac{3}{8} D \right) - D^2 C^{-3} I = 0$$

$$\frac{\partial \Phi(x)}{\partial x} - \mu \Phi(x) \rightarrow \frac{dI}{d\xi} - D^2 C^{-3} I = 0$$

giving a set of four first-order simultaneous differential equations which may be numerically integrated by standard methods. The boundary conditions are

$$v(0) = 0 \quad \rho(0) \rightarrow \infty \quad \epsilon(0) = 0 \quad p(0) = p_0 \quad \Phi(0) = 0 \quad \text{and} \quad \Phi(\infty) = \Phi_0 \quad (\text{S.115})$$

The value of the pressure at the target p_0 is unknown and must be found as an eigenvalue of the solutions satisfying the two point boundary conditions.

Bibliography

Abbott, I H and A E von Doenhoff (1959), *Theory of wing sections: including a summary of airfoil data.* Dover, New York.

Ackeret, J (1925), "Air forces on airfoils moving faster than sound". *Z. Flugtech. Motorluftschiffahrt,* 16, 72–74. Translated as N.A.C.A. Technical Memorandum 317 (1925).

Ackroyd, J A D, B P Axcell, and A I Ruban (2001), *Early developments of modern aerodynamics.* Butterworth Heinemann, Oxford.

Afanas'ev, I V, V M Krol, O N Krokhin, and I V Nemchinov (1966), "Gasdynamic process in the heating of a substance by laser radiation". *Appl. Math. Mech.,* 30, 1218–1225.

Anderson, J D Jr (2007), *Fundamentals of aerodynamics,* fourth edition. McGraw-Hill, New York.

Ashley, H and M Landhahl (1986), *Aerodynamics of wings and bodies.* Dover, New York.

B (1890), "The velocities of particles". *Nature,* 250–251. A summary of the extensive photographic experiments of Mach and co-workers on projectile motions.

Barenblatt, C I (1996), *Scaling, self-similarity and intermediate asymptotics.* Cambridge Texts in Applied Mathematics, Cambridge University Press, Cambridge.

Batchelor, G K (1967), *An introduction to fluid dynamics.* Cambridge Mathematical Library, Cambridge University Press, Cambridge.

Becker, R (1922), "Stosswelle und Detonation". *Z. Phy.,* 8, 321–362. Translated as N.A.C.A. Technical Memoranda 505 and 506 (1929).

Bennett, CO and J E Myers (1982) *Momentum, heat and mass transfer,* third edition, McGraw-Hill, New York.

Introductory Fluid Mechanics for Physicists and Mathematicians, First Edition. Geoffrey J. Pert.
© 2013 John Wiley & Sons, Ltd. Published 2013 by John Wiley & Sons, Ltd.

Bethe, H A (1942), *On the theory of shock waves for an arbitrary equation of state.* Report 545, Office of Scientific Research and Development. Included in Johnson and Chéret (1998, pp. 421–498).

Birkhoff, G (1955), *Hydrodynamics.* Dover, New York.

Birkhoff, G, D P MacDougall, E M Pugh, and G Taylor (1948), "Explosives with lined cavities". *J. Appl. Phys.*, 19, 563–582.

Blasius, H (1908), "Boundary layers in fluid with small friction". *Z. Math. Phys.*, 56, 1–37. Translated in Ackroyd *et al.* (2001, pp. 107–144), also issued as N.A.C.A. Technical Memorandum 1256 (1950).

Boys, C V (1893), "On electric spark photographs; or photography of flying bullets, etc by the light of the electric spark". *Nature*, 47, 440–446.

Bradley, J N (1962), *Shock waves in physics and chemistry.* Methuen, London.

Brushlinskii, K V and Ja M Kazhdan (1963), "On auto-models in the solution of certain problems of gas dynamics". *Rus. math. surv.*, 18, 1–22.

Buckingham, E (1914), "On physically similar systems; illustrations of the use of dimensional equations". *Phys. Rev.*, 14, 345–376.

Busemann, A (1930), "Compression shocks in two-dimensional gas flows". In *Vorträge aus der Gebiete der Aerodynamik und Verwandter Gebiete*, Springer, Berlin. Translated as N.A.C.A., Technical Memorandum 1199 (1949).

Carrier, G F (1951), "Foundations of high speed aerodynamics." Dover, New York.

Caruso, A, B Bertotti, and P Giupponi (1966), "Ionization and heating of solid material by means of a laser pulse." *Nuovo Cimento*, 45, 176–189.

Chandrasekhar, S (1981), *Hydrodynamic and hydromagnetic stability.* Dover, New York.

Cobine, J D (1985), *Gaseous conductors.* Dover, New York.

Colburn, A P (1933). *Trans. Am Inst Chem Eng.*, 29, 174–210.

Colebrook, C F (1939), "Turbulent flow in pipes with particular reference to the transition region between smooth and rough pipes". *Journal of the Instituion of Civil Engineers (London)*, 11, 133–156.

Courant, R, K Friedrichs, and H Lewy (1928), "On the partial difference equations of mathematical physics". *Math. Ann.*, 100, 32–74. Translated in *IBM J., 11, 215–234* (March, 1967).

Courant, R and K O Friedrichs (1948), *Supersonic flow and shock waves.* Interscience, New York.

Courant, R and D Hilbert (1962), *Methods of mathematical physics: Vol II, Partial differential equations*. Interscience, New York.

Dittus, F W and L M K Boelter (1930), "Heat transfer in automobile radiators of the tubular type". *Univ. California Publ. Eng.*, 2, 443–461. Also issued as *Int. Commun. Heat and Mass Transfer*, 12, 3–22 (1985).

Drazin, P G and W H Reid (1981), *Hydrodynamic stability*. Cambridge University Press, Cambridge.

Durand, W F (1934), "Mathematical aids". In *Aerodynamic theory volume 1* (W F Durand, ed.), Springer, Berlin.

Earnshaw, S (1860), "On the mathematical theory of sound". *Philos. Trans.*, 150, 134–158.

Fauquignon, C and F Floux (1970), "Hydrodynamic behaviour of solid deuterium under laser heating". *Phys. Fluids*, 13, 386–391.

Fishenden, M and O A Saunders (1950), *An introduction to heat transfer*. Oxford University Press, Oxford.

Glassman, I and R A Yetter (2008), *Combustion*, fourth edition. Academic Press, Burlington, MA.

Glauert, H (1947), *The elements of aerofoil and airscrew theory*, second edition. Cambridge University Press, Cambridge.

Gnielinski, V (1975), "New equations for heat and mass transfer in turbulent pipe and channel flow", *Int Chem Eng*, 16, 359–367.

Guderley, G (1942), "Powerful spherical and cylindrical compression shocks in the neighbourhood of the centre of the sphere and the cylinder". *Luftfartforschung*, 19, 302–312. Translation issued by the *Ministry of Aircraft Production*, R.T.P. translation no. 1118, undated.

Helmholtz, H (1858), "On integrals of the hydrodynamic equations which express vortex motion". *Crelle's journal für die Reine und Angewandte Mathematik*, 55, 25–55. Translated in *Philos. Mag.*, 33, 485–512 (1867).

Hobson, E W (1911), *A treatise on plane trigonometry*, third edition. Cambridge University Press, Cambridge.

Hopf, E (1969/70), "On the right weak solution of the Cauchy problem for a quasilinear equation of first order". *J. Math. Mech.*, 19, 483–487.

Hugoniot, H (1887), "Memoire sur la propagation des mouvements dans les corps et spécialement dans les gaz parfaits (première partie)". *J. de l'Ec. Polytech.*, 57, 3–97.

Hugoniot, H (1889), "Memoire sur la propagation des mouvements dans les corps et spécialement dans les gaz parfaits (deuxième partie)". *J. de l'Ec. Polytechn.*, 59, 1–125. Translated in Johnson and Chéret (1998, pp. 161–358).

Hunter, C (1960), "On the collapse of an empty cavity in water". *J. Fluid Mech.*, 8, 241–263.

Incropera, F P, D P DeWitt, T L Bergmann and A S Levine (2007) *Fundamentals of heat and mass transfer*, Wiley, New York.

Johnson, J N and R Chéret (1998), *Classic papers in shock compression science*. Springer, New York.

Katz, J and A Plotkin (2001), *Low-speed aerodynamics*. Cambridge Aerospace Series, Cambridge University press, Cambridge.

Kays, W M (1966) *Convective heat and mass transfer*. McGraw-Hill, New York.

Kidder, R E (1975), "Theory of homogeneous isentropic compression and its application to laser fusion". *Nucl. Fusion*, 14, 53–60.

Kolmogorov, A N (1941a), "Dissipation of energy in the locally isotropic turbulence". *Dokl. Akad. Nauk. SSSR*, 32, 16–18. Translated in *Proc. R. Soc. A*, 434, 15–17, 1991.

Kolmogorov, A N (1941b), "The local structure of turbulence in incompressible viscous fluid for very large Reynolds numbers". *Dokl. Akad. Nauk. SSSR*, 30, 299–303. Translated in *Proc. R. Soc. A*, 434, 9–13, 1991.

Kutta, W M (1902), "Auftriebskrafte in strömenden Flüssigkeiten". *Illus. aeronaut. Mitt.*, 6, 133–135. Translated in Ackroyd *et al.* (2001, pp.70–76).

Lamb, H (1932), *Hydrodynamics*, sixth edition. Cambridge University Press, Cambridge.

Landau, L D and E M Lifshitz (1959), *Fluid mechanics*. Course of Theoretical Physics, Pergamon, London.

Langhaar, H L (1980), *Dimensional analysis and the theory of models*. Krieger, Huntington, NY.

Lax, P D (1954), "Weak solutions of nonlinear hyperbolic equations and their numerical computation". *Commun. Pure Appl. Math.*, 7, 159–193.

Lin, C C (1955), *Theory of hydrodynamic stability*. Cambridge Monographs on Mechanics and Applied Mathematics, Cambridge University Press, Cambridge.

Mach, E and L Salcher (1887), "Photographische Fixierung der durch Projektile in der Luft eingeleiten Vorgange". *Sitzungsber. Akad. Wiss. Wien*, 95, 764–780.

Martinelli, R C (1947), *Trans A S M E*, 60, 947–959.

McAdams, W H (1973), *Heat transmission*, third edition. McGraw-Hill, New York.

Meyer, T (1908), *Über zweidimensionale Bewegungsvorgänge in einem Gas, das mit Überschallgeshwingdigkeit strömt*. PhD thesis, Göttingen University. Included in Carrier (1951, pp. 50–89).

Milne-Thomson, L M (1968), *Theoretical hydrodynamics*, fifth edition. Macmillan, London.

Moody, L F (1944), "Friction factors for pipe flow." *Trans. ASME*, 66, 671–684.

Nemchinov, I V (1964), "Dissipation of a heated mass of gas in a regular regime". *Zh. Prik. Mekh. Tekh. Fiz.*, 5, 18–29. Translation by Sandia Corp.

Oleinik, O A (1963a), "Construction of a generalized solution of the Cauchy problem for a quasi-linear equation of first order by the introduction of 'vanishing viscosity"'. *Am. Math. Soc. Trans. Ser. 2*, 33, 277–283.

Oleinik, O A (1963b), "Uniqueness and stability of the generalized solution of the Cauchy problem for a quasi-linear equation". *Am. Math. Soc. Trans. Ser. 2*, 33, 285–290.

Oseen, C W (1910), "Über die Stokessche Formel und über die verwandte Aufgabe in der Hydrodynamik". *Ark. Mathe., Astron. Fy.*, 6.

Poisson, S D (1808), "A paper on the theory of sound." *J. d'Éc. Polytech.*, 14, 319–392. Translated in Johnson and Chéret (1998, pp. 1–69).

Prandtl, L (1904), "Über Flüssigkeitsbewegung bei sehr kleine Reibung". In *Proceedings of the 3rd International Mathematical Congress, Heidelberg*, Leipzig. Translated in Ackroyd *et al.* (2001, pp. 77–87).

Prandtl, L (1921), "Applications of modern hydrodynamics to aeronautics". Report 116, N.A.C.A.

Prandtl, L and O G Tietjens (1957), *Applied hydro- and aerodynamics*. Dover, New York.

Proudman, I and J R A Pearson (1957), "Expansions at small Reynolds numbers for the flow past a sphere and around a circular cylinder". *J. Fluid Mech.*, 2, 237–262.

Quinn, B Keyfitz (1971), "Solutions with shocks: an example of an l_1-contractive semigroup". *Commun. Pure Appl. Math.*, 24, 125–132.

Raizer, Yu P (1977), *Laser-induced discharge phenomena*. Consultants Bureau, New York.

Rankine, W J M (1870), "On the thermodynamic theory of waves of finite longitudinal disturbances". *Philos. Trans. R. Soc.*, 160, 277–280.

Rayleigh, Lord (1876), "On the resistance of fluids." *Philos. Mag.*, 430–441.

Rayleigh, Lord (1899), "On the viscosity of argon affected by temperature." *Proc. Roy.Soc/*, 66, 68-74.

Rayleigh, Lord (1910), "Aerial plane waves of finite amplitude". *Proc. R. Soc. A*, 84, 247–284.

Rayleigh, Lord (1945), *Theory of sound*, second edition. Dover, New York.

Reichenbach, H (1983), "Contributions of Ernst Mach to fluid mechanics". *Annu. Rev. Fluid Mech.*, 15, 1–28.

Reynolds, O (1875), "On the extent and action of the heating surface of steam engines." *Proc Manchester Lit Phil Soc*, 14, 81–85. Also published in Scientific papers 1, 81–85, Cambridge University press (1903).

Reynolds, O (1883), "An experimental investigation of the circumstances which determine whether the motion of water shall be direct or sinuous, and of the law of resistance in parallel channels". *Philos. Trans. R. Soc.*, 174, 935–982.

Reynolds, O (1895), "On the dynamical theory of incompressible viscous fluis and the determination of the criertion." *Phil Trans Roy Soc*, 186, 123–164.

Riemann, G F B (1860), "Uber die Fortpflanzung ebener Luftwellen von endlicher Schwingtungsweite". *Abh. Ges. Wiss. Göttingen, Math.-Phy.*, 8, 43–65. Translated in Johnson and Chéret (1998, pp. 109–128).

Salas, M D (2007), "The curious events leading to the theory of shock waves". *Shock Waves*, 16, 477–487. Paper given at the *17th Shock Interaction Symposium, Rome, 2006* available at ntrs.nasa.gov/archive/nasa/casi.ntrs.nasa.gov/20060047586_2006228914.pdf.

Schlichting, H (1968), *Boundary layer theory*, sixth edition. McGraw-Hill, New York.

Sedov, L I (1959), *Similarity and dimensional methods in mechanics*. Academic Press, New York.

Stanyukovich, K P (1960), *Unsteady motion of continuous media*. Pergamon Press, London.

Stokes, G G (1845), "On the theories of the internal friction of fluids in motion and of the equilibrium and motion of elastic bodies". *Trans. Cambridge Philos. Soc.*, 8, 287–305. Also published in *Mathematical and Physical Papers*, 1, 75–129, Cambridge (1901).

Stokes, G G (1848), "On a difficulty in the theory of sound". *Philos. Mag.*, 33, 349–356. Included in Johnson and Chéret (1998, pp. 71–79).

Stokes, G G (1851), "On the effect of the internal friction of fluids on the motions of pendulums". *Trans. Cambridge Philo. Soc.*, 9, 8–106. Also published in *Mathematical and Physical Papers*, 3, 1–114, Cambridge (1901).

Suits, C G and H Poritsky (1939), "Application of heat transfer data to arc charac-teristics". *Phys. Rev.*, 55, 1184–1191.

Taylor, G I (1910), "The conditions necessary for discontinuous motion in gases". *Proc. R. Soc. A*, 84, 371–377.

Taylor, G I (1941), "The formation of a blast wave by a very intense explosion". Technical Report RC-210, Civil Defence Research Committee. Published as *Proc. R. Soc. A*, 201, 159–186, 1950.

Taylor, G I and J W Maccoll (1935), "The mechanics of compressible flow". In *Aero-dynamic theory volume 3* (W F Durand, ed.), 210–250, Springer, Berlin.

Thomas, L H (1944), "Note on Becker's theory of the shock front". *J. Chem. Phys.*, 12, 449–453.

Thomson, W (1869), "On vortex motion." *Trans. R. Soc. Edinburgh*, 25, 217–260.

Tritton, D J (1988), *Physical fluid dynamics*. Oxford University Press, Oxford.

van Dyke, M (1975), *Perturbation methods in fluid mechanics*. Parabolic Press, Stanford, CA.

van Dyke, M (1982), *Album of fluid motion*. Parabolic press, Stanford.

von Karman, T (1939) *Trans A S M E*, 61, 705–710.

von Karman, T and J M Burgers (1935), "General aerodynamic theory - perfect fluids". In *Aerodynamic theory volume 2: Perfect fluids* (W F Durand, ed.), Springer, Berlin.

von Karman, T and N B Moore (1932), "Resistance of slender bodies moving with supersonic velocities, with special reference to projectiles". *Trans. Am. Soc. Mech. Eng.*, 54, 303–310. Included in Carrier (1951, pp. 142–149).

von Mises, R (2004), *Mathematical theory of compressible flow*. Dover, New York.

von Neumann, J (1941), "The point source solution". Report AM 9, National Defense Research Committee Div B. Issued as Chapter 2 of Los Alamos report LA2000 (1958).

Welty, J R, C L Wicks, and R L Wilson (1984), *Fundamentals of momentum, heat and mass transfer*, third edition. John Wiley & Sons, Inc., New York.

Weyl, H (1944), "Shock waves in arbitrary fluids". Applied Mathematics panel note 12, National Defense Research Committee. Subsequently published in *Commun. Pure Appl. Math.*, 2, 103–122, 1944; Included in Johnson and Chéret (1998, pp. 498–519).

Whitcomb, R T (1956), *"A study of the zero-lift drag-rise characteristics of wing-body combinations near the speed of sound"*. Report 1273, N.A.C.A., 1956. Originally issued as research memorandum RM L52H08, 1952.

Whitcomb, R T and L R Clark (1965), "An aerfoil shape for efficient flight at super-critical Mach numbers". Technical Report TMX-1109, NASA, 1965.

Winteron, R H S (1998), "Where did the Dittus and Boetler equation come from"? *Int. J. of Heat and Mass Transfer*, 41, 809–810, 1998.

Zel'dovich, Ya B and Yu P Raizer (1967), *Physics of shock waves and high temperature phenomena*. Academic Press, New York.

Zhukovskii, N E (1906), "On annexed vortices." *Transaction of the physical section of the Imperial society of the friends of natural science, Moscow*, 13, 12–25. Translated in Ackroyd *et al.* (2001, pp. 88–106).

Index

Absolute instability, 114
Ackeret's formulae for lift and drag, 347
Acoustic reflection, 214
Acyclic flow, 35
Adiabatic compression
    planar, 287
Adiabatic flow, 3
Adiabatic lapse rate, 12
Aeolian tones, 74
Aerodynamic centre, 303
Aerofoils, 298–315
    application of conformal transforms,
        309
    Blasius's equation, 309
    cylinder mapping
        lift and pitching moment, 312
    focus, 303
    Kutta condition, 299
    Kutta–Zhukovskii lift formula, 301
    lift coefficient, 302
    mapping a circular cylinder, 310
    pitching moment, 302
    thin wing
        lift, 304
    thin wing lift, 308
    thin wing moment, 308
Angle of attack, 302
Attached shock, 285

Bernoulli's equation
    strong form, 32
    weak form, 16–17
Blasius's power law distribution, 135

Blasius's solution, 144
Blast wave, 397
Boundary conditions, 36
    Dirichlet condition, 36
    mixed condition, 36
    Neumann condition, 36
Boundary layer separation, 156
Boundary layers, 139–156
    applied pressure gradient, 149
    Blasius's solution for a flat plate,
        144
    momentum integral method, 146
    Prandtl's equations, 141
    Tollmein–Schlichting instability, 151
    turbulent boundary layer, 152
Boussinesq approximation, 195
Burn waves, 364

Capillary waves, 96
Cavitation, 63
Centred self-similar flows, 395–414
    blast wave, 397
    collapse of a void, 412
    convergent shock wave, 407
    spherical collapse, 402
Chapman–Jouget hypothesis, 376
Chapman–Jouget process, 370–373
    polytropic gas, 371
Characteristics
    steady two-dimensional flow, 231
Characteristics in compression, 241
Collapse of a shell, 412
Colliding jets, 418

Introductory Fluid Mechanics for Physicists and Mathematicians, First Edition. Geoffrey J. Pert.
© 2013 John Wiley & Sons, Ltd. Published 2013 by John Wiley & Sons, Ltd.

Complex function methods, 52
Compressible flow, 221
    around a sharp corner, 235
    continuous solutions, 224–240
    correction for lift and drag, 344
    discontinuous solutions, 241–294
    dissipationless approximation, 209
    improved correction for lift, 347
    perturbation methods, 341
    perturbation potential, 343
    small disturbances, 211
    steady flow around a corner, 235
    supersonic flow around an aerofoil, 347
    uniqueness theorem, 222
    weak discontinuities, 223
Compressible gas pipe flow, 20
Conformal transforms, 69
Connectivity, 36
Conservation law form, 9
Conservation of angular momentum
    Eulerian frame, 10
    Lagrangian frame, 8
Conservation of energy
    Eulerian frame, 11
    Lagrangian frame, 8
Conservation of entropy
    Eulerian frame, 11
    Lagrangian frame, 8
Conservation of mass
    Eulerian frame, 8
    Lagrangian frame, 6
Conservation of momentum
    Eulerian frame, 9
    Lagrangian frame, 7
Control of separation, 163
Convected heat transfer, 175
Convective instability, 114
Couette flow, 81
Critical density, 20
Critical Mach number, 351
Critical pressure, 20
Critical sound speed, 20
Crocco's equation, 31
Cylindrical sound waves, 217

d'Alembert's paradox, 63
de Laval nozzle, 21
Deflagrations, 363–377
    in a closed tube, 367
    structure, 373
Deflagrations and detonations, 363–377
    permitted conditions, 370
Density, 2
Detached shock, 286
Detonations, 367–377
    structure, 375
Dimension analysis, 86
Dimensionless products, 91
Displacement thickness, 143
Dissipationless flow, 3
Distortion tensor, 77
Doublet sheets, 45
Doublet sources, 43
Downwash velocity, 320
Drag, 68, 161
    Stokes' sphere, 82
    cylinder, 63
    form drag, 161
    induced drag, 161
    skin drag, 161
Drag divergent Mach number, 352
Drag in high-speed compressible flow, 350
Drag in ideal flow, 70
Drag in supersonic flight, 351
Dry adiabatic lapse rate, 13

Electric arc
    positive column, 205
Elliptic loading, 329
Environmental lapse rate, 15
Equation of continuity, 9
Euler's equation, 7
Eulerian frame, 4, 8–11

First axis of aerofoil, 302
Flame speed, 365
Flow around streamlined bodies, 295
Flow round a corner, 66
Flow velocity, 2

Fluid dynamic equations
    generalised form, 290
Fluid instabilities, 102, 106
    Kelvin–Helmholtz instability, 104
    non linear instability, 115
    Rayleigh–Taylor instability, 103
    Richtmyer–Meshkov instability, 422
    stability of laminar shear flow, 112
    Tollmein–Schlichting instability,
        151
Fluid particle, 1
Fluid point, 1
Force on a body in ideal flow, 66
Forced convection, 176–193, 205
Form drag, 71, 161
Free convection, 193
    heated cylinder, 205
    heated horizontal plate, 200
    parallel horizontal plates, 201
    vertical plate, 197–200
Friction and heat transfer, 182–188
    Colburn's modification, 188
    Martinell correction, 185
    Prandtl–Taylor correction, 183
    Reynolds analogy, 182
    von Karman's correction, 184
Friction factor, 132
Fully developed turbulence, 121
Fundamental equations, 3

Generalised fluid dynamic equations
    characteristic solution, 291
    hyperbolic form, 290
    weak solution, 292
Generation of turbulence, 119
Glauert's equation, 61, 306
Grashof number, 194
Gravity waves, 97
    energy transmission, 98
Guderley's problem, 407

Heat transfer coefficient, 176
Heat transfer rates in flowing fluids,
        178–182
    across a pipe, 179
    along a pipe, 178

    heat exchangers, 180
Helmholtz's theorems, 28
Hodograph plane, 223
Homogeneous compressible flow
    compression, 390–395
      shells, 393
      spheres, 390
    expansion, 386–390
Homogeneous turbulence, 121
Horseshoe vortex pattern, 316
Hugoniot plot, 245
    with energy release, 368
Hugoniot relation, 245
Hydrodynamic condition, 3
Hydrostatic head, 418
Hydrostatic stability, 15
Hydrostatics, 12–16
    equilibrium fluid, 12
    lapse rate, 12
    stability, 15
Hypersonic wing, 359

Ideal flow, 2, 3
Ideal flow around a plate, 71
Ideal fluid flow, 25–74
Incompressible flow, 33–34
    drag at high speed, 350
Incompressible fluid, 18
Induced drag, 161, 325
Induced velocity, 38
Inertial confinement fusion, 414–416
Infinitesimal volume, 1
Intermediate asymptotics, 385
Irrotational flow, 31–33
    Crocco's equation, 31
    velocity potential, 32
Irrotational incompressible flow, 35
    around a corner, 66
    Laplace's equation, 35
    Rankine ovals, 49
    sphere, 48
    two dimensions, 51–74
      complex functions, 52
      flow around a body, 55
      flow around a cylinder, 62–65
      flow around a thin wing, 59
      free vortex, 54

Irrotational incompressible flow
        (*continued*)
        Rankine vortex, 54
        source, 52
        source and doublet sheets, 55
        tied vortex, 53
        vortex, 52
    uniqueness, 35
Isothermal discontinuity, 261
Isothermal flow, 4
Isothermal rarefaction, 230

Jouget's rule, 370

Karman vortex street, 73
Kelvin's theorem, 26
Kelvin's wedge, 100
Kelvin–Helmholtz instability, 104
Kolmogorov distribution, 125
Kutta condition, 299
Kutta–Zhukovskii lift formula, 301

Lagrangian frame, 4–8
Laminar thermal boundary layer, 188
Laminar wake, 163
Lapse rate, 12–15
    dry adiabatic lapse rate, 13
    environmental lapse rate, 15
    moist saturated adiabatic lapse rate,
        13
Laser–matter breakdown, 377–382
    gases, 381
    solids, 378
        deflagration model, 379
        self-regulating model, 380
Laser–plasma interaction, 377
Lift
    Kutta–Zhukovskii, 68, 301
    rotating cylinder, 65
    thin wing, 59, 308
Lifting line theory, 320
Linearised theory of supersonic flight in
        three dimensions, 354
Linearised theory of supersonic flight in
        two dimensions, 347
Logarithmic mean temperature, 181

Magnus effect, 65
Martinelli correction, 185
Method of matched asymptotics, 169
Mixed mean fluid temperature, 186
Modelling, 88
Moist saturated adiabatic lapse rate, 13
Momentum thickness, 143
Moody plot, 132
Munroe effect, 22

Natural convection, 193–205
Navier–Stokes equation, 80
Newton's law of cooling, 178
Newtonian viscosity, 75
Nonlinear instability, 115
Nusselt number, 177

Oblique shocks, 277–287
    high Mach number, 281
    low Mach number, 279
    shock polar, 282
Orr–Sommerfeld equation, 112

Panel method
    three dimensions, 330
    two dimensions, 314
Pelton wheel, 418
Permutation symbol, 5
Perturbation methods for compressible
        flow, 341
Pitching moment
    thin wing, 308
Pitching moment of aerofoil, 302
Pitot tube, 417
Plane Poiseuille flow, 81
Polytropic gas, 17
Prandtl number, 177, 194
Prandtl's boundary layer equations, 141
Prandtl's distribution law for turbulent
        flow through a duct , 130
Prandtl's equation, 327
    Fourier series solution, 327
Prandtl's mixing length model, 136
Prandtl–Glauert correction, 344
Prandtl–Meyer flow, 235
Pressure, 2
Pressure relieving effect on sphere, 63

Rankine ovals, 49
Rankine vortex, 54
Rankine–Hugoniot equations, 242
Rankine–Hugoniot relations
    with energy release, 366
Rarefactions
    centred rarefaction, 226
    limit velocity, 226
    reflection, 228
    steady two dimensional, 231
    time dependent, 224
Rate of dilation, 76
Rayleigh number, 195
Rayleigh–Bénard instability, 203
Rayleigh–Taylor instability, 103
Rectilinear vortex, 29, 39
Reynolds' ink drop experiment, 117
Reynolds' stress, 126
Richtmyer–Meshkov instability, 422
Riemann invariants, 225
Riemann's solution, 224
Rotation velocity, 76
Rough wall, 129

Sears–Haack body, 358
Self-similarity, 89, 383–386
Separation, 156–163
    turbulent boundary layer, 166
Shallow water waves, 106
Shaped charge flow, 22
Shear flow stability, 112
Ship wave drag, 99
Shock adiabat, 245
Shock front structure, 254–267
    gas shocks, 256
    real gases, 264
    supported by heat transfer, 260
    weak shocks, 261
Shock polar, 282
Shock separation, 351
Shock stall, 351
Shock tube
    theory, 269
Shock tubes, 267–271
Shock waves, 241, 242
Shvab–Zel'dovich deflagration, 374
Similarity, 88

Simple waves, 223
Sinks, 42
Skin drag, 161
Slots and flaps, 163
Sound barrier, 352
Sound waves, 211–218
    cylindrical, 217
    energy, 213
    plane, 212
    reflection, 214
    spherical, 215
Source sheets, 43
Sources, 42
Spherical sound waves, 215
Stagnation density, 18
Stagnation pressure, 18
Stagnation sound speed, 18
Stokes equation, 82
Stokes' flow, 82
Stokes streamfunction, 34
Strain rate, 76
    longitudinal, 76
    transverse, 76
Streamfunction, 33
Streamlined bodies, 295
Streamlined flow, 41
Streamlines, 16
Streampipes, 22
Stress, 77
Strouhal number, 74
Supercritical wing, 354
Supersonic flow, 219, 221
Supersonic flow interactions, 271–277
    overtaking interactions, 275–277
    shock overtaking contact surface,
        276
    shock overtaking rarefaction, 276
    shock overtaking shock, 276
    shock collision, 274
    shock reflection at a wall, 271
Swept wings, 350

Temperature, 2
Thermal boundary layer, 188–193
    laminar, 188
    turbulent, 192

Thin wing lift, 304
Tollmein–Schlichting instability, 151
Torque, 77
Trailing vortices, 316
Transonic flight, 351
Turbulent boundary layer
        flat smooth plate, 152
        power law distribution, 154
Turbulent flow, 117–138
        Blasius's power law approximations,
            135
        Kolmogorov distribution, 125
        Kolmogorov's theory, 123
        law of the wall, 127
        Reynolds' ink drop experiment, 117
        Reynolds' stress, 126
        rough wall, 129
        smooth wall, 127
        through a duct, 129
        von Karman similarity model, 127
Turbulent pipe flow
        empirical relations, 132
Turbulent thermal boundary layer, 192
Turbulent wake, 168
Two-dimensional steady flow
        characteristic invariants, 232
        characteristics, 231

Uniqueness, 36
Units, 86

Velocity defect, 130
Velocity potential, 32
Velocity profile near a wall, 129
Vena contracta, 19
Venturi, 418
Viscous fluid equations, 78–81
        energy, 79
        entropy, 80
        momentum, 79
Viscous stress, 78
Viscous sub-layer, 128
von Karman boundary integral method
        for a flat plate, 146
von Karman distribution law for
        turbulent flow through a duct ,
        130
von Karman ogive, 357

von Karman similarity model of shear
        flow, 127
Vortex loop, 55
        induced velocity, 40
Vortex rows, 72–74
Vortex sheet, 40
Vortices, 29–31
        Rankine vortex, 54
        simple vortex, 29, 53
        Vortex sheet, 29
Vorticity, 27
        rotation, 76

Wake, 158, 163–169
        laminar, 163
        turbulent, 168
Wave drag, 351
        ship, 99
Waves in incompressible fluids, 93–102
        capillary waves, 96
        gravity waves, 97
        shallow water waves, 106
        stratified fluid, 108
        surface waves, 94
            contact discontinuity, 102
            free boundary, 96
            infinite fluid, 102
Waves in stratified fluid, 108
Weak solutions, 293
Whitcomb area rule, 358
Winds
        downwash velocity, 320
Wing loading, 327
Wing sections, 296
Wings, 315–331
        elliptic loading, 329
        force on wing, 319
        induced drag, 325
        lifting line theory, 320
        velocity at surface, 318
        wake, 323

Zhukovskii aerofoil profile, 333
Zhukovskii aerofoils, 332, 339
        Karman–Treffetz profile, 334
        Theodorsen's solution, 338
        von Mises profile, 336
        Zhukovskii profile, 333